Springer Series in Geomechanics and Geoengineering

Series editor

Wei Wu, Universität für Bodenkultur, Vienna, Austria
e-mail: wei.wu@boku.ac.at

About this Series

Geomechanics deals with the application of the principle of mechanics to geomaterials including experimental, analytical and numerical investigations into the mechanical, physical, hydraulic and thermal properties of geomaterials as multiphase media. Geoengineering covers a wide range of engineering disciplines related to geomaterials from traditional to emerging areas.

The objective of the book series is to publish monographs, handbooks, workshop proceedings and textbooks. The book series is intended to cover both the state-of-the-art and the recent developments in geomechanics and geoengineering. Besides researchers, the series provides valuable references for engineering practitioners and graduate students.

More information about this series at http://www.springer.com/series/8069

Euripides Papamichos · Panos Papanastasiou
Elena Pasternak · Arcady Dyskin
Editors

Bifurcation and Degradation of Geomaterials with Engineering Applications

Proceedings of the 11th International Workshop on Bifurcation and Degradation in Geomaterials, Dedicated to Hans Muhlhaus, Limassol, Cyprus, 21–25 May 2017

Springer

Editors
Euripides Papamichos
Department of Civil Engineering
Aristotle University of Thessaloniki
Thessaloniki
Greece

Elena Pasternak
School of Mechanical and Chemical Engineering
University of Western Australia
Crawley, WA
Australia

Panos Papanastasiou
Department of Civil and Environmental Engineering
University of Cyprus
Nicosia
Cyprus

Arcady Dyskin
School of Civil, Environmental and Mining Engineering
University of Western Australia
Crawley, WA
Australia

ISSN 1866-8755 ISSN 1866-8763 (electronic)
Springer Series in Geomechanics and Geoengineering
ISBN 978-3-319-56396-1 ISBN 978-3-319-56397-8 (eBook)
DOI 10.1007/978-3-319-56397-8

Library of Congress Control Number: 2017935827

Printed on acid-free paper

This Springer imprint is published by Springer Nature
The registered company is Springer International Publishing AG
The registered company address is: Gewerbestrasse 11, 6330 Cham, Switzerland

Preface

The 11th International Workshop on Bifurcation and Degradation in Geomaterials (IWBDG 2017) was held in Limassol, Cyprus, during the period May 21–May 25, 2017. Interests on localization and instabilities in the field of geomechanics date back to the early 1980s when the first International Workshop on Localization of Soils was organized in Karlsruhe, Germany, in February 1988. The first event created such enthusiasm and interest in the fundamental aspects of bifurcation theory to soils, that a second workshop followed suit in Gdansk, Poland, in September 1989. The topic later extended to rock mechanics at the third international workshop in Aussois, France, in September 1993. The interest grew steadily, and the scope was expanded to instabilities and degradations in geomaterials at the workshops that were held in Gifu, Japan, September 1997; Perth, Australia, November 1999; Minneapolis, USA, June 2002; Crete, Greece, June 2005; Lake Louise, Canada, May 2008; Porquerolles Island, France, May 2011; and Hong Kong, May 2014.

The IWBDGs have grown in size and scope, since their inception, covering more and wider areas of geomaterials and geomechanics including modern trends. The scientific program of the IWBDG 2017 includes 11 keynote lectures and 66 oral communications. The lectures cover a wide range of topics including advances in instabilities, localized and diffuse failure, multiscale phenomena, multiphysics, micromechanical modeling, and other related topics. A major goal was to promote the link between theoretical and applied mechanics with engineering applications in traditional and in emerging fields, such as geomechanics for the energy and the environment. The quality of the contributed papers has benefited greatly from the expertise of numerous colleagues who served as referees and to whom we are very grateful.

The IWBDG 2017 is dedicated to Prof. Hans Muhlhaus, on the occasion of his 70th birthday, in recognition of his contributions and his achievements in the field of Bifurcation and Degradation in Geomaterials. The papers presented in special session in honor of Prof. Muhlhaus are included in these proceedings.

Hans shaped the world of Geomechanics, introducing the theory of bifurcations, localisations and instabilities and higher order theories. He brought and drummed in the mathematical rigor and 'objectivity' into the engineering community. His paper "The thickness of shear bands in granular media", Geotechnique, 1987, co-authored by the late Prof. Ioannis Vardoulakis, is the second-most cited paper since 1970 of this journal. It has a significant impact on the development of constitutive theories for granular materials. Such theories are crucial for a fundamental understanding of the origins of failure of geo-structures-natural or man-made-and the dynamic properties of earthquakes. More recent contributions to the field include his papers on shear banding, discrete element simulations of granular assemblies and his theoretical and computational work on instabilities in sand erosion and fluid-solid interactions. With more than 300 journal and conference papers and more than 8000 citations Professor Mühlhaus is a leading researcher in the area who determined its shape and direction for years to come.

We thank all the authors and especially the authors of the plenary and invited papers for their timely submission and participation in IWBDG 2017, the Session Developers for putting together exciting and stimulating sessions, the reviewers of the papers, and the members of the Scientific and Organizing Committees for their support. We also express our sincere appreciation to the technical and financial cosponsors: the University of Cyprus, the Aristotle University of Thessaloniki, the University of Western Australia, the Cyprus Tourism Organization, ALERT-Geomaterials, SINTEF Petroleum Research, MTS Systems, Themeliotechniki Cyprus LTD. and Dreamscape Conferences.

Thessaloniki, Greece — Euripides Papamichos
Nicosia, Cyprus — Panos Papanastasiou
Crawley, Australia — Elena Pasternak
Crawley, Australia — Arcady Dyskin

Committees

Organizing Committee

Euripides Papamichos, Chair, Aristotle University of Thessaloniki, Greece
Panos Papanastasiou, Chair, University of Cyprus
George Manolis, Aristotle University of Thessaloniki, Greece
Yiannis Dafalias, National Technical University of Athens, Greece and University of California, Davis, USA
Dimitris Loukidis, University of Cyprus
Ioannis Ioannou, University of Cyprus

Organizing Committee of Muhlhaus Tribute

Arcady Dyskin, University of Western Australia
Elena Pasternak, University of Western Australia

Advisory Board

R.I. Borja, USA
J. Desrues, France
F. Nicot, France
F. Darve, France
H.-B. Muhlhaus, Australia
R. Wan, Canada

International Scientific Committee

Contents

Part I
Experimental Modeling of Instabilities Localization and Degradation

Localisation Precursors in Geomaterials?

J. Desrues, E. Andò, P. Bésuelle, G. Viggiani, L. Debove, P. Charrier and J.B. Toni

Abstract Strain localisation in soils and rocks has been studied extensively for the last 40 years or so. On the experimental side, a large number of these studies have been devoted to the experimental observation of localised deformation in laboratory element tests like biaxial (plane strain) and triaxial tests. 2D and 3D imaging techniques and image analysis methods have been used to characterize the onset and subsequent development of strain localisation. In the recent years, these techniques and methods have improved dramatically, allowing considerably more accurate measurement of displacement and strain field in the laboratory specimens. It is time to have a second look, with these new glasses, at some decades-old results, to assess what can be confirmed and what should be reconsidered.

1 Introduction

Strain localisation in soils and rocks has been studied extensively in Laboratoire 3SR, Grenoble, for the last 40 years or so [4, 6, 7]. It was shown, using strain field measurement methods e.g. stereophotogrammetry, that shear banding can take place in both contractive and dilative specimens, under either drained or undrained conditions. Complex localisation patterns have been revealed in axisymmetric triaxial tests using medical X-ray scanners to perform tomographic studies allowing for the first time to have a look inside the specimens rather than simply at their outer surface [1, 5]. These studies, using imaging techniques that were new at that time, already opened new windows on the early mechanisms of strain localisation in laboratory specimens: it became clear that the initiation of strain localisation in specimens undergoing a loading process was not properly characterised by the naked eye observation of the specimen during the test, which let see only final gross patterns that can be rather different from the early ones.

J. Desrues (✉) · E. Andò · P. Bésuelle · G. Viggiani · L. Debove · P. Charrier · J.B. Toni
Institute of Engineering, Université Grenoble Alpes, CNRS, Grenoble INP, 3SR, 38000 Grenoble, France
e-mail: jacques.desrues@3sr-grenoble.fr

E. Papamichos et al. (eds.), *Bifurcation and Degradation of Geomaterials with Engineering Applications*, Springer Series in Geomechanics and Geoengineering, DOI 10.1007/978-3-319-56397-8_1

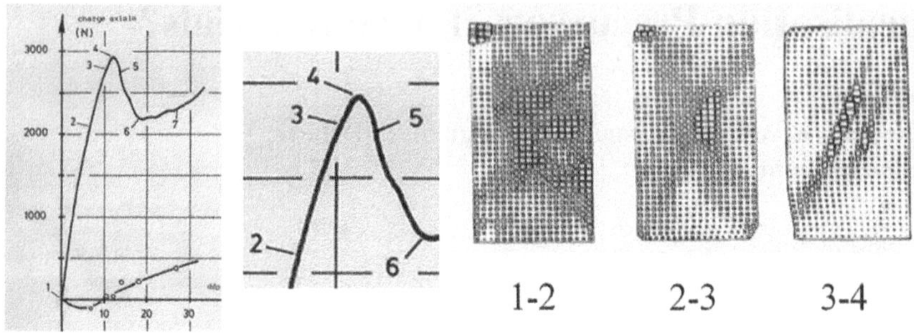

Fig. 1 Incremental deviatoric strain field in biaxial test SHF06 on dense Hostun RF sand: well structured shear bands are observed at the very end of the pre-peak regime. After Desrues [4]

Biaxial tests on sand performed using setups that allowed a direct observation and photography of the front face of parallelepipedic specimens subjected to a plane strain deformation processes [4, 6, 7] indicated that strain localisation starts to develop at or before the stress peak of the global response. Incremental deviatoric strain fields, characterised at the best resolution that was available by that time (i.e. up to the years 2000) were revealing a typical change from strain field distribution from what appeared as a more or less homogeneous, or diffuse heterogeneous mode, to a shear band mode, as illustrated in Fig. 1. Triaxial tests on the same sand, performed under tomographic survey using a medical X-ray scanner, revealed for the first time complex shear band patterns, observable as dilatancy bands but only after the stress peak (Fig. 2). Identical observations were reported by other teams in the world [1, 10].

In recent years, imaging techniques and image analysis methods have improved dramatically, allowing considerably more accurate measurement of displacement and strain field in the laboratory specimens. High resolution computed tomography (CT), associated with 3D volume digital image correlation (DIC), are the new glasses that have changed our observation capabilities. Nowaday, performing in-situ (i.e. inside a tomograph) triaxial tests on 70 mm specimens of Hostun sand with a voxel size as small as 50 μm is possible. With such a resolution, one single grain among the 14 millions contained in the 538 cm^3 volume of the specimen contains on average $6^3 = 216$ voxels. This is not enough to perform a exhaustive survey of the kinematics of the individual grains, as made successfully on 10 mm diameter triaxial specimens of the same sand using our CT scanner (discrete analysis, Andò et al. [2]); but it is enough to perform high resolution DIC in terms of continuum kinematics, allowing for a completely new insight of the pre-peak deformation of granular media in axisymmetric triaxial tests, as will be shown in this paper.

This study was motivated by intriguing observations made recently on several experimental setups developed in our lab, using high resolution imaging techniques on different granular materials, soils and rocks. Andò et al. [2] show pre-peak deformation patterns in small triaxial specimens analysed using discrete DIC (Fig. 3). Bésuelle and Lanatà [3] report the same kind of observations in plane strain

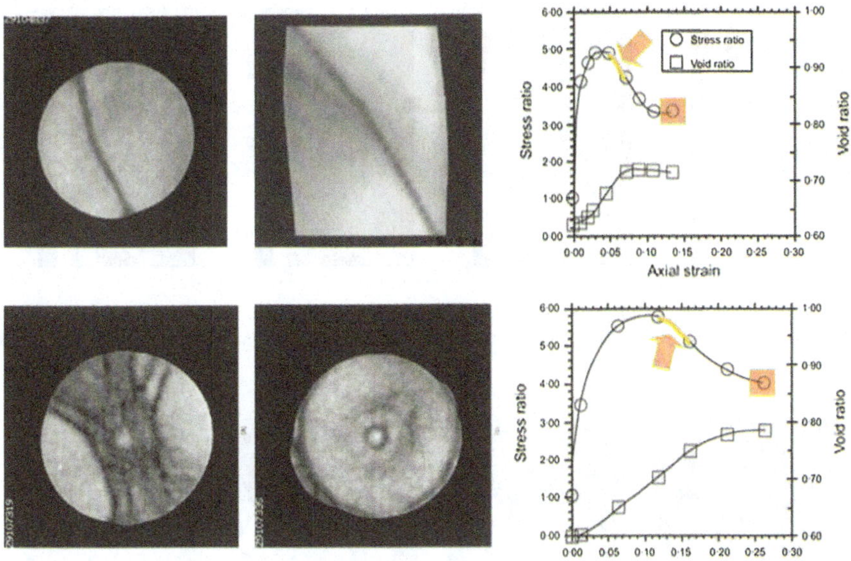

Fig. 2 Complex shear band patterns in triaxial tests on dense Hostun RF sand: observed shear bands in the post-peak regime. After Desrues et al. [5]

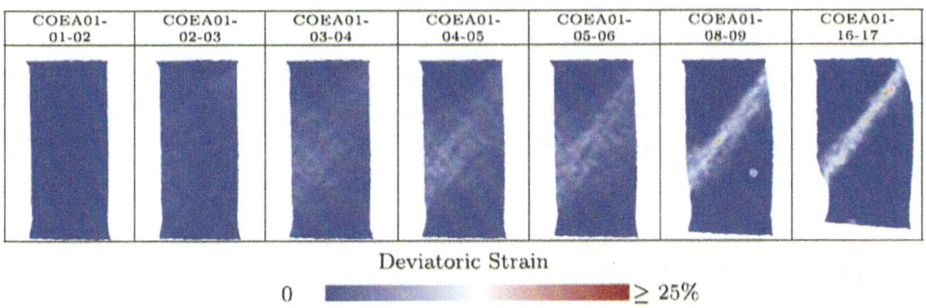

Fig. 3 Incremental deviatoric strain fields in small-size triaxial tests on caicos ooids observed using micro-tomography and discrete DIC. After Andò et al. [2]

compression tests on rocks, using DIC to analyse 2D high resolution digital images of a side of the specimen, taken during the test through a transparent window (Fig. 4). Hall et al. [8] report results suggesting the presence of early localised events in the $1\gamma2\epsilon$ apparatus in 3SR lab, Grenoble (Fig. 5). Lebouil et al. in Institut de Physique de Rennes, France find also some early localised deformation bands in glass beads, using a different testing setup and imaging technique based on laser light diffusion within the first millimeter of the specimen [9].

The present study is performed on standard scale laboratory specimens, subjected to classical axisymmetric triaxial tests with various test conditions in terms of initial void ratio, end lubrication, slenderness ratio. The results can be compared directly

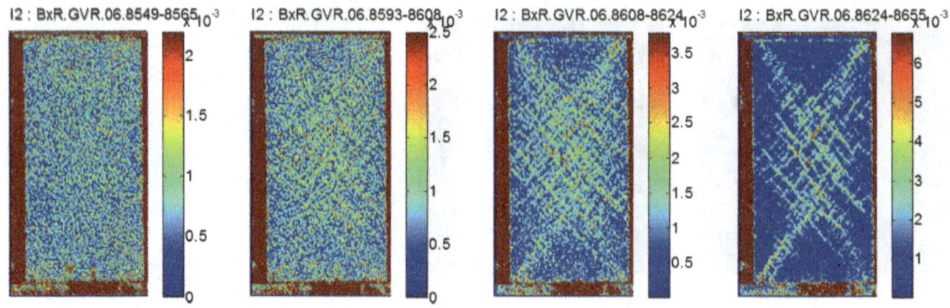

Fig. 4 Plane-strain compression test at 20 MPa confining pressure: incremental fields, between photographs numbered on top of each picture, of the second invariant of the incremental strain tensor. After Bésuelle and Lanatà [3]

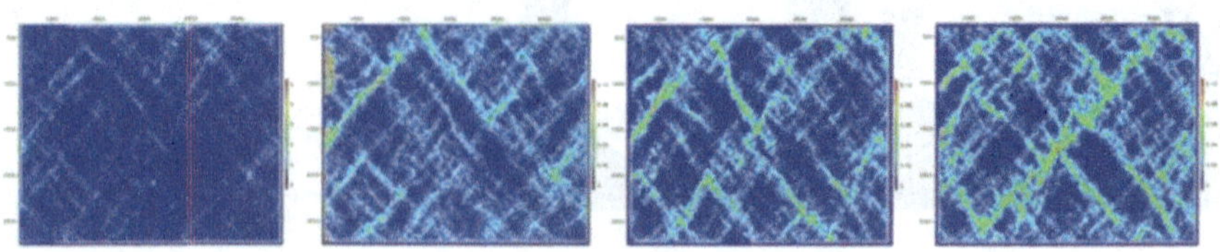

Fig. 5 Incremental deviatoric strain fields in biaxial tests on 2D granular media (Scheenebeli material). After Hall et al. [8]

with our 25 years old tomographic results published in 1996 [5]. The goal is to check whether early localisation events can be detected in classical test conditions, and if yes to try to characterise in detail these events.

2 Materials and Methods

Hostun NH31 sand produced by SIBELCO-France is used in this study. HN31 is the present denomination of Hostun RF sand that was used in the past in our lab. The characteristics of the two sands are the same.

The micro-tomograph used in the study was designed and build by RX-Solution, Annecy, France.

The DIC code Tomowarp2 used is an in-house code developed by Tudisco et al. [11].

The triaxial setup was designed specifically for this study. The pressure cell is made of PVC, it can sustain up to 600 kPa (limited pressure in order to avoid too thick cell walls, which would absorb to much of the X-ray flux). The loading frame allows platen displacement speed in the range 1–25 μm per second. The nominal specimen diameter and length are respectively 70 mm and 140 mm (or 70 mm for short specimens).

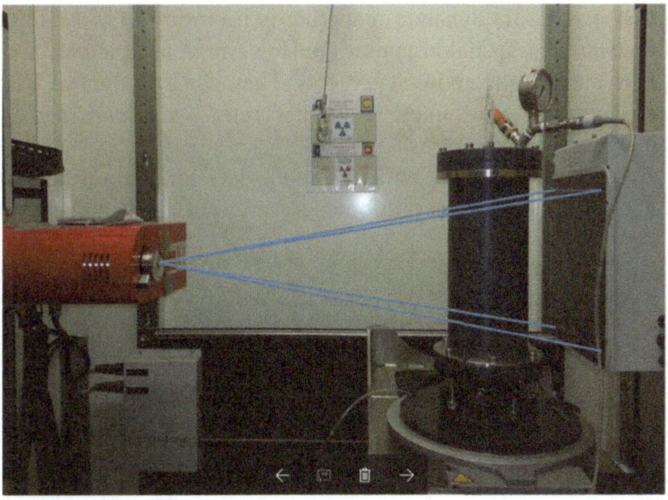

Fig. 6 "Tomotriax Φ70" setup in the X-ray scanner at 3SR

Figure 6 shows the triaxial setup ready for operation inside the CT scan. From left to right one can see the X-ray source, the rotation table with the triaxial cell fixed on it, and the detector. The cell is placed close to the detector because the field to be imaged is large (specimen in its possible laterally expanded configuration).

3 Results and Analysis

Although the test campaign is still running, and despite the very time-consuming data processing steps (tomographic image acquisition and reconstruction, digital image correlation, strain field computation, resulting 3D images exploration et rendering) to perform on 3D images as big as 20 Gbytes each, the picture that emerges already is the following: early localisation events are present almost from the beginning of the loading in all the tests, including those performed on loose specimens. These events are numerous, and for each of them the geometrical extension (length of the band) is significant with respect to the size of the specimen: typically from one third to one half of the specimen's length, often more. Some of these events tend to grow in intensity as the global specimen's deformation progresses, some tend to decay. At 20% global axial strain, a few shear bands can be identified clearly, concentrating all the deformation process, and visible on the specimen's surface with the naked eye. Still, the direct observation at the stress peak (about 6–8% axial strain) did not reveal any visible localisation.

Illustrating these results as 2D images is challenging, because the very subtle and numerous early localised events become confused when projected onto a plane to produce a 2D picture; 3D dynamic rendering is necessary to separate the events, and to understand their spatial organisation. However, a few pictures are proposed in this paper to support the above statements. They illustrate the test TT70-HN31-001 performed on a dense specimen, slenderness ratio 2, non-lubricated ends, tested under 100 kPa confining pressure.

Figure 7 shows the sequence of incremental deviatoric strain fields presented as 3D volumes with transparency in order to reveal the inner organisation of the

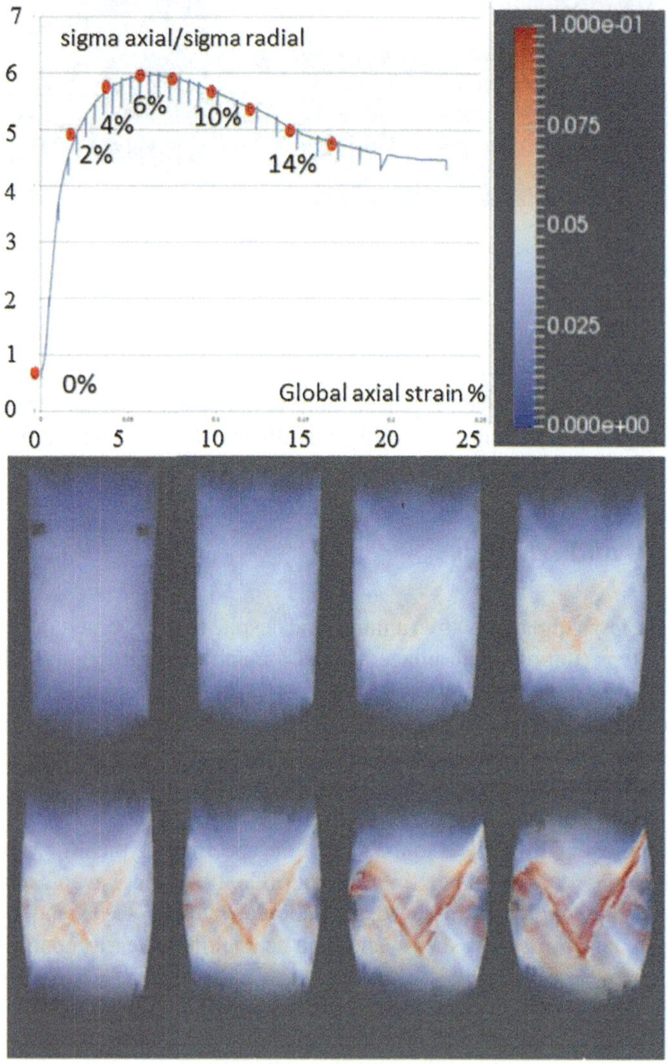

Fig. 7 Incremental deviatoric strain fields in axisymetric triaxial test on Hostun HN31 sand observed using micro-tomography and continuum DIC

deformation events. Each image corresponds to a 2% increment of global axial strain. As indicated by the lookup table, dark blue is for zero and intense red or 10% deviatoric strain. Although difficult to observe clearly on such 2D projections, early events can be detected already in the first increment.

In other tests, not presented in this paper, it was observed that such pre-peak structuration of the deformation process occurs in all tests performed so far, including dense and loose specimens, long and short, end-lubricated or not.

In the authors' belief, the observed localised events cannot be considered as simple local slips between grain clusters: their typical length is two decades above the mean grain size of the tested sand.

4 Conclusions

Diffusely distributed strain localisation events have been observed from the early stages of triaxial tests on sand, using X-ray micro tomography and Digital Image Correlation. These finding seems to be rather general: dense and loose specimens, long and short, end-lubricated or not. These events have all the characteristics of shear bands. Should them be considered as precursors of strain localisation, or as already achieved localised events? The theoretical interpretation of these findings presents a number of interesting questions to be studied in future work.

Acknowledgements The research leading to these results has received funding from the European Research Council under the European Union's Seventh Framework Program FP7-ERC-IDEAS Advanced Grant Agreement No 290963 (SOMEF). The laboratory 3SR is part of the LabEx Tec 21 (Investissements d'Avenir - grant agreement No ANR-11-LABX-0030).

References

1. Alshibli, K., Sture, S., Costes, N., et al.: Assessment of localized deformations in sand using x-ray computed tomography. Geotech. Test. J. GTJODJ **23**(3), 274–299 (2000)
2. Andò, E., Hall, S.A., Viggiani, G., Desrues, J., Bsuelle, P.: Grain-scale experimental investigation of localised deformation in sand: a discrete particle tracking approach. Acta Geotechnica **7**(1), 1–13 (2012)
3. Bésuelle, P., Lanatà, P.: A new true triaxial cell for field measurements on rock specimens and its use in the characterization of strain localization on a vosges sandstone during a plane strain compression test. Geotech. Test. J. **39**(5), 877–890 (2016)
4. Desrues, J.: La localisation de la dformation dans les matriaux granulaires, 1984. Thse de Doctorat es Sciences
5. Desrues, J., Chambon, R., Mokni, M., Mazerolle, F.: Void ratio evolution inside shear bands in triaxial sand specimens studied by computed tomography. Gotechnique **46**(3), 529–546 (1996)
6. Desrues, J., Lanier, J., Stutz, P.: Localization of the deformation in tests on sand sample. Eng. Fract. Mech. **21**, 909–921 (1985)
7. Desrues, J., Viggiani, G.: Strain localization in sand: an overview of the experimental results obtained in Grenoble using stereophotogrammetry. Int. J. Numer. Anal. Methods Geomech. **28**(4), 279–321 (2004)

8. Hall, S.A., Wood, D.M., Ibraim, E., Viggiani, G.: Localised deformation patterning in 2d granular materials revealed by digital image correlation. Granul. Matter **12**(1), 1–14 (2010)
9. Le Bouil, A., Amon, A., Sangleboeuf, J.-C., Orain, H., Bsuelle, P., Viggiani, G., Chasle, P., Crassous, J.: A biaxial apparatus for the study of heterogeneous and intermittent strains in granular materials. Granul. Matter **16**(1), 1–8 (2014)
10. Tatsuoka, F., Sakamoto, M., Kawamura, T., Fukushima, S.: Strength and deformation characteristics of sand in plane strain compression at extremely low pressures. Soils Found. **26**(1), 65–84 (1986)
11. Tudisco, E., and, E., Cailleteau, R., Hall, S.A.: Tomowarp2 : a local digital volume correlation code. Submitted to Software XC (2017)

Water Saturation Induced Strength Degradation of Callovo-Oxfordian Claystone

Zaobao Liu, Jianfu Shao and Ying Xu

Abstract It is necessary to investigate the effect of initial water content on the strength behaviors of the Callovo-Oxfordian (COx) argillite since the construction activities of underground radioactive waste repositories can induce a desaturation and a resaturation process of the hosted rock. The present work is devoted to an experimental characterization of the water induced strength degradation the COx argillite under constant strain rate loadings. Argillite samples of initial relative humidity (RH) of 98% are firstly tested at confining pressure of 4 MPa, 8 MPa and 12.4 MPa to derive a strength criterion. Then another group of tests on argillite samples with different saturation realized by relative humidity (dry, 76 and 85%) are carried out to quantify the water content induced strength degradation in claystone. It is found that both the peak and residual stress and the failure strain are correlated with the humidity level of the claystone. The results give the implications that the desaturation and re-saturation of argillite will exert influences on its surroundings in the underground repositories. Special attention should be paid to minimize the swelling effect of clay minerals and oxidation of pyrite inclusions in the argillite with water presence.

Keywords Argillite · Damage · Failure · Water degradation · Radioactive waste disposal

1 Introduction

Geological disposal is considered feasible worldwide for disposal of high-level radioactive waste. Claystone, due to their low permeability, self-sealing ability and the absence of major natural fractures, have been selected as the hosted rock for

---

Z. Liu (✉) · J. Shao · Y. Xu
Laboratory of Mechanics of Lille, University of Lille, 59655 Villeneuve D'Ascq, France
e-mail: zaobao.liu@polytech-lille.fr

Y. Xu
School of Mines, China University of Mining and Technology, Xuzhou 221116, China

E. Papamichos et al. (eds.), *Bifurcation and Degradation of Geomaterials with Engineering Applications*, Springer Series in Geomechanics and Geoengineering, DOI 10.1007/978-3-319-56397-8_2

underground radioactive waste repositories in France. In this context, the Callovo-Oxfordian (COx) argillite, has been extensively studied during the last decades in France in the framework of the underground research laboratory CIGEO [1] to characterize its thermo-hydro-mechanical behaviors.

In the construction of underground nuclear wastes repositories, the initially saturated claystone will be subjected to various couplings such as excavation induced loadings, water and gas flow, desaturation and resaturation processes after sealing. Due to the presence of clay minerals especially the smectite, the mechanical behavior of the claystone is very sensitive to its water content [2, 3]. Thus, it is necessary to investigate the effect of RH level on the mechanical behaviors of the COx clayey rock. Experimental investigations have been reported on the mechanical and hydro mechanical behaviors of the claystone in partially saturated conditions [2–6]. It is found that the strength decreases and the clayey rock becomes more ductile when RH level increases. The initial elastic modulus also decreases with water content while Poisson's ratio is slightly affected by water content. These effects are thought inherently related to the change of microstructure in the clayey rock [7]. It is thought the desaturation and resaturation process of the clayey rock may modify the distance between the clay platelets, leading to modification of the mechanical properties of clay aggregates and thus the clayey rock [8]. Moreover, damage is found more important in claystone with lower moisture values than higher ones. The shear bands and shearing factures created during triaxial compression are related to moisture levels of the claystone [9].

This study quantifies strength degradation of the COx claystone induced by increasing saturation degree. Triaxial compression tests of three confining pressures are firstly carried out to obtain a strength criterion. The pre-peak and post-peak strength of the COx claystone of different saturation is investigated.

2 Materials, Experimental Device and Method

The tested material is cored from the underground research laboratory in Meuse/Haute-Marne of France in the Callovo-Oxfordian argillite formation. The clayey rock in microscale is constituted of three major minerals, i.e. quartz, calcite and clay minerals [10, 11]. The pores of the argillite are mainly in nanoscale [1, 12].

The rock cores of argillite, although well packaged are sometimes fractured in process of transportation and conservation due to unexpected damage as well as the oxidation induced cracking of highly active inclusions. We have encountered many fractures in the original argillite rock cores although they are protected firstly by a plastic jacket, then by a concrete confinement, and eventually by a rigid plastic package with constrains at both ends.

Preparation of argillite samples with relative large diameter is difficult since the argillite is easily damaged and fractured in process of cutting and drilling in laboratory. Argillite samples with large diameters are more likely to be influenced by the inclusions. We have to use air as coolant since the argillite is prone to splinter when contacted with water without confining pressure. Thus, the drilling of samples will produce around the drilling rig a lot of dusty particles which are very fine and harmful to the health of people. And the drilling will sometimes induce specimen cracking if there are certain pre-existing cracks or weak zone in the rock cores. Sometimes the cracks can go through the specimens, which make the samples useless. Moreover, once some expected samples are drilled from a sound rock core without fractures, they are then subjected to the tests of realizing varying water contents by placement and conservation in a small container where the RH is maintained by different brines. Due to the low permeability of argillite, the desaturation and resaturation processes in the container will take a very long time depending on the sample size. At the sample time, the desaturation or resaturation generates a gradient of water content inside the sample. Due to this gradient, there is kinematic incompatibility of swelling or shrinkage strain leading to creation of local tensile stresses. Such tensile stresses may become significant in large samples and then responsible for micro cracks. Thus, many samples are cracked due to swelling of the clay minerals as well as oxidation of organic inclusions in the container with high RH levels. Therefore, special attention should be paid to the swelling of clay minerals and oxidation of active inclusions in the claystone. In the present study, the samples are protected by a scotch tape at both ends and finally they are in a good manner for usage in tests.

The samples used in this study are drilled from the same rock core with a diameter about 37 mm and a length about 74 mm. Both sample ends are polished to be perpendicular to the sample axis once they are drilled and cut to agree with the expected size. Then, all samples are conserved in a closed container in which the RH is maintained at a given level, say 76, 85 and 98%. The temperature around the container is kept 23 °C by a central air conditioner. The samples are kept in humidity controlled containers and will not be used for mechanical tests until their mass values stabilize. The dry sample is heated before test progressively in an oven at 50 °C (2 days), 80 °C (2 days) and 105 °C (3 days) to minimize the preheating-induced cracks. The RH is taken as 0% for the dry sample, which assumes that no water is retained after successive heating.

The experimental program is devoted to characterize the damage and failure behaviors of the argillite under constant strain rate loading conditions. All the tests are realized in an autonomous and auto-compensated hydro-mechanical testing system patented to the University of Sciences and Technologies. The testing system consists of three independent loading components, respectively for deviator stress loading, confining pressure application, and interstitial pressure generation, which are assembled independently in the triaxial cell. The monitoring and acquisition of pressure/stresses, fluid pressure and/or flow rate, displacements or strain is realized by some specific transducers and recorded by a data acquisition center.

The confining pressure is applied and maintained by an ISCO D260 series pump which has a precision of 0.1 MPa, and the interstitial pressure is maintained by a gas supplier controlled by a manometer which has a precision of 0.1 MPa. When applying confining pressure in the apparatus, the samples are subjected to a hydrostatic stress state. The deviator is loaded by a rigid INSTRON mechanical loading machine. The loading and unloading of the deviatoric stress are controlled by a constant axial strain rate of $\dot{\varepsilon}_1 = 0.5 \times 10^{-5}$/s. A pair of LVDT and a radius ring measure the deformation. The axial strain ε_1 is thus obtained by the ratio between the LVDT displacement and the sample length, and the radius strain ε_3 is obtained by the ratio between the ring deformation and the sample diameter. Volumetric strain ε_v of the samples in the mechanical tests are calculated by

$$\varepsilon_v = \varepsilon_1 + 2\varepsilon_3 \tag{1}$$

The deviatoric stress q in the sample is calculated by

$$q = \sigma_1 - \sigma_3 = (F - F_f)/A \tag{2}$$

where, F is the applied force; F_f is the piston friction about 0.6 KN in the tests.

3 Mechanical Curves

The strain and deviatoric stress throughout the constant strain rate loading are shown in Fig. 1a–c for the COx clayey rock samples of the RH = 98% under three confining pressures and shown in Fig. 1d–f for the samples of three different RH values under Pc = 12 MPa.

The peak strength values of the samples are related to the RH values of the samples and will be discussed later. The axial strain at macro failure of the sample of dried, with RH = 76%, 85% and 98% are respectively 1.19%, 2.22%, 2.23% and

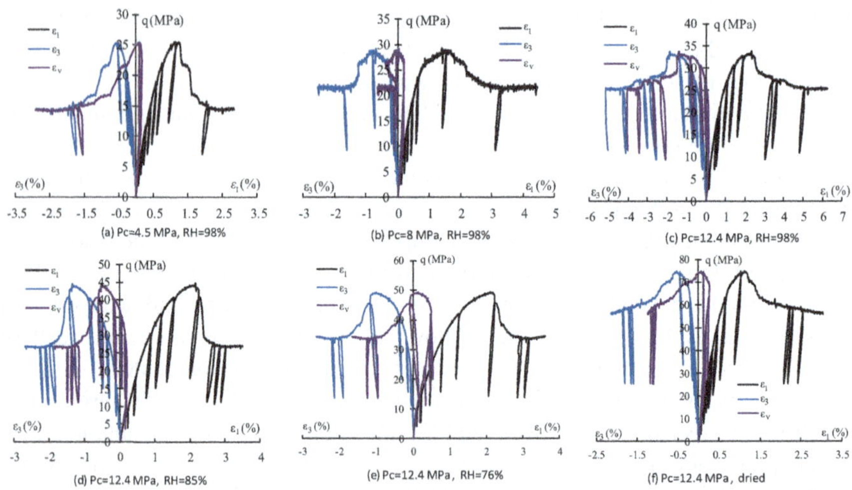

Fig. 1 Strain versus stress of claystone with different RH values

2.51%. It suggests that the strains at failures of the samples are also influenced by the RH. Samples with bigger values of RH seem to be able to bear larger deformation than the ones with lower relative humidity. Thus, high RH level can enhance the strains of the argillite samples.

Moreover, evolution of volumetric strain of the samples is influenced by its initial RH as indicated in Fig. 1, especially the onset of dilation, i.e. the turning point at which the volume strain turns from compaction to dilation in the strain-stress curves. The volumetric strain stops compacting (onset of dilation) at the deviator of q = 68.4 MPa for the dry sample (RH = 0), and respectively about q = 42.8 MPa, 26.1 MPa and 13.5 MPa for samples with RH = 76, 85 and 98% as shown in Fig. 1. Hence, the stress onset of dilation of the argillite is closely related to its RH values. The COx claystone with higher RH values arrives at its dilation onset at a much smaller stress than those with lower ones.

4 Strength Degradation

The detail strength properties of the claystone are shown in Fig. 2. As shown in Fig. 2a, the relation between the strength and mean stress can be well quantified by a linear function, which indicates a linear strength criterion is suitable to describe the strength behavior of the claystone of RH = 98%.

According to the evaluation results in Fig. 2a, one can calculate the cohesion c = 7.24 MPa and the friction angle φ = 20.3°, for the COx claystone at initial saturation degree of RH = 98%.

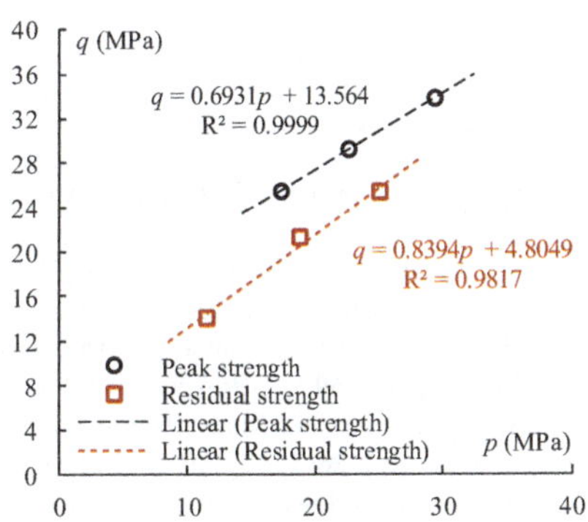

(a) q-p relation

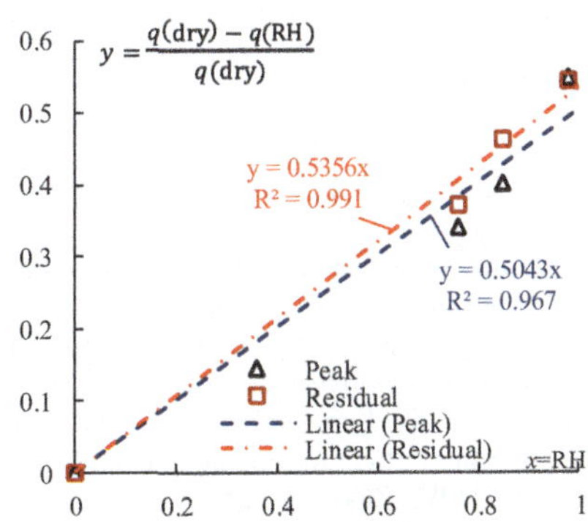

(b) q-RH relation

Fig. 2 Strength of claystone under different Pc and saturation degree

For the degradation in mechanical strength induced by the RH, we define

$$y = \frac{q(\mathrm{dry}) - q(\mathrm{RH})}{q(\mathrm{dry})},\ x = RH - 0\%(dry) = RH \tag{3}$$

One can have a linear relationship that can describe the strength degradation properties of the claystone induced by increasing initial saturation degree at Pc = 12 MPa

$$y = mx \tag{4}$$

The degradation coefficient m is respectively 0.5043 and 0.5356 for the peak and residual strength as indicated in Fig. 2b. The linear degradation is probably induced by the weakening mechanism that the layers of the illite and smectite are easier to collapse and fall down with more water content.

5 Conclusion

Initial saturation degree has important effects on strength and deformation of the COx claystone. High initial saturation weakens argillite mechanical strength but enhances its deformation. The onset of volume dilation advances in the COx claystone with higher saturation during triaxial loading. Linear functions can quantify the claystone saturation induced strength degradation. The degradation may be induced by water weakening of interlayered structures of illite and smectite.

References

1. Andra (2012) Recherche et développement à l'Andra pour les projets de stockage. Andra, Paris
2. Yang, D., Chanchole, S., Valli, P., Chen, L.: Study of the anisotropic properties of argillite under moisture and mechanical loads. Rock Mech. Rock Eng. **46**(2), 247–257 (2013)
3. Zhang, F., Xie, S.Y., Hu, D.W., Shao, J.F., Gatmiri, B.: Effect of water content and structural anisotropy on mechanical property of claystone. Appl. Clay Sci. **69**, 79–86 (2012)
4. Bemer, E., Longuemare, P., Vincké, O.: Poroelastic parameters of Meuse/Haute Marne argillites: effect of loading and saturation states. Appl. Clay Sci. **26**(1–4), 359–366 (2004)
5. Hu, D.W., Zhang, F., Shao, J.F., Gatmiri, B.: Influences of mineralogy and water content on the mechanical properties of argillite. Rock Mech. Rock Eng. **47**(1), 157–166 (2013)
6. Zhang, F., Hu, D.W., Xie, S.Y., Shao, J.F.: Influences of temperature and water content on mechanical property of argillite. Eur J Environ Civ En **18**(2), 173–189 (2013)
7. Bornert, M., Valès, F., Gharbi, H., Nguyen Minh, D.: Multiscale full-field strain measurements for micromechanical investigations of the hydromechanical behaviour of clayey rocks. Strain **46**(1), 33–46 (2010)

8. Robinet, J.-C.: Minéralogie, porosité et diffusion des solutés dans l'argilite du Callovo-Oxfordien de Bure (Meuse, Haute-Marne, France) de l'échelle centimétrique à micrométrique. Université de Poitiers, Poitiers (2008)
9. Liu, Z.B., Shao, J.F.: Moisture effects on damage and failure of Bure claystone under compression. Geotechnique Letters **6**(3), 182–186 (2016)
10. Andra (2005) Dossier 2005: CDROM Argile—plaquettes, synthèse, tomes, référentiels et glossaire
11. Liu, Z., Xie, S., Shao, J., Conil, N.: Effects of deviatoric stress and structural anisotropy on compressive creep behavior of a clayey rock. Appl. Clay Sci. **114**, 491–496 (2015)
12. Liu, Z., Shao, J., Xie, S., Secq, J.: Gas permeability evolution of clayey rocks in process of compressive creep test. Mater. Lett. **139**, 422–425 (2015)

Propagation and Evolution of Strain Localization in Clay

Dunja Perić and Xingdong Wu

Abstract This research focuses on propagation and evolution of strain localization in clay. To this end, an undrained plane strain compression test program was performed to investigate the effects of the past stress history and strain rates on the strain localization response of kaolin clay. Results of a single test (T3) are presented herein. They include the overall global stress/pore pressure versus axial strain response, as well as indicators of propagation and evolution of strain localization. The results indicate that the response consisted of the three distinct phases: (1) a homogenous deformation, (2) an inception and propagation of strain localization, and (3) an evolution of strain localization within the shear band. The Lagrange strain tensor was used to obtain local volumetric and shear strains developed within the shear band during the evolution stage. It was found that local volumetric and shear strains reached 80% and 200% respectively, which corresponds to a global axial strain increment of only 1.7%. Furthermore, the shear band exhibited a decreasing tendency to compress, thus resulting in its incremental dilatancy angle varying from initially −24° to a final value of −6°.

1 Introduction

Strain localization in granular materials such as sands has been extensively investigated in the past 30 years, both physically and mathematically, thus resulting in significant advances in the related knowledge. This study was undertaken to

---

The original version of this chapter was revised: Incorrect author name has been corrected. The erratum to this chapter is available at 10.1007/978-3-319-56397-8_78

D. Perić (✉) · X. Wu (✉)
Department of Civil Engineering, Kansas State University, 2111 Fiedler Hall, Manhattan, KS 66506-5000, USA
e-mail: peric@ksu.edu

X. Wu
e-mail: chinawu@ksu.edu

E. Papamichos et al. (eds.), *Bifurcation and Degradation of Geomaterials with Engineering Applications*, Springer Series in Geomechanics and Geoengineering, DOI 10.1007/978-3-319-56397-8_3

further elucidate propagation and evolution of strain localization in undrained plane strain tests on slurry consolidated kaolin clay.

2 Experimental Program

An experimental program was designed and performed to investigate effects of a stress history and loading rate on development of strain localization in clay samples. To this end lightly and slightly heavily overconsolidated samples were tested in compression at two tenfold different axial strain rates Perić and Hwang [1]. Prismatic clay specimens were trimmed from slurry consolidated samples. They were subsequently mounted on the bottom platen inside a biaxial cell. The bottom platen was placed on the top of a linear bearing or sled, thus enabling an uninhibited development of strain localization. All clay samples were subjected to a mixed control undrained loading whereby the axial and horizontal out of plane strain rates were controlled along with the rate of the horizontal in plane total stress.

The plane strain apparatus was heavily internally instrumented to facilitate close monitoring of ensuing failure mechanisms. The internal instrumentation included seven LVDTs and eight load cells. External instrumentation included three pore pressure transducers for measurements of top and bottom pore pressures, and of a volume change of a specimen during the consolidation phase.

3 Results

Although a total of six undrained plane strain compression experiments were carried out only one of them (T3) is discussed herein. The corresponding sample was consolidated to a mean effective stress of 55.2 kPa in the initial stage of a biaxial test. This state corresponds to the overconsolidation ratio of 2.91 based on the past maximum mean effective stress applied during the slurry consolidation. An overall response of the clay sample observed in test T3 is depicted in Figs. 1 through 2,

Fig. 1 Deviatoric stress versus global axial strain for test T3

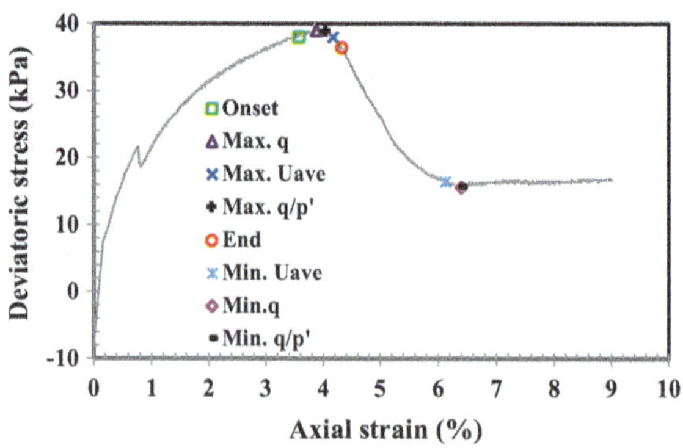

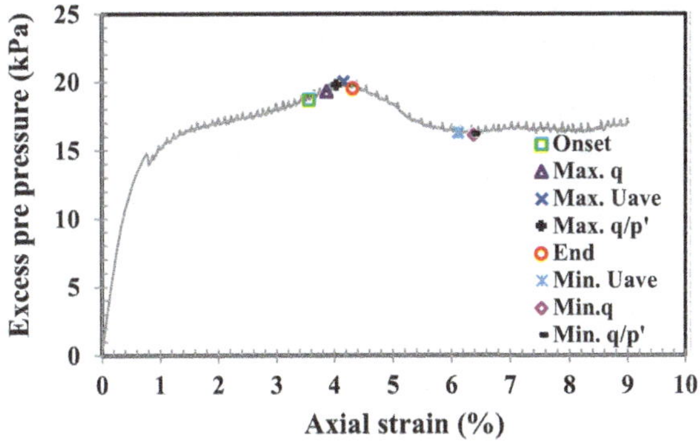

Fig. 2 Excess pore pressure versus global axial strain for test T3

which show plots of the deviatoric stress and average excess pore pressure versus global axial strain. The average excess pore pressure is the average value of the top and bottom measurements. All stresses were obtained from the boundary measurements of forces and a chamber pressure. Furthermore, it is noted that in test T3 the global axial strain rate was about tenfold decreased (from 3.3 %/hr to 0.33%/hr) at about value of global axial strain of 0.75%.

Measurements obtained from internal LVDTs indicated that the response of the sample was homogenous prior to the onset of strain localization, thus indicating that a global or sample level response nearly coincides with a pointwise or local response. This phase was followed by the first cluster of closely spaced events, which commences with the onset of strain localization. It also contains the maximum values of deviatoric stress, average pore pressure, and stress ratio (deviatoric to mean effective stress), as well as the end of a shear band propagation (Figs. 1 and 2). Thus, the maximum values of deviatoric stress, average pore pressure and stress ratio occurred during the propagation of the shear band. The second cluster of closely spaced events emerged during the further evolution of the shear band. It contains the minimum values of the average excess pore pressure, deviatoric stress and stress ratio (Figs. 1 and 2). It is noted that the deviatoric stress herein is a generalized deviatoric stress, thus including effects of all three principal stresses.

Evolution of local strains is presented next. A shear band is assumed to be inextensible during the evolution phase simply because a change in length due to an elastic unloading is negligible in comparison to the length of the shear band. It is also assumed that the entire shear band deforms uniformly, thus undergoing the displacement pattern shown in Fig. 3, which depicts only a small portion of the total length of the shear band. Furthermore, the orientation was obtained from the photographs and assumed to be constant after the end of propagation when computing local strains.

In the present computation elastic unloading outside the shear band is neglected. Thus, horizontal and vertical displacements of the upper boundary of the shear band are equal to the displacements of the sled and of the top sample surface respectively. The Lagrange strain is used to compute local volumetric and shear strains based on the displacements of the top boundary of the shear band.

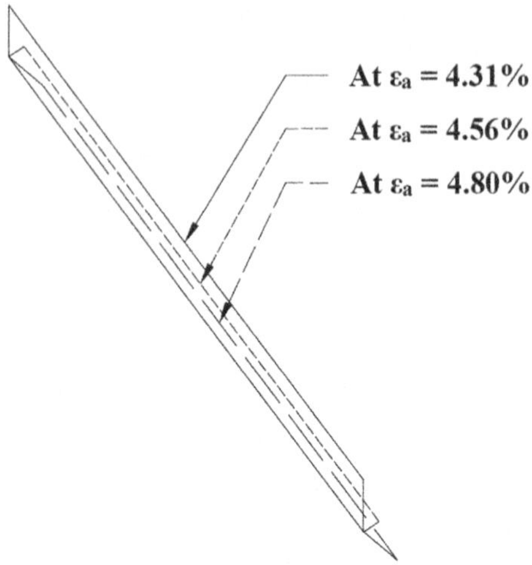

Fig. 3 Un-deformed and deformed configurations of the shear band at different values of global axial strain for test T3

Consequently, shear (ε_s) and volumetric strains (ε_v) of the shear band respectively are given by:

$$\varepsilon_s(t) = \frac{u_T(t)}{t_E} \cos \psi(t) \tag{1}$$

and

$$\varepsilon_v(t) = -\frac{u_T(t)}{t_E} \sin \psi(t) \tag{2}$$

where t is time, u_T is a magnitude of a top shear band boundary displacement, while t_E is the thickness of the shear band that corresponds to the end of the propagation stage. In this particular test t_E was rather small and equal to 0.33 mm. Finally, a total dilatancy angle (ψ) is given by:

$$\psi(t) = \theta_B - \tan^{-1}\left[\frac{u_{Tz}(t)}{u_{Ty}(t)}\right] \tag{3}$$

where u_{Tz} and u_{Ty} are vertical and horizontal displacements of the top shear band boundary respectively, while θ_B is the angle between the shear band and horizontal direction. It is noted that this definition results in negative dilatancy angle for a contracting shear band.

Figure 4 shows local volumetric and shear strains computed in accordance with Eqs. (1) and (2).

Figure 5 depicts total and incremental dilatancy angles computed in accordance with Eq. (3).

Fig. 4 Local volumetric and axial strains for test T3

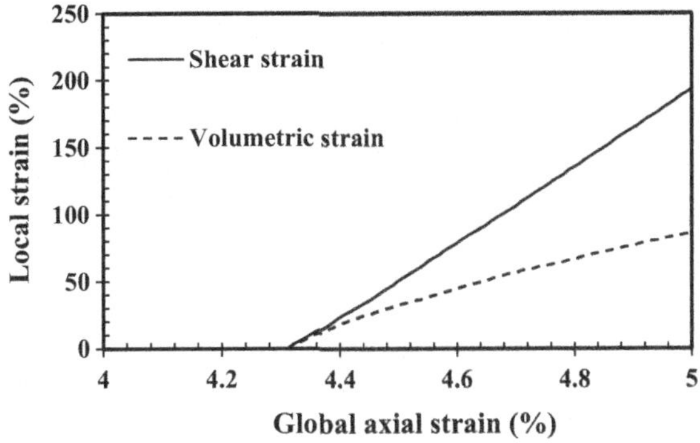

Fig. 5 Total and incremental shear band dilatancy angles for test T3

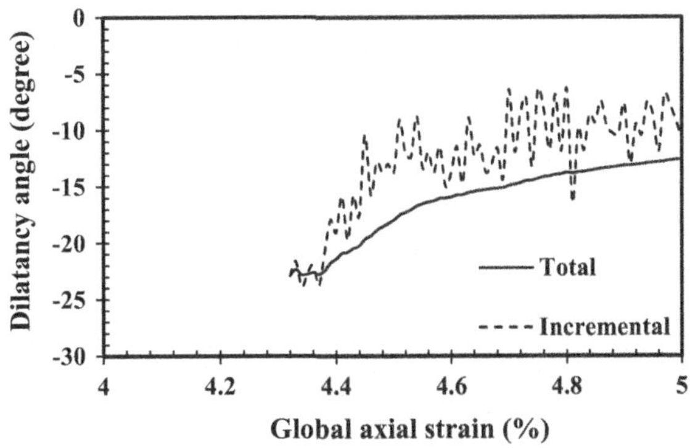

4 Conclusions

Results of an undrained plane strain compression test on clay sample having an overconsolidation ratio of 2.91 are presented. They indicate that three distinct phases of response exist: (1) a homogeneous deformation, (2) onset of strain localization, which is followed by shear band propagation, and (3) shear band evolution. Furthermore, strains are highly localized during the third stage, thus resulting in values of local volumetric and shear strains of 80% and 200% respectively, which correspond to the global axial strain of 5%. Furthermore, a dilatancy angle of the shear band increases from −24° to −6° during the evolution stage, thus indicating a decreasingly contractant shear band.

Reference

1. Perić D, Hwang, C.: Experimental investigation of plane strain behaviour of Georgia kaolin. In: Pande, G.N., Pietruszczak, S. (eds.) Numerical Models in Geomechanics—NUMOG VIII, Proceedings of the 8th International Symposium on Numerical Models in Geomechanics, pp. 93–98 (2002)

Emergence of Strain Localization in Porous Rocks Characterized by Full-Field Measurement in Plane Strain Condition

Pierre Bésuelle and Patrizia Lanatà

Abstract The development of a new true triaxial cell which contains a device to observe the specimen under loading for full field measurement, has allowed studying the emergence of strain localisation in rock specimens. Two porous rocks have been investigated, a sandstone and a clayey rock. Both show similarities in the strain localization process, with an early diffuse strain localization pattern well before the stress peak that evolves progressively toward a reduced number of shear bands pattern. This final pattern, being usually reported in the literature, is associated to the stress peak and initiation of strain softening.

1 Introduction

This study attempts to review recent experimental observations on two porous rocks about the progressive evolution from diffuse to localized deformation. Spatial descriptions of physical quantities and the time evolution of these fields are needed for this aim, in contrast to more conventional measurement techniques based on global measurement by transducers positioned outside of the loading device and/or at the specimen boundaries (no field measurement) or on *post mortem* characterization (no time evolution). Such techniques which are referred to as full-field measurements became more and more popular during the last two decades. They are potentially well adapted to detecting, during laboratory tests, the transition from the initial (quasi-) homogeneous regime to the localized regime.

P. Bésuelle (✉) · P. Lanatà
Université Grenoble Alpes, CNRS 3SR, 38000 Grenoble, France
e-mail: pierre.besuelle@3sr-grenoble.fr

E. Papamichos et al. (eds.), *Bifurcation and Degradation of Geomaterials with Engineering Applications*, Springer Series in Geomechanics and Geoengineering, DOI 10.1007/978-3-319-56397-8_4

2 A True Triaxial Cell for Full Field Measurement and Studied Rocks

A true triaxial apparatus was developed at Laboratoire 3SR in Grenoble (France), with the aim of characterizing the onset of localization and the post-localization regime in rocks [3]. The device can apply three independent stresses along the three space directions of a prismatic rock specimen, and allows its visualization under loading. A simplified schematic of the apparatus is shown in Fig. 1. The specimens surfaces, perpendicular to the principal stress (direction 1) and intermediate compressive stress (direction 2), are in contact with rigid platens, which are moved by two perpendicular pistons, as in some triaxial setups developed previously [4, 6]. The two surfaces of the specimen, perpendicular to the minor stress (direction 3), are free to deform because the stress is applied by a confining fluid (by means of a soft silicon membrane). Moreover, one of the two surfaces perpendicular to the intermediate stress (direction 2) is in contact with a hard window, in order to observe the specimen under loading and take photographs. The intermediate stress can be reg-

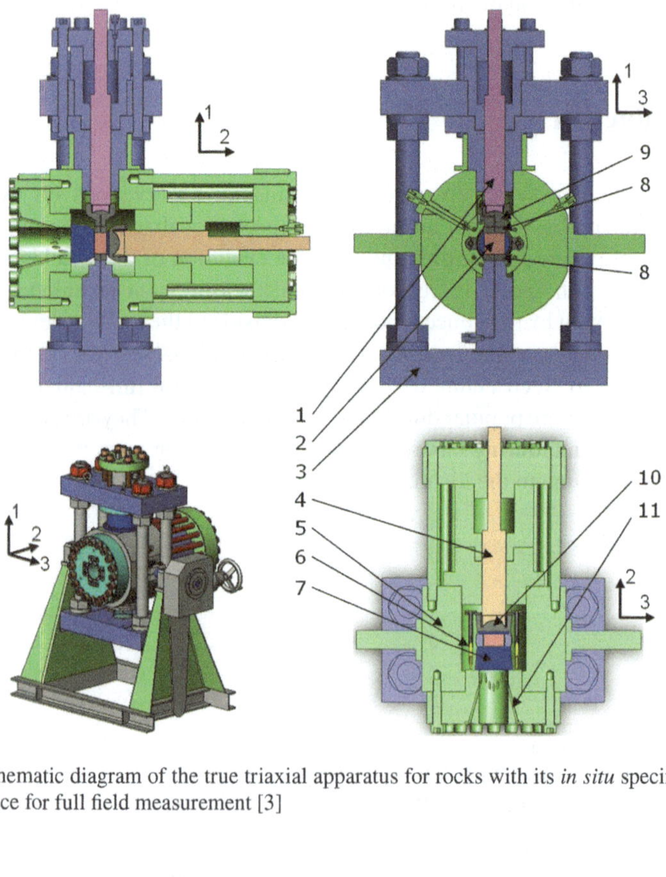

Fig. 1 Schematic diagram of the true triaxial apparatus for rocks with its *in situ* specimen observation device for full field measurement [3]

ulated to ensure that there is no deformation along this direction, thus allowing the application of plane strain loading, which implies that the kinematic of the surface in contact with the window is representative of the kinematic of the whole specimen, up to strain localization and beyond that level of stress.

Photographs of the visible surface of the specimen were taken throughout the loading with a high resolution camera. DIC analysis can thus be carried out on the resulting images to compute the displacement and strain fields. It is worth noting that to apply DIC, the observed specimen must have a random pattern, which is here artificial.

The size of the specimen is 50 mm in the axial direction, 30 mm in the direction of intermediate stress and 25 or 50 mm in the direction of minor stress, which corresponds to slenderness ratios (the ratio between the height and width) of two and one, respectively. The loading cell can apply a maximum confining pressure of 100 MPa, while the axial and the horizontal pistons can apply forces of 500 and 700 kN, respectively. These values correspond to a differential stress, with respect to the confining pressures of 670 and 530 MPa, respectively, for a specimen with a slenderness ratio of two, and half of these values for a 50 mm width specimen.

The first tested rock is a natural red sandstone coming from the Vosges mountains, Eastern France [2]. It is a pink quartz sandstone (quartz = 93%), a few percent of which is feldspar and white mica. The sandstone is poorly cemented, its cohesion is due to the interpenetration between the grains. Its porosity is about 22%. The mean grains size is about 0.30 mm. Its uniaxial compression strength is about 35 MPa.

The second tested rock is a Callovo-Oxfordian argillite, which has been considered as a potential host rock for radioactive waste disposal facilities, from the ANDRA Underground Research Laboratory (URL) at approximately 500 m below the ground surface. It is a sedimentary rock composed of phyllosilicates, tectosilicates, carbonates and pyrite [7]. Clay particles are clustered into aggregates, which are globally oriented along the stratification, whereas the other mineral inclusions have no preferential orientation. Its uniaxial compressive strength is about 20 MPa.

3 Selected Results

The mechanical behavior and failure by strain localization in the Vosges sandstone was extensively studied through axisymetric compression and extension tests [2]. The recent work of [5] has extended the analysis to plane strain compression tests. Two tests were selected: test BxR_GVR_06 was performed at a lateral stress of 20 MPa and test BxR_GVR_11 at 50 MPa. From [2], we know that the behavior of the rock at the lowest mean stress is quite brittle, the mean stress dependency on the deviatoric stress peak envelope is positive (cohesive-frictional behavior) and the shear bands are dilative. At the highest mean stress, the behavior was more ductile, the mean stress dependency on the deviatoric stress peak envelope was negative (cap surface), the shear bands were compactive and the inclination of the bands with respect to the axial loading direction was also higher than for a lower mean stress [1].

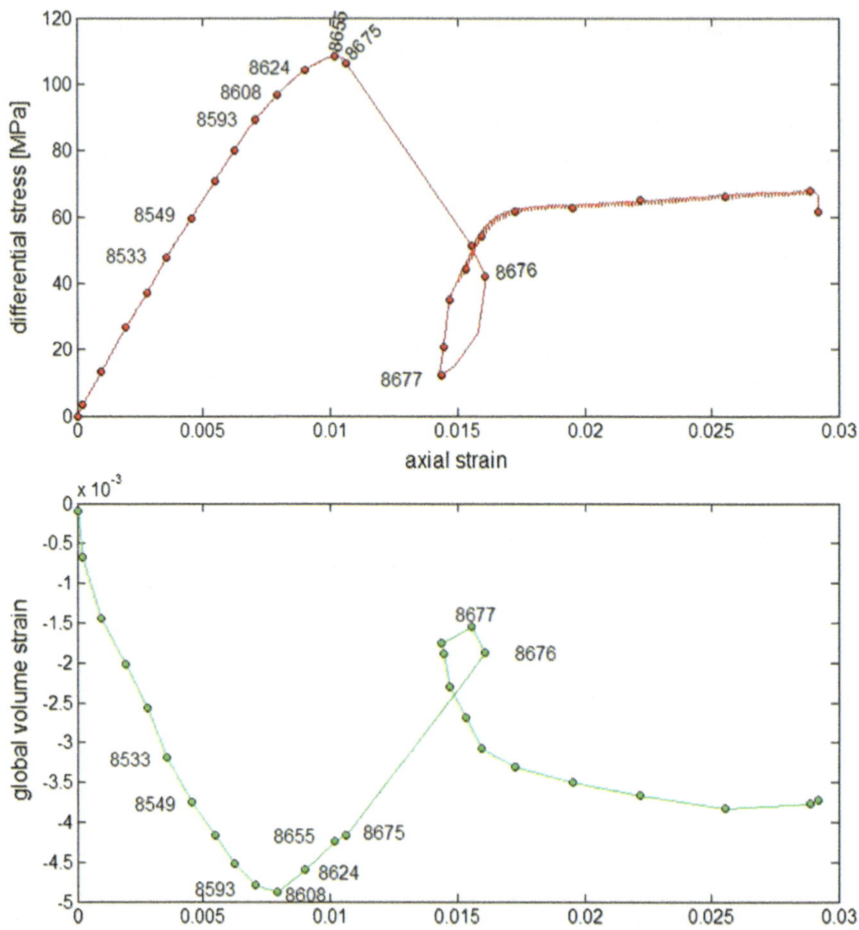

Fig. 2 Test BxR_GVR_06 [5]: (*top*) differential stress ($\sigma_1 - \sigma_3$) versus axial strain and (*bottom*) volume strain versus axial strain; the numbers noted on each curve are the selected photographs numbers

Figure 2 shows the stress-strain response and global volumic curve measured for the test BxR_GVR_06. There was first a quasi linear part of the stress versus strain curve, followed by an incurvation up to a stress peak and a progressive and then sudden strain softening. The stress level stabilized at the end of the test. The global volume strain compacted at the beginning and then dilated. The maximum compaction corresponds more or less to the loss of linearity of the upper curve. At the end of the test, the volume strain rate vanished.

Some photographs were selected during the test (Fig. 2): the incremental strain fields between these images are presented in Fig. 3. The strain field was quite homogeneous during the linear part of the stress versus strain curve (increment 8533–8549). The small fluctuations in both shear and volume strains correspond to the

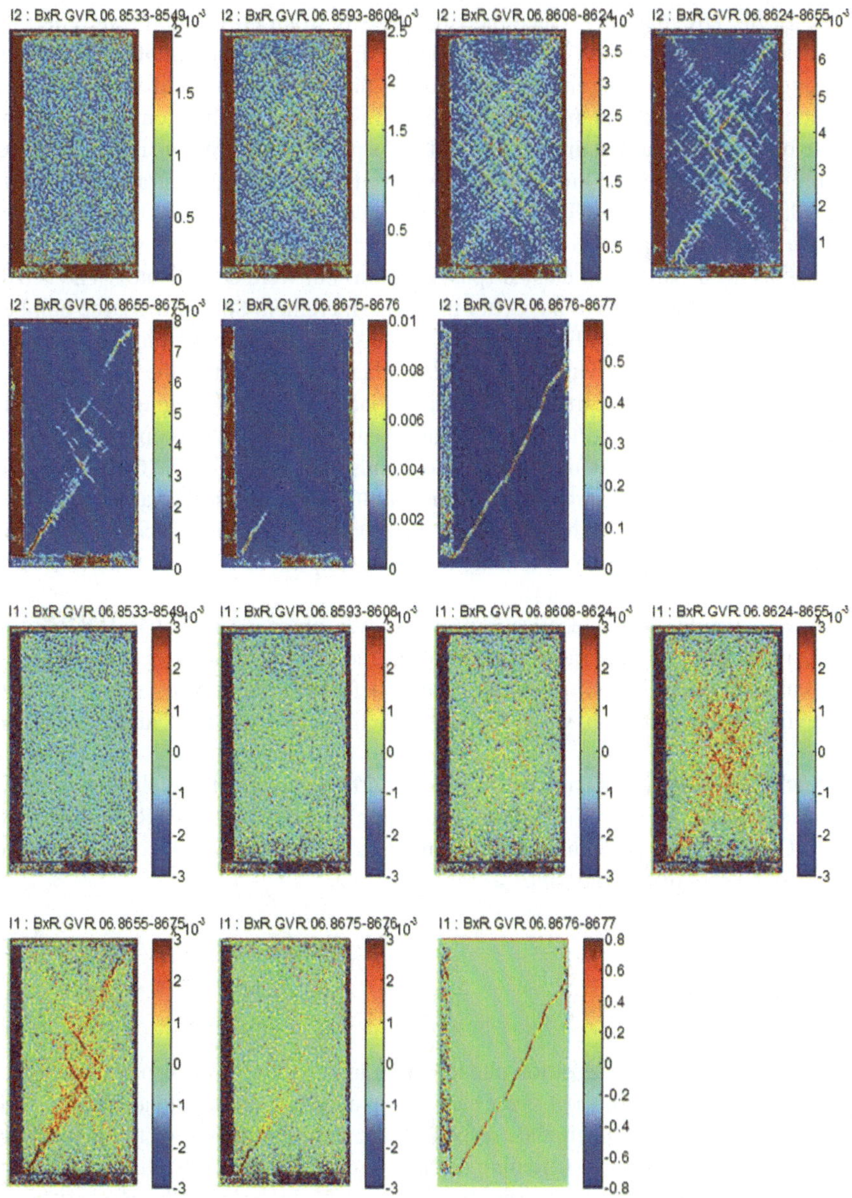

Fig. 3 Test BxR_GVR_06 [5]: (the two first lines) incremental second strain tensor invariant (shear strain) fields; the numbers at the *top* of pictures correspond to the photographs numbers. (the two last lines) incremental first strain tensor invariant (volume strain) fields

noise of the DIC measure. Once the stress versus strain curve significantly left its quasi-linear regime, which corresponded to a significant reduction of the specimen volume compaction (increment 8593–8608), the shear strain field was no longer homogeneous. We distinguished a dense, diffuse strain localization band pattern: these bands were parallel and conjugated. This pattern was more visible in the next increment, 8608–8624. A similar pattern was visible in the volume strain field, showing dilatancy in the bands. This step corresponded to dilatancy in the global volume. Before the stress peak (increment 8624–8655), the complex pattern was less dense and some shear bands still existed (plastic strain) but were de-activated. The active shear bands continued to dilate. After the peak (increment 8655–8675), the number of active bands significantly decreased and the shear and volume dilatancy became increasingly concentrated in a decreasing number of bands. Then, the testing device, due to a too small axial stiffness, could not control a stable softening of the specimen. During the abrupt softening, one main shear band started to propagate from the bottom left corner and dominated the other bands (increment 8675–8676). The band was fully propagated through the specimen at 8677 and was still dilating. The dilatancy decreased later (not shown). The inclination of the early bands before the stress peak, with respect to the axial loading direction, was about 45°, which decreased to around 30° at the end of the test.

A very similar strain localization process was observed in test BxR_GVR_11 performed at 50 MPa lateral stress. However, the volume strain was compactant in the bands. The inclination of the early bands before the stress peak, with respect to the axial loading direction, was about 45°, which stayed around 45° at the end of the test.

Such a progressive localization process with an early diffuse shear band pattern was also observed in the Callovo-Oxfordian clay rock. Tests BxR_COx_06 and BxR_COx_13 were performed at a lateral stress of 12 MPa, which corresponded to the in situ stress of the material.

4 Conclusion

Until now, the strain localization phenomenon in geomaterials was understood as a propagation or abrupt birth of shear bands from a diffuse strain field associated with the stress peak of the specimen's response. The strain softening of the material in the bands induced a global softening of the specimen. Historically, this aspect came from observations of the specimens after the tests, in which relatively simple patterns of shear bands were generally observed. This was also enforced by the theoretical approach of shear band analysis using bifurcation theory. This approach is generally quite consistent with experimental results at the stress peak.

The development of more efficient imaging tools and methods for full-field measurement has allowed better quantification of the deformation with better space and time resolutions and at a smaller scale. The recent results on several porous rocks suggest that the transition from a diffuse regime to a localized regime is more progressive than what is recalled above. A complex pattern of numerous parallel and conjugated shear bands appears well before the stress peak. These diffuse shear-band patterns can generate volume strain, dilatancy or compaction, depending on the material and stress state. Some bands are progressively de-activated, and close to the stress peak, only a few shear bands stay active which concentrates the deformation. They are generally prone to strain softening, inducing global strain softening. This early diffuse shear band pattern questions also the nature of plastic strain in the studied materials. Can it be homogeneous or is it invariably spatially organized under shear bands pattern?

References

1. Bésuelle, P.: Evolution of strain localisation with stress in a sandstone: brittle and semi-brittle regimes. Phys. Chem Earth Part A **26**(1–2), 101–106 (2001)
2. Bésuelle, P., Desrues, J., Raynaud, S.: Experimental characterisation of the localisation phenomenon inside a Vosges sandstone in a triaxial cell. Int. J. Rock Mech. Min. Sci. **37**(8), 1223–1237 (2000)
3. Bésuelle, P., Lanatà, P.: A new true triaxial cell for field measurements on rock specimens and its use in the characterization of strain localization on a Vosges sandstone during a plane strain compression test. Geotech. Test. J. **39**(5), 1–12 (2016)
4. Haimson, B., Chang, C.: A new true triaxial cell for testing mechanical properties of rock, and its use to determine rock strength and deformability of Westerly granite. Int. J. Rock Mech. Min. Sci. **37**(1–2), 285–296 (2000)
5. Lanatà, P.: Full-field experimental characterization of mechanical behaviour and failure in a porous rock in plane strain compression: homogeneous deformation and strain localization. Ph.D. thesis, University of Grenoble, France (2014)
6. Mogi, K.: Effect of the triaxial stress system on the failure of dolomite and limestone. Tectonophysics **11**, 111–127 (1971)
7. Robinet, J.-C., Sardini, P., Coelho, D., Parneix, J.-C., Prêt, D., Sammartino, S., Boller, E., Altmann, S.: Effects of mineral distribution at mesoscopic scale on solute diffusion in a clay-rich rock: example of the Callovo-Oxfordian mudstone (Bure, France). Water Resour. Res. **48**, W05554 (2012)

Localization in Plane Strain Compression of Fluid-Saturated Rock

Roman Y. Makhnenko and Joseph F. Labuz

Abstract Localization in terms of a bifurcation from a homogeneous pattern of deformation is predicted to be different if the boundary conditions are drained or undrained. This phenomenon is evaluated in plane strain compression experiments with water-saturated Berea sandstone, where corresponding poroelastic and inelastic properties were measured. Non-uniformity of lateral deformation and clustering of acoustic emission events were used as the experimental evidence of localization, which occurred soon after the peak stress in both drained and undrained cases. These observations are in reasonable agreement with an analysis that predicted negative values of the critical hardening modulus (localization post-peak) for fluid-saturated sandstone under plane strain loading conditions.

1 Introduction

Elastic, inelastic, and strength characteristics of fluid-saturated rock depend on the rate of deformation and boundary conditions. Moreover, studies show that the conditions for localization of deformation can also be different for specimens tested in drained and undrained regimes Rudnicki and Rice [6], Rudnicki [5]. Even though localization may result from inhomogeneities or stress concentrations, an alternative point of view is that this phenomenon is a bifurcation from a smoothly varying pattern of deformation, where the constitutive description of homogeneous deformation can admit a solution that is compatible with boundary conditions for further homogeneous deformation, but corresponding to non-uniform deformation in a planar zone Rudnicki and Rice [6]. In the following, this approach is compared with

R.Y. Makhnenko (✉)
Department of Civil and Environmental Engineering,
University of Illinois at Urbana-Champaign, Champaign, IL, USA
e-mail: romanmax@illinois.edu

J.F. Labuz
Department of Civil, Environmental, and Geo-Engineering,
University of Minnesota – Twin Cities, Minneapolis, MN, USA

E. Papamichos et al. (eds.), *Bifurcation and Degradation of Geomaterials with Engineering Applications*, Springer Series in Geomechanics and Geoengineering, DOI 10.1007/978-3-319-56397-8_5

two experimental methods of detecting the onset of localization: one is based on the deviation of uniform deformation of the specimen and the other is based on clustering of acoustic emission (AE) events. Experiments were performed on water-saturated Berea sandstone to determine its poroelastic and inelastic properties and to observe localization in drained and undrained plane strain compression.

2 Background

Rudnicki and Rice [6] suggested a constitutive description that is compatible with boundary conditions for homogeneous deformation but non-uniform deformation in a planar zone. The specific constitutive relations for (confined and compressed) brittle rock can be written in terms of a hardening modulus h. For the case of changing mean stress P (P' = Terzaghi's effective mean stress), h is related to the tangent modulus $h_{tan} = (d\tau - \mu dP')/(d\gamma - d\tau/G)$, where γ = shear strain, τ = shear stress, μ = friction coefficient, and G = shear modulus. The RR analysis predicts a critical value of the hardening modulus at which localization occurs:

$$h_{crit} = \frac{G(1+\nu)}{9}\left[\frac{(\beta-\mu)^2}{(1-\nu)} - \frac{1}{2}\left(\sqrt{3}\eta+\beta+\mu\right)^2\right] \tag{1}$$

Here β is the dilatancy factor and $\eta = \sqrt{3}\cdot s_2/\tau$ with s_2 being the intermediate principal value of the deviator stress tensor. This criterion is valid for both dry and drained deformation (constant pore pressure). The localization of deformation happens when the condition $h = h_{crit}$ is met (bifurcation point).

If the drained response is rate-independent, then the alternative limit of undrained deformation is also rate-independent and a similar bifurcation analysis can be applied. Rudnicki [5] proposed a form of the constitutive relation for undrained response, which is obtained from the drained response (Eq. 1) by substitutions that involve Skempton's B coefficient and undrained Poisson's ratio ν_u. The following assumptions are involved: (i) the elastic portion of strain increments can be described by linear, isotropic poroelasticity; (ii) the role of the pore pressure p in the inelastic strain increments is included by replacing the mean stress P by Terzaghi's effective mean stress P'; (iii) the inelastic increment in the apparent porosity is equal to the inelastic volume strain increment. From Rudnicki [5], the critical undrained hardening modulus for localization H_{crit} can be written as:

$$H_{crit} = \frac{G(1+\nu_u)}{9}\left[\frac{(1-B)^2(\beta-\mu)^2}{(1-\nu_u)} - \frac{1}{2}\left(\sqrt{3}\eta+(1-B)(\beta+\mu)\right)^2\right] \tag{2}$$

3 Methods

A Vardoulakis and Goldscheider [7] type plane strain apparatus was used for this study and modified for fluid-saturated rock testing of 100 × 87 × 44 mm specimens Makhnenko and Labuz [2]. The apparatus combines the positive features of a constitutive (plane strain) compression test, such that the two-dimensional material behavior can be evaluated, and a direct shear test, such that the characteristics of the shear band can be measured. Lateral displacement is measured at two positions, which allows an evaluation of non-uniform deformation and the onset of localization. Additionally, eight AE sensors are used to monitor microcracking (Fig. 1). AE locations can be determined to evaluate the onset of localization. For example, a criterion based on the value of the fractal dimension of the events positions has been proposed Iverson et al. [1]. However, a simple criterion is used in this study: if 90% of continuously located AE events with accuracy $r_a = 3$ mm are situated within sphere of $3r_a$ diameter (=9 mm or 3 standard deviations), the deformation is described to be localized (Fig. 1b).

Another method of determining localization is to detect when the deformation of the specimen is no longer homogeneous. Two LVDTs are positioned to measure the lateral displacement of the specimen (Fig. 1a). Typically, the output of the two LVDTs is changing at a uniform rate up to the peak axial stress, and just after the peak, one is approximately constant and the other continues to increase (Fig. 2a). This behavior is related to the formation of the shear band, when one part of the specimen slides and the other part remains stationary. The point where the lateral displacements change rate is taken as the onset of localization.

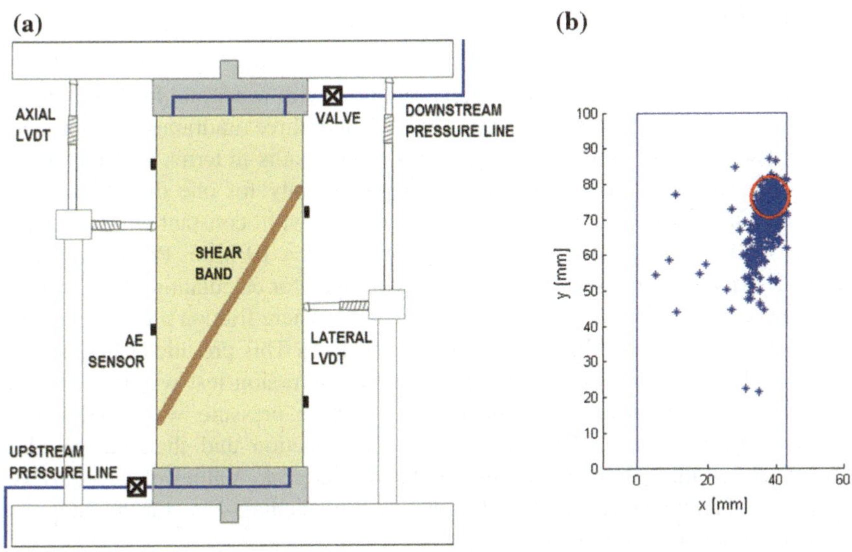

Fig. 1 **a** Sketch of the experimental setup and **b** AE events clustering in the sandstone

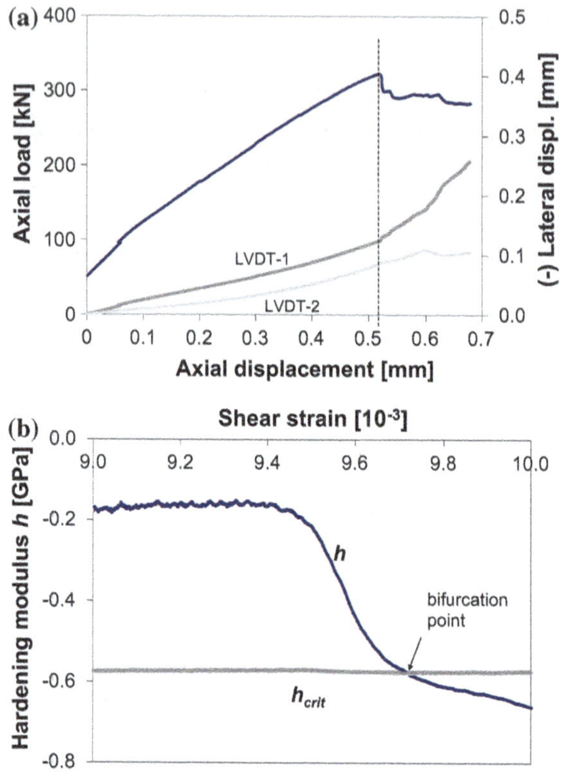

Fig. 2 Bifurcation point predicted by **a** onset of non-homogeneous deformation (*dashed line*) and **b** critical hardening modulus method for dry plane strain experiment

4 Results

Specimens of Berea sandstone were fully saturated with water and loaded to failure under plane strain conditions. Three drained and three undrained compression experiments were performed and exhibited similar results in terms of detecting the onset of localization, so the data here is reported only for one drained and one undrained test. The drained test was performed with constant (i) pore pressure = 5 MPa and (ii) minimum principal stress σ_3 = 10 MPa. Precise measurements of axial and lateral stresses and strains allowed for calculation of the material parameters: $G = 4.6$ GPa, $\mu = 0.79$, and $\beta = 0.41$, where friction μ and dilatancy β are reported at the peak Makhnenko and Labuz [3, 4]. This provided calculation of h_{crit} from Eq. (1). The undrained plane strain compression test was performed at σ_3 = const = 10 MPa and the initial value of pore pressure was 3 MPa. Pore pressure increased to 7 MPa during elastic compaction and then decreased to 4.5 MPa when the specimen was dilating before the peak. Values of $B = 0.61$ and $\nu_u = 0.35$ Makhnenko and Labuz [4] were used to calculate H_{crit} for the undrained plane strain compression test.

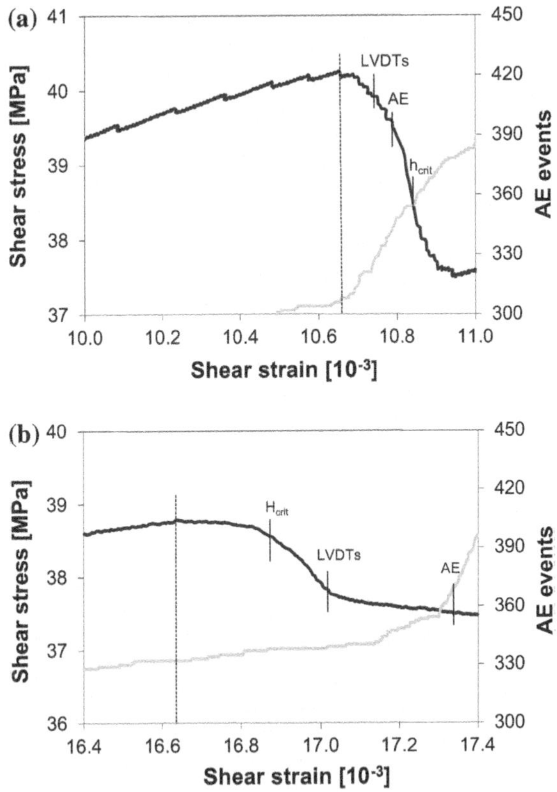

Fig. 3 Localization points determined by three methods: non-uniform lateral deformation (LVDTs), $h = h_{crit}$, and clustering of acoustic emission (AE) for **a** drained and **b** undrained plane strain compression tests. *Vertical dotted line* designates the shear strain at peak stress

Additionally, onsets of non-uniform rock deformation and AE clustering were obtained from the proposed methods. The results of the three localization criteria for drained and undrained plane strain compression experiments are presented in Fig. 3. The response of the undrained specimen is significantly less brittle, which is explained by the increase in effective minimum principal stress σ_3' caused by dilatant pore pressure decrease around and after axial peak stress Makhnenko and Labuz [3].

5 Discussion and Conclusions

The onset of localization in water-saturated Berea sandstone was determined by three methods: non-uniformity of specimen deformation detected by lateral LVDTs, clustering of AE events, and a critical value of the hardening modulus computed from measured elastic and inelastic material properties. In general, for both drained and undrained compression tests, all three methods provided localization happening soon after the peak axial stress was reached. In the case of drained compression, all the prediction methods for the onset of localization are consistent (lie within 0.15×10^{-3} shear strain) whereas in the undrained test AE clustering was delayed by 0.5×10^{-3} shear strain and the AE rate is significantly lower. It should be noted that the detection of the onset of localization with AE locations is sensitive to the chosen criterion. Also, stress and directional dependency of some material parameters (e.g. B, ν, and G), as well as their degradation during material damage, are not considered here, but influence the reported values of the critical hardening modulus.

Some discrepancy in the results of observed (from non-uniformity of deformation and AE) and predicted (from the critical value of the hardening modulus) localization also could be explained by the use of a relatively low effective minimum principal stress (below 10 MPa), where for the drained test, the failure plane was quite steep, and for the undrained case, the failure mode exhibited features associated with axial splitting.

References

1. Iverson, N., Kao, C.-S., Labuz, J.F.: Clustering analysis of AE in rock. J Acoustic Emission **25**, 364–372 (2007)
2. Makhnenko, R., Labuz, J.: Plane strain testing with passive restraint. Rock Mech. Rock Eng. **47**(6), 2021–2029 (2014)
3. Makhnenko, R.Y., Labuz, J.F.: Dilatant hardening of fluid-saturated sandstone. J Geophys Res Solid Earth **120**, 909–922 (2015)
4. Makhnenko, R.Y., Labuz, J.F.: Elastic and inelastic deformation of fluid-saturated rock. Phil Trans R Soc A **374**, 20150422 (2016)
5. Rudnicki, J.W.: Localization in undrained deformation. In: Proceedings of the 4th Biot Conference on Poromechanics, Columbia University, New York (2009)
6. Rudnicki, J.W., Rice, J.R.: Conditions for the localization of deformation in pressure-sensitive dilatant materials. J Mech Phys Solids **23**, 371–394 (1975)
7. Vardoulakis, I., Goldscheider, M.: Biaxial apparatus for testing shear bands in soils. In: Proceedings of 10th International Conference on Soil Mechanics and Foundation Engineering, pp. 819–824. Balkema, Rotterdam (1981)

3D Particle-Scale Displacement Gradient to Uncover the Onset of Shear Bands in Sand

Andrew M. Druckrey and Khalid A. Alshibli

Abstract In situ synchrotron micro-computed tomography (SMT) scans of a specimen composed of uniform silica sand were acquired at multiple strains during a triaxial compression test. Individual sand particles were identified and tracked through multiple strain increments. The paper presents the concept of displacement gradient to expose the inception of shear bands (strain localization) in sheared sand. Each particle's neighboring particles were identified and translation fields for all particles were calculated. The second order norm between the particle translation vector and neighboring particles translation vectors were averaged, resulting in a relative displacement value for each particle. The relative displacement concept is effective to uncover the onset of localized shearing within sheared granular materials.

1 Introduction

Granular materials are composed of discrete particles that translate and rotate against neighboring particles when they are sheared. Shearing of granular materials localizes into zones of intensive shearing known as shear bands. There is extensive literature about experimental measurements and numerical modelling of factors that influence the onset and properties of shear bands. In some cases, failure is not manifested through a single well-defined shear band which makes it difficult to identify the mode of failure. Investigations of mechanisms leading to strain localization into shear bands in granular material experiments have been investigated using discrete element method (DEM) [9–11], digital image correlation (DIC) [7, 12], x-ray computed tomography (CT) [3, 4, 6, 8, 13], as well as theoretical and numerical methods [1, 2, 14]. Traditional particle kinematics alone do not uncover intermittent strain localization during hardening regime or the evolu-

A.M. Druckrey · K.A. Alshibli (✉)
Department of Civil & Environmental Engineering, University of Tennessee,
325 John Tickle Building, Knoxville 37996, TN, USA
e-mail: alshibli@utk.edu

E. Papamichos et al. (eds.), *Bifurcation and Degradation of Geomaterials with Engineering Applications*, Springer Series in Geomechanics and Geoengineering, DOI 10.1007/978-3-319-56397-8_6

tion of strain localization into a single shear band at failure. Few researchers have investigated intermittent local shear bands formation during hardening phase using 2D DEM or DIC techniques.

2 Experiment

A silica sand with grain size between US sieves #40 (0.429 mm) and #50 (0.297 mm) was used in the experiment. A small triaxial apparatus described in Hasan and Alshibli [8] and Druckrey and Alshibli [4] was setup in beamline 13BMD of Advanced Photon Source (APS), Argonne National Laboratory (ANL), Illinois, USA. The specimen is cylindrical and initially measures about 10 mm in diameter by 20 mm in height. The experiment was conducted on a very dense dry specimen

Fig. 1 PSR versus axial strain and volumetric strain versus axial strain

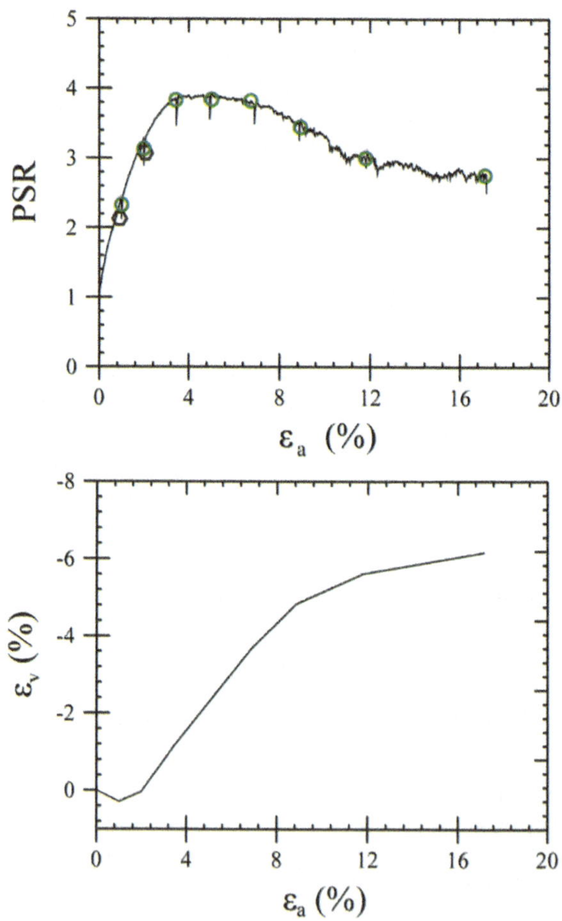

under drained condition at 400 kPa confining pressures. SMT images were collected at multiple compression stages of the experiment. An initial image was acquired, the top end plate was moved at a constant compressive displacement rate of 0.2 mm/min to a predetermined target displacement, loading was paused and the next image was acquired. The spatial resolution of the images is 11.18 μm/pixel. Compressive Load and displacement measurements were also collected during each of the experiments and the principal stress ratio (PSR = σ_1/σ_3) and volume change (ε_v) versus nominal axial strain (ε_a) relationships are presented in Fig. 1. Markers in Fig. 1 label axial strains at which SMT scans were acquired.

3 Image Processing and Particle Tracking

Image processing was conducted using the procedure described in Druckrey et al. [5]. Grayscale SMT images were initially filtered and binarized, and particles were separated using a combination of watershed, distance transform, and numerical reconstruction algorithms that removed small areas of contact between particles. With particles no longer in contact, each particle was assigned a unique label number and contacts were restored. SMT technique produced high quality grayscale images that resulted in accurate particle identification after processing. Images at each axial strain increment were then quantified using the code described in Druckrey et al. [5] and particle centroid, volume, surface area, short axis length, intermediate axis length, long axis length, and contact information were extracted and saved for each particle. In addition, orientation of the particle long, intermediate, and short axes were computed and saved in output files for further analysis.

An incremental particle tracking code was developed to track individual particles using particle morphology data, similar to the particle tracking code in Druckrey and Alshibli [4]. The magnitude of particle displacement was then calculated as the distance between the centroid of a particle in the first SMT image (x_1, y_1, z_1) and the centroid of the matching particle in the second SMT image (x_2, y_2, z_2). Particle rotations were also calculated based on differences in long axis orientations between the increments. Applying this particle kinematics analysis to the entire specimen at multiple strains is similar to 3D digital volume correlation. The particle kinematic measurements are displayed in Fig. 2 in the plane parallel to specimen shear band at critical state, displaying the shear band at its narrowest point of view. Particle displacements are normalized for each strain increment by the axial displacement of the specimen top plate (i.e., global compression). Strain hardening for this experiment took place between ε_a = 0–4.9%, peak PSR was reached between ε_a = 4.9–6.9%, followed by softening between ε_a = 6.9–11.8%, and critical state transpires thereafter (Fig. 1). The specimen failed via a single well-defined shear band that initiated near the peak PSR ($\varepsilon_a \sim 4.9\%$) and is easily identified from particles displacements and rotations during the critical state ($\varepsilon_a > 11.8\%$) of the experiment (Fig. 2). Particle rotations do not exhibit localization during earlier stages of the experiment. Particle displacements show a nucleation of the shear

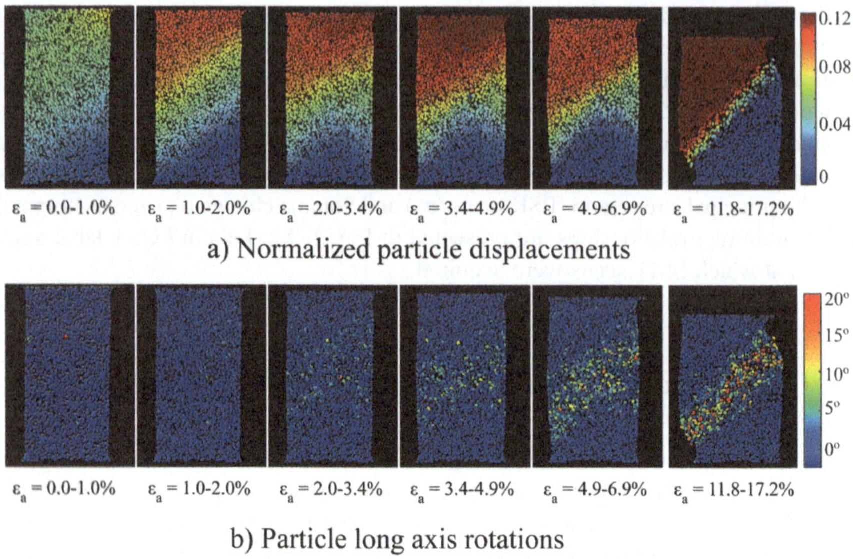

Fig. 2 Maps of particle displacements and rotations based on SMT measurements

band, but are not definitively localized until post-peak strain softening stage. Such conventional particle kinematics has been widely used to analyze and quantify the failure mode of granular materials, but mask many intricacies and lesser intermittent localized strains (local shear bands, LSB) during hardening that lead to failure, especially if a specimen fails through bulging.

Intricate zones of localized strain within a specimen that are not exposed by particle kinematics can be mined from the data by comparing a particle's displacement vector with that of all of its neighboring particles. The concept of relative particle displacement branches from kinematic displacements and relates a particle displacement vector (δ) to that of all neighboring particles in contact with that particle (δ_1, δ_2, ..., δ_n), using the second order norm of vector differences:

$$rd_i = norm(\delta - \delta_i) \tag{1}$$

$$RD = \frac{1}{n}\sum_{i=1}^{n} rd_i \tag{2}$$

where rd_i is the magnitude of relative displacement for a single contacting neighboring particle and n is the number of contacting particles. The overall magnitude of relative displacement (RD) is the average of all individual relative displacement vectors. RD not only differentiates vertical displacements, but localized displacement in any 3D direction. Relative displacement better captures the shear between particles, exposes small localized strains within a granular material, and allows for quantification and visualization of local shear bands (LSB) throughout the experiment as well as analysis of mechanisms that contribute to the onset and growth of the major shear band if any. Figure 3 displays the Incremental relative displacements for the experiment (slices are taken normal to final shear band for better visualization). Zones of localized strain, not necessarily shear bands, sporadically developed as early as the first strain increment (ε_a = 0–1%). Then at ε_a = 1–2%, strains localized into multiple LSB with thicknesses of 1d_{50} to 3d_{50}, where d_{50} is the mean particle size (0.36 mm). The LSB exhibited preferred orientations parallel and perpendicular to the inclination of the of final major shear band that would develop at a higher strain. The LSB became more developed and more defined in the third strain increment (ε_1 = 2–3.4%), where multiple cross-hatched LSB develop in multiple directions. These multiple intersecting LSB typically extend over the entire width of the specimen in conjugate directions with orientations similar to the final shear band. The LSB began to retreat and merge into a zone intensive shearing near the center of the specimen at the end of strain hardening phase of the experiment

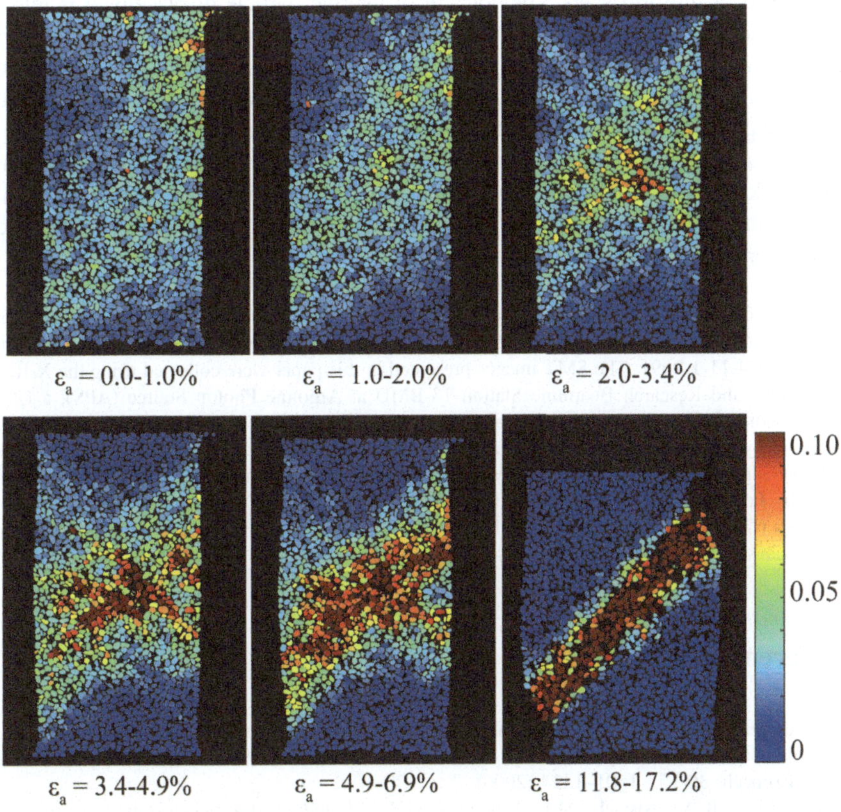

Fig. 3 Normalized relative displacements in the loading direction. *Color* scale shows relative displacements normalized by the specimen global axial compression imposed by the *top* end plate

(ε_1 = 3.4–4.9%). The zone of highly disordered particle relative displacements near the end of hardening is still in the general direction of the final shear band, although LSB in other directions are still growing from the zone of disarray. During the peak PSR phase (ε_1 = 4.9–6.9%), most LSB have merged into a large zone of disordered relative displacements that has an overall inclination of the final major shear band, while few minor LSB were still extend from the zone of intensive shearing. The final major shear band fully develops during the softening phase of the experiment and continues through the critical state phase. It has a thickness of $9d_{50}$ to $10d_{50}$. Particles within the final major shear band exhibit a high relative displacement values which uncovers an interesting experimental evidence that intensive shearing and interaction takes place between particles within the shear band.

4 Conclusions

3D SMT was used to capture the evolution of failure mode of a triaxial specimen composed of uniform sand. The concept of relative particle displacement is introduced and used to visualize and quantify intricate zones of localized strains and local shear bands that cannot be exposed when one uses particle translation and rotation alone. Local shear bands in triaxial experiments can be better analyzed and quantified in 3D using the second order norm of differences in displacement vectors. This technique exposes more intricate strain localizations than conventional particle kinematics approach. RD technique uncovered the onset of multiple local shear bands that develop during the hardening phase and eventually evolve into fewer persistent shear band(s).

Acknowledgements This material is partially funded by the US National Science Foundation (NSF) under Grant No. CMMI-1266230 and Office of Naval Research (ONR) grant No. N00014-11-1-0691. The SMT images presented in this paper were collected using the X-Ray Operations and Research Beamline Station 13-BMD at Argonne Photon Source (APS), a U.S. Department of Energy (DOE) Office of Science User Facility operated for the DOE Office of Science by Argonne National Laboratory under Contract No. DE-AC02-06CH11357. We acknowledge the support of GeoSoilEnviroCARS (Sector 13), which is supported by the National Science Foundation—Earth Sciences (EAR-1128799), and the Department of Energy, Geosciences (DE-FG02-94ER14466). We thank Dr. Mark Rivers of APS for help in performing the SMT scans.

References

1. Alsaleh, M.I., Voyiadjis, G.Z., Alshibli, K.A.: Modelling strain localization in granular materials using micropolar theory: mathematical formulations. Int. J. Numer. Anal. Methods Geomech. **30**(15), 1501–1524 (2006)
2. Alshibli, K.A., Alsaleh, M.I., Voyiadjis, G.Z.: Modelling strain localization in granular materials using micropolar theory: numerical implementation and verification. Int. J. Numer. Anal. Methods Geomech. **30**(15), 1525–1544 (2006)

3. Andò, E., Hall, S., Viggiani, G., Desrues, J., Bésuelle, P.: Grain-scale experimental investigation of localised deformation in sand: a discrete particle tracking approach. Acta Geotech. **7**(1), 1–13 (2012)
4. Druckrey, A., Alshibli, K.: 3D behavior of sand particles using X-ray synchrotron micro-tomography. In: Geo-Congress 2014 Technical Papers, pp. 2814–2821 (2014)
5. Druckrey, A.M., Alshibli, K.A., Al-Raoush, R.I.: 3D characterization of sand particle-to-particle contact and morphology. Comput. Geotech. **74**, 26–35 (2016)
6. Hall, S.A., Bornert, M., Desrues, J., Pannier, Y., Lenoir, N., Viggiani, G., Bésuelle, P.: Discrete and continuum analysis of localised deformation in sand using X-ray μCT and volumetric digital image correlation. In: Géotechnique, vol. 60, pp. 315–322 (2010)
7. Hall, S.A., Muir Wood, D., Ibraim, E., Viggiani, G.: Localised deformation patterning in 2D granular materials revealed by digital image correlation. Granular Matter **12**(1), 1–14 (2009)
8. Hasan, A., Alshibli, K.: Three dimensional fabric evolution of sheared sand. Granular Matter **14**(4), 469–482 (2012)
9. Jiang, M., Zhu, H., Li, X.: Strain localization analyses of idealized sands in biaxial tests by distinct element method. Front. Archit. Civil Eng. China **4**(2), 208–222 (2010)
10. Kuhn, M.R., Bagi, K.: Alternative definition of particle rolling in a granular assembly. J. Eng. Mech. **130**(7), 826–835 (2004)
11. Oda, M., Iwashita, K.: Study on couple stress and shear band development in granular media based on numerical simulation analyses. Int. J. Eng. Sci. **38**(15), 1713–1740 (2000)
12. Rechenmacher, A.L.: Grain-scale processes governing shear band initiation and evolution in sands. J. Mech. Phys. Solids **54**(1), 22–45 (2006)
13. Sjödahl, M., Siviour, C.R., Forsberg, F.: Digital volume correlation applied to compaction of granular materials. Procedia IUTAM **4**, 179–195 (2012)
14. Walker, D., Tordesillas, A., Rechenmacher, A.: Transmission of kinematic information in dense granular systems: local and nonlocal network sensing. Acta Geotech. **8**(5), 547–560 (2013)

Load Transmission by Granular Materials

Danuta Lesniewska

Abstract The first attempts to recognize mechanisms of load transmission in granular materials using photo-elastic technique were undertaken in the first half of the 20th century. Integrated photo-elasticity was employed for further experimental studies on big granular assemblies. This paper contains new proposal of non-standard analysis of photo-elastic images of deforming granular material. This proposal is based on defining some basic components of 2D Fourier transform of an image.

1 Introduction

The first attempts to recognize mechanisms of load transmission in granular materials using photo-elastic technique were undertaken in the first half of the 20th century. Integrated photo-elasticity was employed for further experimental studies on big granular assemblies. Works of Dantu [2], Wakabayashi [10], Drescher and De Jong [4], Drescher [3], Allersma [1], Majmudar and Behringer [6], Oda and Kazama [8] and others led to the widely accepted concept of 'force chains' bearing most of the load put on a granular material.

It was suggested by Oda and Kazama that physical mechanism of a granular material's deformation is related to the 'force chains' buckling. No substantial progress on other possible physical mechanisms were made since then and also creation and evolution of the whole 'force chains' network was not studied.

Recent developments of digital photography and image analysis created new tools to trace deformation and 'force chains' network. Lesniewska and Muir Wood [5] and Muir Wood and Lesniewska [7] found that photo-elastic images of 'force chains' are not suitable for the analysis using standard image analysis tools. The problem lies in the specific nature of 'force chains', being very thin and elongated structures, difficult to be captured by any image segmentation procedures.

D. Lesniewska (✉)
Koszalin University of Technology, Koszalin, Poland
e-mail: danuta.lesniewska@tu.koszalin.pl

E. Papamichos et al. (eds.), *Bifurcation and Degradation of Geomaterials with Engineering Applications*, Springer Series in Geomechanics and Geoengineering, DOI 10.1007/978-3-319-56397-8_7

This paper proposes some new way of non-standard analysis of photo-elastic images of deforming granular material. The proposal is based on defining some basic components of 2D Fourier transform of an image. These components represent different features of 'force chain' network and can help to look for characteristic material lengths, like average dimensions of 'force chains' mesh, average characteristic distance between neighbouring 'force chains', average length of 'force chains' at different stages of deformation. The idea of the proposed approach comes from the analogy between poly-crystals and granular materials and between Lauegrams, used to measure internal structure of crystals, and 2D Fourier transforms.

2 Experimental Setup

The paper presents series of 2D Fourier transforms of photo-elastic transmission images, representing large (consisting of $\sim$1e + 6 grains) granular assembly. The assembly can be classified as pseudo 2D, because it is placed in a narrow rectangular box, enforcing plane strain condition on macro (sample) scale, but individual grains are spherical and form several layers, none of them perfectly plane. The assembly deforms under varying boundary conditions (boundary load and boundary displacement).

The level of external load is selected in such a way, that photo-elastic effect (enforced birefringence) in granular material is easy to observe. Principle of integrated photo-elasticity is used (transparent grains immerged in immersion liquid having the same refractive index as grains' material). The analysis refers to photo-elastic images of granular samples, which cover the whole sample ('macro' scale, the area of an image equal to $\sim$250 $\times$ 200 individual grains—glass granules of $D_{50} \sim 1$ mm). Similar test arrangement was detaily described by Lesniewska and Muir Wood [5] and Muir Wood and Lesniewska [7].

The photographs from one of the tests, presented in Fig. 1 (column A), were taken using RGB scientific grade camera and later transformed to grayscale images to calculate Fourier transforms. Specific type of circular polariscope was used, equipped with rectangular polarizer and analyser and thus giving much better quality of photo-elastic images of granular sample under load.

3 Applications of 2D Fourier Transforms

Fourier analysis is a well known analytical tool transforming a sample of data points (i.e. brightness of image pixels) into a mathematically equivalent sum of component sine waves of different frequencies and amplitudes [9]. The amplitudes of each of the component sine waves in the Fourier transform expresses the relative contributions of each frequency of variation to the periodicity of the original data.

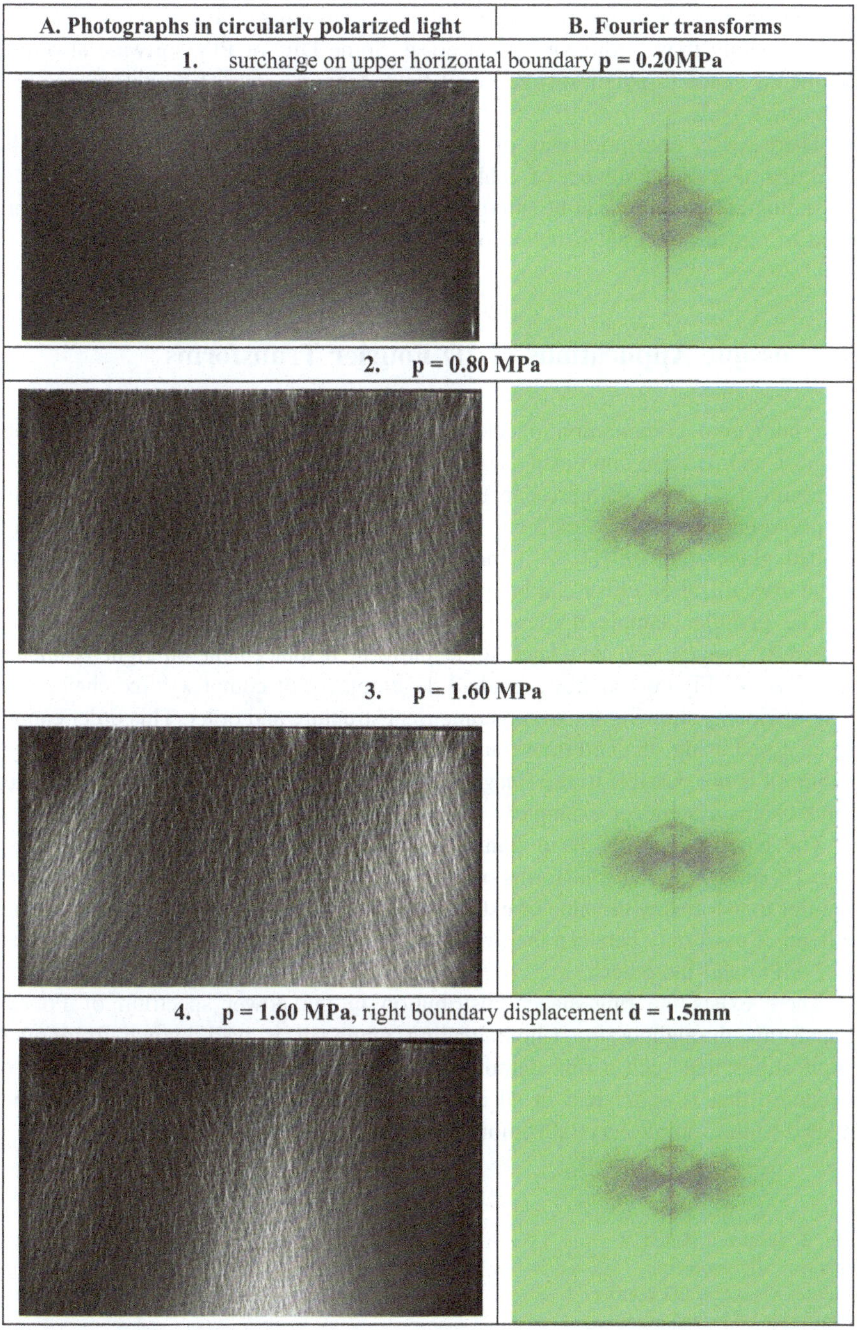

Fig. 1 *Column A* Force chain system development during uniform loading of granular sample (glass granules $D_{50} \sim 1$ mm) on *top* boundary (rows 1–3) and after 1.5 mm displacement of *right* boundary (row 4). *Column B* Corresponding 2D Fourier transforms

Fourier transforms are the basic tool in constructing different types of image filters, mainly those reducing optical noise. Some DIC or PIV software also uses frequency space to obtain displacement fields from a series of digital images of a deforming body.

There exists one more way of utilizing 2D Fourier transforms—they can be regarded as a direct subject of analysis themselves, similarly to Laue patterns in crystallography, which can be interpreted as 'physical Fourier transforms', and are used to measure internal structure of crystals or nanostructures [9].

4 Possible Applications of 2D Fourier Transforms

The numerical computation of the Fourier transforms presented in this paper (Figs. 1 and 3) were done using DaVis8 software.

Figure 1 (column A) represents some typical experimental data—2D projection of photo-elastic response of 2 cm thick granular sample, consisted of uniformly graded glass granules ($D_{50} \sim 1$ mm). Due to dimensional uniformity of grains some crystallization effect can be expected.

The granular sample undergoes growing external loading on top boundary (Fig. 1A, images 1–3) and later some uniform displacement on right boundary (Fig. 1A, 4). Figure 1A shows gradual development of complex force chains system, obviously showing traces of geometrical structure and order. This order cannot be analysed using standard image analysis tools, however, due to its highly discreet nature (it is not possible to use image segmentation to find characteristic features of force chains system for example).

The other possibility is to transform the images into frequency spectrum—Fig. 1B shows Fourier transforms of images shown in Fig. 1A. Some evolution of Fourier transforms with value of external loading is also observable, but significant difference exist only between the first phase of relatively low value of external load (0.2 MPa) and the others.

There exists the ring-shaped distribution in the power spectrum of Fourier transforms throughout the whole loading range—it is well known from other applications that such a ring demonstrates periodicity at intermediate spatial frequencies, that is equivalent in all directions in the image (this is most probably related to the sample 'crystallization').

Fig. 2 Idealized sketch of characteristic geometric features visible in 2D Fourier transforms of loaded granular material (Fig. 1)

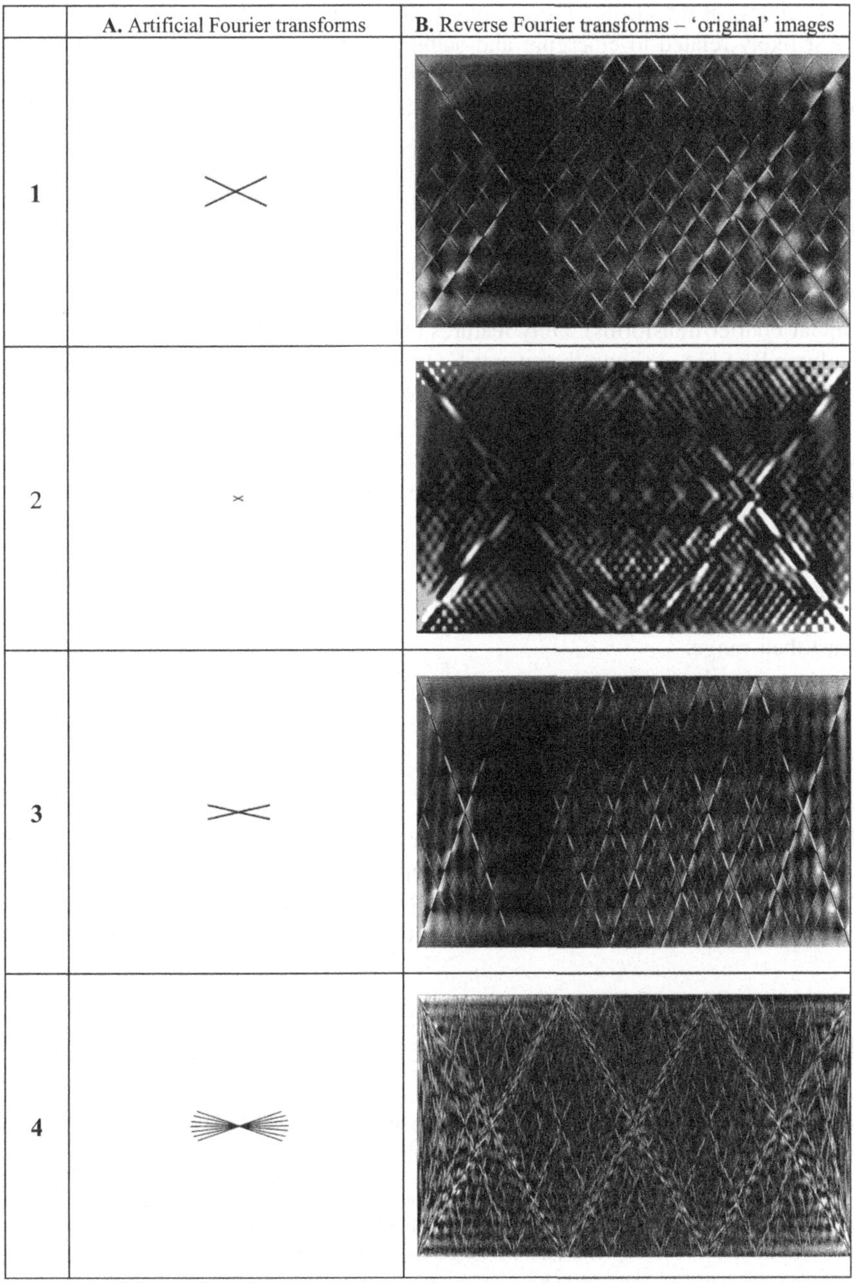

Fig. 3 Idealized components of Fourier transforms (A) and their reverse (B), showing corresponding force chain networks

In course of the loading process other objects appear in Fourier power spectrum, most likely related to emerging and stabilizing force chain network. As the geometric features visible in Fig. 1B (rows 2–4) are quite regular (they are sketched in Fig. 2) it is possible to recognize these directly related to force chains and their arrangement.

It can be assumed that Fig. 2 consists of a ring-shaped area and a double fan, which can have complicated structure, but is most probably related to anisotropic force chain network, visible in Fig. 1A.

Figure 3 shows the result of simple back-analysis based on Fig. 2. Column A of Fig. 3 contains artificial images of features which could be included in Fig. 2 (artificial Fourier transforms). Only features characterizing the double fan are studied. Column B of Fig. 3 shows the reverse Fourier transforms of the images from column A. Figure 3B shows that the information on the force chain network geometry is really coded in the double fan. The fact that the extent of the fan is quite well defined can give the limiting range of force chains directions within a granular sample (Fig. 3B 1 and 3). Also some information on changing length of force chains can be deduced from the dimensions of double fan (compare Fig. 3B 1 and 2). Figure 3B, row 4, explains the complexity of force chain networks: the double fan may cover a range of possible force chains directions. Such different directions are present in Fig. 1A, but using the Fourier transforms of Fig. 1B it is possible to determine at least their range.

5 Conclusion

Direct analysis of Fourier transforms can help to find characteristic material lengths, which are necessary to model granular material behaviour: average dimensions of 'force chains' mesh, average characteristic distance between neighbouring 'force chains', average length of 'force chains' at different stages of deformation.

References

1. Allersma, H.G.B.: Optical analysis of stress and strain in photoelastic particle assemblies (1987). Ph.D. thesis, Delft Technical Univ. Delft, The Netherlands
2. Dantu, P.: Contribution à l'étude mécanique et géométrique des milieux pulvérulents. In: Proceedings of 4th International Conference on Soil Mechanics and Foundation Engineering, vol. 1, pp. 144–148. Butterworths Scientific Publications, London (1957)
3. Drescher, A.: An experimental investigation of flow rules for granular materials using optically sensitive glass particles. Géotechnique **26**(4), 591–601 (1976)
4. Drescher, A., Josselin, De Jong, G.D.: Photoelastic verification of a mechanical model for the flow of a granular material. J. Mech. Phys. Solids **20**, 337–351 (1972)
5. Lesniewska, D., Muir Wood, D.: Observations of stresses and strains in a granular material. J. Eng. Mech. **135**, 1038–1054 (2009)

6. Majmudar, T.S., Behringer, R.P.: Contact force measurements and stress-induced anisotropy in granular materials. Nature (London) **435**, 1079–1082 (2008)
7. Muir Wood, D., Lesniewska, D.: Stresses in granular materials. Granular Matter **13**, 395–415 (2011)
8. Oda, M., Kazama, H.: Microstructure of shear bands and its relation to the mechanisms of dilatancy and failure of dense granular soils. Géotechnique **48**(4), 465–481 (1998)
9. Prum, T., Torres, R.H.: A Fourier tool for the analysis of coherent light scattering by bio-optical nanostructures. Integr. Comp. Biol. **43**, 591–602 (2003)
10. Wakabayashi, T.: Photoelastic method for determination of stress in powdered mass. In: Proceedings of 7th National Conference on Applied Mechanics, Japan, pp. 153–158 (1957)

Periodic Stick-Slip Deformation of Granular Material Under Quasi-static Conditions

Danuta Lesniewska, Magdalena Pietrzak, Michal Nitka and Jacek Tejchman

Abstract Stick-slip motion is generally observed in situations involving dry friction. It is commonly present in granulates and results from interactions at micro-scale. Analysing the data base collected during a series of small scale model tests the evidence was found of periodic deformation mode, induced within the granular material by a series of very small displacement increments. Image analysis was used to study the phenomenon. It was found that there exist minimum displacement of the external boundary necessary to produce 'plastic localization', being a function of the external load and grain coarseness.

1 Introduction

Stick-slip motion is generally observed in situations involving dry friction. It is commonly present in granulates and results from interactions at micro-scale. If Coulomb-Euler laws of friction are employed and it is acknowledged that grains exhibit also elastic behaviour, it is possible to explain qualitatively stick-slip motion using simple model of an interrupted oscillator [1].

Analysing the data base collected during a series of small scale model tests (performed on glass granules of $D_{50} \sim 1$ mm and coarse Borowiec sand) the evidence was found of periodic deformation mode induced within the granular material by a series of very small ($\sim$1/10 of grain diameter) quasi-static external displacement increments. Image analysis by geoPIV-RG and DaVis8 software was used to study the phenomenon.

D. Lesniewska (✉) · M. Pietrzak
Koszalin University of Technology, Koszalin, Poland
e-mail: danuta.lesniewska@tu.koszalin.pl

M. Nitka · J. Tejchman
Gdansk University of Technology, Gdańsk, Poland

E. Papamichos et al. (eds.), *Bifurcation and Degradation of Geomaterials with Engineering Applications*, Springer Series in Geomechanics and Geoengineering, DOI 10.1007/978-3-319-56397-8_8

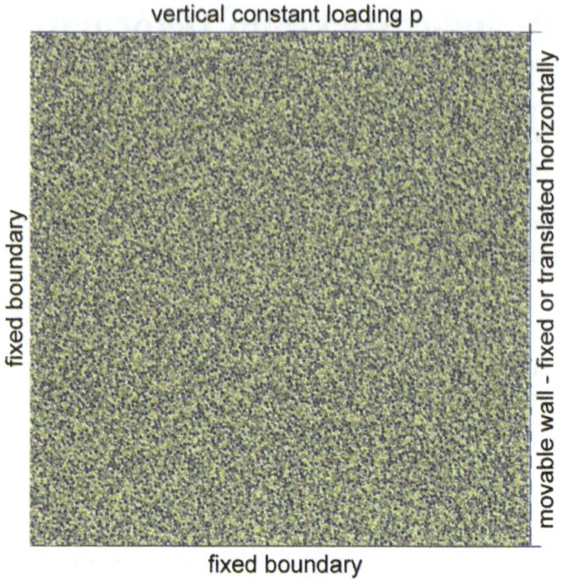

Fig. 1 DEM sample, simulating conditions of real test for one layer of grains

2 Experiments

The paper presents results of PIV (particle image velocimetry) analysis performed on images of large (consisting of ~1e + 6 grains) granular assembly. The assembly can be classified as pseudo 2D, because it is placed in a narrow rectangular box, enforcing plane strain condition on macro (sample) scale. Individual grains are spherical and form several layers, none of them perfectly plane. The assembly deforms under varying boundary conditions (boundary load and boundary displacement, like in Fig. 1).

PIV analysis refers to the images covering the area equal to ~250 × 200 individual grains (glass granules of $D_{50} \sim 1$ mm). Similar test arrangement was described in detail by Lesniewska and Muir Wood [2], Muir Wood and Lesniewska [3] Lesniewska et al. [4] and Pietrzak and Lesniewska [5]. RGB scientific grade camera was used to record the tests. The resolution of camera was ~2500 × 2000 pixels. One average grain was represented by ~10 × 10 pixels.

3 DEM Simulations

In order to simulate real tests, the discrete element method (DEM) was used. The open source 3D code YADE, developed at Grenoble University [6], was employed. So called soft-particles approach was accepted (i.e. particles deformation represented by their overlap). At first stage of the simulation by YADE, interaction

Table 1 Two sets of parameters used in DEM simulations

Number of set	Young modulus (kPa)	Poisson ratio	Angle of internal friction (°)	Rotation resistance	
1	7.5e08	0.3	35	0.02	0.01
2	2.4e10	0.3	20	0.3	0.005

forces are computed for elements (spheres) in contact. In the second step, Newton second law of dynamics is used to compute for each element the resulting acceleration, next integrated to find new element position. To reduce the calculation time, only one layer of grains was modelled.

Two sets of parameters were used, listed in Table 1. In order to increase the rolling resistance of perfect spheres, contact moments were introduced [6]. The normal force contributed to the rolling resistance only. The increments of contact moments were calculated by means of rolling stiffness.

4 Discussion of Results

The results of PIV analysis, performed both on real images (photographs taken during the experiments) and on artificial images, created during DEM simulations (Fig. 1) are presented in Fig. 2. The figure contains, in case of real, experimental images, maps of volumetric and shear strains for four subsequent right boundary displacement increment $\Delta d \sim 0.1$ mm (columns A and B). Column C of Fig. 2 shows shear strain maps obtained for DEM simulations for the same loading conditions and boundary displacements.

4.1 Cyclic Behaviour

Figure 2A, B show the characteristic sequence of incremental (representing one displacement step of moving boundary) strain maps, typically repeating itself throughout a whole model test. The last one in sequence appears a distinct strain localization (Fig. 2A, B, row 4). Each localization incident is proceeded by several displacement steps, during which shear and volumetric strains are much smaller and relatively uniformly distributed inside the wedge shaped region, adjacent to the moving boundary (Fig. 2A, B, rows 1–3).

It can be seen from Fig. 2A that relatively uniform deformation within the wedge (rows 1–3) is mostly of dilative nature (blue colour), while strong localization event causes contraction (red and yellow colours) to appear within the wedge and dilation prevails inside the shear band (row 4). Thus the granular material within the wedge undergoes alternate transition from weakening to

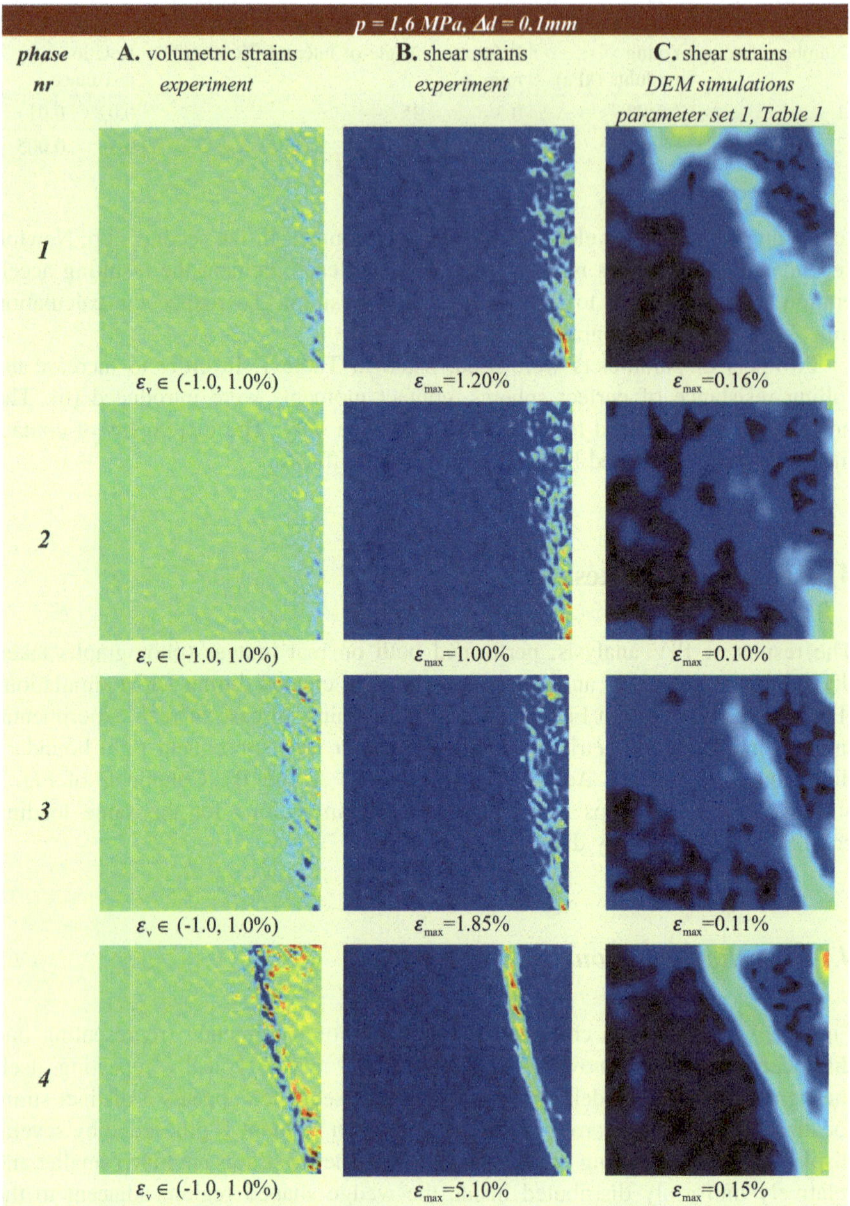

Fig. 2 Single cycle of incremental deformation behind the model wall, visualized by PIV analysis: A, B—volumetric and shear strain maps obtained for four subsequent wall displacement increments (model tests), C—corresponding shear strain maps obtained from DEM simulation

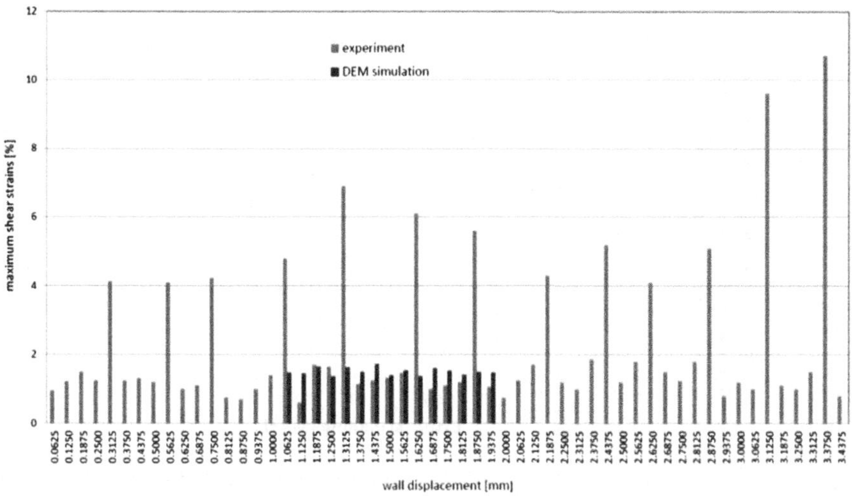

Fig. 3 Relation between maximum shear strain within shear band and wall displacement in one of the model tests (*grey*) compared to the results of corresponding DEM simulation (parameters set 1, *black*). Each displacement increment Δd equal to 0.0625 mm, p = 1.6 MPa

strengthening states during the wall movement. These alternate phases of deformation could be called 'elastic relaxation' and 'plastic localization' of strains. 'Elastic relaxation' phase produces small and dispersed shear strains (stick), while 'plastic localization' significant and localized shear strains (slip).

A thorough analysis of deformation fields from many tests suggested that also within a shear band occur cyclic changes. A simple measure of deformation —maximum value of shear strains within shear band was selected to demonstrate this periodic behaviour. It is presented in Fig. 3, which shows (in grey colour) the relation between maximum shear strain and the value of boundary displacement in one of the model tests. Each of localization incidents can be easily defined in Fig. 3 by a peak value of shear strains. The distance between adjacent peaks (a period) is between 3 and 5 unit displacement increments (Δd ~ 0.1 mm) in Fig. 3, mostly 4 increments (~4 increments on average). The finding that there exists, for certain test conditions, a minimum value of boundary displacement, necessary to activate strain localization, is important to arrange properly experiments in which image analysis is planned to be used. Too big boundary displacement may cause deformation outside the range of PIV (DIC—digital image correlation) applicability [7], while too small may not activate strain localization at all.

Figure 3 shows the minimum difference between peak shear strains and background level of diffused deformation about 100%, while maximum difference reaches 400%.

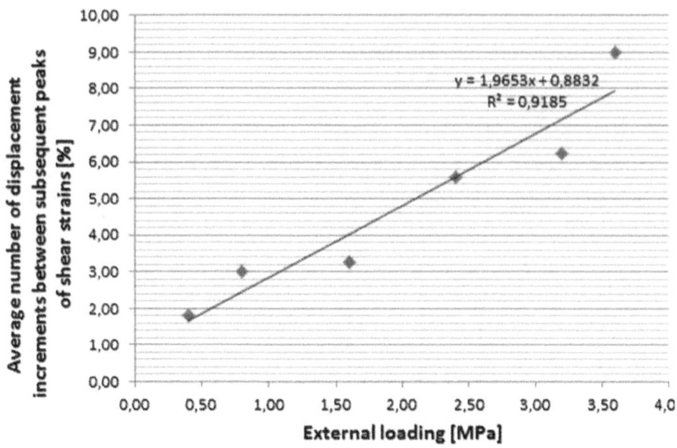

Fig. 4 Relation between the average number of displacement increments $\Delta d = 0.0625$ mm (length of the 'cycle') and the value of external loading

4.2 Minimum Displacement

It was found that the minimum displacement of the external boundary (model of rigid retaining wall), necessary to produce 'plastic localization' depends on the value of the external load (the higher the load, the greater the value of minimum displacement, what seems to be quite obvious) and is also a function of grain coarseness (glass granules in immersion liquid, dry glass granules and sand were compared). The lack of space does not allow to present the results obtained for course Borowiec sand and dry granules—Fig. 4 shows synthesis of some 20 tests performed on glass granules submerged in immersion liquid (clove oil). Each point in Fig. 4 is an average calculated from 2 or 3 tests. The linear relation between length of the 'cycle' and the external load is obtained.

4.3 Comparison with DEM Simulations

Only some selected results of DEM simulation are presented in this paper, namely maps of shear strains produced by PIV for pairs of images created by DEM (Fig. 2C) and maximum values of shear strains (Fig. 3, in black colour). The resolution of DEM images was lower than the resolution of camera images, so more continuous picture of shear strain maps was obtained for DEM.

The difference in resolution does not explain however that only qualitative similarity between experimental data and simulation can be observed in Fig. 2, as neither the geometry of shear band nor the range of shear strains are close. DEM simulation gave definitely smaller strains than the experiment. This is clearly visible in Fig. 3, which also does not show an indication of cyclic behaviour, so clear in case of the real test. Figure 2C could suggest some cyclic changes in shear strain maps, however it is much less convincing than Fig. 2A, B.

The reasons for this discrepancy may be numerous, as the DEM model is built in 1:1 scale and contains a lot of grains and some parameters which cannot be determined experimentally, leaving the only available trial and error method.

First the influence of material parameters was checked (set 2, Table 1), however without much success—the shear band was still located far from the real one. Further research on the topic is continued.

5 Conclusions

A linear correlation between minimum displacement activating strain localization and the level of external load was found for granular samples. Finding physical limits on boundary displacement have an influence on experimental and numerical testing, as the proper choice of external deformation increment is crucial to obtain valid results, if image analysis is planned to be employed [7].

References

1. Duran, J.: Sands, Powders and Grains. An Introduction to Physics of Granular Materials. Springer (2000)
2. Lesniewska, D., Muir Wood, D.: Observations of stresses and strains in a granular material. J. Eng. Mech. **135**, 1038–1054 (2009)
3. Muir Wood, D., Lesniewska, D.: Stresses in granular materials. Granular Matter **13**, 395–415 (2011)
4. Lesniewska, D., Niedostatkiewicz, M., Tejchman, J.: Experimental study on shear localization in granular materials within combined strain and stress field. Strain **48**, 430–444 (2012)
5. Pietrzak, M., Lesniewska, D.: Failure evolution in granular material retained by rigid wall in active mode. Studia Geotechnica et Mechanica **XXXIV**(4) 69–77 (2012)
6. Kozicki, J., Donze, F.V.: A new open-source software developed for numerical simulations using discrete modeling methods. Comput. Methods Appl. Mech. Eng. **197**, 4429–4443 (2008)
7. Stanier, S., Dijkstra, J., Lesniewska, D., Hambleton, J., White, D., Muir, Wood D.: Vermiculate artefacts in image analysis of granular materials. Comput. Geotech. **72**, 100–113 (2016)

Part II
Constitutive Modeling and Multi-physical Coupling for Localization, Instabilities and Degradation of Geomaterials

Modeling Physico-Chemical Degradation of Mechanical Properties to Assess Resilience of Geomaterials

T. Hueckel, M. Ciantia, B. Mielniczuk, M.S. El Youssouffi and L.B. Hu

Abstract It is widely accepted that critical properties of geo-materials that play a key role in failure of earth-structures undergo often a substantial evolution induced by non-mechanical processes and variables. That includes: hydro-thermal fracture, thermal collapse, chemical mass removal or accretion (dissolution or precipitation), chemical shrinkage/swelling, drying shrinkage, capillary force evolution during pore water phase change. The properties affected are: strength in all its manifestation, compressibility, permeability, thermal conductivity, to mention just a few. The physical processes involved are either natural or engineered. Their phenomenology is per se a conundrum, as often they constitute a series of parallel or sequential processes. A review of several phenomena leading to geomaterial degradation, and methodology is presented to deal with multi-physical couplings in constitutive modeling. In plasticity, the central constitutive function is a hardening rule. Also in this case, phenomenological observations indicate a chemo-mechanical, two-way coupling. Other degradation phenomena discussed include drying—cracking, and or the role of suction induced hardening in unsaturated materials.

T. Hueckel (✉)
Duke University, Durham, NC, USA
e-mail: hueckel@duke.edu

M. Ciantia
Imperial College London, London, UK

B. Mielniczuk · M.S. El Youssouffi
Montpellier 2 University, Montpellier, France

L.B. Hu
University of Toledo, Toledo, USA

E. Papamichos et al. (eds.), *Bifurcation and Degradation of Geomaterials with Engineering Applications*, Springer Series in Geomechanics and Geoengineering, DOI 10.1007/978-3-319-56397-8_9

1 Introduction

One of a most common misconceptions in structural mechanics and geo-mechanics is to perform failure analysis assuming that the properties of the materials involved are those of a “nominal” material, i.e. a material as it was at the moment of its manufacturing, or of the placement of the structure into the soil mass. However, over the recent years it became widely obvious that critical properties of materials that play a key role in failure undergo invariably a substantial evolution induced by non-mechanical processes and variables, referred to in general as “material weathering”, or “material wear”. Such processes may be spontaneous, engineered, or “inflicted” (by error, or erroneous, or unpredicted usage). The mechanisms being part of the above mentioned processes include: hydro-thermal fracture, thermal collapse, chemical (biological) mass removal or accretion (dissolution or precipitation), chemical shrinkage/swelling, drying shrinkage and cracking, capillary force evolution during pore water phase change. Many of these processes are seasonally, or daily, cyclic, and their recurrence may count from hundreds to million times, with accumulated irreversible strain, damage and change in conductivities.

The properties affected are: strength in all its manifestations, deformability, permeability, thermal conductivity, to mention just a few. The physical processes involved are not well identified, and are subject of inquiry per se. Their phenomenology is a conundrum, as often they constitute a series of parallel of sequential physic-chemical processes of which we have only fragmentary knowledge.

The currently accepted practice required by law, standards and codes, and philosophy of legislation supporting them to deal with such processes are limited exclusively to the concept of factor of safety, which is a single number, to reduce the allowable load. However, it is widely perceived that the depth of considerations to arrive at the values of safety factor are disproportionally simplistic given the enormous progress in computational capabilities, on one hand, and in understanding of the behavior and of the properties of the materials. While several branches of engineering evolved to base the design on the long-term performance analysis, many others (including civil engineering) rarely even consider long-term processes.

However, there is a growing social expectation to provide infrastructure that would be resilient in the circumstances, not only of the routine loading, but also exceptional loading, including extreme and accidental exposure. This expectation seems to derive from less than satisfactory performance of the infrastructure, urban systems, as well as individual structures during recent hurricanes Katrina, and Sandy, repeated flooding across EU and US, including the coastal flooding, the tsunami disaster in Fukushima, the overall deterioration of roads and bridges in the US, the inundation of cities, which may be, or not, related to climate change.

With the new technologies entering the common practice, or being studied, like CO_2 sequestration, fracking, methane hydrate exploitation, nuclear waste disposal, geo-thermal energy exploitation, etc., this leads to the expectation, that they be designed as resilient, which implies that as many as possible scenarios of loading of lesser probability of occurrence be examined, and long-term performance analyses be conducted. That clearly implies multi-physics performance analysis, and knowledge and thorough investigation of phenomena potentially taking place.

The paper focuses on a series of phenomena of geomaterial degradation (short and long term) and methods to model them via thermo—and chemo—plasticity and —elasticity, emphasizing their multi-physics couplings and multi-scale description. The very meaning of degradation implies a gradual loss of engineering qualities, and we will leave aside similar, interesting that are of scope of this specific definition.

2 The Essence of Time-Dependent Degradation and Resilience Assessment

Classical constitutive modeling knows only one trick to deal with a time-dependent irreversible evolution of geomaterials, which is a visco-elastic or -plastic creep. However, it becomes increasingly clear that a generic creep, with its explicit dependence on time is of very limited use in providing any controllability of the process. What is being sought are the mechanisms behind the time dependence and the variables that allow the processes to be controlled, as much as it is possible.

Degradation is viewed as processes that are chemical or physical in nature, usually, but not necessarily developing over long term. Often these processes are not at all, or poorly, controlled, frequently even ignored, intentionally, or not, typically because of lack of imagination that their negative outcome could be quantified, predicted and often prevented.

Four types of processes can be considered as examples of degradation: chemical reactions, non-reactive chemical phenomena, and mineral transformation, affecting material properties, and finally, water phase-transition effect in drying cracking. Two of them are discussed below.

In the first category, such the degradation processes are caused commonly by chemical reactions with the environment, enhanced by temperature and humidity variations, linked either to diurnal, seasonal cycles, or related to the climate change. They may be explicitly cyclic, but often are related to natural irregular environmental variations. It is also highly relevant in the phenomena of sinkholes, CO_2 sequestration technology, and in nuclear waste disposal. In the latter area it has recently become a point of scrutiny, because of the corrosion reaction induced by bacteria at the contact between steel canisters and clay. The chemical reactions are dissolution of minerals, their precipitation, mineral transformations, as in natural analogue for nuclear waste disposal [1], are typically characterized by reaction rate equations, which explicitly make them time-dependent. An additional time effect may come from diffusion/advection rate dependence.

The non-reactive chemo-mechanical coupling is characterized by an instantaneousness of chemical change, like water salinity change, leading to swelling/shrinkage, or inundation of pore water with aromatic contaminants [2]. The latter has been reported to induce a highly stress dependent changes in hydraulic conductivity in clay based landfill liners. The phenomena involved are osmosis, ion exchange, to mention a few. A time-dependence in these phenomena comes from a short-term diffusion of ions or water.

The third group of degradation phenomena addressed is physico-mechanical in nature and regards the effects of phase transition of soil water, namely that of drying-cracking. Drying-cracking is an effect critically affecting almost any type of soil structure, from landfill liners, to nuclear waste disposal barriers, to levees. The basic physical phenomena involved are evaporation of pore water, ensuing soil shrinkage, evolution of capillary forces and menisci, and air entry, and macroscopic cracking. They tend to occur at least three scales. The time-dependence results from the process of evaporation and diffusive-advective transport of water toward the interface with the air.

To make earth structures resilient in extreme conditions, one needs to conduct a performance assessment. That requires identifying scenarios under which the material and earth structures will be functioning. Within such scenarios it is critical to thoroughly understand the complexity of phenomena and mechanisms developing, usually over long-term. That includes determining which variables and properties control the phenomena involved. The performance assessment requires developing constitutive models (most commonly coupled) for the phenomena and mechanisms and implement them within numerical codes.

3 Modeling Framework for Degradation and Resilience Assessment

In what follows we limit ourselves to chemically, thermally, or phase-transition induced degradation of geomaterials. That leaves outside processes of degradation due to earthquakes, physical wear, etc. We also focus on degradation of mechanical and hydraulic properties of geomaterials.

There are countless chemical processes and reactions that may impact the mechanical properties of geomaterials. The most important in the present context are removal or addition of mass of minerals within the pore space. This may occur through dissolution of precipitation. The result of such mass loss/gain for the mechanical properties is a strength increase or decrease, direct chemical strain, usually volumetric swelling or compaction. Also other properties of relevance may be affected, such as hydraulic and thermal conductivity. Another type of change is in ionic content of pore fluid at the interface with the materials solids. In materials that are electrically charged, as clays, it produces swelling or shrinkage, resulting often in a change in strength and compressibility, as well as in hydraulic conductivity.

The causes of the mentioned processes are diverse. They may result directly from the technology, as injection of water or vapor, as in fracking, or in geothermal technologies, injection of CO_2 in technologies related to its geological sequestration, or specific chemicals, for instance acids, as in fracking. Some processes that are of interest are natural processes as a part of diagenesis, like pressure solution. Similar processes are involved in subsidence due to oil or gas extraction, such as resulting compaction.

To account for the role of chemical, thermal or physical changes in the mechanical material response one needs to formulate a framework of balance laws and a set of constitutive hypotheses, taking into account the following features of geomaterials. First, one needs to recognize that in many cases, the above changes occur because of the two- (or three-) phase nature of geomaterials. Second, such changes often consist of interaction (of all sorts) between the phases, including mass, or momentum, or heat exchange. Next, specific mechanisms and corresponding variables need to be identified and mathematically described to couple the mechanical properties of the soil solids to their chemistry, and in particular to the evolution and possibly transport of chemical species. Finally, the changes, and corresponding mechanisms can occur or be described at different geometrical or temporal scales, the fact that needs to note that to describe total free energy in two-phase, multi-species reactive porous medium be incorporated in our framework.

To start with let us consider the material free energy, in which the work by total stress, σ during the process of deformation $d\varepsilon$, and heat must be supplemented by the work of chemical potential $\mu_{k,\ K}$ (mass based, [J/g]) during addition of mass, $m_{k,K}$ of species of both phases, (k are indices for the species, K are for phases); $\mathfrak{H}$ being entropy, and T absolute temperature

$$d\Psi = \sigma d\varepsilon - \mathfrak{H}\delta T + \sum_{k,K} \mu_{k,K} m_{k,K} \tag{1}$$

Species may be entities of any kind, pore water, adsorbed water, other fluid species, minerals, ions, even entire rock. Mass of species added to, or removed from the system and energy associated with the addition/removal (i.e. chemical potential $\delta\mu_{kK} = \frac{\delta p_K}{\rho_{kK}} + \frac{RT}{m_k^{(M)}}\delta(\ln x_{kK})$ are the only variables related to the chemical processes. In this definition, $R = 8.31451$ J/molK is the universal gas constant, h is entropy, T [K] the absolute temperature and $m_k^{(M)}$ is the molar mass of the species k, e.g. for free pore water $m_w^{(M)} = 18$ g, p_K is pressure or mean stress in phase K, assumed as equal in all the species of the given phase, whereas x_{kW} are molar fractions of the species k in phase W, which is actually the mass variable. We will use both the changes in mass and or chemical potentials in reaction equations, as well as to quantify the effect that reactions make on the mechanical properties of soil/rock.

The central challenge is to derive constitutive properties of the solid phase, which may be loosing mass, inclusive of the adsorbed water. For that task we shall adopt a strategy proposed originally by Heidug and Wong [3]. The free energy of the solid phase is represented by the difference between the total energy of the whole system of porous medium and that of free pore water, the latter classically defined per unit volume of the fluid phase as

$$\delta\Psi_W = \mu_{wW}\delta m_{wW} + \sum \mu_{sW}\delta m_{sW} - \mathfrak{H}_W\delta T - p_W\delta v_W \tag{2}$$

The first term describes molecular water mass change, the second term, with the summation over the components m_{sW} implies all changes: in ionic content in free water, as well as mineral relative mass changes, being dissolved into, or precipitated from the liquid phase, or finally transported away, the last term is the mechanical work associated with the water volume content change.

Following Heidug- Wong and assuming that solid and fluid have the same temperature, the free energy of the reactive solid phase becomes

$$\delta\Psi_S = \delta\Psi - \delta\Psi_W = \sigma\delta\varepsilon + \sum_{k,S} \mu_{kS}\delta m_{kS} + p_W\delta v_W - (\mathfrak{H} - \mathfrak{H}_W)\delta T \tag{3}$$

Clearly, the key variable relatives to chemical effects are mass changes. Depending on the process it is necessary/convenient to identify a reversible part of the change of $m_{kS}{}^{el}$ (mainly in close systems), or not, which means to say that all mass changes are irreversible (mainly in open systems). In the former case, ignoring temperature changes and the effect of fluid volume change, the elastic part of free energy of the reactive solid expressed via the effective stress, σ' and employing Legendre transform, $\Psi_M{}^{el}$, becomes

$$\delta\Psi_M^{el} = \varepsilon^{el}\delta\sigma' - \sum_{k,S} \mu_{k,S} m_{kS}^{el}; \sigma' = \sigma - pI; \varepsilon^{el} = \frac{\partial\Psi_M}{\partial\sigma'}; \tag{4}$$

Hence, the chemo elasticity relationships take the form (in vector notation)

$$\boldsymbol{\varepsilon}^{el} = \left(-\beta(m_{kS}^{el})m_{kS}^{el} + \kappa(m_{kS}^{el})\, ln\frac{p'}{p_o} \right)\frac{I}{3} + \frac{s}{2G(m_{kS}^{el})} \tag{5}$$

where $\boldsymbol{s}$ is the stress deviator, while G is the shear modulus, κ is a bulk modulus and β is the coefficient of chemical expansion. Thus, the primary task is to identify which of the species when changing mass affect most the material moduli, and strength and what is the relationship between the variable. On the other hand constitutive laws need to be established for rates of mass change individual minerals or ions, $\dot{m}_{kS}^{el}$. As for the chemical reactions, most of the rates are well known from rate constants determined for specific reactions. However, additional coupling may arise in the case for instance of the variable specific surface areas, for example as in dissolution from a micro-cracking rock. One such case will be discussed below.

Equally, if not more important is the description of the effects of chemical reactions, thermal and physical processes on the irreversible straining and failure of geomaterials. The main focus is on irreversible chemo-plastic strain and the accumulated mass removal (isothermal) effect on the material strength as framed within the second thermodynamic law leading to a bound on dissipation function

$$\dot{D} = \boldsymbol{\sigma}' \dot{\boldsymbol{\varepsilon}}^{irr} + \sum_{k,S} \mu_{kS} \dot{m}_{kS}^{irr} \geq 0 \tag{6}$$

One possible set of constitutive equations is as spelled below:

$$\delta\varepsilon^{irr} = \delta\Lambda \frac{\partial f}{\partial \boldsymbol{\sigma}'}; f = f\left(\boldsymbol{\sigma}', p_c', M\right) = 0; p_c' = p_c'\left(\varepsilon_v^{irr}, \int \dot{m}_{kS}^{irr} dt\right); M = M\left(\int \dot{m}_{kS}^{irr}\right) \tag{7}$$

which imply the fulfillment of the yield condition, $f = 0$, but the "size" of which, p_c' depends on the accumulated mass removed, as does ratio of its semi-axes, or aspect ratio, M. Experience suggests that both the yield locus characteristics, but most of all the chemical dependence are both geo-material, as well as geochemical process specific.

4 Degradation of Coastal Calcarenite Subject to Water Inundation

Large segments of marine coasts in Southern Europe and elsewhere are formed of calcarenite, which is notoriously prone to degradation and to rock falls causing extensive damage and endangering human life and wellbeing. Calcarenite is more than 90% carbonate rock. When exposed to water it may lose instantly up to 60% of its dry strength, Fig. 1 (short-term effect) [4]. On the other hand, when continuously saturated, calcarenite weakens due to the dissolution of the grains and bonds over the long-term with a much lower rate [5]. It is known that strength of carbonate rocks depends on its volume fraction filled with calcite [6].

The short-term loss and regain of strength is seen as a result of suspension and re-sedimentation of solid mineral into and from the fluid [4]. It is a reversible process for a small number of wetting-drying cycles in a closed system [7]. However, for a larger number of cycles the dry strength may not be fully recovered, while the strength in wet conditions remains about the same over short-term [8]. The long-term changes are seen as irreversible. In open systems, specific transport conditions play a crucial role.

On the micro-scale the current findings are: (i) Grains and bonds in the solid structure are formed of calcite microcrystals of 5 μm; (ii) about 50% of the pore volume are macro-pores 25–250 μ ca, while the rest are micro-pores ($25 < r < 0.01$ μ); (iii) Two distinct types of bonds, strong and weak, are observed in SEM. The strong bonds are bridges of calcite formed in a diagenetic process, while the weak bonds are formed by a mixture of the calcite "powder" and sea water salts that settle into a meniscus during evaporation; it is re-suspended when

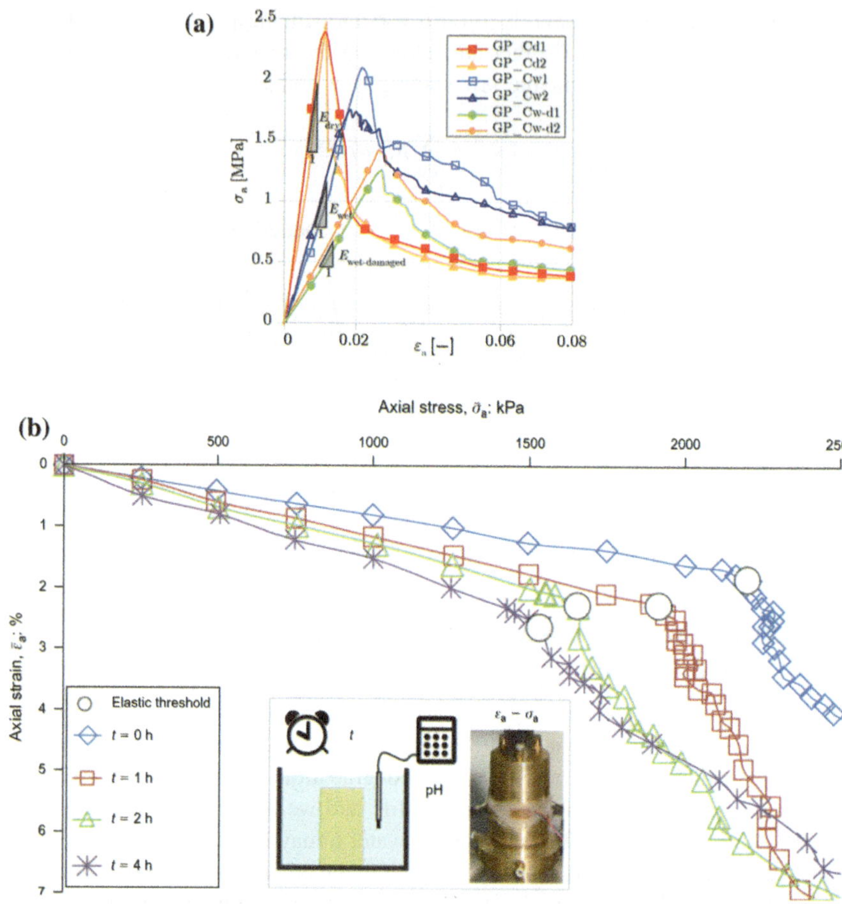

Fig. 1 Tests on wetted calcarenite: **a** uniaxial compression on dry (Cd1&2), wetted (Cw1&2) and acidified (Cw-d1&2)-from Ciantia et al. 2014; **b** Oedometer test with saturation of 0, 1, 2 and 4 h —from Ciantia and Hueckel [4]

water refills the pores [9]; (iv) the chemical composition of the calcarenite of concern is: 98% $CaCO_3$ and some traces of Al and Mg); (v) 99% of the porosity of calcarenite is connected; (vi) the initial specific surface area results to be 2.7 m^2/g.

In what follows, we concentrate on the long-term interaction of calcarenite with water. The material is considered as continuously saturated and drained. The main component of calcarenite, calcite ($CaCO_3$) dissolves in water diminishing the dimensions of bonds and grains (and increasing the porosity), hence reducing the material strength at the macro-scale [7]. A meso-scale scenario centers around a meso-pore surrounded by deformable rock matrix with water penetrating radially from the fluid-rock inter-face and effecting dissolution of the mineral. If the material is stressed at yielding, the developing irreversible micro-cracks induce an increase in the specific surface area. As the reaction rate per volume of fluid is proportional to the surface area at the fluid/solid interface [10], it follows that the rate is coupled to the amount of the mechanical damage. The formulation for dissolution and specific surface area increase is developed at the micro-scale whereas the phenomena of damage described above are formulated at a meso-scale level of a pore. Finally, quantities from these two scales are re-calibrated to the macro-scale, at which continuum mechanics constitutive models are formulated.

The rate equations are different for basic and acidic environment, after Sjöberg [10]. In acidic conditions it depends on pH of the fluid alone [11].

$$\begin{aligned} \dot{\xi} &= -k_b A^* M^m \left\{ C - \left[Ca^{2+}\right]^{1/2} \left[CO_3^{2-}\right]^{1/2} \right\} \quad \text{for } 7.5 < \text{pH} < 14 \\ \dot{\xi} &= -k_a A^* M^m \left[H^+\right] \quad \text{for } 4 < \text{pH} < 7.5 \end{aligned} \tag{8}$$

where $0 \leq \xi \leq 1$ is the reaction progress, and its time derivative is the reaction rate (denoted by superposed dot), k_a and k_b are reaction constants for acid or base, whereas the reacting surface area A^* is calculated per unit volume of reacting fluid, M^m is the molar mass of calcium carbonate, and the quantities in brackets denote concentrations of respective species. Following Hu and Hueckel [12] the variable surface area, which includes that of generated micro-cracks is linked to the variable of dilatant volumetric irreversible strain, $\varepsilon_{\mathrm{v}}^{irr} < 0$, however when the latter are compactant, the reactive surface area is assumed as remaining constant. Hence,

$$A^* = f(\varepsilon_{\mathrm{v}}^{irr}) = \begin{cases} A_0^*(1 + \phi|\varepsilon_{\mathrm{v}}^{irr}|) & \text{if} \quad \varepsilon_{\mathrm{v}}^{irr} < 0 \\ A_0^* & \text{if} \quad \varepsilon_{\mathrm{v}}^{irr} \geq 0 \end{cases} \tag{9}$$

where ϕ is a constant, while A_0^* is effective surface area of pre-existing voids.

To describe the water sensitivity of strength for calcarenite, the simplest approach is through rigid hardening plasticity with chemical softening. A linearized plane stress yield locus, in terms of radial and circumferential stress components reads

$$f = \begin{cases} \frac{1}{2}(\sigma_\vartheta - \sigma_r) - \frac{\alpha}{2}(\sigma_r + \sigma_\vartheta) - \left[1 + \gamma\varepsilon_{\mathrm{v}}^{irr} - \delta\xi\right] \frac{\tan\varphi_1 - \tan\vartheta_1}{\tan\vartheta_1 + 1} \sigma_{01}^* = 0 & ; \quad \sigma_\vartheta \geq \sigma_r \tan\varphi_1 \\ \frac{1}{2}(\sigma_r - \sigma_\vartheta) - \frac{\alpha}{2}(\sigma_r + \sigma_\vartheta) - \left[1 + \gamma\varepsilon_{\mathrm{v}}^{irr} - \delta\xi\right] \frac{\tan\vartheta_2 - \tan\varphi_2}{\tan\vartheta_2 + 1} \sigma_{02}^* = 0 & ; \quad \sigma_\vartheta \leq \sigma_r \tan\varphi_2 \\ \sigma_r - \sigma_{02}^*\left[1 + \gamma\varepsilon_{\mathrm{v}}^{irr} - \delta\xi\right] = 0 & ; \quad \sigma_r \tan\varphi_2 \leq \sigma_\vartheta \leq \sigma_r \\ \sigma_\vartheta - \sigma_{02}^*\left[1 + \gamma\varepsilon_{\mathrm{v}}^{irr} - \delta\xi\right] = 0 & ; \quad \frac{\sigma_r}{\tan\varphi_1} \leq \sigma_r \leq \sigma_\vartheta \end{cases} \tag{10}$$

where α, δ, γ, ϑ_i, φ_i are all constants.

With the restriction that despite the chemical softening, the positive strain hardening function, h is always dominant

$$h = 1 + \gamma\varepsilon_{\mathrm{v}}^{irr}\delta\xi > 0 \tag{11}$$

The associated flow rule is adopted, and as a result there is an irreversible chemo-plastic strain rate component [12–14].

We employ these constitutive laws to consider a meso-scale representative elementary volume (REV) of a single, plane strain, axisymmetrical pore between four bonded grains and under a constant external isotropic compression, and chemical degradation.

The solution for such pore features three zones: dilatant, critical and compactant, evolving as the chemical reaction progresses at constant external radial stress at B, Fig. 2. Eventually as the critical zone *CD* disappears, the solution ceases to exists, which is interpreted as a chemically induced instability; for details see Ciantia and Hueckel [4].

The meso-scale behavior has been up-scaled, using a periodic representation and a macroscopic chemo-plasticity. The macroscopic prediction of reaction progress and degradation of the yield limit over a year is shown against the experimental results in Figs. 3 and 4. Such a tool can be used for resilience assessment for coastal structures.

5 Degradation of Geomaterials via Drying-Cracking

Drying of soils contributes to (often critical) degradation of earth structures such as levees, dykes, earth dams, clay liners, clay buffers and backfills in nuclear waste disposal. Sometimes, drying is enhanced by a simultaneous heating. Drying and cracking of soils and like materials appear to develop following an involved scenario, which consists of several phases: (i) shrinkage of the solid upon evaporation induced suction, or total tensile stress build up, if the soil is constrained against displacement. Effective stress in the presence of high suction and low total tensile stress results to be compressive; (ii) air entry when the water-air interface undergoes instability; (iii) total tensile stress concentration at the tip of the air entry finger; (iv) development of tensile crack, as due to stress magnification, the latter becomes larger than suction, and the effective stress becomes tensile. Peron et al. [15] observed in MPS that first to dry are the larger pores, which consistently become smaller contributing to the macroscopic shrinkage of soil. The larger pores closure and practically soil shrinkage cease, simultaneously with the air entry. However, the smaller pores continue to evaporate, but their contribution to shrinkage is minor.

Modeling of the above phenomena requires considerations at several scales. Hu et al. [16, 17] simulated evaporation from a porous system seen as a bunch of parallel deformable tubes of two sizes, including a moment of air entry, point ***a*** for large pores, and point ***b*** for smaller pore, shown in Fig. 5a. Air entry occurs at a point when a decreasing meniscus radius becomes smaller than the largest pore throat radius, the latter depending on deformability of the vessels, reaching a critical

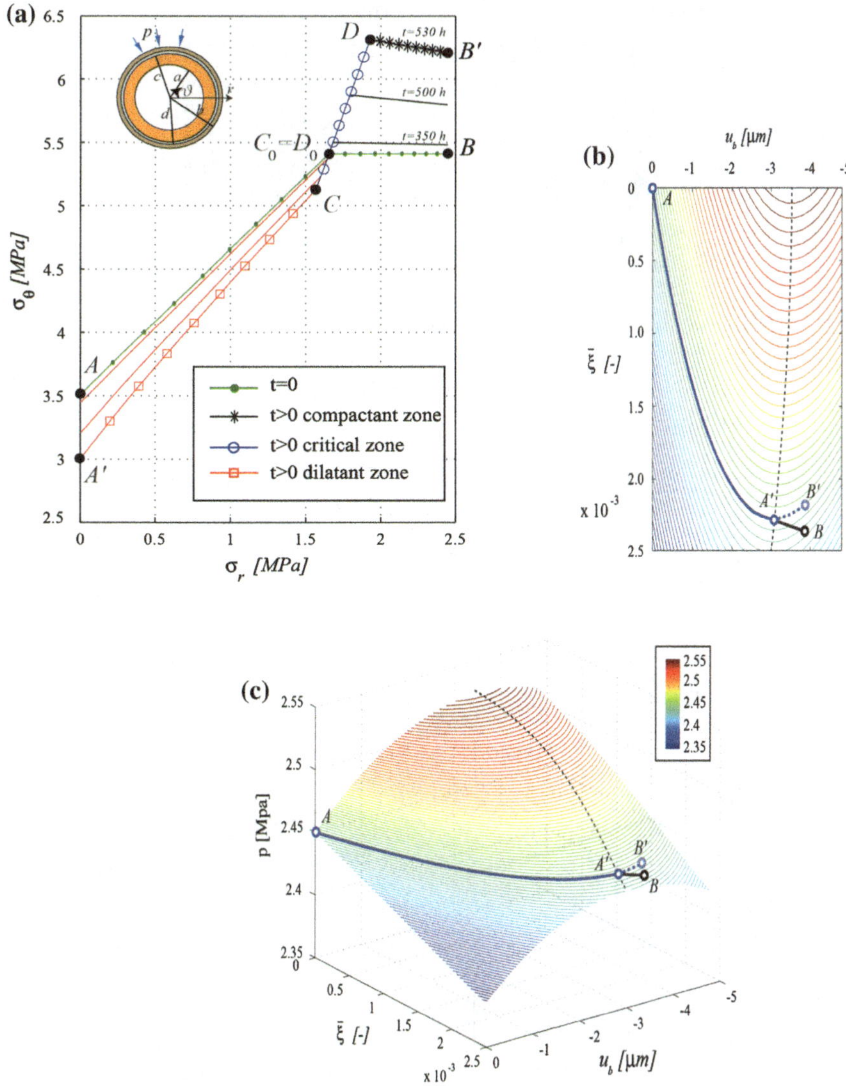

Fig. 2 Stress components (**a**) and external displacement (**b** and **c**) as functions of reaction progress. In (**b**) and (**c**), segment *AA'* visualizes a response along an actual path u_b/b versus dissolved mass. Segment *A'B*, shows an unstable response to an arbitrary path δt > 0 from *A'*. *(see Ciantia and Hueckel [4] for detailed explanations)*

pressure difference between fluid, p, and atmosphere, p_a. It was observed that air entry occurs within 1/1000 of a sec producing a well-articulated singularity of the external surface, Fig. 5.

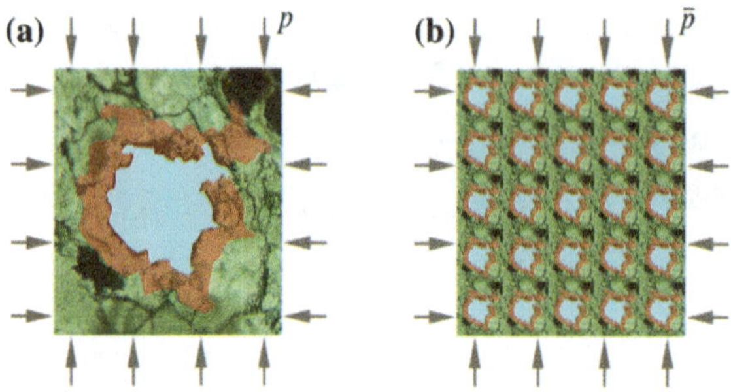

Fig. 3 Micro- and macro-scale medium, (**a**) and (**b**)

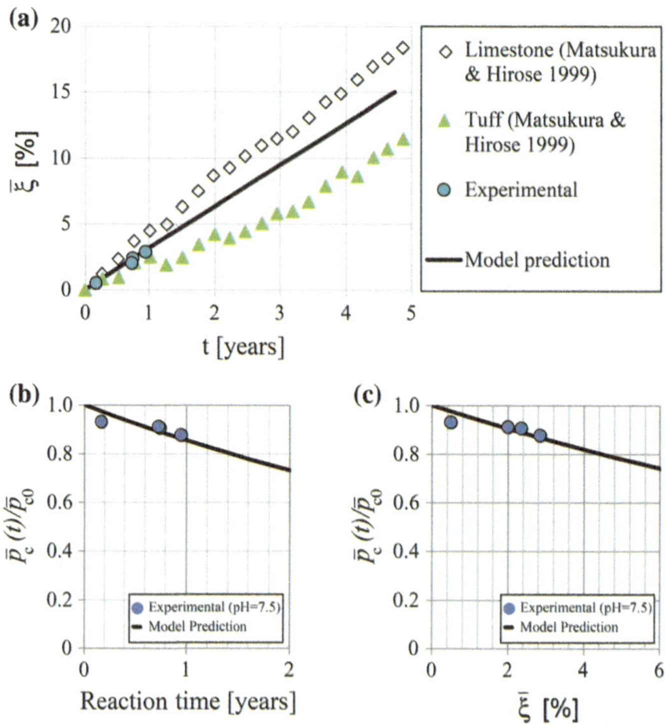

Fig. 4 Reaction progress in time (**a**)—results for a limestone and tuff are shown for comparison, Degradation of yield limit size, p_c' (**b** and **c**)

We postulate that at the tip of that singularity, a feeble total tensile stress induced by a constrained shrinkage becomes amplified by an order of magnitude, and changes the sign of compressive effective stress to tensile and its value easily becomes larger than tensile strength. Using the principles of linear fracture mechanics, the stress at the external boundary of the tubes, which are placed near

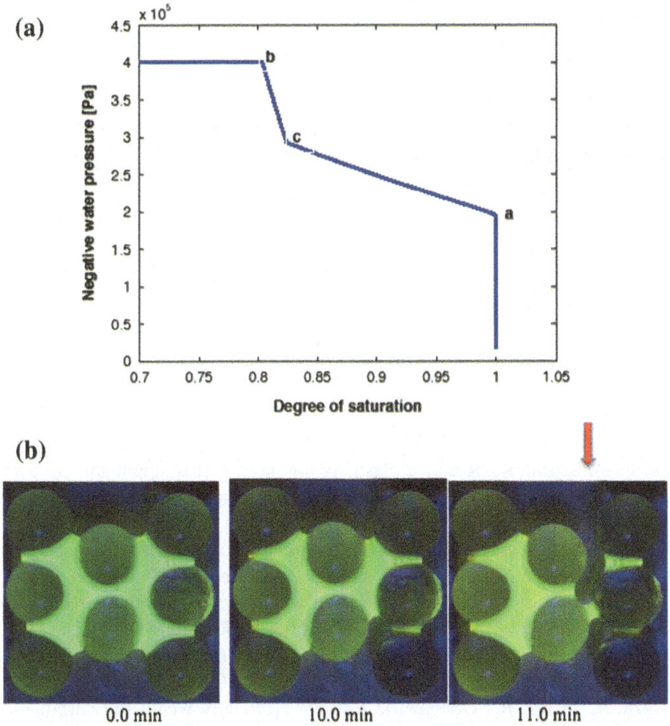

Fig. 5 **a** Simulated soil-water characteristic curve obtained via pressure weighing procedure for the systems of the two modes of vessels versus Degree of Saturation; **b** air entry into a set of glass spheres immersed in water

the flaw tip, with the curvature radius of r_c can be calculated as the far-field (macro-scale) stress at crack initiation as (stress is positive in tension)

$$\sigma_x = \frac{K_{I\,crit} + \frac{1}{2} p \sqrt{\pi r_c}}{B\sqrt{\pi c}} \tag{12}$$

where $K_{I\,crit}$ is the critical stress intensity factor, c is the depth of the flaw, B is a constant equal to 1.12 and p is the fluid pressure inside of the pore [18]. Once a crack is generated at a most stressed air entry point, it propagates further at a modest speed controlled by the rate of continuous evaporation (for details see Hueckel et al. [19]. The critical time to cracking depends on the water advection to the interface with the atmosphere to generate the suction required to trigger the air entry.

6 Conclusions

Outline of modeling is presented for degradation of geomaterials due to time dependent chemical or physical processes [20]. Further progress in such modeling is necessary to provide tools to assessment of resilience of geomaterials and earth structures. The time dependence of chemical softening is inherited from the characteristic rates of reactions, and that of drying cracking from advective transport through deformable porous skeleton. The above findings require a thorough experimental verification.

References

1. Hueckel, T., Pellegrini, R.: Reactive plasticity for clays. Part II: application to a natural analog of long-term geomechanical effects of nuclear waste disposal. Eng. Geol. **64**(2–3), 195–216 (2002)
2. Fernandez, F., Quigley, R.M.: Controlling the destructive effect of clay-organic liquid interactions by application of effective stresses. Can. Geotech. J. **28**, 388–398 (1991)
3. Heidug, W.K., Wong, S.W.: Hydration swelling of water absorbing rocks: a constitutive model, Int. Jnl. Numer. Anal. Methods Geomech. **20**, 402–430 (1996)
4. Ciantia, M., Hueckel, T.: Weathering of stressed submerged calcarenites: chemo-mechanical coupling mechanisms. Geotechnique **63**(9), 768–785 (2013)
5. Castellanza, R., Nova, R.: Oedometric tests on artificially weathered carbonatic soft rocks. J. Geotech. Geoenvironmental Eng. ASCE **130**(7), 728–739 (2004)
6. Hueckel, T., Pellegrini, R., Del Olmo, C.: Constitutive properties of thermo-elasto-plastic behavior of deep carbonatic clays. Int. J. Numer. Anal. Methods Geomech. **22**(7), 549–576 (1998)
7. Castellanza, R., Parma, M., Pescatore, V., Silvestro, G.: Model footing load tests on soft rocks. Geotech. Test. J. **32**(3), 262–272
8. Andriani, G.F., Walsh, N.: The effects of wetting and drying, and marine salt crystallization on calcarenite rocks used as building material in historic monuments. Geolo. Soc. London, Spec. Publ. **271**, 179–188 (2007)
9. Delenne, J.Y., Soulié, F., El Youssoufi, M.S., & Radjai, F.: Compressive strength of an unsaturated granular material during cementation. Powder Technol. **208**, 308–311 (2011)
10. Sjöberg, E.L.: A fundamental equation for calcite dissolution kinetics. Geochim. Cosmochim. Acta **40**, 441–447 (1976)
11. Sjöberg, E.L., Rickard, D.T.: The effect of added dissolved calcium on calcite dissolution kinetics in aqueous solutions at 25 °C. A fundamental equation for calcite dissolution kinetics. Chem. Geol. **49**, 405–413 (1985)
12. Hu, L., B., Hueckel, T.: Coupled chemo-mechanics of intergranular contact: toward a three-scale model. Comput. Geotech. **34**, 306–327 (2007)
13. Hueckel, T.: Reactive plasticity for clays during dehydration and rehydration. Part I: concepts and options. Int. J. Plast. **18**(3), 281–312 (2002)
14. Hueckel, T.: Reactive plasticity for clays during dehydration and rehydration. Part I: concepts and options. Int. J. Plast. **18**(3), 281–312 (2002)
15. Peron, H., Hueckel, T., Laloui, L., Hu, L.B.: Fundamentals of desiccation cracking of fine-grained soils: experimental characterisation and mechanisms identification. Can. Geotech. J. **46**, 1177–1201 (2009)

16. Hu, L.B., Peron, H., Hueckel, T., Laloui, L.: Desiccation shrinkage of non-clayey soils: multi-physics mechanisms and a microstructural model. Int. J. Numer. Anal. Methods Geomech. **37**(12), 1761–1781 (2013)
17. Hu, L.B., Peron, H., Hueckel, T., Laloui, L.: Desiccation shrinkage of non-clayey soils: a numerical study. Int. J. Numer. Anal. Methods Geomech. **37**(12), 1782–1800 (2013)
18. Scherer, G.: Crack-tip stress in gels. J. Non-Cryst. Solids **144**, 210–216 (1992)
19. Hueckel, T., Mielniczuk, B., El Youssoufi, M.S., Hu, L.B., Laloui, L.: A three-scale cracking criterion for drying soils. Acta Geophys. **62**(5), 1049–1059 (2014)
20. Loret, B., Hueckel, T., Gajo, A.: Chemo-mechanical coupling in saturated porous media: elasto-plastic behaviour of homoionic expansive clays. Int. J. Solids Struct. **39**, 2773–2806 (2002)

Cyclic Elastoplastic Constitutive Model for Soils Based on Non-linear Kinematical Hardening Rule with Volumetric and Deviatoric Kinematical Parameters

Fusao Oka and Sayuri Kimoto

Abstract The aim of this paper is to present cyclic constitutive models for sandy and clayey soils based on the kinematical hardening rule with the strain-degradation, and the applicability of the models to the practically important problem such as the of the levee during earthquakes. For sandy soils, we have been developed a cyclic elasto-plastic model based on the kinematical hardening rule with two yield surfaces for the change of the stress ratio and the mean effective stress considering the degradation of the yield surface. From the simulation by the present model, we have found the strong non-associativity leads to the large decrease in the mean effective stress during the cyclic deformations under undrained conditions while the model with associated flow rule is not. This result is quite important because the mean effective stress becomes almost zero at the full liquefaction state. We have found that the model can well reproduce the cyclic behavior of soil.

1 Introduction

In the present model, firstly, we have developed a new cyclic model using a single kinematical hardening yield surface for both the change in the stress ratio and the mean effective stress based on the cyclic model. Then in order to take account of the non-associativity that comes from the friction of soils, we have introduced a non-associativity parameter [1]. The proposed model is based on the non-associative flow rule including the associative one. The model is derived based on the non-linear kinematical hardening rule which is advocated by Armstrong and

F. Oka (✉)
Emeritus of Kyoto University, Lab.3 138-1 Association for Disaster Prevention Research, Tanakaasukai-Cho, Sakyo-Ku, Kyoto 606-8226, Japan
e-mail: oka.fusao.38x@st.kyoto-u.ac.jp

S. Kimoto
Department of Civil and Earth Resources Engineering, Kyoto University, Kyoto 615-8540, Japan

E. Papamichos et al. (eds.), *Bifurcation and Degradation of Geomaterials with Engineering Applications*, Springer Series in Geomechanics and Geoengineering, DOI 10.1007/978-3-319-56397-8_10

Frederick [2] with two hardening parameters for both the volumetric and deviatoric strains. From the simulation, we have found the strong non-associativity leads to the large decrease, i.e. almost zero in the mean effective stress during the cyclic deformations under undrained conditions.

2 Cyclic Elastoplastic Constitutive Model

Kimoto et al. [3] proposed an elasto-plastic cyclic model for sandy soils based on the non-linear kinematical hardening rule. The model includes two yield surfaces; one is for the change of stress ratio and the other is for the change of the mean effective stress. In the present model, we have proposed a single yield function for the change in the stress ratio and the mean effective stress.

2.1 Total Strain Increments

The total strain rates decomposes into elastic and plastic ones $d\varepsilon_{ij}^{E}$ and $d\varepsilon_{ij}^{P}$ as:

$$d\varepsilon_{ij} = d\varepsilon_{ij}^{E} + d\varepsilon_{ij}^{P} \tag{1}$$

The strain dependency of the elastic shear modulus is given by

$$G^{E} = \frac{G_{0}^{E}}{1 + \alpha_{e}(\gamma_{ap}^{P*}/\gamma_{r}^{E*})^{r_{e1}}} \left(\frac{\sigma'_{m}}{\sigma'_{m0}} \right)^{r_{e2}} \tag{2}$$

where r_{e1}, r_{e2}, α_{e} are material parameters, G_{0}^{E} is the initial value of G^{E}, γ_{ap}^{P*} is the accumulated plastic deviatoric strain increment after the stress ratio reaches M_{m}^{*} and γ_{r}^{E*} is a referential strain.

The elastic volumetric modulus is given by the similar manner as Eq. (2) while the initial one is given by the swelling index κ.

2.2 Overconsolidation Surface

We use an overconsolidation surface for defining the boundary between the normally consolidated region ($f_b \geq 0$)and overconsolidated region ($f_b < 0$)in the stress space, which is given by

$$f_b = \bar{\eta}^*_{(0)} + M^*_m \ln\left(\frac{\sigma'_m}{\sigma'_{mb}}\right) = 0,\ \bar{\eta}^*_{(0)} = \left\{\left(\eta^*_{ij} - \eta^*_{ij(0)}\right)\left(\eta^*_{ij} - \eta^*_{ij(0)}\right)\right\}^{1/2},\ \eta^*_{ij} = s_{ij}/\sigma'_m \tag{3}$$

where σ'_m is the mean effective stress, s_{ij} is the deviatoric stress tensor, M^*_m is the stress ratio $\eta^* = (\eta_{ij}\eta_{ij})^{1/2}$ at the critical state which is called the stress at the phase change (maximum compression), $\eta^*_{ij(0)}$ is the value of η^*_{ij} after the end of consolidation, σ'_{mb} is the parameter depending on the plastic volume strain.

For the soil that exhibits the strong structural change, σ'_{mb} may change due to the change of the plastic shear strain as well as the plastic volumetric plastic strain. We can describe the strain softening induced by the plastic strain by the decrease of σ'_{mb} as:

$$\sigma'_{mb} = \left\{\sigma'_{mbf} + \left(\sigma'_{mbi} - \sigma'_{mbf}\right)\exp(-\beta z)\right\}\exp\left(\frac{1+e_0}{\lambda-\kappa}v^p\right),\ z = \int dz = \int \left(d\varepsilon^p_{ij} d\varepsilon^p_{ij}\right)^{1/2} \tag{4}$$

where σ'_{mbf} is the final vale of σ'_{mbi}, β is a parameter of the rate of its reduction, σ'_{mbi} is the initial value of σ'_{mb}, σ'_{m0} is the initial value of mean effective stress e is the void ratio, λ is the compression index, κ is the swelling index and v^P is the plastic volumetric strain. In general, σ'_{mbi} of the isotropic consolidated soil corresponds to the preconsolidation stress σ'_{m0}. Since, for sand, it depends on the structural anisotropy and the compaction and aging etc., σ'_{mbi} does not necessary coincides with σ'_{m0}. Hence, σ'_{mbi} can be determined by the characteristics of volumetric strain and the quasi-overconsolidation ratio $OCR^* (= \sigma'_{mbi}/\sigma'_{m0})$ is defined.

2.3 Yield Surface

The yield function is assumed to be a non-linear function of the kinematical hardening parameters of the stress ratio and the mean effective stress, and is given by

$$f_y = \bar{\eta}^*_\chi + C_{ns}\tilde{M}^*\left(\ln\frac{\sigma'_{mk}}{\sigma'_{ma}} + \left|\ln\frac{\sigma'_m}{\sigma'_{mk}} - y^*_m\right|\right) = 0 \tag{5}$$

$$\bar{\eta}^*_\chi = \left\{\left(\eta^*_{ij} - \chi^*_{ij}\right)\left(\eta^*_{ij} - \chi^*_{ij}\right)\right\}^{1/2} \tag{6}$$

where σ'_{mk} is the unit of the stress, y^*_m and z_m are the kinematical hardening parameters for the change of the mean effective stress and σ'_{ma} is a parameter for the structural change, χ^*_{ij} is the nonlinear kinematical hardening for the stress ratio change. In the present paper, the unite of the stress is kiropascal (kPa). The parameter C_{ns} is controlling the non-associativity which is less than 1.0; $C_{ns} = 1.0$ for the associative flow rule and $C_{ns} < 1.0$ for non-associative flow rule. In the present model, we assume $C_{ns} = 1.0$ before the states reaches the phase transformation line and $C_{ns} < 1.0$ after the states reaches the phase transformation line.

χ_{ij}^* has a dimension of η_{ij}^* and is called back stress. The nonlinear hardening rule, an evolutional equation of χ_{ij}^* is given by

$$d\chi_{ij}^* = B^* \left(A^* de_{ij}^P - \chi_{ij}^* d\gamma^{P*} \right) \tag{7}$$

where A^* and B^* are the material parameters and relates to the stress ratio at failure M_f^* and the non-dimensional initial plastic shear modulus G^P as:

$$A^* = M_f^*, \quad B^* = \frac{G^P}{M_f^*} \tag{8}$$

In addition, B^* follows the evolutional law as:

$$dB^* = C_f \left(B_1^* - B^* \right) d\gamma^{P*} \tag{9}$$

where C_f and B_1^* are material parameters. The another law can be used with a referential strain γ_r^{P*} as:

$$B^* = B_0^* / (1 + \gamma_{ap}^{P*} / \gamma_r^{P*}) \tag{10}$$

The softening due to the internal structural change can be described by the same equation as for the overconsolidation surface and is given by

$$\sigma'_{ma} = \frac{\left\{ \sigma'_{mbf} + \left(\sigma'_{mbi} - \sigma'_{mbf} \right) \exp(-\beta z) \right\}}{\sigma'_{mbi}} \sigma'_{mai} \tag{11}$$

where σ'_{mai} coincides with the initial mean effective stress. For the kinematical hardening associated with the change of the mean effective stress is given by

$$y_m^* = y_{m1}^* + y_{m2}^* \tag{12}$$

$$dy_{m1}^* = B_2^* \left(A_2^* d\varepsilon_v^p - y_{m1}^* \left| d\varepsilon_v^p \right| \right) \tag{13}$$

$$dy_{m2}^* = B_3^* d\varepsilon_v^p \tag{14}$$

2.4 Plastic Potential Function

The plastic potential function is given by

$$g=\bar{\eta}_{\chi}^{*}+\tilde{M}^{*}\left(\ln\frac{\sigma'_{mk}}{\sigma'_{mp}}+\left|\ln\frac{\sigma'_{m}}{\sigma'_{mk}}-y_{m}^{*}\right|\right)=0 \quad (15)$$

where σ'_{mp} is a constant and $\tilde{M}^{*}$ is given by

$$\tilde{M}^{*}=\begin{cases} M_{m}^{*} & f_b \geq 0 \ (Normally\ consolidated\ region) \\ \left(\sigma_{m}^{*}/\sigma'_{mb}\right)^{n_2}M_{m}^{*} & f_b<0 \ (Overconsolidated\ region) \end{cases} \quad (16)$$

where n_2 is a material parameter. σ_{m}^{*} the value of the mean effective stress at the intersection point between the surface passing through the present stress point which is similar to the overconsolidation surface and the anisotropic stress axis. $\tilde{M}^{*}$ which controls the direction of the plastic strain increment depends on the stress state and is smaller than the stress ratio at the phase transformation M_{m}^{*}. $\tilde{M}^{*}=A_{cm}M_{m}^{*}$ after $\tilde{M}^{*}$ reaches M_{m}^{*}.

In the present model, we use the following generalized flow rule including 4th order tensor H_{ijkl} [3] to more precisely reproduce the dilatancy characteristics.

$$d\varepsilon_{ij}^{P}=H_{ijkl}\frac{\partial g}{\partial\sigma'_{kl}}, \quad H_{ijkl}=a\delta_{ij}\delta_{kl}+b(\delta_{ik}\delta_{jl}+\delta_{il}\delta_{jk}) \quad (17)$$

where a, b are material parameters depending on the stress and strain.

The stress-dilatancy relationship is given by

$$\frac{dv^{P}}{d\gamma^{P*}}=D^{*}\left(\tilde{M}^{*}-\eta_{\chi}^{*}\right), \quad D^{*}=\frac{3a}{2b}+1 \quad (18)$$

where D^{*} is a so-called dilatancy coefficient and controls the plastic volumetric strain associated with the shearing. In the case of $a=0$, the relationship corresponds to the conventional one. D^{*} is given by

$$D^{*}=D_{0}^{*}(\tilde{M}^{*}/M_{m}^{*})^{n} \quad (19)$$

where D_{0}^{*}, n are material parameters.

3 Simulation Results

In order to show the applicability of the present model, the response of the loose Toyoura sand under the undrained simple shear conditions. The material parameters used in the simulations are as follows:

$e_0 = 0.772,\ \lambda = 0.0091,\ \kappa = 0.0052,\ OCR^* = 1.0,\ G_0^E/\sigma'_{m0} = 1023.6,\ M_f^* = 0.99,$ $M_m^* = 0.707,\ B_0^* = 40889.0,\ B_1^* = 54.5,\ C_f = 0.0,\ \gamma_r^{P*} = 0.002,\ \gamma_r^{E*} = 0.012,\ r_{e2} = 1.0,$ $\alpha_e = 1.0, r_{e1} = 1.0, D_0^* = 0.85, n = 5.1, C_d = 2000.0, \beta = 0.0, n_2 = 0.5\ \sigma'_{mbf}/\sigma'_{mbi} = 1.0,$ $A_2^*, B_2^* = 0.0,\ 0.0,\ B_3^* = (1+e_0)/(\lambda - \kappa) = 204.4,\ A_{cm} = 1.0,\ C_{ns} = 0.01$

Equation (12) was used for the calculation. Figure 1 shows calculated results well reproduce experimental results.

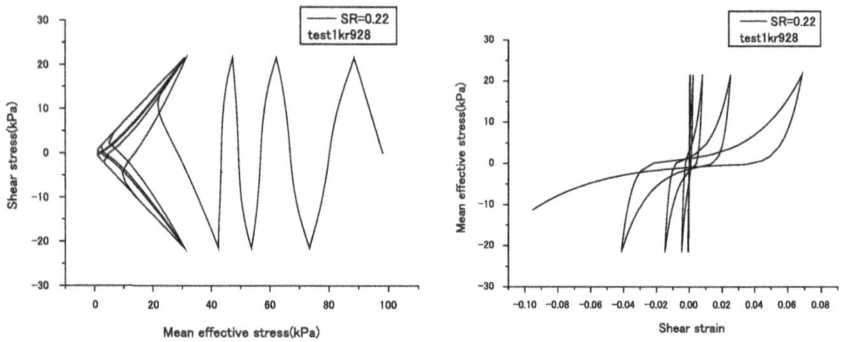

(a) Calculated results

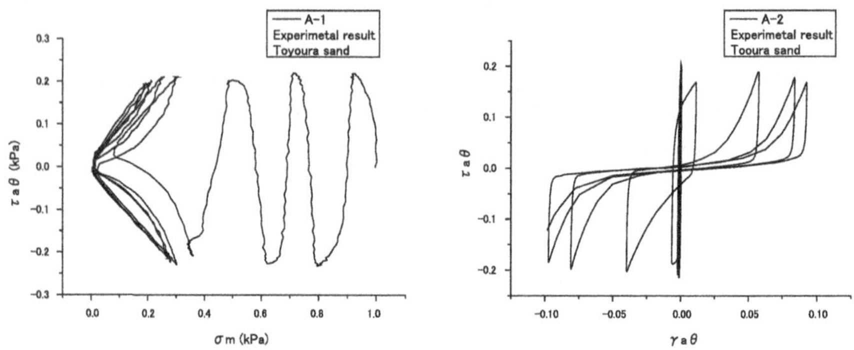

(b) Experimental results

Fig. 1 Stress path and stress-strain relation

4 Conclusions

Cyclic elasto-plastic constitutive models have been developed based on the kinematical hardening theory which includes two kinematical hardening parameters for volumetric and shear strains. We have introduced non-associativity parameter that controls the non-associative behavior of sandy soil which is important for the simulation of instable behavior such as liquefaction [4]. Numerical simulation shows the model can well reproduce the cyclic behavior of soils. In the future, we will present the parametric study and the application of this model to the boundary value problem including instability problem such as liquefaction.

References

1. Oka, F., Yashima, A., Tateishi, A., Taguchi, Y., Yamashita, S.: A cyclic elasto-plastic constitutive model for sand considering a plastic-strain dependence of the shear modulus. Geotechnique **49**(5), 661–680 (1999)
2. Armstrong, P.J., Frederick, C.O.: A mathematical representation of the multiaxial Bauschinger effect. Berkeley Nuclear Laboratories 1966. Technical Report RD/B/N 731 (1966)
3. Kimoto, S., Shahbodagh Khan, B., Mojtaba Mirjalili, M., Oka, F.: A Cyclic Elasto-viscoplastic constitutive model for clay considering the nonlinear kinematic hardening rules and the structural degradation. Int. J. Geomech. ASCE **15**(5): A4014005-14 (2015)
4. Oka, F., Kimoto, S.: Effect of the non-associativity of plasticity model on the cyclic behavior of geomaterials. In: Proceedings of 13th Asia-Pacific Symposium on Engineering Plasticity and Its Applications, 4–8 Dec (2016)

Theoretical Prediction of Strain Localization in Anisotropic Sand by Non-coaxial Elasto-Plasticity

Maosong Huang, Zhouquan Chen, Xilin Lu and Xiaoqiang Gu

Abstract Due to the interaction of stress state and soil micro-structure, the onset of strain localization in anisotropic sand is different from that in isotropic sand. In order to accurately predict the onset of strain localization in inherent anisotropic sands under multi-dimensional stress condition, a state-dependent critical state constitutive model was proposed. The anisotropic critical state line was modified by incorporating micro-structure information, termed as fabric anisotropy. The model was shown to be able to capture influences of loading direction and intermediate principal stress ratio on stress-strain relationships and volumetric characteristics. Through the integration of the rate-form stress-strain relationship, bifurcation analysis was employed to predict the onset of strain tests. The results showed that the major principal strain at the bifurcation points increases with the deposition angle, while the stress ratio decreases with the angle varying from 0° to around 60° and increases afterwards in plain strain tests. Overall, the predicted shear stress at bifurcation points compare well with the stress peak in experiments.

Keywords Theoretical prediction · Strain localization · Anisotropic sands · Noncoaxial plasticity · Fabric anisotropy

1 Introduction

In the bifurcation analysis proposed by Rudnicki and Rice [1], the influence of anisotropy of materials on the strain localization was not taken into account sufficiently. The previous works mainly focused on the isotropic constitutive model and the improvement of the theory which leaded to a non-coaxial plastic flow rule. Based on the extended form of this theory in the 3D stress space by Qian et al. [2],

M. Huang (✉) · Z. Chen · X. Lu · X. Gu
Department of Geotechnical Engineering, Tongji University, Shanghai 20092, China
e-mail: mshuang@tongji.edu.cn

E. Papamichos et al. (eds.), *Bifurcation and Degradation of Geomaterials with Engineering Applications*, Springer Series in Geomechanics and Geoengineering, DOI 10.1007/978-3-319-56397-8_11

an anisotropic generalized Mohr-Coulomb model accounting for the microstructure characteristics of granular materials was introduced to predict the strain localization of cross-anisotropic sand [3]. However, there was also an anisotropic model considering state-dependence which derived from the original works by Li and Dafalias [4], and no attempt was made to apply this model to the bifurcation analysis. An important concept was encompassed in this models, which is termed as the anisotropic critical state lines (ACSL). This concept refers to a scale-valued anisotropic state variable A, which is defined as the first joint isotropic invariant of the fabric tensor and a properly defined loading direction tensor. And the location of the critical state line (CSL) in the void ratio e versus mean effective stress p space is argued to be strongly affected by the anisotropic state variable A. Based on the anisotropic model, this paper intends to modify the constitutive model by Li and Dafalias [4] by revising the function forms of ACSL and developed a corresponding non-coaxial model to predict the onset of shear bands with comparison to plane strain tests on Toyoura sands further.

2 Anisotropic State Variable and Critical State Line

In the constitutive model developed by Li and Dafalias [4], the anisotropy of sand mainly depends on the anisotropic state variable A. The variable is related with all key aspects of the model, such as the location of CSL, the forms of plastic modulus and the dilatancy function, etc. And the fabric tensors which characterize the micro structure of granular material can be defined as

$$F_{ij}^{'} = \frac{1}{3+\Delta}\begin{bmatrix} 1-\Delta & 0 & 0 \\ 0 & 1+\Delta & 0 \\ 0 & 0 & 1+\Delta \end{bmatrix} \tag{1}$$

where Δ is the vector magnitude proposed by Curray [5], which indicates the magnitude of the anisotropy of the preferred orientation of the particles.

The loading direction tensor is another component of the anisotropic state variable A, which is characterized by the normalized stress $\hat{\sigma}_{ij}$ and expressed as

$$\hat{\sigma}_{ij} = \frac{M_c g(\theta)}{R} r_{ij} + \delta_{ij} \tag{2}$$

where M_c is the critical stress ratio under triaxial compression. Then, the tensor r_{ij} is defined as $r_{ij} = s_{ij}/p$, and $R = \sqrt{3r_{ij}r_{ji}/2}$, is the second invariant of r_{ij}. In addition, the lode angle θ is defined as $\theta = -[\sin^{-1}(9r_{ij}r_{jk}r_{ki}/2R^3)]/3$.

The interaction between the fabric anisotropy and the loading direction can be characterized by the new stress tensor $\widehat{T}_{ij}$, which is formulated by

$$\widehat{T}_{ij} = \frac{1}{6}(\widehat{\sigma}_{ik}F_{kj}^{-1} + F_{ik}^{-1}\widehat{\sigma}_{kj}) = \widehat{p}(\widehat{r}_{ij} + \delta_{ij}) \tag{3}$$

where F_{ij}^{-1} is the inverse of the fabric tensor F_{ij}. $\widehat{p} = \widehat{T}_{ii}/3$ is the mean stress, $\widehat{r}_{ij} = \widehat{s}_{ij}/\widehat{p}$ is the ratio of deviatoric stress. The related second and third invariants can be expressed as $\widehat{R} = \sqrt{3\widehat{r}_{ij}\widehat{r}_{ji}/2}$ and $\widehat{\theta} = -[\sin^{-1}(9\widehat{r}_{ij}\widehat{r}_{jk}\widehat{r}_{ki}/2\widehat{R}^3)]/3$, respectively. With combination of two invariants, the anisotropic state variable can be redefined as

$$A = \frac{\widehat{R}}{M_c g(\widehat{\theta})} - \left(\frac{\widehat{R}}{M_c g(\widehat{\theta})}\right)_c \tag{4}$$

where $\widehat{R}/(M_c g(\widehat{\theta}))$ can be regarded as a function corresponding to the vector magnitude Δ, the lode angle θ, and the angle α which is defined as an inclination between the deposition direction and the loading direction. The term $(\widehat{R}/(M_c g(\widehat{\theta})))_c$ is the function value under triaxial compression, which is a constant and do not vary with the loading condition.

The basic assumption of ACSL, which was proposed in the critical state model of sand accounting for the inherent anisotropy [4], is that the CSL in the *e*-*p* plane depends on the loading conditions and the fabric anisotropy. The experimental work of Yang et al. [6], to some extents, establish the correlation between the inherent fabric anisotropy and the CSL. According to this method, the concept of ACSL can be verified. Hence, a revised form of ACSL is defined as

$$e_c = e_\Gamma(A) - \lambda_c(p/p_a)^\xi = e_{\Gamma 0} + k_\Gamma A \exp\left[\frac{b}{b+c}\sin(2\alpha)\right] - \lambda_c(p/p_a)^\xi \tag{5}$$

where $e_{\Gamma 0}$, k_Γ, c, λ_c and ξ are material constants, $e_{\Gamma 0}$ is the critical-state void ratio for triaxial compression at intercept $p = 0$, p_a is a reference pressure (101 kPa). The influence of A on CSL is governed by the function $e_\Gamma(A)$, which determine the location of CSL. The location of the revised ACSL is mainly affected by the term $A\exp[\frac{b}{b+c}\sin(2\alpha)]$ and its variation of various intermediate stress ratio b with α is shown in Fig. 1.

Fig. 1 Variation of $A\exp\left[\frac{b}{b+c}\sin(2\alpha)\right]$ with α

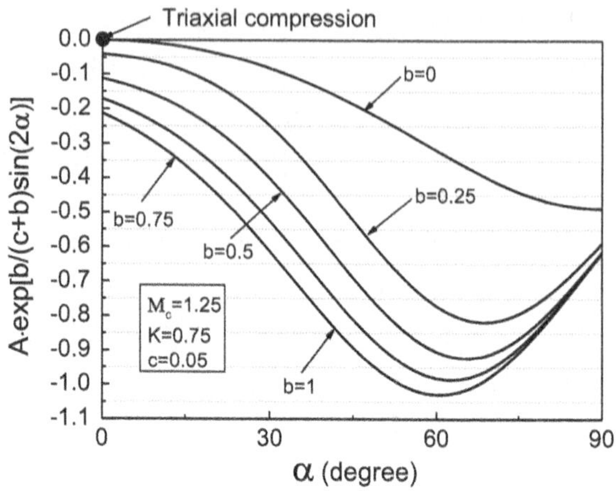

3 Non-coaxial Anisotropic Model and Bifurcation Prediction

According to the yield vertex theory by Rudnicki and Rice [1], the total plastic strain increment $d\varepsilon_{ij}^{p}$ can be composed into the coaxial term $d\varepsilon_{ij}^{pc}$ and non-coaxial term $d\varepsilon_{ij}^{pn}$, that is

$$d\varepsilon_{ij}^{p} = d\varepsilon_{ij}^{pc} + d\varepsilon_{ij}^{pn} \tag{6}$$

Combination of the stress-strain relationship including the non-coaxial strain increment and the consistency condition of the yield function, the elasto-plastic modulus matrix always can be formulated as follows [3]

$$D_{ijkl}^{ep} = D_{ijkl}^{e} - D_{ijmn}^{e}\left(\frac{\frac{\partial Q}{\partial \sigma_{mn}}\frac{\partial f}{\partial \sigma_{st}}}{H_p + \frac{\partial f}{\partial \sigma_{pq}} D_{pquv}^{e}\frac{\partial f}{\partial \sigma_{uv}}} + \frac{H_t}{H_t + 2G} C_{mnst}^{np}\right) D_{stkl}^{e} \tag{7}$$

where D_{ijmn}^{e} is the elastic modulus matrix, H_p is the hardening modulus and C_{ijkl}^{np} is the non-coaxial term

$$C_{ijkl}^{np} = \frac{1}{H_t}\left(\frac{\delta_{ik}\delta_{jl} + \delta_{il}\delta_{jk}}{2} - \frac{\delta_{kl}\delta_{ij}}{\delta_{mn}\delta_{mn}} - \frac{s_{ij}s_{kl}}{2J_2} - \frac{S_{ij}S_{kl}}{S_{mn}S_{mn}}\right) \tag{8}$$

In Eq. (7), the yield function f and the plastic potential function Q are given as respectively

$$f = q - Mg(\theta)p = 0 \tag{9}$$

$$Q = q + \frac{\mathrm{B}}{1-B} M_d p g(\theta)[1 - (\frac{p}{p_o})^{B-1}] = 0 \tag{10}$$

where M is the stress ratio in triaxial compression, d_0 is a model parameter, and $M_d = M_c e^{m\psi}$ is the phase transformation stress ratio, $\psi = e - e_c$ is a state parameter proposed by Been and Jefferies [7], m is a model parameter. $B = d_0/M_{cs}$ and p_o is the size parameter of the potential surface.

In this study, the hardening rule is determined by the incremental equation

$$d\varepsilon_s^p = \frac{pM}{hG(M_{cs}e^{-n\psi} - M)} dM \tag{11}$$

where $\varepsilon_s^p = \sqrt{2e_{ij}^p e_{ij}^p/3}$ and $e_{ij}^p = \varepsilon_{ij}^p - \delta_{ij}\varepsilon_{ij}^p/3$. The constant n is a model parameter and h is related to the anisotropic state variable, defined as

$$h = (h_1 - h_2 e)(1 + k_h A) \tag{12}$$

where e is the void ratio, and h_1, h_2 and k_h are model parameters.

According to the bifurcation theory, the condition for the onset of strain localization in the small strain condition is

$$\det(A_{ik}) = \det\left[n_j D_{ijkl}^{ep} n_l - \frac{1}{2}(\sigma_{ik} + \sigma_{jk}n_j n_i - \sigma_{ij}n_j n_k - \sigma_{jl}\delta_{ik}n_j n_l) \right] = 0 \tag{13}$$

where A_{ik} is referred to as the acoustic tensor; n_i is the unit vector normal to the shear band; and D_{ijkl}^{ep} is elasto-plastic modulus tensor. More details about the process can refer to these previous works, such as Huang et al. [8].

4 Results and Discussion

A series of drained plane strain tests were performed with different deposition angle α by Oda et al. [9]. And the experimental data is employed to compare with the prediction results of strain localization based on the revised anisotropic critical state model. The model parameters in simulations are shown in Table 1 and mostly refer to those by Yang et al. [6].

Figure 2 shows the stress-strain relationships for e = 0.67 at a confining pressure $\sigma_3 = 2.0\,\mathrm{kg/cm^2}$ (1 kg/cm^2 = 98.1 kPa) when α = 0°, 30°, 45°, 60° and 90°, as well as the predicted results in bifurcation analysis. On the one hand, the simulation results can demonstrate the ability of the model to describe the characteristics of anisotropic behavior reasonably. On the other hand, the onsets of shear

Table 1 Model parameters for Toyoura sand

Elastic parameters	Critical state parameters	Dilatancy parameters	Hardening parameters	Anisotropic parameters	Non-coaxiality parameters
$G_0 = 80$	$M_{cs} = 1.38$	$d_0 = 0.68$	$h_1 = 3.15$	$\Delta = 0.26$	$H_t = 8500$ kPa
$\nu = 0.25$	$K = 0.75$	$m = 3.5$	$h_2 = 3.05$	$c = 0.05$	
	$e_{\Gamma 0} = 0.934$		$n = 1.35$		
	$\lambda_c = 0.019$		$k_h = 0.5$		
	$\xi = 0.7$				
	$k_\Gamma = 0.15$				

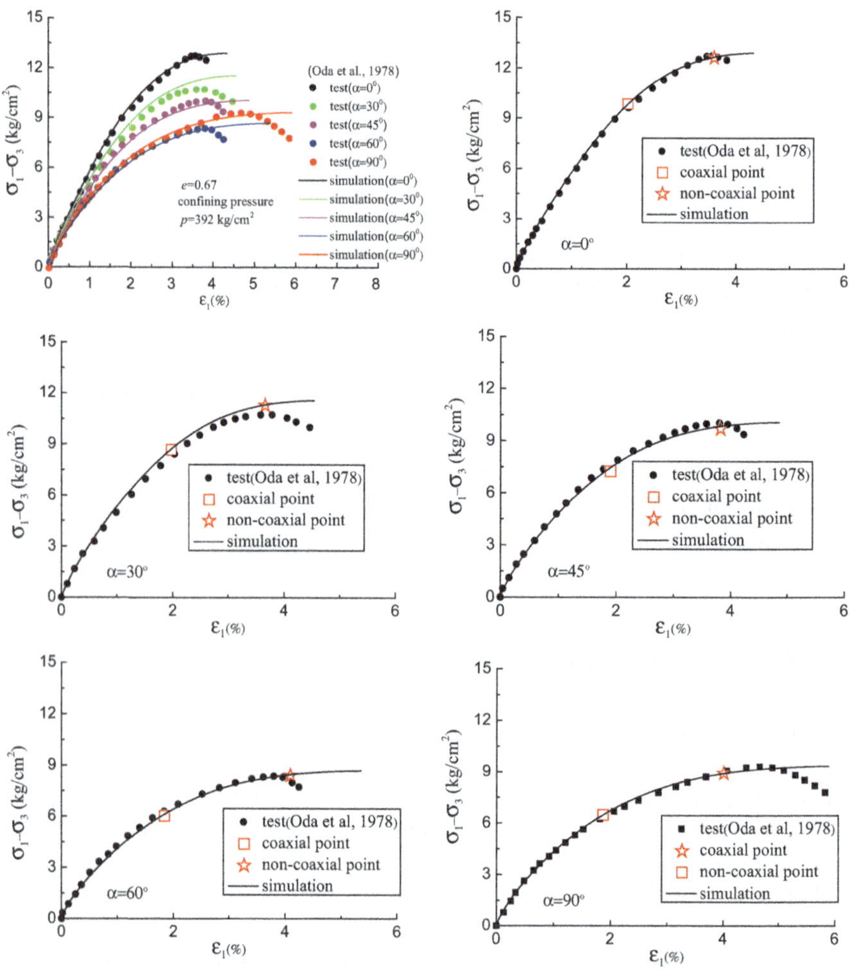

Fig. 2 Stress-strain relationships and bifurcation point in plane strain tests

band are both predicted by the coaxial and non-coaxial model. It can be seen from the figures that the predicted peak stress ratio is underestimated based on coaxial model and that based on non-coaxial model can agree well with the test data.

5 Conclusions

The failure point predicted by the bifurcation analysis was markedly affected by the peak strength of the constitutive model. In the anisotropic critical state model, the peak strength is dependent on the state parameters. Since the variation of state parameter ψ with α is determined by the anisotropic critical state line function, the peak strength inherit the trend as the same as that of ψ altered by the modified function, in which the stress ratio descend with deposition angle ranging from 0° to around 60° and ascend afterwards. Thus, the prediction of the strain localization also shows the same trend and is consistent with the tests results.

Acknowledgements This work was financially supported by the National Natural Science Foundation of China (Grant No. 11372228).

References

1. Rudnicki, J.W., Rice, J.R.: Conditions for the localization of deformation in pressure-sensitive dilatant materials. J. Mech. Phys. Solids **23**, 371–394 (1975)
2. Qian, J.G., Yang, J., Huang, M.S.: Three-dimensional noncoaxial plasticity modeling of shear band formation in geomaterials. J. Eng. Mech. **134**, 322–329 (2008)
3. Lu, X., Huang, M., Qian, J.: The onset of strain localization in cross-anisotropic soils under true triaxial condition. Soils Found. **51**(4), 693–700
4. Li, X.S., Dafalias, Y.F.: Constitutive modeling of inherently anisotropic sand behavior. J. Geotech. Geoenviron. Eng. **128**, 868–880 (2002)
5. Curray, J.R.: The analysis of two-dimensional orientation data. J. Geol. **64**(2), 117–131 (1956)
6. Yang, Z.X., Li, X.S., Yang, J.: Quantifying and modeling fabric anisotropy of granular soils. Geotechnique **58**(4), 237–248 (2008)
7. Been, K., Jefferies, M.G.: A state parameter for sands. Geotechnique **35**(2), 99–112 (1985)
8. Huang, M., Lu, X., Qian, J.: Non-coaxial elasto-plasticity model and bifurcation prediction of shear banding in sands. Int. J. Numer. Anal. Methods Geomech. **34**, 906–919 (2010)
9. Oda, M., Koishikawa, I., Higuchi, T.: Experimental study on anisotropic shear strength of sand by plane strain test. Soils Found. **18**(1), 25–38 (1978)

Modeling Sand Behavior Under Partially Drained Stress Paths

M.S. Yaghtin and A. Lashkari

Abstract Soil behavior is conventionally studied under fully drained or undrained conditions. However, general drainage with complete control on both pore volume and pore water pressure leads to a wide spectrum of behaviors which may not be delimited by drained and undrained ones. Herein, a critical state bounding surface plasticity model is employed to simulate partially drained behavior of sand samples sheared over a wide domain of drainage conditions. Using the concept of loss of uniqueness, onset of instability under partial drainage is studied.

1 Introduction

In the laboratory, undrained triaxial and direct simple shear tests have been applied to study the liquefaction of sands. Traditionally, it is assumed that behavior of saturated sands is bounded by those of undrained and fully drained tests, and behavior during and soon after earthquake is essentially undrained. However, results of shaking table tests have revealed that some volume changes during earthquakes are inevitable due to high permeability of sands. Depending on drainage condition, five potential responses for loose sands are schematically illustrated in Fig. 1. Type 1 is fully drained behavior in which excess pore water pressure is zero and total and effective stresses are equal. Drainage is prohibited and soil volume is kept fixed in the undrained response (i.e., type 2 of behavior). Types 3 to 5 are technically called partially drained behaviors in which soil shears with simultaneous changes in pore volume and pore water pressure. Soil behavior is bounded by drained and undrained ones in type 3. Nevertheless, soil load bearing structure excessively weakens in type 4, and gradually strengthens in type 5 of the behavior.

M.S. Yaghtin · A. Lashkari (✉)
Faculty of Civil and Environmental Engineering,
Shiraz University of Technology, Shiraz, Iran
e-mail: lashkari@sutech.ac.ir; lashkari_ali@hamyar.net

E. Papamichos et al. (eds.), *Bifurcation and Degradation of Geomaterials with Engineering Applications*, Springer Series in Geomechanics and Geoengineering, DOI 10.1007/978-3-319-56397-8_12

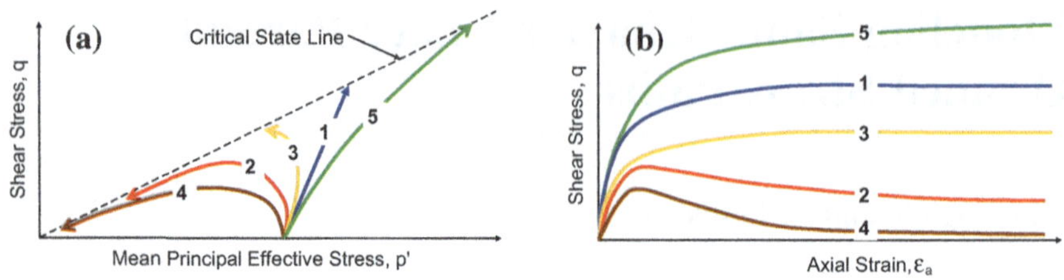

Fig. 1 Sand behavior as a function of drainage state: **a** stress path; **b** shear stress versus strain

A plasticity model is employed to simulate the behavior of sand under general drainage. Using the concept of loss of uniqueness, onset of instability is predicted.

2 Constitutive Model

Golchin and Lashkari [6] and Lashkari and Golchin [7] put forward a sand model with elastic-plastic coupling in which, elastic strains are obtained from energy function to conserve energy in closed loop elastic stress paths. Coupling of elasticity and plasticity enables the model to consider evolving anisotropic elastic moduli. Explicit form of the model constitutive equations in triaxial space is:

$$\begin{Bmatrix} \dot{p} \\ \dot{q} \end{Bmatrix} = \begin{bmatrix} D_{pp} & D_{pq} \\ D_{qp} & D_{qq} \end{bmatrix} \begin{Bmatrix} \dot{\varepsilon}_v \\ \dot{\varepsilon}_q \end{Bmatrix} \tag{1}$$

where p $[=(\sigma_a^{'}+2\sigma_r^{'})/3]$ and q $[=\sigma_a^{'}-\sigma_r^{'}]$ are, respectively, mean principal effective stress and shear stress in which $\sigma_a^{'}$ and $\sigma_r^{'}$ are axial and radial effective stresses. $\varepsilon_v\ [=\varepsilon_a+2\varepsilon_r]$ and $\varepsilon_q\ [=2/3(\varepsilon_a-\varepsilon_r)]$ are volumetric and shear strains where ε_a and ε_r are axial and radial strains. D_{pp}, D_{pq}, D_{qp}, and D_{qq} are calculated from:

$$D_{pp}=K-\left[\frac{1+\Gamma,_{q\chi}\,\chi,_{\alpha}\,K_p}{pK_p+(1+\Gamma,_{q\chi}\chi,_{\alpha}K_p)\,[3G-J\eta+(J-K\eta)d]}\right](J+Kd)(J-K\eta) \tag{2a}$$

$$D_{pq}=J-\left[\frac{1+\Gamma,_{q\chi}\,\chi,_{\alpha}\,K_p}{pK_p+(1+\Gamma,_{q\chi}\chi,_{\alpha}K_p)\,[3G-J\eta+(J-K\eta)d]}\right](J+Kd)(3G-J\eta) \tag{2b}$$

$$D_{qp}=J-\left[\frac{1+\Gamma,_{q\chi}\,\chi,_{\alpha}\,K_p}{pK_p+(1+\Gamma,_{q\chi}\chi,_{\alpha}K_p)\,[3G-J\eta+(J-K\eta)d]}\right](3G+Jd)(J-K\eta) \tag{2c}$$

$$D_{qq}=3G-\left[\frac{1+\Gamma,_{q\chi}\,\chi,_{\alpha}\,K_p}{pK_p+(1+\Gamma,_{q\chi}\chi,_{\alpha}K_p)\,[3G-J\eta+(J-K\eta)d]}\right](3G+Jd)(3G-J\eta) \tag{2d}$$

In Eqs. (2a)–(2d), Γ is the energy function of Einav and Puzrin [4]:

$$\Gamma(p,q,\chi)=\frac{p^{2-\chi}-(2-\chi)p\,p_0^{1-\chi}}{\overline{K}\,(2-\chi)(1-\chi)p_{ref}^{1-\chi}}+\frac{q^2}{6\overline{G}\,p_{ref}^{1-\chi}p^{\chi}}-\frac{q_0(2q\,p_0-\chi q_0\,p)}{6\overline{G}\,p_{ref}^{1-\chi}p_0^{1+\chi}} \quad (3)$$

where p_0 and q_0 are, respectively, p and q at zero elastic strains and p_{ref} (= 101 kPa) is a reference pressure. $\overline{K}=K_0\,(e_g-e)^2/(1+e)$ and $\overline{G}=G_0\,(e_g-e)^2/(1+e)$ in which K_0 and G_0 are soil parameters and e_g = 2.97 is a reasonable assumption for sands with angular particles. K, G, and J are, respectively, the elastic bulk, shear, and shear-volumetric coupling moduli:

$$K=\frac{\Gamma,_{qq}}{\Gamma,_{pp}\Gamma,_{qq}-\Gamma,_{pq}\Gamma,_{qp}}=\frac{1}{1-\frac{\chi(\chi-1)}{2}\eta^2\left(\frac{\overline{K}}{3\overline{G}}\right)}\overline{K}\,p_{ref}\left(\frac{p}{p_{ref}}\right)^{\chi} \quad (4a)$$

$$G=\frac{1/3\,\Gamma,_{pp}}{\Gamma,_{pp}\Gamma,_{qq}-\Gamma,_{pq}\Gamma,_{qp}}=\frac{1}{1-\frac{\chi(\chi-1)}{2}\eta^2\left(\frac{\overline{K}}{3\overline{G}}\right)}\left(\overline{G}+\frac{\chi(\chi+1)}{6}\eta^2\overline{K}\right)p_{ref}\left(\frac{p}{p_{ref}}\right)^{\chi} \quad (4b)$$

$$J=-\frac{\Gamma,_{pq}}{\Gamma,_{pp}\Gamma,_{qq}-\Gamma,_{pq}\Gamma,_{qp}}=\frac{1}{1-\frac{\chi(\chi-1)}{2}\eta^2\left(\frac{\overline{K}}{3\overline{G}}\right)}\chi\eta\overline{K}\,p_{ref}\left(\frac{p}{p_{ref}}\right)^{\chi}=\chi\eta K \quad (4c)$$

where $\Gamma,_{pp}=\partial^2\Gamma/\partial p\partial p$, $\Gamma,_{pq}=\Gamma,_{qp}=\partial^2\Gamma/\partial p\partial q$ and $\Gamma,_{qq}=\partial^2\Gamma/\partial q\partial q$ and $\eta=q/p$ is stress ratio. χ in Eqs. (3) and (4a)–(4c) varies through:

$$\chi=\chi(\alpha)=\chi_{min}+\frac{(\chi_{max}-\chi_{min})}{1+B\,(\alpha^b/\alpha-1)} \quad (5)$$

where χ_{min} = 0.45, χ_{max} = 0.95 and B = 0.10 are practical estimations for various sands (see [8]). In Eqs. (2a)–(2d), $\chi,_{\alpha}$ and $\Gamma,_{q\chi}$ are the compact forms for $\partial\chi/\partial\alpha$ and $\partial^2\Gamma/\partial q\partial\chi$, respectively. K_p and d in Eqs. (2a)–(2d) are, respectively, plastic hardening modulus and dilatancy function:

$$K_p=h_0\exp(-c_h\,e)\cdot\frac{\alpha^b-\alpha}{\alpha-\alpha_{in}}\left(\frac{p_{ref}}{p}\right)^{1-\chi};\quad d=A\left[2-\sqrt{\frac{<\alpha^d-\alpha>}{(1+c)M}}\right](\alpha^d-\alpha) \quad (6)$$

where h_0, c_h and A are parameters and c = 0.70–0.75. α is back-stress ratio which is linked to stress ratio through $f=\eta-\alpha-m=0$ in triaxial compression wherein m = 0.001 here. In Eq. (6), α^b and α^d are back-stress ratios associated with bounding and dilatancy surfaces (see [3]):

$$\alpha^b = M\ \exp(-n^b\ \psi) - m \quad ; \quad \alpha^d = M\ \exp(n^d\ \psi) - m \tag{7}$$

n^b and n^d are soil parameters, M is slope of critical state line in the q versus p plane and ψ is state parameter:

$$\psi = e - e_{cs} = e - \left[e_0 - \lambda\,(p/p_{ref})^{\xi}\right] \tag{8}$$

e_0, λ and ξ are soil parameters and e_{cs} is critical void ratio corresponding to the current p.

3 Criterion of Instability Under Partial Drainage Condition

Following the pioneering works of Borja [2] and Andrade et al. [1] based on bifurcation theory, loss of uniqueness or stability in triaxial space requires:

$$[[\dot{p}]]\ [[\dot{\varepsilon}_v]] + [[\dot{q}]]\ [[\dot{\varepsilon}_q]] = 0 \tag{9}$$

where $[[\dot{p}]]$, $[[\dot{q}]]$, $[[\dot{\varepsilon}_v]]$ and $[[\dot{\varepsilon}_q]]$ are, respectively, jumps in p, q, ε_v and ε_q due to duplicate solutions for velocity field. In partially drained tests, volumetric and axial strain rates are linked through (see [5]):

$$\zeta = \frac{\dot{\varepsilon}_v}{\dot{\varepsilon}_a} \tag{10}$$

Using Eq. (10) together with ordinary algebra, one has:

$$\dot{\varepsilon}_v = \left(\frac{\zeta}{1 - \zeta/3}\right)\dot{\varepsilon}_q \tag{11}$$

Implementing Eqs. (1) and (11) in Eq. (9) gives the following criterion for loss of uniqueness under partial drainage condition in triaxial tests:

$$\zeta^2 D_{pp} + (1 - \zeta/3)^2 D_{qq} + \zeta(1 - \zeta/3)\,(D_{pq} + D_{qp}) = 0 \tag{12}$$

4 Simulation of Sand Behavior Under Different Drainage Conditions

Fraser River sand, a uniform angular to sub-angular sand with low to medium sphericity, is an alluvial deposit from Fraser River delta in British Columbia, Canada [5]. Using the p model, behaviors of four Fraser River sand samples with different initial states under fully drained condition are simulated and predictions are illustrated against corresponding experimental data in parts "a" and "b" of Fig. 2. Applying identical parameters, behaviors of three loose Fraser River sand samples sheared under undrained condition are simulated and drawn with experimental data in parts "c" and "d" of Fig. 2. Model parameters in all simulations are provided in Table 1. In Fig. 2, the model captured the general trend of sand behavior associated with density, confining stress, and drainage condition and the observed discrepancies may be attributed to uncertainty in critical state line of Fraser River sand.

The behaviors of six Fraser River sand samples (e_{in} = 0.808, p_c = 200 kPa) sheared under partially drained condition with ζ = + 1.0, +0.40, +0.20, 0.0, −0.10, −0.40 and −1.0 are shown in parts "a" and "b" of Fig. 3 wherein tendency towards contraction and continuous strain hardening gradually turns into contraction and strain softening as ζ varies from +1.0 to −1.0. Certain states for instability and associated Instability Line in experimental data are included in the Figs. 3a, b. Without changing the parameters given in Table 1, behaviors of samples under

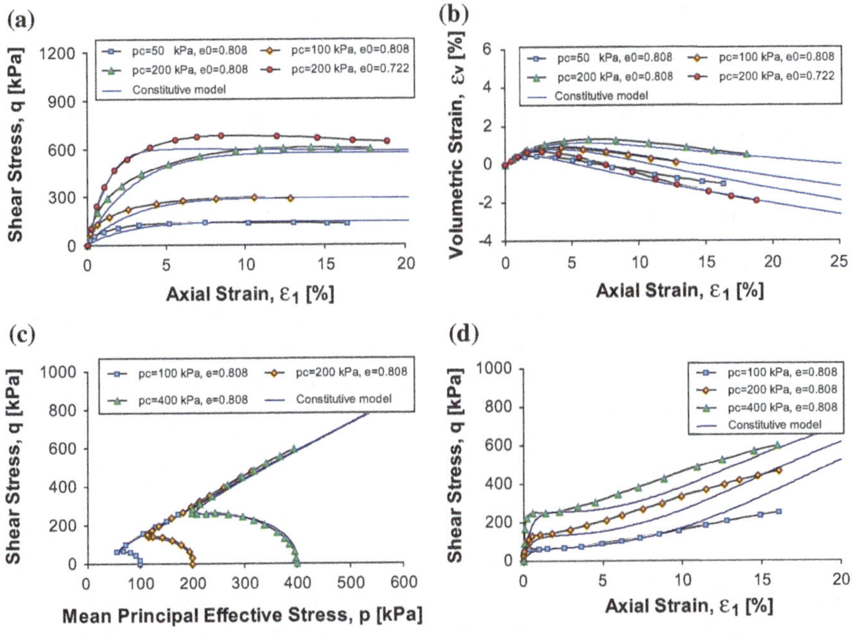

Fig. 2 Simulation of Fraser River sand behavior for: **a** and **b** four tests under fully drained

Table 1 Model parameters used in simulation of Fraser River sand behavior

Elasticity		Critical state line				Dilatancy		Plastic modulus		
G_0	K_0	M	e_0	λ	ξ	A	n^d	h_0	c_h	n^b
45.0	41.25	1.45	0.98	0.05	0. 70	0.40	0.80	2.5×10^6	13.5	0.30

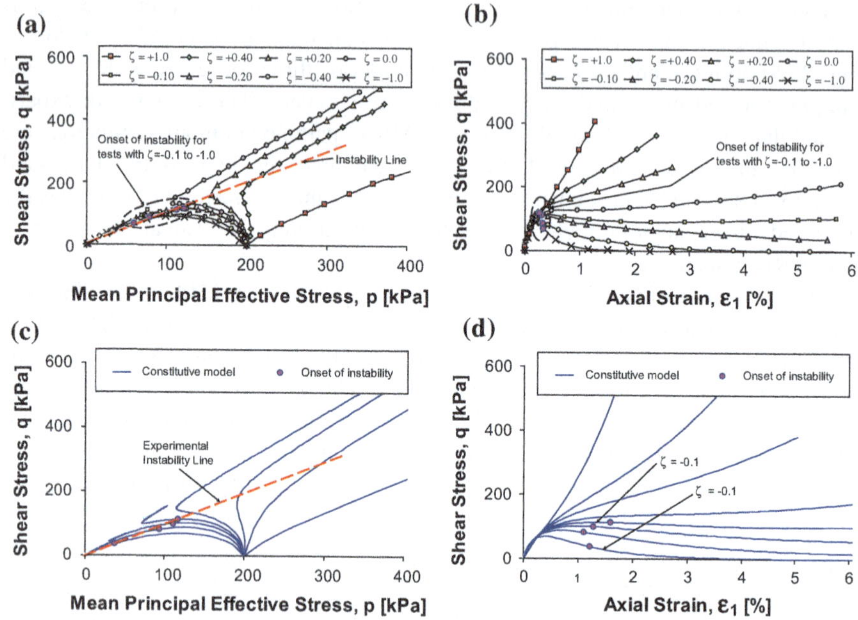

Fig. 3 Simulation of Fraser River sand behavior under partially drained condition: **a**, **b** experimental data for q versus p and q versus ε_1, **c** and **d** predicted behaviors

partially drained stress paths are simulated (see Fig. 3a, b). Similar to experimental data, the tendency toward contraction increases as ζ evolves from +1.0 to −1.0 in the model simulations. Besides, instability instigates beyond the peak shear strength in strain softening regime of the behavior for tests with $\zeta = -0.10$ to −1. Except for the sample with $\zeta = -1$, predicted stress states for onset of instability in samples with z = −0.1 to −0.4 are similar to those in experiments. However as shown in Fig. 3c, experimental failure line fits the predicted states reasonably.

5 Conclusions

Recent experimental findings emphasize that generalized drainage of medium loose and loose sands may lead to certain states more damaging than those of undrained condition. A state-dependent constitutive model was used for simulation of the behavior of sand samples u fully drained, undrained, and partially drained shear. Under partial drainage, it was observed that positive $\zeta = \dot{\varepsilon}_v/\dot{\varepsilon}_a$ ratio results in improvement of load bearing structure; however, negative ζ causes progressive debilitation of soil structure. Onset of instability in partially drained tests was predicted using the concept of loss of uniqueness. It was shown that instability occurs once post-peak stress path intersects the failure line.

References

1. Andrade, J.E., Ramos, A.M., Lizcano, A.: Criterion for flow liquefaction instability. Acta Geotech. **8**, 525–535 (2013)
2. Borja, R.I.: Computational modeling of deformation bands in granular media. II. Numerical simulations. Comput. Methods Appl. Mech. Eng. **193**, 2699–2718 (2004)
3. Dafalias, Y.F., Manzari, M.T.: Simple plasticity sand model accounting for fabric change effects. ASCE J. Eng. Mech. **130**(6), 622–634 (2004)
4. Einav, I., Puzrin, A.M.: Pressure-dependent elasticity and energy conservation in elastoplastic models for soils. J. Geotech. Geoenviron. Eng. **130**(1), 81–92 (2004)
5. Eliadorani, A.: The response of sands under partially drained states with emphasis on liquefaction. Ph.D. thesis. The University of British Columbia, Canada (2000)
6. Golchin, A., Lashkari, A.: A critical state sand model with elastic-plastic coupling. Int. J. Solids Struct. **51**, 2807–2825 (2014)
7. Lashkari, A., Golchin, A.: On the influence of elastic-plastic coupling on sands response. Comput. Geotech. **55**, 352–364 (2014)
8. Lashkari, A.: Prediction of flow liquefaction instability of clean and silty sands. Acta Geotech. **11**, 987–1014 (2016)

Prediction of Sand Instability Under Constant Shear Drained Paths

M.J. Alipour and A. Lashkari

Abstract The unstable behavior of soil slopes under water infiltration is usually studied by means of constant shear drained triaxial tests, wherein mean principal effective stress decreases gradually, while shear stress is maintained fixed and full drainage is allowed. A certain criterion for loss of stability under triaxial constant shear drained paths is obtained here. A state-dependent plasticity model is employed to simulate behavior of Changi sand samples under constant shear drained path. Comparison of model simulations with experimental data indicates that, sand behavior under triaxial drained and undrained paths, as well as triaxial constant shear drained tests, can be simulated realistically using a unique set of parameters.

1 Introduction

Instability in soil mechanics is usually attributed to abrupt generation of irreversible strains, and inability to support external loads [4]. Recent experimental findings have revealed that sands may become unstable under Constant Shear Drained (CSD) paths in which mean principal effective stress declines gradually under constant shear stress while complete drainage is allowed (e.g., [2, 4, 5]). In practice, experimental findings have indicated that water infiltration may eventuate in calamitous failure of slopes. It is suggested that the behavior of slope elements during water infiltration under fixed external loads can be investigated effectively in triaxial CSD tests (e.g., [4]).

A critical state plasticity model is employed here to simulate the behavior of sand under CSD. Using the concept of loss of uniqueness, general condition for onset of instability under CSD is obtained. It is shown that the plasticity model can reasonably simulate instability of sands under constant shear drained path.

M.J. Alipour · A. Lashkari (✉)
Faculty of Civil and Environmental Engineering, Shiraz University of Technology, Shiraz, Iran
e-mail: lashkari@sutech.ac.ir; lashkari_ali@hamyar.net

E. Papamichos et al. (eds.), *Bifurcation and Degradation of Geomaterials with Engineering Applications*, Springer Series in Geomechanics and Geoengineering, DOI 10.1007/978-3-319-56397-8_13

2 Constitutive Model

The constitutive model of Golchin and Lashkari [8], an extension of the bounding surface model by Dafalias and Manzari [6], is adopted here. Elastic moduli are obtained from a Gibbs energy function to guarantee conservation of energy in any arbitrary closed loop elastic path. Elastic-plastic coupling permits stress-induced evolution of elastic moduli, when soil state steps in the elastoplastic regime of behavior. In triaxial space, constitutive model is:

$$\begin{aligned}\begin{Bmatrix} \dot{p} \\ \dot{q} \end{Bmatrix} &= \begin{bmatrix} D_{pp} & D_{pq} \\ D_{qp} & D_{qq} \end{bmatrix} \begin{Bmatrix} \dot{\varepsilon}_v \\ \dot{\varepsilon}_q \end{Bmatrix} \\ &= \left[\begin{pmatrix} K & J \\ J & 3G \end{pmatrix} - \frac{\varphi}{pK_p + \varphi\,[3G - J\eta + (J - K\eta)d]} \right. \\ &\quad \left. \times \begin{pmatrix} (J+Kd)(J-K\eta) & (J+Kd)(3G-J\eta) \\ (3G+Jd)(J-K\eta) & (3G+Jd)(3G-J\eta) \end{pmatrix} \right] \begin{Bmatrix} \dot{\varepsilon}_v \\ \dot{\varepsilon}_q \end{Bmatrix}\end{aligned} \tag{1}$$

where $p\,[=(\sigma_a^{'}+2\sigma_r^{'})/3]$ and $q\,[=\sigma_a^{'}+\sigma_r^{'}]$ are, respectively, mean principal effective stress and shear stress in which $\sigma_a^{'}$ and $\sigma_r^{'}$ are axial and radial effective stresses. $\varepsilon_v\,[=\varepsilon_a+2\varepsilon_r]$ and $\varepsilon_q\,[=2/3(\varepsilon_a-\varepsilon_r)]$ are volumetric and shear strains in which ε_a and ε_r are axial and radial strains. In Eq. (1), Γ is the Gibbs free energy function of Einav and Puzrin [7]:

$$\Gamma(p,q,\chi) = \frac{p^{2-\chi} - (2-\chi)p\,p_0^{1-\chi}}{\overline{K}\,(2-\chi)(1-\chi)p_{ref}^{1-\chi}} + \frac{q^2}{6\overline{G}\,p_{ref}^{1-\chi}p^{\chi}} - \frac{q_0(2q\,p_0 - \chi q_0\,p)}{6\overline{G}\,p_{ref}^{1-\chi}p_0^{1+\chi}} \tag{2}$$

where p_0 and q_0 are p and q at zero elastic strains, respectively. $\overline{K}=K_0\,(e_g-e)^2/(1+e)$ and $\overline{G}=G_0\,(e_g-e)^2/(1+e)$ in which K_0 and G_0 are parameters and e_g = 2.17 is suitable for sands with sub-rounded to well-rounded grains. p_{ref} (= 101 kPa) is a reference pressure and $\eta = q/p$ is stress ratio. K, G, and J are, respectively, the hyperelastic bulk, shear, and shear-volumetric coupling moduli:

$$K = \frac{\partial^2\Gamma/\partial q\partial q}{\det \mathbf{H}} = \frac{1}{1-\frac{\chi(\chi-1)}{2}\eta^2\left(\frac{\overline{K}}{3\overline{G}}\right)}\overline{K}\,p_{ref}\left(\frac{p}{p_{ref}}\right)^{\chi} \tag{3a}$$

$$G = \frac{1/3\;\partial^2\Gamma/\partial p\partial p}{\det \mathbf{H}} = \frac{1}{1-\frac{\chi(\chi-1)}{2}\eta^2\left(\frac{\overline{K}}{3\overline{G}}\right)}\left(\overline{G} + \frac{\chi(\chi+1)}{6}\eta^2\,\overline{K}\right)p_{ref}\left(\frac{p}{p_{ref}}\right)^{\chi} \tag{3b}$$

$$J = \frac{-\,\partial^2\Gamma/\partial p\partial q}{\det \mathbf{H}} = \frac{1}{1-\frac{\chi(\chi-1)}{2}\eta^2\left(\frac{\overline{K}}{3\overline{G}}\right)}\chi\,\eta\,\overline{K}\,p_{ref}\left(\frac{p}{p_{ref}}\right)^{\chi} \tag{3c}$$

where 2×2 Hessian matrix $\mathbf{H}$ is defined by:

$$\mathbf{H} = \begin{pmatrix} \partial^2 \Gamma/\partial p \partial p & \partial^2 \Gamma/\partial p \partial q \\ \partial^2 \Gamma/\partial q \partial p & \partial^2 \Gamma/\partial q \partial q \end{pmatrix} \tag{4}$$

χ in Eqs. (2) and (3a)–(3c) varies through:

$$\chi = \chi(\alpha) = \chi_{min} + \frac{(\chi_{max} - \chi_{min})}{1 + B\ (\alpha^b/\alpha - 1)} \tag{5}$$

where $\chi_{min} = 0.45$, $\chi_{max} = 0.95$ and $B = 0.10$ are realistic boundaries for various sands. $\varphi = 1 + \Gamma,_{q\chi}\chi,_{\alpha}K_p$ where $\chi,_{\alpha}$ and $\Gamma,_{q\chi}$ are, respectively, represent $\partial\chi/\partial\alpha$ and $\partial^2\Gamma/\partial q\partial\chi$. Plastic modulus ($K_p$) and dilatancy function (d) are defined by:

$$K_p = h_0(1 - c_h\, e) \cdot \frac{\alpha^b - \alpha}{|\alpha - \alpha_{in}|} \left(\frac{p_{ref}}{p}\right)^{1-\chi} \quad ; \quad d = A(\alpha^d - \alpha) \tag{6}$$

where h_0, c_h and A are soil parameters. α is back-stress ratio which is linked to stress ratio through $f = \eta - \alpha - m = 0$ in triaxial compression wherein $m = 0.001$ here. According to Dafalias and Manzari [6], α^b and α^d are calculated through:

$$\alpha^b = M\ \exp(-n^b\ \psi) - m \quad ; \quad \alpha^d = M\ \exp(n^d\ \psi) - m \tag{7}$$

in which, M is slope of critical state line in the q versus p plane and n^b and n^d are soil parameters. ψ is state parameter that is calculated by:

$$\psi = e - e_{cs} = e - \left[e_0 - \lambda\, (p/p_{ref})^{\xi}\right] \tag{8}$$

e_0, λ and ξ are soil parameters and e_{cs} is critical void ratio at p.

3 Instability Under CSD

Following the pioneering works of Borja [3] and Andrade et al. [1] based on bifurcation theory, loss of uniqueness or stability in multiaxial space requires $[[\dot{\sigma}]] : [[\dot{\varepsilon}]] = 0$ in which $[[\dot{\varepsilon}]]$ is jump in strain rate tensor due to duplicate solution for velocity. $[[\dot{\sigma}]]$ is jump in effective stress tensor related to $[[\dot{\varepsilon}]]$. Under triaxial compression CSD in which $[[\dot{q}]] = 0$ holds, the above criterion becomes:

$$[[\dot{p}]]\ [[\dot{\varepsilon}_v]] = 0 \tag{9}$$

where $[[\dot{p}]]$, $[[\dot{q}]]$ and $[[\dot{\varepsilon}_v]]$ are, respectively, jumps in p, q and ε_v due to duplicate solutions for velocity field. Implementing Eq. (1) in Eq. (9) and rearrangement of terms gives the following criterion for loss of uniqueness under CSD:

$$[[\dot{p}]]^2 \, D_{qq} = 0 \tag{10}$$

4 Simulation of Sand Behavior Under Conventional Triaxial Paths

Data of triaxial tests on Changi sand, a marine dredged silica sand for land reclamation in Singapore, are adopted for evaluation of the constitutive model. The basic properties of Changi sand are presented in Wanatowski and Chu [9]. For loose and very loose Changi sand samples, the model predictions are illustrated versus experimental data of drained, and K-consolidated undrained triaxial tests in Fig. 1, respectively. The model parameters used in simulations are given in Table 1. In Fig. 1, the plasticity model takes into account the common trend of Changi sand behavior associated with density, confining stress, and drainage condition.

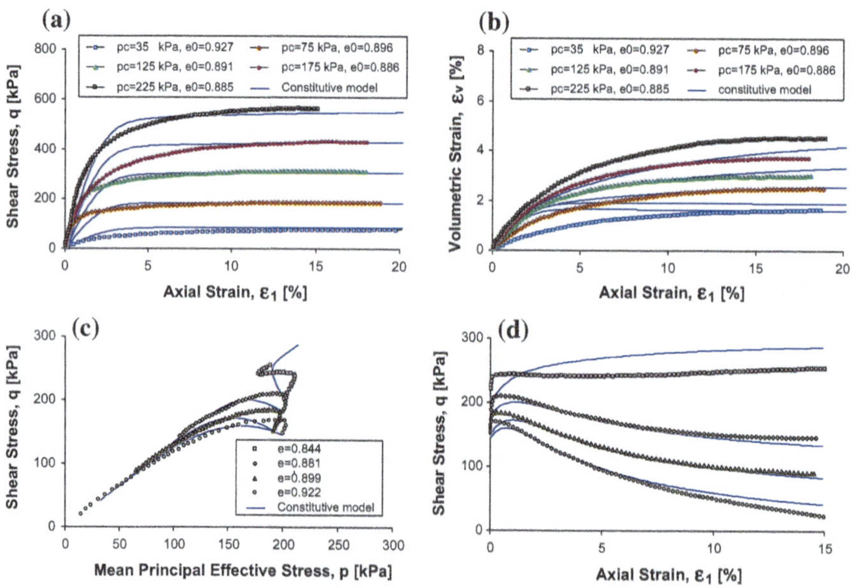

Fig. 1 Simulation of the behavior of Changi sand: **a** and **b** q versus ε_1 and ε_v versus ε_1 curves for five conventional triaxial drained tests, **c** and **d** stress paths and q versus ε_1 curves for four anisotropically consolidated undrained tests (data from [9])

Table 1 Model parameters in simulation of Changi sand behavior

Elasticity		Critical state line				Dilatancy		Plastic modulus		
G_0	K_0	M	e_0	λ	ξ	A	n^d	h_0	c_h	n^b
70	54.4	1.353	0.9345	0.0578	0.5913	1.6	2.4	1132	0.9756	0.80

5 Simulation of Sand Behavior Under CSD

The constitutive model is applied to simulate behavior of initially very loose ($e_0 = 0.945$) and very dense ($e_0 = 0.657$) samples of Changi sand under CSD. In laboratory, the samples were sheared along conventional drained triaxial stress path up to q = 165 kPa (for very loose sample) and q = 300 kPa (for very dense sample) after isotropic consolidation under $p_c = 150$ kPa. Consequently, samples were subjected to CSD by continuously decreasing mean principal effective stress under relatively fixed shear stress, until $\varepsilon_1 = 8–10\%$ was attained. Without changing the model parameters in Table 1, soil behavior in the CSD tests are simulated and depicted against data in Fig. 2. Parts "a" to "d" of Fig. 2 indicate that instability occurs prior to complete mobilization of critical state stress ratio for the loose sample; however, the dense one becomes unstable at peak stress ratio, whereat decrease in mean principal effective stress halts. Besides, axial strain is generated rapidly once soil becomes unstable in both loose and dense samples.

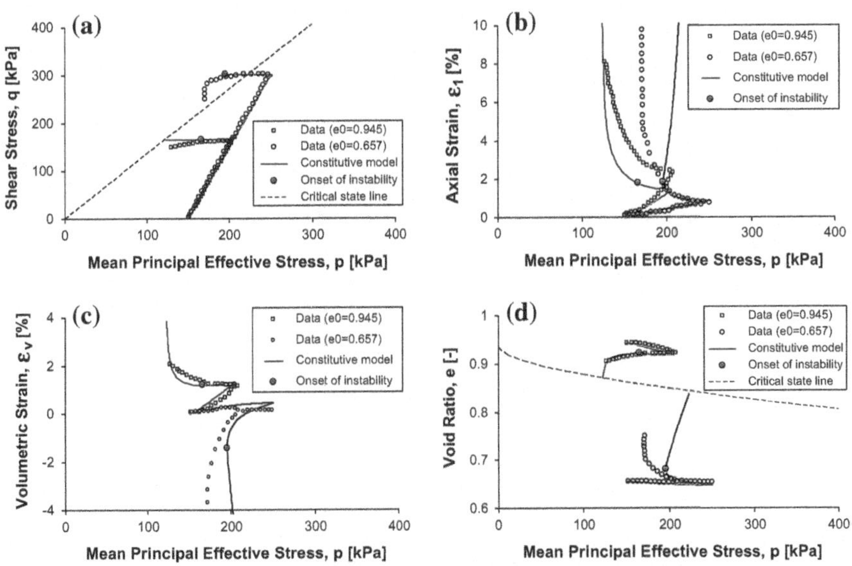

Fig. 2 Simulation of behavior for very loose ($e_0 = 0.945$) and very dense ($e_0 = 0.657$) Changi sand samples under constant shear drained stress paths: **a** stress path, **b** axial strain versus mean principal effective stress, **c** volumetric strain versus mean principal effective stress and **d** void ratio versus mean principal effective stress (data taken from [4])

6 Conclusions

Loose granular soils are susceptible to abrupt loss of stability when subjected to constant shear drained (CSD) paths. In this paper, a critical state bounding surface plasticity model was applied to predict instability of a very loose sample of Changi sand under CSD. When subjected to CSD, it was observed that loose sands become unstable at certain states beyond which strains are generated faster, prior to complete mobilization of critical state stress ratio (i.e., friction angle). It is found that the state-dependent model can be applied to simulate conventional drained and undrained triaxial tests, as well as CSD test using a single set of parameters.

References

1. Andrade, J.E., Ramos, A.M., Lizcano, A.: Criterion for flow liquefaction instability. Acta Geotech. **8**, 525–535 (2013)
2. Azizi, A., Imam, R., Soroush, A., Zandian, R.: Behavior of sands in constant deviatoric stress loading. In: Oka, F., Murakami, A., Kimoto, S. (eds.) Prediction and Simulation Methods Geohazard Mitigation, pp. 319–324 Taylor & Francis Group (2009)
3. Borja, R.I.: Computational modeling of deformation bands in granular media. II. Numerical simulations. Comput. Methods Appl. Mech. Eng. **193**, 2699–2718 (2004)
4. Chu, J., Leroueil, S., Leong, W.K.: Unstable behavior of sand and its implication for slope stability. Can. Geotech. J. **40**(5), 873–885 (2003)
5. Chu, J., Wanatowski, D.: Instability of loose sand in plain strain. J. Geotech. Geoenviron. Eng. **134**(1), 136–142 (2008)
6. Dafalias, Y.F., Manzari, M.T.: Simple plasticity sand model accounting for fabric change effects. J. Eng. Mech. **130**(6), 622–634 (2004)
7. Einav, I., Puzrin, A.M.: Pressure-dependent elasticity and energy conservation in elastoplastic models for soils. J. Geotech. Geoenviron. Eng. **130**(1), 81–92 (2004)
8. Golchin, A., Lashkari, A.: A critical state sand model with elastic-plastic coupling. Int. J. Solids Struct. **51**, 2807–2825 (2014)
9. Wanatowski, D., Chu, J.: Static liquefaction of sand in plane strain. Can. Geotech. J. **44**, 299–313 (2007)

Model Prediction of Static Liquefaction in Unsaturated Sands

Xilin Lu, Maosong Huang and Jiangu Qian

Abstract The initiation of static liquefaction of unsaturated sand was studied by an extended Mohr-Coulomb elasto-plasticity hardening model. The second-order work was used to check the initiation point of static liquefaction under undrained shear. The results showed that the mechanical behavior of unsaturated sands will be changed by the compressibility of gas. The critical stress ratio at the liquefaction instability point decreases with the saturation degree.

1 Introduction

Static liquefaction has been widely studied in saturated soil [1, 11]. Other than the density of sands, the consolidation state [7], and the stress state [6], the drainage condition [17], and the initial degree of saturation [5] also showed influence on the instability of sand. The existence of gas makes the unsaturated sand compressible and restricts the buildup of excess pore-water pressure [14]. Undrained triaxial compression tests have shown that loose unsaturated sands produce static liquefaction only when the initial saturation degree exceeds a critical value [8]. In order to describe the mechanical behavior of unsaturated soil, a suitable constitutive model is needed. Most of the current models are based on models proposed for saturated soil. Wheeler [15] proposed a conceptual model to calibrate the behavior of unsaturated soils containing large gas bubbles. Grozic et al. [9] proposed a constitutive model for gassy sand based on an existing model. Sultan and Garziglia [13] presented a constitutive model for gassy soil based on Cam-Clay model.

X. Lu (✉) · M. Huang · J. Qian
Key Laboratory of Geotechnical and Underground Engineering of Ministry of Education, Tongji University, Shanghai 200092, China
e-mail: xilinlu@tongji.edu.cn

X. Lu · M. Huang · J. Qian
Department of Geotechnical Engineering, Tongji University, Shanghai 200092, China

E. Papamichos et al. (eds.), *Bifurcation and Degradation of Geomaterials with Engineering Applications*, Springer Series in Geomechanics and Geoengineering, DOI 10.1007/978-3-319-56397-8_14

This paper studied the onset of static liquefaction in unsaturated sand by proposing a material state-dependent elastoplasticity model. The fluid pressure change in unsaturated sand under undrained loading was obtained. The stress-strain relationship under the undrained triaxial condition was formulated by the proposed model. The ability of the model in predicting static liquefaction was validated by comparing to existing experiments.

2 Constitutive Modeling of the Unsaturated Sand

The constitutive model for unsaturated sand was built on an existing material state dependent Mohr-Coulomb hardening model. The yield function and potential plastic function are

$$\begin{cases} F = q - Mp' = 0 \\ Q = q + M_d p' \ln\frac{p'}{p_0} = 0 \end{cases} \tag{1}$$

where the mean effective pressure is $p' = \sigma_{ii}/3 - u_{gw}$, the equivalent shear stress is $q = \sqrt{3J_2} = \sqrt{3s'_{ij}s'_{ij}/2}$, the deviatoric stress is $s'_{ij} = \sigma'_{ij} - \delta_{ij}p'$ and δ_{ij} is the Kronecker delta, M is the stress ratio, and M_d is the stress-dilatancy.

The liquid phase is assumed continuous and gas exists as discrete bubbles embedded in liquid. The effective stress of unsaturated sands is formulated as

$$\boldsymbol{\sigma}' = \boldsymbol{\sigma} - u_{gw}\mathbf{1} \tag{2}$$

The change of pore fluid pressure is [3]

$$\Delta u_{gw} = \frac{u_g + p_{at}}{(1 - S_r)n}\Delta\varepsilon_v \tag{3}$$

The material-dependent peak stress ratio and stress-dilatancy is

$$M_f = M_{cs}\exp(-n_b\psi) \tag{4a}$$

$$M_d = M_{cs}\exp(n_d\psi) \tag{4b}$$

where $\psi = e - e_c$ is material state parameter [4], e_c is void ratio at critical state, n_b and n_d are peak stress ratio and dilatancy parameters.

The critical state line is defined [16] by

$$e_c = e_{c0} - \lambda_c\left(\frac{p}{p_{at}}\right)^{\xi} \tag{5}$$

where e_{c0}, λ_c and ξ are the material parameters, $p_{at} = 101.3$ kPa is atmospheric pressure.

The evolution of M follows the hyperbolic law

$$M = M_f \frac{\varepsilon_s^p}{A + \varepsilon_s^p} \tag{6}$$

After derivation by elastoplasticity theory, the rate form constitutive relationship can be obtained. The obtained relationship is same with Pietruszczak and Pande [12] if the surface tension force of air bubble ignored. This model is based on deviatoric hardening; it may not suitable for soil of volumetric hardening. Besides, the effect of pressure on the water-gas pressure difference is not considered in the model.

3 Criteria for the Onset of Static Liquefaction

The condition for material instability of soil is

$$d^2w = d\boldsymbol{\sigma}' : d\boldsymbol{\varepsilon} \leq 0 \tag{7}$$

where d^2w is the second-order work, $d\tilde{\boldsymbol{\sigma}}'$ is effective stress increment and $d\tilde{\boldsymbol{\varepsilon}}$ is the corresponding strain increment.

In the case of fully saturated sands under a undrained triaxial condition, the criterion for static liquefaction is [2]

$$\mathrm{D}_{pq}^{ep}(2,2) = 0 \tag{8}$$

Since the volumetric strain is not zero in unsaturated sand when subjected to undrained loading, Eq. (8) cannot be used in the case of unsaturated sand, and the initiation of static liquefaction predicted by Eq. (7) in unsaturated sand will be different from the saturated sand.

4 Model Prediction

The undrained triaxial tests of Ottawa sand [10] were simulated. The size of specimen was 100 mm in height and 50 mm in diameter; the confining pressure was 100 kPa. The initial saturation degrees of specimen were 94.5, 96.3, 98.1, 99.2, and 100%; the initial void ratios of the specimens were 0.774, 0.773, 0.773, 0.770, and 0.769. After determining material parameters directly from experiments and choosing suitable fitting parameters, the material parameters adopted in the simulation were $G_0 = 50$, $\nu = 0.15$, $M_{cs} = 1.2$, $n_b = 1.1$, $n_d = 3.5$, $e_{c0} = 0.815$,

$\lambda_c = 0.063$, $\xi = 0.3$, $A = 0.002$. An explicit integration method was used to integrate the rate-form constitutive relationship under undrained condition. The predicted stress-strain behavior and stress path are shown in Fig. 1. In the case of low initial saturation degree, the predicted stress path is close to that of drained triaxial test at the initial stage, and then approaches to a stress path of undrained triaxial test. Although there is discrepancy of predicted results from experimental data, the trend of the undrained behavior can be well captured. The second order work was calculated during the integration of the rate-form stress-strain relationship. The results are shown in Fig. 2a, the increase of the initial saturation degree induces the second order work falling below zero rapidly. The stress ratio corresponding to the instability points are shown in Fig. 2b, it decreases with the initial saturation, and the predicted results compare very well with the experiments.

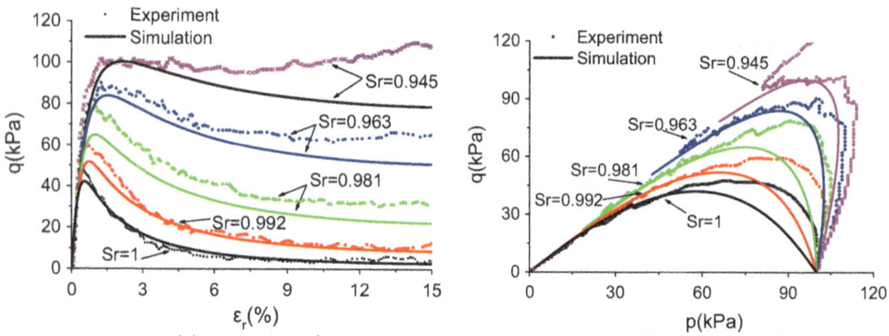

(a) stress-strain (b) stress path

Fig. 1 Stress-strain relationships in a undrained test [10]

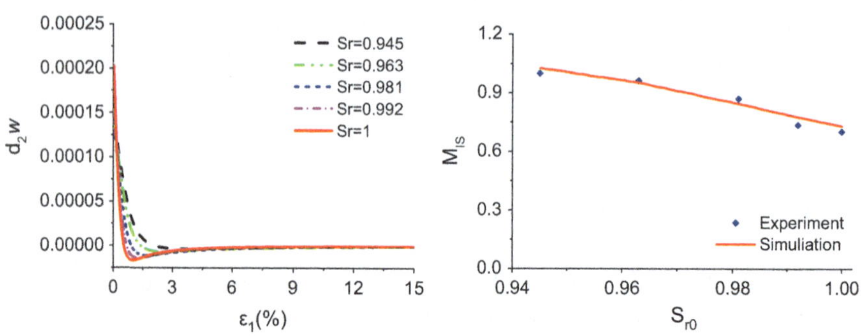

(a) second order work (b) stress ratio at instability point

Fig. 2 The predicted initiation of static liquefaction

5 Conclusion

A material state-dependent Mohr-Coulomb hardening elasto-plasticity model was proposed to calibrate the stress-strain relationship of unsaturated sand. The mechanical behavior under undrained condition was shown to rely on the initial saturation ratio. The onset of static liquefaction under undrained condition was predicted by the second-order work. The decrease of initial saturation degree results in compressibility of soil and induces higher shear stress. It also reduces the stress ratio at the instability point.

Acknowledgements The financial supports by the National Natural Science Foundation of China (No. 11372228 and No. 41672270) are gratefully acknowledged.

References

1. Andrade, J.E.: A predictive framework for liquefaction instability. Géotechnique **59**(8), 673–682 (2009)
2. Andrade, J.E., Ramos, A.M., Lizcano, A.: Criterion for flow liquefaction instability. Acta Geotech. **8**(5), 525–535 (2013)
3. Atigh, E., Byrne, P.M.: Liquefaction flow of submarine slopes under partially undrained conditions: an effective stress approach. Can. Geotech. J. **41**(1), 154–165 (2004)
4. Been, K., Jefferies, M.G.: A state parameter for sands. Géotechnique **35**(2), 99–112 (1985)
5. Buscarnera, G., Nova, R.: Modelling instabilities in triaxial testing on unsaturated soil specimens. Int. J. Numer. Anal. Meth. Geomech. **35**(2), 179–200 (2011)
6. Chu, J., Wanatowski, D.: Instability conditions of loose sand in plane strain. J. Geotech. Geoenviron. Eng. ASCE **134**(1), 136–142 (2008)
7. Fourie, A.B., Tshabalala, L.: Initiation of static liquefaction and the role of K0 consolidation. Can. Geotech. J. **42**(3), 892–906 (2005)
8. Grozic, J.L., Robertson, P.K., Morgenstern, N.R.: The behaviour of loose gassy sand. Can. Geotech. J. **36**(3), 482–492 (1999)
9. Grozic, J.L.H., Imam, S.M.R., Robertson, P.K., Morgenstern, N.R.: Constitutive modeling of gassy sand behaviour. Can. Geotech. J. **42**(3), 812–829 (2005)
10. He, J., Chu, J.: Undrained responses of microbially desaturated sand under monotonic loading. J. Geotech. Geoenviron. Eng. ASCE **140**(5), 04014003 (2014)
11. Lu, X., Huang, M.: Static liquefaction of sands under isotropical and K0 consolidated undrained triaxial conditions. J. Geotech. Geoenviron. Eng. ASCE **141**(1), 04014087 (2015)
12. Pietruszczak, S., Pande, G.N.: Constitutive relations for partially saturated soils containing gas inclusions. J. Geoeng. ASCE **122**(1), 50–59 (1996)
13. Sultan, N., Garziglia, S.: Mechanical behaviour of gas-charged fine sediments: model formulation and calibration. Géotechnique **64**(11), 851–864 (2015)
14. Vega-Posada, C.A., Finno, R.J., Zapata-Medina, D.G.: Effect of gas on the mechanical behavior of medium-dense sands. J. Geotech. Geoenviron. Eng. ASCE **140**(11), 04014063 (2014). doi:10.1061/(ASCE)GT.1943-5606.0001163

15. Wheeler, S.J.: A conceptual model for soils containing large gas bubble. Geotechnique **38**(3), 389–397 (1988)
16. Yang, J., Li, X.S.: State-dependent strength of sands from the perspective of unified modeling. J. Geotech. Geoenviron. Eng. ASCE **130**(2), 186–198 (2004)
17. Yang, X.L., Yan, R.M.: Collapse mechanism for deep tunnel subjected to seepage force in layered soils. Geomech. Eng. **8**(5), 741–756 (2015)

Microscale Analysis of the Effect of Suffusion on Soil Mechanical Properties

Rodaina Aboul Hosn, Cong Doan Nguyen, Luc Sibille, Nadia Benahmed and Bruno Chareyre

Abstract Suffusion is a particular internal erosion process that can lead to important disorders in water retaining structures such as embankment dams and levees. It causes modifications in the soil micro-structure and may modify the mechanical behaviour of the soil leading to deformations at the macroscopic scale. Therefore, the aim of this study is to investigate the consequences of internal erosion on the mechanical properties of the soil. We present such an investigation through numerical and experimental approaches. For the experimental approach, a newly developed suffusion test apparatus is used while for the numerical approach, a model is established based on the discrete element method (DEM) with a one-way fluid-solid coupling.

Keywords Internal erosion · Suffusion · Shear strength · Erosion tests · Discrete element method

1 Introduction

Suffusion is a particular internal erosion process involving the selective erosion of fine particles within the matrix of coarse particles. Although the consequences of suffusion on the mechanical properties of the soil have been investigated through experimental [3, 9] and numerical studies [7, 8], they are not yet completely described. Therefore, with the objective to establish a clear relation between the removal of fine particles and the mechanical properties of eroded soils, this paper presents such an analysis following both experimental and numerical approaches. Effects on the hydraulic and the mechanical properties of the eroded soils were noticed in

R.A. Hosn (✉) · L. Sibille · B. Chareyre
Grenoble Alpes University, 3SR Laboratory, 38400 Grenoble, France
e-mail: rodaina.aboulhosn@3sr-grenoble.fr

R.A. Hosn · C.D. Nguyen · N. Benahmed
IRSTEA, 13182 Aix-en-Provence, France

E. Papamichos et al. (eds.), *Bifurcation and Degradation of Geomaterials with Engineering Applications*, Springer Series in Geomechanics and Geoengineering, DOI 10.1007/978-3-319-56397-8_15

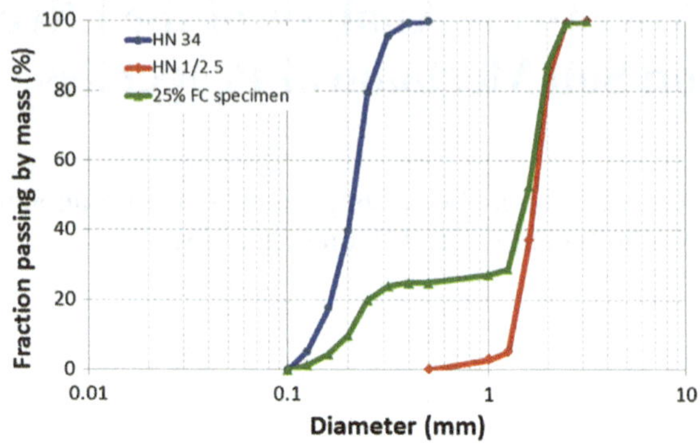

Fig. 1 Particle size distribution of Hostun sand

experiments using a newly developed suffusion test apparatus. To reproduce these effects numerically, an extraction procedure, based on the discrete element method with a one-way fluid-solid coupling, is defined.

2 Experimental Approach

2.1 The Material Used

A mixture of coarse and fine silica sand, (HN 1/2.5 and HN34, respectively), is used. The soil samples were prepared with an initial relative density of 40% and a fines content (FC) of 25%. Figure 1 shows the particle size distribution of the used sand. The potential of internal instability is assessed following the methods proposed by [4, 5] which indicated that the studied soil is vulnerable to internal erosion.

2.2 Experimental Setup and Procedure

Suffusion tests were carried out using a newly developed suffusion test apparatus made up of a cylindrical Plexiglas cell, 70 mm in internal diameter, to fit the pedestal triaxial cell. The cell is connected to a water supply system and fines collector. The soil samples were reconstituted by moist tamping. Then, they were saturated by flushing CO_2 first, followed by de-aired water in an upward direction at very low flow rate to prevent the heave phenomenon. Thereafter, the samples were subjected to the erosion test by flushing water in a downward direction. In this test, the flow rate is increased by steps. The corresponding hydraulic gradient, i, for each flow rate is determined from the pressure gradient measured by two pressure transducers installed at the upper and lower parts of the suffusion cell. Similarly, the eroded mass is collected and weighted. Note that the flow rate is increased until the erosion of fine

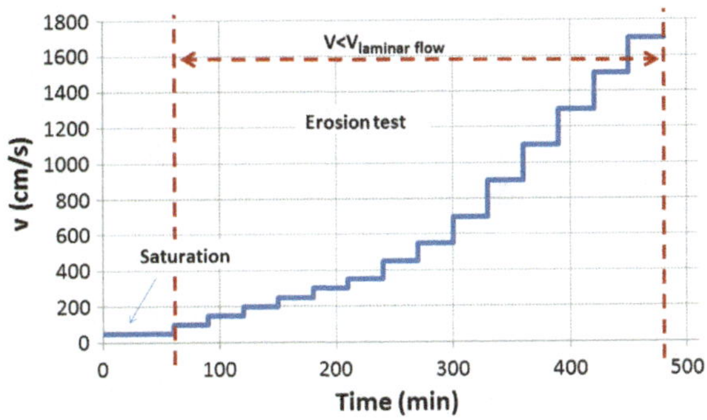

Fig. 2 Illustration of the hydraulic loading during internal erosion

particles is triggered, and then kept constant until no eroded fines are observed. The test is then stopped when reaching the maximum laminar flow velocity. The increasing flow rate process is illustrated in Fig. 2.

Once the internal erosion process is done, the soil samples are recovered and frozen to preserve the microstructure. Afterwards, they are installed in the triaxial cell, confined at 100 kPa, and then sheared under drained conditions at a constant strain rate of 1% per minute to investigate the stress-strain behavior of eroded soil samples.

2.3 Results and Discussion

The variation of the hydraulic gradient (i), the hydraulic conductivity (k) and the eroded mass (M_e) are illustrated in Figs. 3 and 4. It is noticed that at low flow velocities (v), the hydraulic gradient is almost negligible. When the flow velocity exceeds 0.06 cm/s, a sudden increase of the hydraulic gradient is observed accompanied with a sudden decrease of permeability. It is worth noting here, that no eroded mass was collected at this stage. Therefore, such a sharp increase in the hydraulic gradient may be attributed to the clogging of the constrictions by the transported fine particles. In

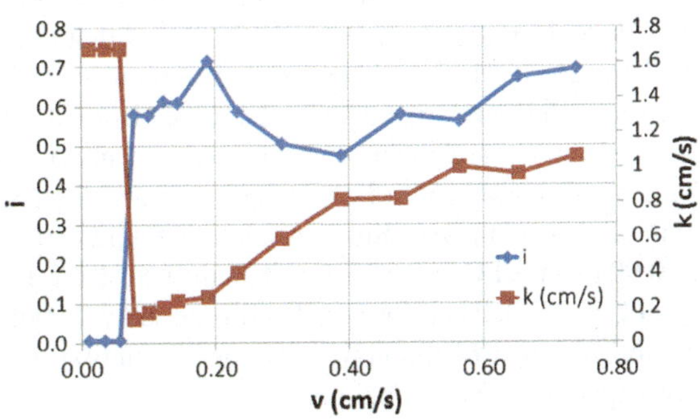

Fig. 3 The variation of the hydraulic gradient with seepage velocity

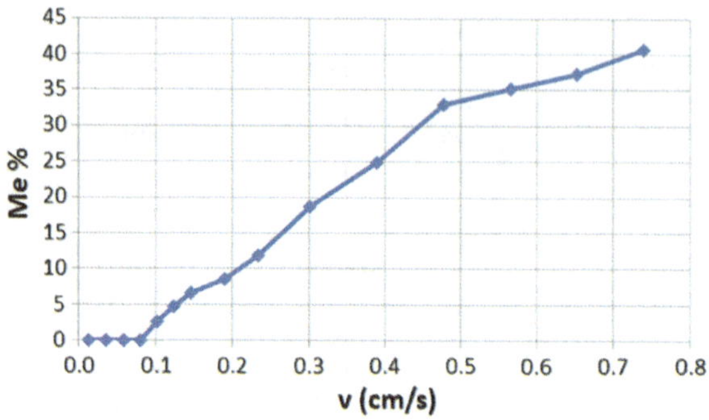

Fig. 4 Ratio of the eroded mass to the initial fine particle mass as a function of the seepage velocity

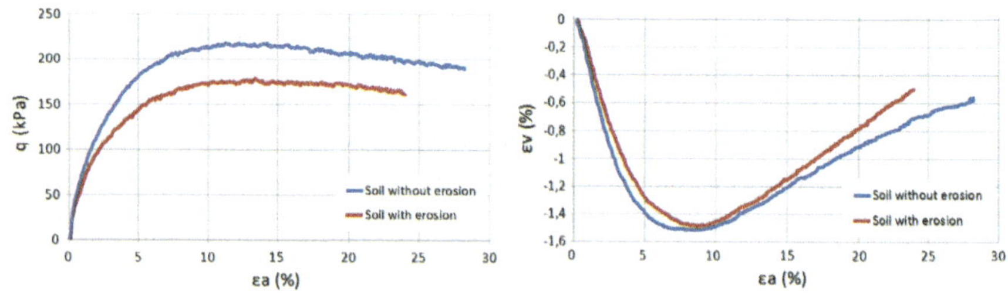

Fig. 5 Responses to drained triaxial compression tests of soil samples with and without erosion

other words, erosion might have occured inside the sample, but particles were not washed out.

With further increase of the flow rate, the hydraulic gradient increases very slightly and some fine grains were collected resulting in a decrease in the hydraulic conductivity. This indicates that the trapped particles were dislodged and transported outside the sample; However, the erosion is not predominant. As the flow velocity increases, the hydraulic properties keep changing as a result of the co-existance of erosion-clogging processes.

Responses to drained triaxial compression tests are illustrated in Fig. 5. The eroded soil shows a lower shear strength compared to the soil without erosion. The peak friction angle decreases from 31.4° to 27.8° and that at the critical state decreases from 29.2° to 26.7°. However, both soil specimens exhibit the same volumetric strains, even though the relative density decreases from 40 to 20% after erosion. Nevertheless, recent researches on reconstituted soil mixtures indicate that the mechanical behavior is mainly controlled by the coarse grains structure; hence by the inter-grains void ratio, e_g. Applying this concept, it is found that e_g of the eroded sample is equal to 0.904 (with FC = 16%), lower than that of the non-eroded sample (e_g = 1.039 and FC = 25%). Therefore, such a less contractant behavior may be explained by the combined effect of the new fines content and the intergranular void ratio. Further investigations are still ongoing.

3 Numerical Approach

Complementary to the experimental approach, a discrete numerical model is defined. The granular assembly is composed of spherical discrete elements. Periodic boundary conditions are adopted and gravity is neglected in all simulations.

The interparticle interaction is modelled by a linear elastic relationship between forces and displacements with a slip Coulomb model. Since spherical particles cause excessive rolling, rolling resistance is taken into account in the contact law. A particle extraction procedure, based on a one way fluid-solid coupling, is defined to mimic the internal erosion (detachment and transport) of fine particles. For the one-way coupling, the DEM-PFV (Pore scale Finite Volume) [2] is followed to solve the interstitial fluid flow and to compute the fluid forces applied on each soil particle.

3.1 Extraction Procedure

The suffusion process can be described accurately by a complete fluid-solid coupling which would represent an important computational cost. Therefore, the following simplified extraction procedure, involving a partial coupling, has been defined [1]. Starting from a granular assembly at a given equilibrium state reached for a given stress state (isotropic or deviatoric), the interstitial fluid flow, generated by applying a global hydraulic gradient, is solved and fluid forces applied on solid particles are computed. Then DEM cycles are iterated under constant fluid forces (this is why the coupling is "one way") to check whether solid particles can retrieve an equilibrium state under the combined action of contact and fluid forces. When the fluid force is not balanced by contact forces, the particle loses its equilibrium and is considered detached. Thereafter, the possibility of detached particles to be transported is checked by comparing their sizes with a controlling constriction size, chosen by following the suggestion of [6] in their filter criterion and corresponds to D_{c35} (the constriction size for which 35% of the constrictions of the granular assembly are finer than this size). Detached particles with a diameter smaller than D_{c35} are considered as eroded particles and are removed from the granular assembly. Finally, DEM cycles are iterated again with the new particle configuration (i.e. without the eroded particles) and without any fluid forces. Due to the particles removal, the granular assembly may deform under the constant applied stress state. Then the simulation is run until a new equilibrium state (if any) is reached. Such a process can be repeated to pursue the erosion development, or a drained triaxial compression test can be performed to evaluate the new mechanical properties of the eroded soil.

Using the particle size distribution of Fontainebleau soil (Fig. 6), the defined extraction procedure is applied by progressively increasing the hydraulic gradient. Figure 7 shows an increase in the eroded mass (M_e), including active particles M_{ea} (i.e., particles participating in the force transfer), with the hydraulic gradient, i. When a little amount of active particles is eroded, the soil shows negligible deformations as

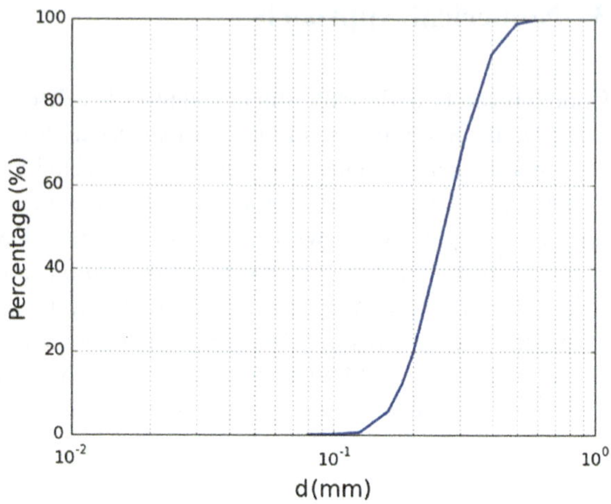

Fig. 6 Particles size distribution of Fontainebleau sand

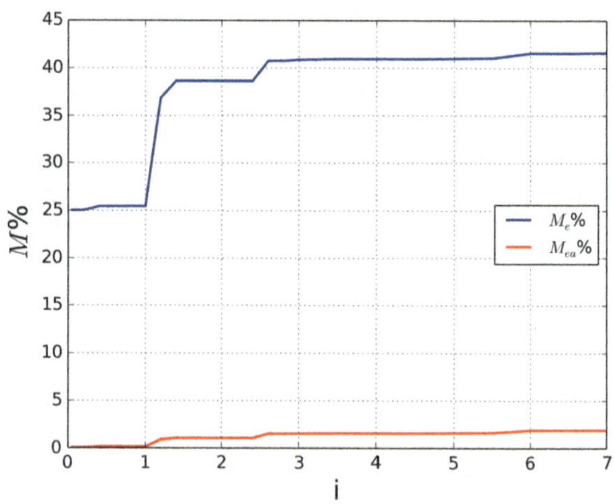

Fig. 7 The variation of the eroded mass with the hydraulic gradient, *i*

demonstrated in Fig. 8. However, as more active particles are eroded, the soil compresses significantly showing larger deformations at the macroscopic scale. Nevertheless, as noticed from Fig. 8, the porosity increases after erosion even if the soil shows large deformations. Therefore, the creation of a more open microstructure is predominant during suffusion.

3.2 *Post-suffusion Properties*

The effect of erosion on the shear strength, obtained by performing drained triaxial compression tests, is shown in Fig. 9. A non-linear non-monotonic relation is found between the maximum friction angle and the percentage of the active eroded mass. It

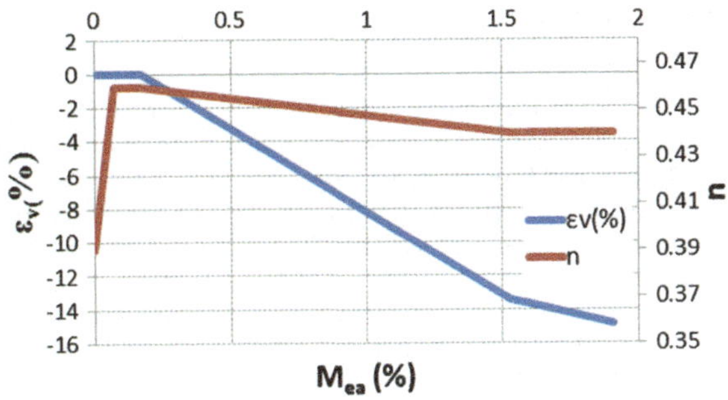

Fig. 8 The variation of the volumetric deformation and the porosity with the active eroded mass, M_{ea}

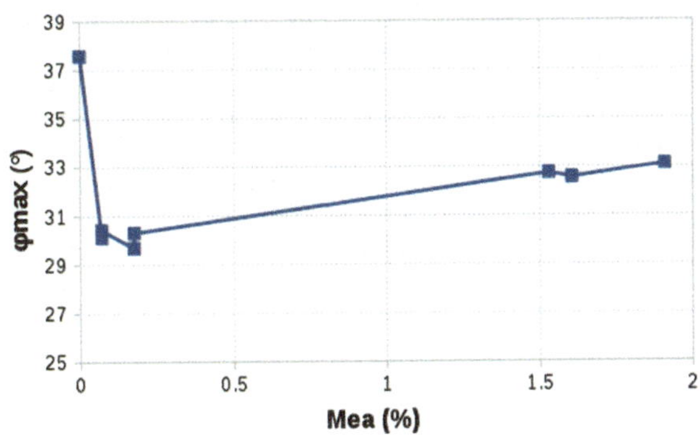

Fig. 9 The variation of the maximum friction angle with the active eroded mass, M_{ea}

is noticed that, at low percentages of active eroded mass, when the soil didn't deform, the shear strength decreases significantly. However, as we continue eroding active particles, the soil deforms and the porosity decreases slightly, as shown previously in Fig. 8, so that the soil recovers part of its strength. But in all cases, the shear strength decreases after erosion similarly to the experimental results (Fig. 5).

4 Conclusion

This paper presents two approaches to describe the effect of suffusion on the mechanical properties of the soil. Experiments were done using a newly developed seepage test apparatus while numerically, an extraction procedure is defined based on a one-way fluid-solid coupling. Performing such erosion tests, effects on the hydraulic and mechanical properties of the soil were noticed with a decrease in the soil shear strength for the studied soil samples. However, the experimental and numerical approaches presented in this paper are currently performed on two different gradings where a simplified one was used numerically to limit the computational cost. In future work, the numerical approach will be reiterated by considering the grading used experimentally, with the objective to reach a description of the shear strength degradation, complementary to the experimental one.

References

1. Aboul Hosn, R., Sibille, L., Benahmed, N., Chareyre, B.: A discrete numerical description of the mechanical response of soils subjected to degradation by suffusion. In: International Conference of Scour and Erosion, ICSE (2016)
2. Chareyre, B., Cortis, A., Catalano, E., Eric, B.: Porescale modeling of viscous flow and induced forces in dense sphere packings. Transp. Porous Media **92**, 473–493 (2012)
3. Ke, L., Takahashi, A.: Experimental investigations on suffusion characteristics and its mechanical consequences on saturated cohesionless soil. Soils Found. **54**, 713–730 (2014)
4. Kenney, T.C., Lau, D.: Internal stability of granular filters. Can. Geotech. J. 22(2), 215-225 (1985)
5. Kezdi, Á.: Soil physics: selected topic (1979)
6. Raut, A.K., Indraratna, B.: Further advancement in filter criteria through constriction-based techniques. J. Geotech. Geoenviron. Eng. **134**(6), 833887 (2008)
7. Scholtès, L., Hicher, P.-Y., Sibille, L.: Multiscale approaches to describe mechanical responses induced by particle removal in granular materials. Comptes Rendus Mécanique **338**, 627–638 (2010)
8. Wood, D., Maeda, K.: Discrete element modelling of soil erosion. In: Fourth International Conference on Scour and Erosion (2008)
9. Xiao, M., Shwiyhat, N.: Experimental investigation of the effects of suffusion on physical and geomechanic characteristics of sandy soils. Geotech. Test. J. **35**, 111 (2012)

A Numerical Model of Internal Erosion for Multiphase Geomaterials

Sayuri Kimoto, Toshifumi Akaki, Benjamin Loret and Fusao Oka

Abstract Internal erosion is the detachment of fine soil particles due to seepage flow, with ensuing increasing porosity, and transport of these particles out of the soil mass. In the present study, firstly we have formulated the constitutive equations of the internal erosion, that is, the erosion criteria and the rate equation of the mass transfer. The driving force for erosion is assumed to be given by the interaction term, i.e. relative velocity between two phases in the equation of motions for the two-phase mixture. Then, field equations to simulate hydro-mechanical behavior due to the internal erosion were derived in the framework of multiphase mixture theory. In addition, laboratory erosion tests using gap-graded sandy soil are simulated by the proposed model and the validity are discussed with respect to the rate of eroded soil mass and the particle size distribution after the erosion test.

1 Introduction

Soil particles in the earth structures are dislodged/extracted and transported by seepage flow when the hydraulic flow is intense enough, which is called the internal erosion. Progressive degradation of soil structure due to the internal erosion may

S. Kimoto (✉)
Department of Civil and Earth Resources Engineering, Kyoto University, Kyoto 615-8540, Japan
e-mail: kimoto.sayuri.6u@kyoto-u.ac.jp

T. Akaki
Department of Civil and Earth Resources Engineering, Kyoto University, Kyoto, Japan

B. Loret
Institut National Polytechnique, Université de Grenoble, Grenoble, France

F. Oka
Emeritus of Kyoto University, Kyoto, Japan

E. Papamichos et al. (eds.), *Bifurcation and Degradation of Geomaterials with Engineering Applications*, Springer Series in Geomechanics and Geoengineering, DOI 10.1007/978-3-319-56397-8_16

lead to the local failure, such as, piping, and may result in cavities in the earth structures, such as earth dams, dikes and levees. These are one of the main causes of the failure of geomaterials. In the present study, firstly we have formulated the constitutive equations of the internal erosion, that is, the erosion criteria and the rate equation of the mass transfer. We assume that the erosion takes place when the hydrodynamical driving force acting on the small volume of mixture is larger than the resistance force over the area. Then, we derived field equations to simulate hydro-mechanical behavior due to the internal erosion in the framework of multi-phase mixture theory. Finally, laboratory erosion tests using gap-graded sandy soil are simulated by the proposed model and the validity is discussed with respect to the rate of eroded soil mass and the particle size distribution after the erosion test.

2 Equations of Motion and Mass Balance Equations

The onset conditions of the internal erosion are derived based on the multi-phase mixture theory. The constituents are the soil particles (*S*), the eroded soil particles (*FS*), and the pore water (*W*). It is assumed that the eroded soil particles moves unified with the pore fluid, and the fluid phase (*F*) is determined as $F = W + FS$. The equation of the balance of the linear momentum of the solid phase is given by,

$$\bar{\rho}^S \ddot{u}_i^S = \frac{\partial \sigma_{ij}^S}{\partial x_j} + \bar{\rho}^S b_i + R_i \tag{1}$$

where $\bar{\rho}^S (= n^S \rho^S)$ is the mass density of the solid phase, n^S is the volume fraction of the solid phase, ρ^S is the mass density of soil particle, $\sigma_{ij}^S \left(= \sigma'_{ij} - n^S P^F \delta_{ij} \right)$ is the partial stress of the solid phase, σ'_{ij} is the skeleton stress, P^F is the average pressure of the fluid phase, $\ddot{u}_i^S$ is the acceleration of the solid phase, b_i is the body force, and R_i is the interaction force term induced by the relative motion between the solid and fluid phases.

The equation of the balance of the linear momentum of the fluid phase is given by the following equation, where

$$\bar{\rho}^F \ddot{u}_i^F = \frac{\partial \sigma_{ji}^F}{\partial x_j} + \bar{\rho}^F b_i - R_i \tag{2}$$

where $\bar{\rho}^F (= n^F \rho^F)$ is the mass density of the fluid phase, $n^F (= n^W + n^{FS})$ is the volume fraction of the fluid phase, ρ^F is the mass density of the fluid phase, $\sigma_{ij}^F (= -n^F P^F \delta_{ij})$ is the partial stress of the fluid phase, $\ddot{u}_i^F$ is the acceleration of the fluid phase. The interaction term R_i in Eqs. (1) and (2) is proportional to the relative velocity between two phases and is given by

$$R_i = n^F \frac{\rho^F g}{k} q_i \tag{3}$$

$$q_i = n^F \left(\dot{u}_i^F - \dot{u}_i^S\right) \tag{4}$$

where $\dot{u}_i^F$ and $\dot{u}_i^S$ are velocities of the fluid phase and the solid phase respectively, γ_w is the unit weight of the water, k is the proportional coefficient (permeability coefficient). Neglecting the acceleration term of Eq. (2) and assuming the space gradient is small, we have

$$R_i = -n^F \left(\frac{\partial P^F}{\partial x_i} - \rho^F b_i\right) = -n^F \rho^F g \frac{\partial h}{\partial x_i} \tag{5}$$

where h is the total head of the fluid phase and Eqs. (3)–(5) expresses Darcy's law. The mass balance equation for each phase is given by

$$\frac{\partial \overline{\rho}^\alpha}{\partial t} + \frac{\partial\left(\overline{\rho}^\alpha \dot{u}_i^\alpha\right)}{\partial x_i} = \dot{\rho}^\alpha \quad (\alpha = S, W, FS) \tag{6}$$

where $\overline{\rho}^{FS}(=n^{FS}\rho^S)$, $\dot{u}_i^{FS} = \dot{u}_i^F$, $\dot{\rho}^{FS} = -\dot{\rho}^S$, and $\dot{\rho}^W = 0$.

3 Constitutive Equation of the Internal Erosion

Firstly we derive the onset conditions of the internal erosion of the uncemented soil. It is assumed that the erosion occur when the driving force acts on the small volume of soil, which is larger than the resistance force (see Fig. 1). In the model, the driving force of the erosion is given by the interaction term R_i in Eqs. (1) and (2). The onset condition in the flow direction n_i is given by

$$VR_i n_i \geq f_1\left(\sigma_n^{'}, \varepsilon^p, S\right) S_{gr} \tag{7}$$

where V is the total volume around the soil particles, S_{gr} is the surface area over which acts the resistance force, f_1 is the resistance force against the erosion per unite surface that may depend on the effective stress $\sigma_n^{'}$ acting in the direction n_i, ε^p is the inelastic strain, S is the other factor such as the saturation of the gas in the case of

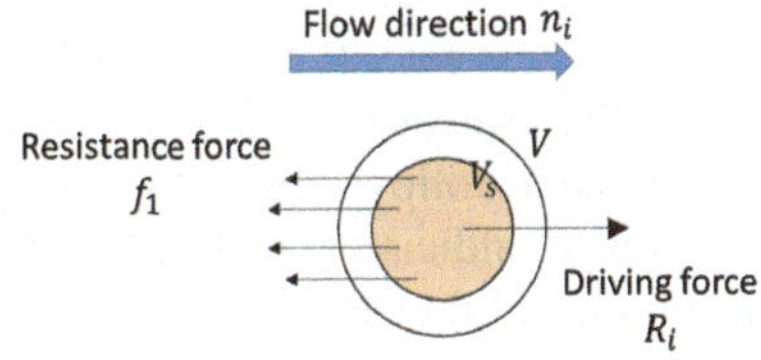

Fig. 1 Driving force and resistance force for the volume V of a set of soil particles in the direction of n_i

the existence of the gas, the initial fine content, clay content, chemical action etc. $S_{gr}/V=4n^S/D$ for the cylindrical volume, $S_{gr}/V=6n^S/D$ for the spherical volume; D is the diameter of soil particle.

For the spherical volume, Eq. (7) becomes

$$R_i n_i - f_1\left(\sigma_n^{'}, \varepsilon^p, S\right)\frac{6n^s}{D} \geq 0 \tag{8}$$

In the case that the permeability is proportional to the square of the diameter as $k=\tilde{k}D^2$, the erosion condition for the particle with the diameter D becomes

$$\therefore q(n_i, D) = q_i n_i - f_1\left(\sigma_n^{'}, \varepsilon^p, S\right)\tilde{k}D\frac{6n^s}{\rho^F g n^F} \geq 0 \tag{9}$$

where Eq. (3) is used. Equation (9) is the onset conditions for the erosion of the particles with the diameter of D in the n_i direction. The direction of the erosion is the direction n_i where $q(n_i, D)$ in Eq. (9) is maximum. When the resistance force is free from the direction, the erosion occurs in the direction of the flow. The resistance to erosion is assume to increase when the overall effective mean stress is large. On the other hand, a large shear strain is considered to have deteriorated the solid skeleton and therefore favored the erosion. These two effects are introduced in the function of the resistance f_1 as

$$f_1 = \frac{k^*}{\tilde{k}}\left(\frac{\sigma_m^{'}}{\sigma_{mr}^{'}}\right)^{m_1}\exp(-m_2\gamma^P) \tag{10}$$

where k^*, m_1, m_2, $\sigma_{mr}^{'}$ are the material parameters. $\sigma_m^{'}$ is the mean effective stress, γ^P is the second invariant of the plastic deviatoric strain. In the present study, $m_1=m_2=0$ is assumed since we do not have enough data to determine.

The rate equation of the erosion of the particle with a diameter of D is given by

$$\dot{\rho}^{FS}(D) = -\dot{\rho}^S(D) = \langle f_2(q(D))\rangle\frac{M(D)-RM_0(D)}{M_T} \tag{11}$$

$$\begin{cases} \langle f_2(q(D))\rangle = 0 & for\ q<0 \\ \langle f_2(q(D))\rangle = f_2(q(D)) = \alpha_{er}(q-q_{cr}) & for\ q\geq 0 \end{cases} \tag{12}$$

where $\dot{\rho}^{FS}(D)(=-\dot{\rho}^S(D))$ is the rate of mass erosion per unit time and unit volume of the particle D, M_T is the total mass of particles, $M(D)$ is the mass of the particle with a diameter of D, $M_0(D)$ is the initial mass of the particle. α_{er} is the coefficient of the erosion rate, R is a material parameter of the ratio of the remaining particles, q_{cr} is the critical value of flow rate for erosion. Equations (11) and (12) can be explained as follows. First, the erosion rate can be non zero only if the criterion for erosion as indication in Eq. (9) via the erosion force q is satisfied. Second the magnitude of the erosion is made function of the erosion force q via the function f_2.

The second factor in Eq. (11) accounts for the particles that have already been eroded.

The governing equations are given by the above equations with the constitutive equation of soil skeleton. The constitutive equation is of the elasto-viscoplastic type. The void ratio dependency of the permeability is assumed to be given by Kozeny-Carman equation. The governing equations are solved by FEM and the unknowns are the displacement of soil skeleton, pore fluid pressure and the concentration c of the eroded soil particles defined by $c = n^{FS}/n^{F}$.

4 Numerical Analysis

In order to validate the analysis method, we have simulated the experiments of the internal erosion tests for the gap-graded sands of the mixture of silica sand No. 3 and No. 8 with a downward flow conducted by [1]. Figure 2 shows the analysis model. The grain size distribution curves before and after the erosion are shown in Fig. 3, and the simulated and the experimental results of the eroded soil volume for the different values of the coefficient α_{er} of the erosion rate is illustrated in Fig. 4. The suffusion (the washing out of the fine) of the fine sand (silica sand No. 8) is well reproduced.

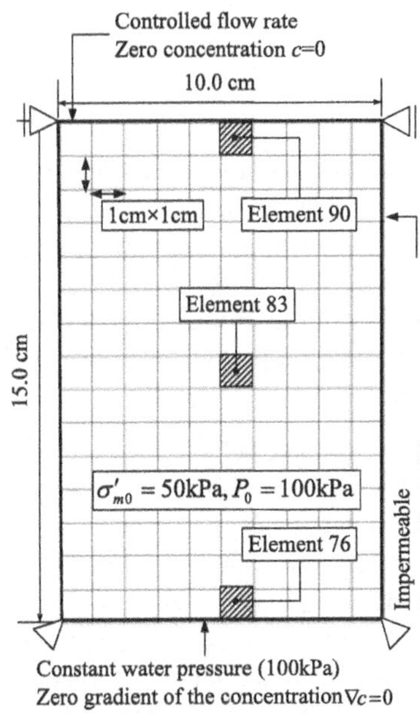

Fig. 2 Finite element mesh and boundary conditions

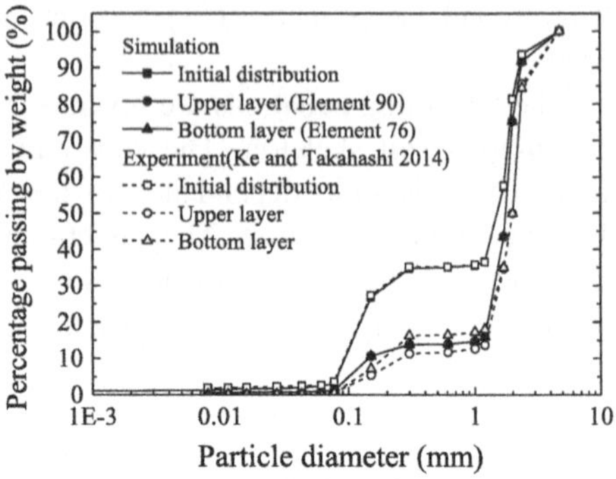

Fig. 3 Grain size distribution before and after erosion

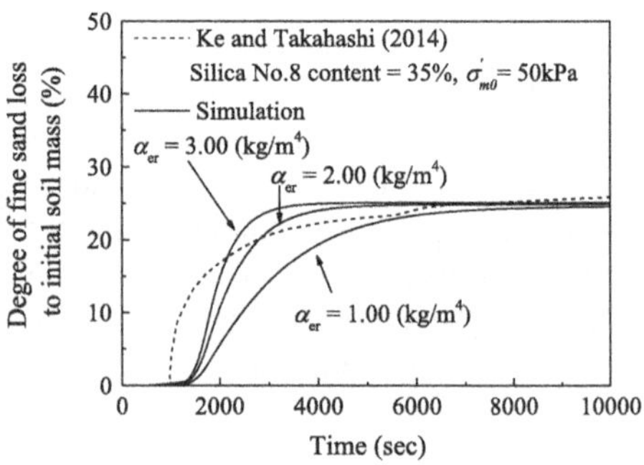

Fig. 4 Simulated and experimental results of the eroded soil

5 Conclusions

In the present study, the constitutive equations of the internal erosion, that is, the erosion criterion and the rate equation of the mass transfer are formulated. Then, the field equations to simulate hydro-mechanical behavior due to the internal erosion were derived in the framework of multiphase mixture theory [2, 3]. In addition, validity of the proposed model was discussed with respect to the rate of eroded soil mass and the particle size distribution after the erosion test. The main conclusions obtained are as follows:

1. A coupled behavior of soil-water interaction and the internal erosion followed by the transportation of eroded particles can be simulated by the proposed model. Change of grain size distribution due to flow out of fine particles observed in the experiments was well reproduced.
2. The simulated mass of eroded soil was compared to that obtained in the experiments. The tendency that the rate of eroded soil increases with increasing flow rate and decreases to zero was reproduced, even though the behavior at an early stage was slightly different from the one obtained in the experiments.

References

1. Ke, L., Takahashi, A.: Experimental investigations on suffusion characteristics and its mechanical consequences on saturated cohesionless soil. Soils Found. **54**(4), 713–730 (2014)
2. Loret, B.: Sand production during hydrate dissociation: constitutive and field equations, Note, 30 Aug 2015
3. Oka, F., Kimoto, S.: Computational modeling of multiphase geomaterials. Taylor & Francis, CRC Press (2012)

Chemically Induced Strain Localization in Geomaterials

Ioannis Stefanou and Jean Sulem

Abstract Deformation bands play an important role in reservoir engineering, geological storage, underwater landslides and slow geological procedures related to creep and aseismic slip. Various mechanisms can be involved at different scales and may be responsible for deformation bands. Mechanical and chemical degradation of the grain skeleton is a softening mechanism that leads to compaction, shear or even dilation band formation. The present study is twofold. On one hand it focuses on the mathematical modeling of chemically induced strain localization instabilities in porous rocks and on the other hand it explores the conditions for their creation. The post localization regime is then studied by numerically integrating the governing equations of the system.

1 Introduction

In a recent paper [1], instabilities in the form of deformation bands triggered by chemical degradation of the solid skeleton have been studied. Chemical dissolution and grain breakage have been considered. The interest of the approach is in the strong chemo-poro-mechanical coupling which was considered. As the stresses and the deformations evolve, the grains of the material break leading to an increase of their specific surface. As the dissolution rate depends upon the area of contact between the reactive fluid and the minerals, dissolution is accelerated by grain fracturing and grain breakage and chemical softening is further enhanced. All these phenomena are observed macroscopically as an overall creep behavior, However, strain localization can occur, affecting drastically the strength and the permeability of the rock formation.

I. Stefanou (✉) · J. Sulem
Laboratoire Navier, CNRS (UMR 8205), ENPC, IFSTTAR, Université Paris-Est, Champs-sur-Marne, France
e-mail: ioannis.stefanou@enpc.fr

E. Papamichos et al. (eds.), *Bifurcation and Degradation of Geomaterials with Engineering Applications*, Springer Series in Geomechanics and Geoengineering, DOI 10.1007/978-3-319-56397-8_17

The effect of chemical dissolution is important in field and in reservoir applications. For instance, the experimental results of [2] showed that the chemical dissolution of a limestone leads to a significant increase of the porosity (from 23% for the intact rock to 27% for the degraded one). According to the same authors, the plastic pore collapse threshold is also reduced and the chemically degraded materials become more collapsible and more ductile due to the increase in porosity and the degradation of the inter-granular cementation. This evidence is corroborated by other authors (e.g. [3–6]) for a class of geomaterials and results in a contraction of the elastic domain only due to chemical reasons (chemical softening). In parallel, in a saturated porous geomaterial, the progressive mechanical damage of the solid skeleton during compaction has as a result the increase of the interface area of the reactants (i.e. of the solution with the solid) and consequently the acceleration of the dissolution rate of the solid phase [7]. Thus, the solid skeleton is degraded more rapidly (mass removal because of dissolution), the overall mechanical properties of the system diminish (contraction of the elastic domain—chemical softening), deformations increase and the solid skeleton is further damaged (intergranular fractures, debonding, breakage of the porous network etc.). Figure 1 schematically shows this positive feedback process, whose stability is not guaranteed. Notice that chemical softening is central for compaction banding in the absence of other softening mechanisms such as mechanical softening.

A two-scale approach was proposed in [1] to model this strongly coupled system. In particular two scales are distinguished in this model. The first one is the micro-scale, where the reaction kinetics and the variation of the specific surface of the solid skeleton constituents due to breakage are described. These quantities are then upscaled to the macro-scopic scale (here defined as the scale of Representative Volume Element—RVE), where the balance and constitutive equations are written. Due to the heterogeneity of the microstructure (e.g. different grain sizes and constituents in the RVE) the dissolution rate is not homogeneous over the RVE. The size of the RVE is a finite statistical quantity that depends upon the geomaterial at hand. The chemical softening rate of the yield surface is a macro-scopic quantity that is related to the average, over the RVE, reaction rate at the grain level. This average procedure (upscaling) naturally introduces a characteristic length (size of the RVE). This approach is inspired from the development of non-local continuum theories. Details of the analysis can be found in [1].

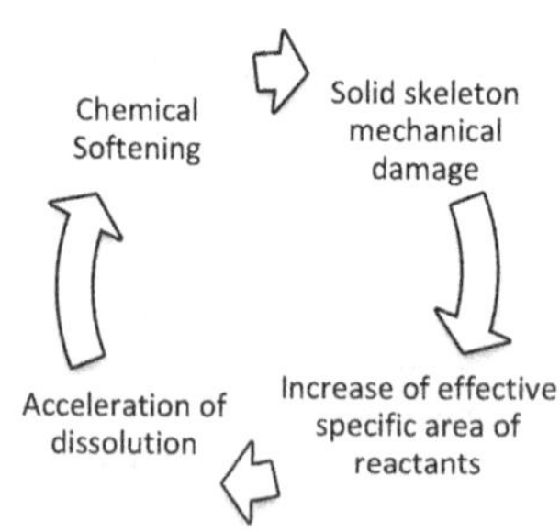

Fig. 1 Positive feedback process due to dissolution and solid skeleton damage (e.g. intergranular fracturing, breakage of the porous network, matrix cracking, grain-matrix debonding etc.)

2 Compaction Banding in Oedometric Compression

In the following we refer to the main results obtained in [1, 8] for oedometric compaction in a carbonate reservoir at 1.8 km depth because of CO_2 injection. The set of reactions that take place because of CO_2 injection is of dissolution type. The conditions for compaction band formation are derived mathematically and investigated in the aforementioned papers. It is shown that there is a region in the $q-p'$ plane where compaction band formation is possible. In other words in this region homogeneous deformation is unstable and the system bifurcates to non-homogeneous solutions corresponding to compaction banding. Figure 2d shows the instability zone for typical chemo-poro-mechanical parameters of the carbonate grainstone considered. Inside the instability region there exists a minimum critical wavelength above which perturbations are unstable (positive growth coefficient). This critical wavelength is related to the thickness of the localization zone and is expressed as a function of the characteristic internal length, ℓ_c. If $\ell_c = 0$ then the system is unstable for any perturbation wave length and the compaction band degenerates to a mathematical plane, as in the classical Cauchy continuum (local formulation).

Initially we assume that the material is in a state of elastic deformation (Point A, Fig. 2d) under the applied total vertical stress (45 MPa). At time $t = 0$, the injection of the CO_2 solution starts. It is assumed that the CO_2 solution is continuously renewed in such a way that practically open flow conditions hold. In field, CO_2

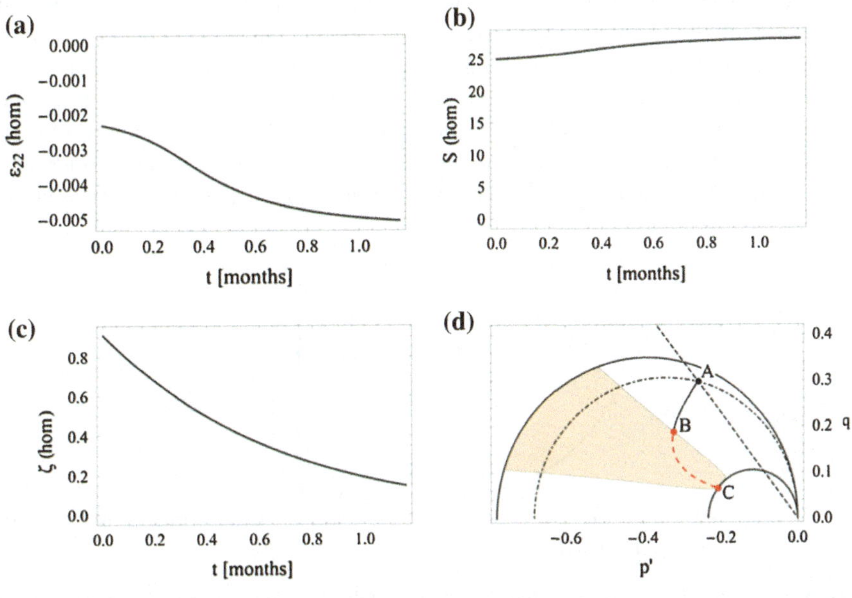

Fig. 2 Evolution in time of **a** deformation, **b** the specific surface, **c** the chemical softening parameter and **d** of the stress path during imposed homogeneous dissolution of a specimen under oedometric conditions

injection open flow conditions would correspond to a zone outside the gas plume, where the formation fluid is saturated with CO_2, but is not in chemical equilibrium with the rock so that carbonate dissolution occurs continuously. Rohmer and Seyedi [9] show that the dissolution front in a reservoir might extend few kilometers around the injection well after 10 years of continuous injection.

The material is progressively degraded due to chemical softening. When the elastic domain envelope reaches point A the material yields, plastic strains are accumulated and solid skeleton damage occurs. This phase of deformation under constant applied loading (i.e. the overburden) corresponds to the creep behavior that is observed due to CO_2 injection [10–12] (see Fig. 2a–c). When the stress state enters to the shaded area (Fig. 2d) homogeneous deformations become unstable and compaction bands are possible.

We emphasize that the bifurcation analysis only gives the conditions for compaction band triggering. In order to assess the evolution of the system and the gradual strain localization inside the band one has to study the post-bifurcation regime. This was performed numerically [8]. When the system is left free to develop non-homogeneous deformations, strain localization in the form of compaction bands can occur. At the beginning of CO_2 injection, the system creeps from point A to point B. No compaction band is triggered and the system behaves as in the case of homogeneous deformation (Fig. 2). Once the stress path crosses the shaded area, non-homogeneous deformations start to grow and a compaction band forms. Figure 3 (left) shows the profile of the vertical deformation at various times. The deformations localize into a narrow band whose thickness depends on the characteristic internal length, ℓ_c. Here we chose $\ell_c = 4$ mm (~20 grains for a typical grain size of 200 μm). In Fig. 3 (right) we present the average vertical (axial, oedometric) deformation over the entire sample, the vertical deformation at the peak of the compaction band and the vertical deformation of a point far from the localization zone. The vertical deformation at the peak of the compaction band is 14 times larger than the vertical deformation far from it. A small initial perturbation was introduced in the numerical system in order to trigger a compaction band in the

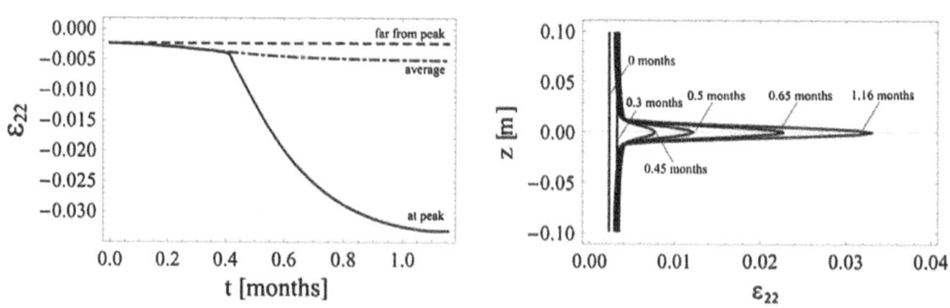

Fig. 3 *Left* Vertical deformation in function of time: at the peak of the compaction band, average over the entire sample and at a point far from the localization zone. *Right* Profile of the vertical deformation at various times

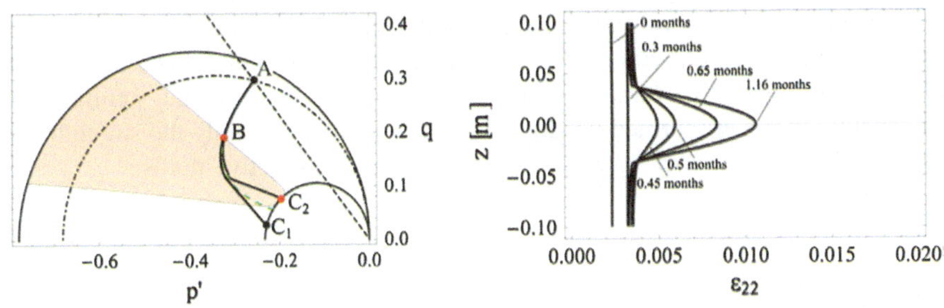

Fig. 4 *Left* Stress paths of the point at the peak of the compaction band (ABC_2) and of a point far from the localization zone (ABC_1). The *dashed line* corresponds to the stress path of the homogeneous deformation (see Fig. 2d). *Right* Profile of the vertical deformation at various times for $\ell_c = 16$ mm (80 grains). The deformations localize into a band which is thicker than in the case of $\ell_c = 4$ mm (20 grains, see Fig. 3)

middle of the sample. Otherwise, tiny numerical errors would cause compaction band triggering anywhere in the sample.

Figure 4 (left) shows the stress path of the point at the peak of the compaction band and the stress path of a point far from the localization zone. In the beginning the stress paths coincide, but after entering in the instability zone they start to diverge due to compaction band formation. It is worth emphasizing that compaction band thickness depends on the chosen characteristic length. This is shown mathematically in [1] and it is corroborated numerically by choosing a different characteristic length, $\ell_c = 16$ mm (~80 grains) in Fig. 4 (right). Finally, the smoothness of the peak of the compaction band is dictated by the internal length, ℓ_c.

3 Conclusions

A strong chemo-mechanical coupling was considered for studying the behavior of rocks under chemically reactive fluids: the material softens in due course of dissolution mechanisms whereas the reaction accelerates with increasing damage (because of the increase of the specific surface of the grain and thus of the area of contact between the reactive fluid and the minerals). A two-scale approach has been proposed in order to account for the heterogeneity of the dissolution process over the RVE. This naturally introduces a 'chemical' material length related to the non-local character of the relation between the softening rate of the yield surface (at the macro-scale) and the dissolution process (at the grain scale, i.e. at the micro-scale). The introduction of this material length is crucial to limit the compaction band thickness to a finite value.

Post-localization robust computations have been performed to simulate the progressive evolution of compaction band under open flow conditions. It is shown that even though the overall deformation creep does not change significantly, compaction bands are formed locally, which can alter significantly the strength and the permeability of a rock formation under injection of reactive fluids.

References

1. Stefanou, I., Sulem, J.: Chemically induced compaction bands: triggering conditions and band thickness. J. Geophys. Res. Solid Earth **119**(2), 880–899 (2014)
2. Xie, S.Y., Shao, J.F., Xu, W.Y.: Influences of chemical degradation on mechanical behaviour of a limestone. Int. J. Rock Mech. Min. Sci. **48**(5), 741–747 (2011)
3. Nova, R., Castellanza, R., Tamagnini, C.: A constitutive model for bonded geomaterials subject to mechanical and/or chemical degradation. Int. J. Numer. Anal. Methods Geomech. **27**(9), 705–732 (2003)
4. Hu, L.-B., Hueckel, T.: Coupled chemo-mechanics of intergranular contact: toward a three-scale model. Comput. Geotech. **34**(4), 306–327 (2007)
5. Ciantia, M.O., Castellanza, R., di Prisco, C.: Experimental study on the water-induced weakening of calcarenites. Rock Mech. Rock Eng. 441–461 (2014)
6. Hu, L.-B.B., Hueckel, T.: Creep of saturated materials as a chemically enhanced rate-dependent damage process. Int. J. Numer. Anal. Methods Geomech. **31**(14), 1537–1565 (2007)
7. Rimstidt, J.D., Barnes, H.L.: The kinetics of silica-water reactions. Geochim. Cosmochim. Acta **44**(11), 1683–1699 (1980)
8. Sulem, J., Stefanou, I.: Thermal and chemical effects in shear and compaction bands. Geomech. Energy Environ. (2016)
9. Rohmer, J., Seyedi, D.M.: Coupled large scale hydromechanical modelling for caprock failure risk assessment of CO2 storage in deep saline aquifers. Oil Gas Sci. Technol.—Rev. l'Institut Français du Pétrole **65**(3), 503–517 (2010)
10. Le Guen, Y., Renard, F., Hellmann, R., Brosse, E., Collombet, M., Tisserand, D., Gratier, J.-P.: Enhanced deformation of limestone and sandstone in the presence of high Pco2 fluids. J. Geophys. Res. **112**(B5), B05421 (2007)
11. Liteanu, E., Spiers, C.J.: Influence of pore fluid salt content on compaction creep of calcite aggregates in the presence of supercritical CO2. Chem. Geol. **265**(1–2), 134–147 (2009)
12. Rutqvist, J.: The geomechanics of CO2 storage in deep sedimentary formations. Geotech. Geol. Eng. **30**(3), 525–551 (2012)

Unstable Creeping in Geomaterials

Claudio di Prisco and Federico Pisanò

Abstract Time plays a dominant role in many areas of geomechanics, such as in the assessment of landslide hazard and territorial vulnerability. In this context, the reliable description of the temporal evolution of both soil mass movements and their dependence on loading rate effects is deemed crucial. From a theoretical standpoint, the time-dependent behaviour of geomaterials can be described by means of viscoplastic constitutive theories, also beneficial to foster the objectivity of numerical results from strain localisation simulations. However, not all the aspects of viscoplastic modelling have been so far clearly acknowledged in the literature. For this reason, in this paper the authors focus on creeping strain localization phenomena under simple shear conditions, discussing, from a theoretical point of view, the mechanical condition to be satisfied for tertiary creep to occur.

1 Introduction

Traditionally, time effects have received less attention in the study of strain localisation problems, although they can affect significantly both triggering and evolution of many geohazards. The present paper aims at highlighting theoretically the influence of both viscosity and loading rate on the strain localization of geomaterils. A suitable tool to investigate time effects in geomaterials is the well-known theory of viscoplasicity, based on the concept of "delayed plastic flow". From a physical standpoint, delayed plasticity embodies in constitutive equations the micro-inertiae of soil particles (in the case of granular materials), as well as irreversible phenomena occurring at the contacts between grains (grain damage induced by indentation, grain crushing, etc.). The approaches available to model the rate-dependency in

C. di Prisco (✉)
Politecnico di Milano, Piazza Leonardo da Vinci 32, 20133 Milano, Italy
e-mail: claudio.diprisco@polimi.it

F. Pisanò
Delft University of Technology, Stevinweg 1, 2628 CN Delft, The Netherlands

E. Papamichos et al. (eds.), *Bifurcation and Degradation of Geomaterials with Engineering Applications*, Springer Series in Geomechanics and Geoengineering, DOI 10.1007/978-3-319-56397-8_18

geomaterials can be classified in two major groups: those allowing the stress state to lie outside the yield surface [1, 2] and those relying on time- or rate-dependent hardening rules (e.g. consistency viscoplasticity—[3–6]). Here, the first-type approach proposed by Perzyna is considered for two reasons: (i) it has been proved successful in capturing the rate-dependence of both fine- and coarse-grained soils [7–10]; (ii) it enables a straightforward convergence to rate-independent elastoplasticity in presence of slow loading processes. Further, viscoplastic models not only enable to capture certain experimental evidences, but also work as a simple regularization technique in the continuum-based simulation of strain localisation [3, 11–13].

In this paper, the interplay of material rate-dependence (viscosity) and loading rate in shear strain localisation problems is critically discussed. In particular, the authors put in evidence, from a theoretical point of view, the equivalence between the theoretical conditions for strain localisation in elasto-plastic solids and unstable/tertiary creep in rate-sensitive materials. The analysis is developed in the framework of the "theory of controllability" and applied to one-dimensional slope stability problems.

2 Theory of Viscoplasticity

Standard Perzyna's viscoplasticity—Perzyna's viscoplastic theory relies upon the assumption of additive reversible (elastic) and unrecoverable (viscoplastic) strain rates. In this framework, viscoplastic strain rates are obtained as [1]:

$$\dot{\boldsymbol{\varepsilon}}^{vp} = \Phi(f)\frac{\partial g}{\partial \boldsymbol{\sigma}} \tag{1}$$

According to Eq. (1), the scalar Φ function (the so-called "viscous nucleus") determines the magnitude of the viscoplastic strain rate tensor, while its direction in the strain rate space is given by the stress gradient of the plastic potential function g. The viscous nucleus function is most commonly assumed to be bilinear and governed by the fluidity parameter. As no plastic consistency needs to be enforced, the stress state is not constrained to lie on the yield locus during plastic loading stages (whence the term "overstress plasticity"). While the viscous nucleus Φ must be a non-negative non-decreasing function of the yield function f, rate-independent plasticity is recovered as the limit of viscoplasticity at vanishing rate-sensitiveness (or infinitely slow loading). It could also be proven that $\Phi \rightarrow \infty$ (infinite plastic strain rate) and $f \rightarrow 0$ (consistency satisfied) as the elasto-plastic limit is approached. Although different formulations are possible, it could be proven that bilinear definition for Φ, imposing $\Phi = 0$ for $f \leq 0$, ensures a viscoplastic response tending to the elasto-plastic limit at vanishing rate-sensitiveness.

In general, we have also to state that, when strain localization is concerned:

- material internal length and characteristic time should both be present in the continuum analysis of geomechanical problems;
- depending on the type of material and loading conditions, the micro-mechanical processes underlying the occurrence of strain localisation are governed by the competitive interplay of spatial and temporal factors;
- in non-local models for geomaterials the relationship between material micro-structure and localisation thickness can be well-defined through non-local material parameters, whereas, when the time factor is also considered via viscoplastic modelling, the material response (and the shear band thickness) is affected by at least three factors at the same time: (i) the analytical definitions of both the yield function and the viscous nucleus, (ii) the values of the calibrated viscous parameters, (iii) the loading rate;
- even in presence of viscoplastic laws, the common lack of non-local enhancement implies the vanishing of the simulated shear band thickness as the loading rate tends to zero—regardless of the viscous nucleus definition.

3 Stability Analysis for Viscoplastic Constitutive Laws

The relation between strain localisation in elastic-plastic media and tertiary creep in elastic visco-plastic continua is theoretically discussed in what follows. For this purpose, the authors resort to the so-called "theory of controllability" [14], originally introduced by Nova et al. [15] to analyse the constitutive response of single potential elastoplastic relationships. Nova's work has been recently reinterpreted in the light of the Hill's stability criterion [16], and extended by this paper's authors to cope with elasto-viscoplastic materials under general loading programmes [17]. A practical application of the viscoplastic controllability theory to the analysis of creep-induced liquefaction in very loose sands is reported in [18]. The same theory is employed hereafter to investigate the inception of tertiary creeps under simple shear (SS) loading conditions, idealising the mechanics of very elongated soil layers (infinite slopes). The stability analysis of ideal infinite slopes is interesting in many theoretical respects. The intrinsic symmetric/kinematic features of the system imply that localisation failure may only occurs in the form of a shear band oriented along the sloping direction (Fig. 1).

As was first discussed in [19] and then in [20], localisation/instability under SS conditions takes place when:

$$D_{11}D_{44} - D_{14}D_{41} < 0 \tag{2}$$

where D_{11}, D_{44}, D_{14} and D_{41} are entries of the incremental elasto-plastic stiffness matrix (subscripts 1 and 4 identify direct stress/strain components normal to the sloping direction and shear components along the slope, respectively). In the case of

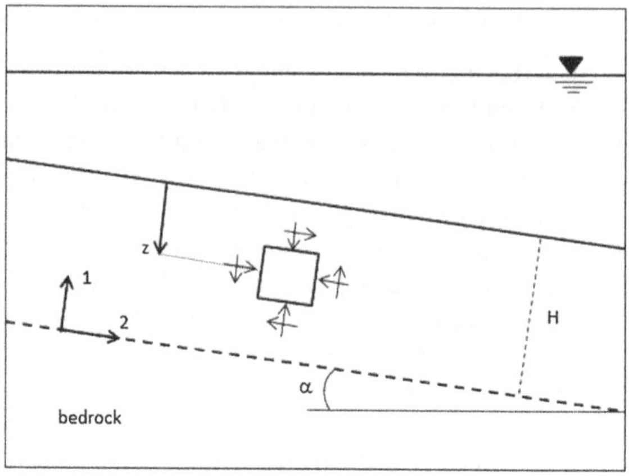

Fig. 1 Geometrical sketch of the infinite slope idealisation

single potential elasto-plastic materials, inequality (2) can be proven to correspond to the following condition on the hardening modulus H:

$$H < H_L = -D_{44}^{el} \frac{\partial f}{\partial \sigma_2} \frac{\partial g}{\partial \sigma_2} \tag{3}$$

where σ_2 is the along-slope normal stress, H_L the instability index defined in [16] and evaluated under SS conditions, D_{44} the elastic shear stiffness, f and g the yield function and the plastic potential, respectively. In [20] the implications of non-associativeness are considered in detail and it is shown that, even in case of perfect plasticity (i.e. $H = 0$), inequality (3) may be fulfilled under certain conditions. In particular, when H = 0, loaclization takes place when either $\frac{\partial g}{\partial \sigma_2}$ or $\frac{\partial f}{\partial \sigma_2}$ are positive.

Pisanò and di Prisco [17] extended to most general loading conditions the concept of tertiary creep and demonstrated that, in the case of creep tests, viscoplastic instability stems from the positiveness of at least one eigenvalue in the following matrix's spectrum:

$$A = \begin{bmatrix} \mathbf{A}_{\alpha\alpha} & \mathbf{A}_{\alpha\beta} \\ \mathbf{A}_{\beta\alpha} & \mathbf{A}_{\beta\beta} \end{bmatrix}, \tag{4}$$

where under SS conditions the single terms become:

$$A_{\alpha\alpha} = -\frac{\partial \Phi}{\partial f}(H - H_L) I_{\alpha\alpha} \tag{5}$$

$$A_{\beta\beta} = -\frac{\partial \Phi}{\partial f}(H - H_L) I_{\beta\beta} - \Phi D_{22} \frac{\partial^2 g}{\partial^2 \sigma_{\beta\beta}} \tag{6}$$

$$A_{\beta\alpha} = 0 \tag{7}$$

$$A_{\alpha\beta} = \Phi\left(\frac{\partial^2 g}{\partial\sigma_\alpha \otimes \partial\sigma_\beta} - C_{\alpha\beta} C_{\beta\beta}^{-1} \frac{\partial^2 g}{\partial\sigma_\beta \otimes \partial\sigma_\beta}\right) \tag{8}$$

in which subscripts α and β indicate controlled stress and strain components (in the case here considered $\alpha = 1, 4$, whereas $\beta = 2, 3$) respectively. In fact, normal and shear stresses are controlled, whereas along the slope and out- of-plane normal strains are imposed to be nil. Moreover, C^{el} is the elastic compliance matrix ($C_{\alpha\beta}$ the partition implied by the specific loading programme), whereas $I_{\beta\beta}$ and $I_{\alpha\alpha}$ are identity matrices and D_{22} stands for the along the slope elastic stiffness.

In Eqs. (5) and (6) the derivative of the viscous nucleus with respect to the yield function is positive by definition, and positive is also the second derivative of the plastic potential with respect to the stress tensor because of its convexity. Thus, Eqs. (5) and (6) impliy that (i) instability takes place when $H = H_L$ and (ii) depending on the current Φ value, the relevant shear strain component accelerates either almost simultaneously or much before than the acceleration of the stresses along the tangential and the out of plane directions, since the second term right of Eq. (6) is strictly negative. It is worth noting that the localisation condition for elasto-plastic constitutive relationships coincides with the condition for unstable creep in rate-sensitive materials. Nevertheless, this by no means implies that instability is expected at the same stress level in the two situations. In fact, both H and H_L in the elasto-viscoplastic case are largely influenced by the time factor, that is by the previous loading and loading rate history.

4 Concluding Remarks

In this paper, the relation between viscoplasticity and strain localisation in geo-materials has been tackled by considering creeping phenomena under simple shear conditions. The use of elasto-viscoplastic constitutive relationships enables the analysis of time effects in localisation processes. For this purpose, Perzyna-type viscoplastic models have been exclusively considered due to their wide popularity in literature and in numerical applications. According to Perzyna's approach, the increment in irreversible strains is governed by the viscous nucleus, which is a function of the yield function and whose definition deeply affects the mechanical response of the material and the inception of instability/localisation.

Under simple shear conditions, the authors have demonstrated the coincidence of the condition for localization to occur, valid for elastic-plastic single potential media, with the condition for unstable creep in viscoplastic Perzyna's type materials to take place.

Although the two conditions are perfectly coincident, both the previous loading rate and the fluidity parameter (or, alternatively, the viscous nucleus definition) severely affects quantitatively the mechanical response of the system, since all the variablesgoverning the instability occurrence (both the current hardening modulus and the instability index) vary with time and are influenced by the rate sensitiveness of the constitutive rule. In particular, this analysis justifies the transition, during the evolution of time, from a stable to an unstable creep response, even in a system where, from a macroscopic point of view, any spatial evolution of the damaged zone is absent.

References

1. Perzyna, P.: Fundamental problems in viscoplasticity. Adv. Appl. Mech. **9**, 243–377 (1966)
2. Duvaut, G., Lions, J.L.: Les inequations en mechanique et en physique. Dunod, Paris (1972)
3. Wang, W.M., Sluys, L.J., de Borst, R.: Viscoplasticity for instabilities due to strain softening and strain-rate softening. Int. J. Numer. Methods Eng. **40**(20), 3839–3864 (1997)
4. Runesson, K., Ristinmaa, M., Mähler, L.: A comparison of viscoplasticity formats and algorithms. Mech. Cohesive-frictional Mater. **4**(1), 75–98 (1999)
5. Heeres, O.M., Suiker, A.S., de Borst, R.: A comparison between the perzyna viscoplastic model and the consistency viscoplastic model. Eur. J. Mech.—A/Solids **21**(1), 1–12 (2002)
6. Carosio, A., Willam, K., Etse, G.: On the consistency of viscoplastic formulations. Int. J. Solids Struct. **37**(48–50), 7349–7369 (2000)
7. di Prisco, C., Imposimato, S., Vardoulakis, I.: Mechanical modelling of drained creep triaxial tests on loose sand. Géotechnique **50**(1), 73–82 (2000)
8. di Prisco, C., Imposimato, S.: Static liquefaction of a saturated loose sand stratum. Int. J. Solids Struct. **39**, 3523–3541 (2002)
9. Borja, R.I., White, J.A.: Continuum deformation and stability analyses of a steep hillside slope under rainfall infiltration. Acta Geotech. **5**(1), 1–14 (2010)
10. Yin, Z.-Y., Chang, C.S., Karstunen, M., Hicher, P.-Y.: An anisotropic elastic–viscoplastic model for soft clays. Int. J. Solids Struct. **47**(5), 665–677 (2010)
11. Needleman, A.: Material rate dependence and mesh sensitivity in localisation problems. Comput. Methods Appl. Mech. Eng. Arch. **67**(1), 69–85 (1988)
12. Loret, B., Prevost, J.: Dynamic strain localisation in elasto-(visco-)plastic solids. Comput. Methods Appl. Mech. Eng. **83**(3), 247–273 (1990)
13. Loret, B., Prevost, J.: On the existence of solutions in layered elasto-(visco-)plastic solids with negative hardening. Eur. J. Mech. A/Solids **10**, 575–586 (1991)
14. Imposimato, S., Nova, R.: An investigation on the uniqueness of the incremental response of elastoplastic models for virgin sand. Mech. Cohesive-Frictional Mater. **3**, 65–87 (1998)
15. Nova, R.: Controllability of the incremental response of soil specimens subjected to arbitrary loading programmes. J. Mech. Behav. Mater. **5**(2), 193–202 (1994)
16. Buscarnera, G., Dattola, G., di Prisco, C.: Controllability, uniqueness and existence of the incremental response: A mathematical criterion for elastoplastic constitutive laws. Int. J. Solids Struct. **48**(13), 1867–1878 (2011)
17. Pisanò, F., di Prisco, C.: A stability criterion for elastoviscoplastic constitutive relationships. Int. J. Numer. Anal. Methods Geomech. **40**(1), 141–156 (2016)
18. Marinelli, F., Pisanò, F., di Prisco, C., Buscarnera, G.: Mechanical interpretation of undrained creep instability in loose sands. Géotechnique, in review (2017)

19. di Prisco, C., Nova, R.: Stability problems related to static liquefaction of loose sand. In: Chambon, R., Desrues, J., Vardoulakis, I. (eds.) Proceedings of the 3rd International Workshop Localisation and Bifurcation Theory for Soils and Rocks, pp. 59–70. Rotterdam, The Netherlands, Balkema (1994)
20. di Prisco, C., Pisanò, F.: An exercise on slope stability and perfect elastoplasticity. Géotechnique **61**(11), 923–934 (2011)

The Effect of Rotational and Isotropic Hardening on the Onset of Compaction Bands

Chara Prassa, Sotiris Alevizos, Manolis Veveakis and Yannis F. Dafalias

Abstract Compaction bands are localized failure patterns that appear in highly porous rock material under the effect of relatively high confining pressure. Being affected mainly by volumetric compression, these bands appear to be almost perpendicular to the most compressive principal stress at a stress state at the so-called "cap" of the yield surface (Issen and Rudnicki, J Geoph Res 105:21529–21536 (2000) [4]). In this study we focus on the mechanism that leads to the onset of compaction bands by using a viscoplasticity model able to describe the post-localization response of these materials. The proposed constitutive framework is based on the overstress theory of Perzyna (Adv Appl Mech 9:243–377 (1966) [7]) and the anisotropic clay plasticity model of Dafalias (Mech Res Commun 13 (6):341–347 (1986) [1]) as modified by Dafalias and Taiebat (Geotechnique 63 (16):1406–1418 (2013) [2]) which provides not only the necessary "cap" of the yield surface, but introduces a rotational hardening mechanism thus taking into account possible anisotropic phenomena. Following the analysis of Veveakis and Regenauer-Lieb (J Mech Phys Solids 78:231–248 (2015) [8]) we identify the compaction bands as "static" cnoidal wave formations in the medium that occur at a post-yield regime and we study the effect of rotational and isotropic hardening on their onset. Moreover, we determine a theoretical lower limit of confining pressure in triaxial compression tests for the compaction bands to develop.

Keywords Compaction bands · Viscoplastic model · Anisotropy · Strain localization · Highly porous rock

C. Prassa (✉) · Y.F. Dafalias
Faculty of Applied Mathematical and Physical Sciences, Department of Mechanics, National Technical University of Athens, 15780 Zographou, Greece
e-mail: chprassa@gmail.com

S. Alevizos · M. Veveakis
School of Petroleum Engineering, UNSW, Kensington, NSW, Australia

Y.F. Dafalias
Department of Civil and Environmental Engineering, University of California, Davis, CA 95616, USA

E. Papamichos et al. (eds.), *Bifurcation and Degradation of Geomaterials with Engineering Applications*, Springer Series in Geomechanics and Geoengineering, DOI 10.1007/978-3-319-56397-8_19

1 Introduction

It is widely recognized that soft rocks display a highly viscous behavior. It is also well known that soft rocks with high porosity succumb to volumetric compression with the development of the so called compaction bands. Compaction bands are planar deformation zones that are oriented perpendicular to the maximum compressible principal stress. Such bands were first observed in the field by Mollema and Antonellini [5] in natural outcrops of sandstone. Oka et al. [6] performed conventional drained triaxial experiments on diatomaceous mudstone and observed that material fails by the formation of either shear bands and/or compaction bands, depending on the confining pressure under which the material is tested. Compaction band formation has been observed by increasing the confining pressure. An interesting finding is that compaction bands in contrast to shear bands are not solitary, but periodic. Issen and Rudnicki [4] admit solutions for compaction bands when the inelastic volumetric deformation is compactive and is associated with a 'cap' on the yield surface. In this study, the formation of compaction bands in highly porous rock material is demonstrated by applying an extension of the anisotropic clay plasticity model of Dafalias and Taiebat [2] to an overstress type of viscoplasticity. This model is suitable for the ensuing analysis because it combines the 'cap' property of a yield surface with the concept of anisotropy. More precisely, since we focus on the behavior of natural soil samples, we assume that the material exhibits a pre-existing anisotropy expressed by a rotated yield surface in stress space, which subsequently can evolve by rotational hardening as the anisotropy evolves during loading. This evolution could explain the observed sharp transition in the localized failure mode from shear bands to compaction bands with the increase of confining pressure. The high viscosity of soft rock is accounted for by adopting an overstress approach based on the concept of Perzyna's overstress theory in conjunction with the yield surface of Dafalias and Taiebat [2]. The viscoplastic constitutive framework was adopted in order to describe the post-localization response of the material.

2 Constitutive Modeling

The anisotropic clay plasticity model of Dafalias and Taiebat [2] (part of SANI-CLAY family) is extended to become viscoplastic of an overstress type. It addresses anisotropy due to fabric orientation of the particulate medium by introducing $\boldsymbol{\alpha}$, which is a dimensionless deviatoric stress-ratio-type tensor-valued variable defining the rotation of Plastic Potential Surface (PPS)/Yield Surface (YS), and whose evolution is defined by the adopted Rotational Hardening (RH) rule. The analysis is made for associative flow rule. The multiaxial stress space expression of SANI-CLAY PPS/YS is given by

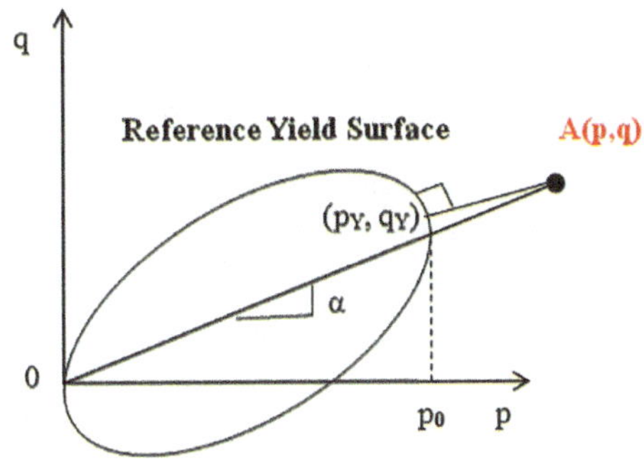

Fig. 1 Schematic diagram of minimum distance of actual stress state from the static yield surface

$$f = g = \frac{3}{2}(s - p\boldsymbol{\alpha}) : (s - p\boldsymbol{\alpha}) - \left[M^2 - \frac{3}{2}\boldsymbol{\alpha} : \boldsymbol{\alpha}\right] p(p_0 - p) = 0 \tag{1}$$

where M is the critical state stress ratio (possibly a function of Lode angle), $\boldsymbol{s}$ is the deviatoric part of the stress $\boldsymbol{\sigma}$ and p_O in triaxial space is the p-coordinate of the intersection of the line with slope $\boldsymbol{\alpha}$ and the YS, defining its size.

The additive decomposition of total strain-rate into elastic and viscoplastic parts is assumed such that $\dot{\varepsilon} = \dot{\varepsilon}^e + \dot{\varepsilon}^{vp}$, where a superposed dot implies the rate. The hypo-elastic strain and stress rate relations are given in terms of an elastic bulk modulus K and shear modulus G (obtained from K and Poisson's ratio ν), by $\dot{\varepsilon}_q^e = \dot{q}/3G, \dot{\varepsilon}_v^e = \dot{p}/K$. The viscoplastic model is derived from the well-known Perzyna's overstress theory [7] by assuming a YS that hardens, while the viscoplastic flow rule is given by a function of the overstress that represents the distance between the actual stress state (p, q) and the YS, measured perpendicularly to it (shortest distance), as shown in Fig. 1 in the triaxial p, q space. The adopted viscoplastic model introduces an original suggestion by proposing to use the two components of the aforementioned distance along p and q axes, as two separate overstress measures to be used for volumetric and deviatoric plastic strain rates, respectively, and define the corresponding overstress functions by:

$$\Phi_1 = \left(\frac{p - p_Y}{p_n}\right)^N, \Phi_2 = \left(\frac{q - q_Y}{q_n}\right)^N \tag{2}$$

where p_Y, q_Y are shown in Fig. 1 and represent the yield values of p and q respectively, and p_n, q_n are reference stresses. The power N of Eq. (4) defines the order of the Perzyna viscoplasticity and is assumed to be equal to $N = 3$. Hence, the viscoplastic flow rule yields the following expressions for the volumetric and deviatoric plastic strain rates in conjunction with Eq. (1) expressed in triaxial space:

$$\dot{\varepsilon}_v^{vp} = \mu H(f^*) \Phi_1 \frac{\partial g}{\partial p} = \mu H(f^*) \left(\frac{\overline{p}}{p_n}\right)^3 p_Y (M^2 - \eta^2) \tag{3a}$$

$$\dot{\varepsilon}_q^{vp} = \mu H(f^*)\Phi_2 \frac{\partial g}{\partial q} = \mu H(f^*)\left(\frac{\overline{q}}{q_n}\right)^3 2p_\gamma(\eta - \alpha) \tag{3b}$$

where $\eta = q/p$ and μ is the viscosity parameter. In this formulation, $\overline{p} = p - p_y$, $\overline{q} = q - q_y$ are the volumetric and the deviatoric components of the overstress and $(f^*) = \frac{3}{2}(s - p\boldsymbol{\alpha}):(s - p\boldsymbol{\alpha}) - \left[M^2 - \frac{3}{2}\boldsymbol{\alpha}:\boldsymbol{\alpha}\right]p(p_0 - p)$ which is the left hand side of Eq. (1), while H is the Heaviside step function that is defined as:

$$H(n) = \begin{cases} 0, & n \leq 0 \\ 1, & n > 0 \end{cases} \tag{4}$$

3 Method of Analysis

The post-localization response is described using the concept of viscoplasticity in order to regularize the equations. The stress equilibrium equations along with a simple power-law-type of overstress function are used in conjunction with the mass balance law following the analysis of Veveakis and Regenauer-Lieb [8].

3.1 Momentum Balance

The equilibrium equations in a 2-D plane strain setting are:

$$\frac{\partial \sigma_{xx}}{\partial x} + \frac{\partial \sigma_{yx}}{\partial y} = 0 \tag{5a}$$

$$\frac{\partial \sigma_{yy}}{\partial y} + \frac{\partial \sigma_{xy}}{\partial x} = 0 \tag{5b}$$

where a 2-D plane strain setting is assumed.

The Levy [3] transformation gives through the Mohr circle the relation between p, q, θ and the biaxial stresses as

$$\sigma_{xx} = p + q\sin 2\theta, \sigma_{yy} = p - q\sin 2\theta, \sigma_{xy} = \sigma_{yx} = q\cos 2\theta \tag{6}$$

where θ is the angle of the slip lines with respect to the principal stresses, $p = (1/2)(\sigma_{xx} + \sigma_{yy})$ and $q = \sigma_{xy}/\cos 2\theta$ are biaxial stresses.

3.2 Mass Balance

The total mass balance expression for any material allowing internal mass transfer is written as Veveakis and Regenauer-Lieb [8]

$$\frac{1}{K'}\frac{\partial \overline{p}}{\partial t} = D_{eff}\left[\frac{\partial^2 \overline{p}}{\partial x^2} + \frac{\partial^2 \overline{p}}{\partial y^2}\right] - \dot{\varepsilon}_n \left(\frac{\overline{p}}{p_n}\right)^3 \quad (7)$$

where p_n and $\dot{\varepsilon}_n = \mu p_Y (M^2 - \eta^2)$ are the boundary values for the applied overstress and volumetric strain rate, respectively. Furthermore, $1/K' = 1/K + \beta_m$, and $\beta_m = (1-\varphi)\beta_\alpha + \varphi\beta_b$ whereas, $D_{eff} = \varphi\beta_\varphi D$ is a generalized mass flow diffusivity, with units [m^2/Pa.s], with D the diffusion coefficient, representing any internal mass exchange process.

3.3 Generalized Slip-Line Field Theory

The equilibrium Eqs. (5a) and (5b) must first be solved using the adopted YS and acquiring the general form $q = q_Y(p)$, in order to determine the slip lines of the problem at hand which correspond to the directions where localized failure could occur [3]. If we assume that the initial YS does not harden, then the characteristics of the corresponding system quasi-linear set of p.d.e's (5a) and (5b) are,

$$\frac{h\sqrt{1-h^2}\cos 2\theta - (-1+h^2)\sin 2\theta}{h+\sin 2\theta} p - 2\frac{h\cos 2\theta + \sqrt{1-h^2}\sin 2\theta}{h+\sin 2\theta} q\theta = 0 \quad (8a)$$

$$\frac{\sin 2\theta - h\left(\sqrt{1-h^2}\cos 2\theta + h\sin 2\theta\right)}{h+\sin 2\theta} p + 2\frac{-h\cos 2\theta + \sqrt{1-h^2}\sin 2\theta}{h+\sin 2\theta} q\theta = 0 \quad (8b)$$

where $h = q_Y'(p)$ being a generalized pressure modulus, which is equal to the slope of the tangent to the yield surface. Notice that this system is hyperbolic if it has two real characteristic directions and turns into parabolic if there is one such direction. The latter corresponds to the limit $|h| \to 1$. We identify this limit as the one where pure compaction ($h = -1$) or pure dilation ($h = 1$) bands are formed.

3.4 Results

The mass balance equation (Eq. 7) was solved along the characteristics of the system, where yield occurs. Introducing the appropriate wave normalization, we have determined analytically the expressions for the minimum reference length for

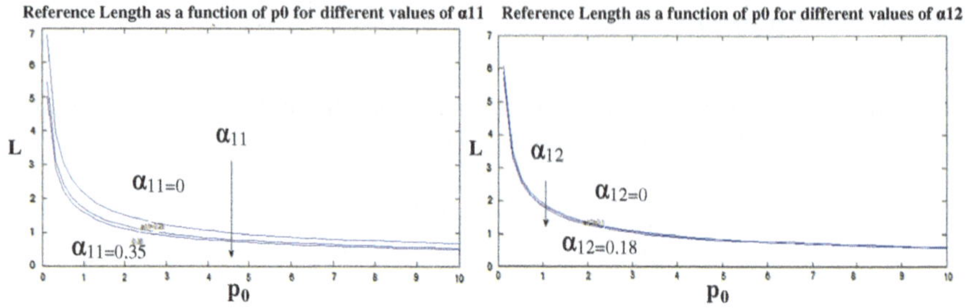

Fig. 2 The effect of the rotational anisotropy tensor components on the onset of compaction bands. With increasing preconsolidation pressure p_0 the required height of the specimen to obtain compaction bands is reduced. *Left* The effect of the diagonal components. Note that a_{22} and a_{11} have the same effects due to self-similarity. *Right* The effect of the off-diagonal component. We observe that it is negligible

the development of one set of compaction bands in terms of the tensor of anisotropy and the isotropic hardening. Our main finding is that higher anisotropy and isotropic hardening can promote the growth of compaction bands (Fig. 2).

4 Conclusions

Using the concept of overstress theory we developed a viscoplastic constitutive framework based on and the yield surface of Dafalias [1] and Dafalias and Taiebat [2]. Following the analysis of Veveakis and Regenauer-Lieb [8] we examined the effect of rotational and isotropic hardening on the onset of compaction bands within the foregoing constitutive setting.

Acknowledgements This research has received funding from the European Research Council under the European Union's Seventh Framework Program FP7-ERC-IDEAS Advanced Grant Agreement n° 290963 (SOMEF).

References

1. Dafalias, Y.F.: An anisotropic critical state soil plasticity model. Mech. Res. Commun. **13**(6), 341–347 (1986)
2. Dafalias, Y.F., Taiebat, M.: Anatomy of rotational hardening in clay plasticity. Geotechnique **63**(16), 1406–1418 (2013)
3. Hill, R.: The Mathematical Theory of Plasticity. Oxford University Press, London (1950)
4. Issen, K., Rudnicki, J.: Conditions for compaction bands in porous rock. J. Geoph. Res. **105**, 21529–21536 (2000)
5. Mollema, P.N., Antonellini, M.A.: Compaction bands: a structural analog for anti-mode I cracks in Aeolian sandstone. Tectonophysics **267**, 209–228 (1996)

6. Oka, F., Kimoto, S., Higo, Y., Ohta, H., Sanagawa, T., Kodaka, T.: An elasto-viscoplastic model for diatomaceous mudstone and numerical simulation of compaction bands. Int. J. Numer. Anal. Meth. Geomech. **35**, 244–263 (2011)
7. Perzyna, P.: Fundamental problems in viscoplasticity. Adv. Appl. Mech. **9**, 243–377 (1966)
8. Veveakis, E., Regenauer-Lieb, K.: Cnoidal waves in solids. J. Mech. Phys. Solids **78**, 231–248 (2015)

Localisation of Deformation for Shearing of a Fault Gouge with Cosserat Microstructure and Different Couplings

H. Rattez, I. Stefanou, J. Sulem, M. Veveakis and T. Poulet

Abstract In this paper, we show the impact of Thermo-Hydro Mechanical couplings (THM) on the stability of a saturated fault gouge under shear. By resorting to Cosserat continuum mechanics, that allows to take into account rotational degrees of freedom, we regularize the problem of localisation and we predict the thickness of a shear band. A linear stability analysis of the homogeneous state is performed and then the system of equations is integrated using a Finite Element (FE) analysis. These analyses can be used for studying the evolution of the thickness of the principal slip zone in a fault under undrained adiabatic shear. Good agreement is found between theoretical predictions and field observations.

1 Introduction

Shear banding is one of the major modes of failure in geomaterials. In particular, it is observed during catastrophic landslides [9] and seismic faults [6] involving mechanisms that occur at several length and time scales. Field observations attest that the width of the band where the shear deformation localizes is narrow, i.e. of millimetric scale or even thinner. Strain localization in narrow bands can be seen as a bifurcation from the homogeneous deformation solution of the underlying mathematical problem, which is favoured by softening behaviour. Mechanical processes (e.g. grain crushing, reduction of internal friction etc.), thermal effects, chemical reactions (e.g. dissolution, dehydration etc.) can induce a softening behaviour.

H. Rattez (✉) · I. Stefanou · J. Sulem
Laboratoire Navier/CERMES, Ecole des Ponts ParisTech, IFSTTAR, CNRS, Université Paris-Est, 77455 Marne la Vallée, France
e-mail: hadrien.rattez@enpc.fr

M. Veveakis · T. Poulet
School of Petroleum Engineering, University of New South Wales, Kensington, NSW 2033, Australia

M. Veveakis · T. Poulet
CSIRO, 11 Julius Avenue, North Ryde, NSW 2113, Australia

E. Papamichos et al. (eds.), *Bifurcation and Degradation of Geomaterials with Engineering Applications*, Springer Series in Geomechanics and Geoengineering, DOI 10.1007/978-3-319-56397-8_20

During seismic slip, a large amount of the accumulated deformation energy is converted into heat. Therefore thermal weakening effects are expected to be crucial.

Here we present a model for the shearing of a saturated fault gouge under THM couplings [8]. Cosserat theory is used for the mathematical description of the mechanical behaviour of the gouge. It allows to regularize the underlying mathematical problem in the softening regime. Cosserat theory is particularly interesting as it can explicitly take into account the granular size.

Adopting an elastoplastic constitutive model for the Cosserat continuum, a linear stability analysis of the homogeneous state of deformation is conducted. The mathematical system is then integrated numerically in order to study the post-bifurcation regime of the problem [4]. The numerical simulations are performed using the FE code called Redback [5]. The numerical analyses permit to verify and to complete the theoretical predictions [8] in the post-bifurcation regime for a fault gouge under THM couplings.

2 Equations of the Model

A Cosserat continuum takes into account six degrees of freedom, i.e. three translations u_i and three rotations ω_i^c, at each point. For the strain tensor γ_{ij}—which is split into its symmetric ε_{ij} and antisymmetric part $\gamma_{[ij]}$—and the curvature tensor κ_{ij} a decomposition into an elastic and a plastic part is assumed.

$$\gamma_{ij} = \gamma_{ij}^e + \gamma_{ij}^p \quad and \quad \kappa_{ij} = \kappa_{ij}^e + \kappa_{ij}^p \tag{1}$$

The stress tensor τ_{ij} is non symmetric in general and is also divided into symmetric σ_{ij} and antisymmetric part $\tau_{[ij]}$. A couple-stress tensor μ_{ij} is introduced which is dual in energy with the curvature κ_{ij}.

The momentum balance equations can be written as follow:

$$\tau_{ij,j} - \rho \frac{\partial^2 u_i}{\partial t^2} = 0 \tag{2}$$

$$\mu_{ij,j} - e_{ijk}\, \tau_{jk} - \rho I \frac{\partial^2 \omega_i^c}{\partial t^2} = 0 \tag{3}$$

where e_{ijk} is the Levi-Civita symbol, ρ is the mass density and I is the micro-inertia.

By assuming that all the mechanical plastic work is converted into heat and that the heat flux is expressed through Fourier's law, we obtain the following diffusion equation for the temperature T:

$$\rho C(\frac{\partial T}{\partial t} - c_{th} T_{,ii}) = \sigma_{ij}\dot{\varepsilon}_{ij}^p + \tau_{[ij]}\dot{\gamma}_{[ij]}^p + \mu_{ij}\dot{\kappa}_{ij}^p \tag{4}$$

where c_{th} is the thermal diffusivity and ρC is the specific heat per unit volume of the material.

The diffusion equation for the pore pressure p is obtained from the fluid mass balance equation:

$$\frac{\partial p}{\partial t} = c_{hy}\, p_{,ii} + \Lambda \frac{\partial T}{\partial t} - \frac{1}{\beta^*} \frac{\partial n^p}{\partial t} \tag{5}$$

where c_{hy} is the hydraulic diffusivity, β^* is the storage capacity, Λ is the undrained thermal pressurization coefficient and n^p is the inelastic porosity change.

Thermal pore fluid pressurization is a destabilizing mechanism as it results in a significant decrease of the effective mean stress and consequently to the reduction of the shear strength. The effective stress tensor is linked to the total stress tensor by $\tau'_{ij} = \tau_{ij} + p\delta_{ij}$.

3 Bifurcation Analysis

Let us consider the problem of undrained adiabatic shearing of an infinite layer as described in Fig. 1. This problem presents an homogeneous solution in terms of stresses, strains, pressure and temperature.

We are interested in determining the conditions for which the above homogeneous solutions become unstable in the Lyapunov sense. The constitutive equations are based on the Muhlhaus-Vardoulakis model [3] which uses a generalized Drucker-Prager yield surface. A non associated plastic flow rule and strain hardening/softening is considered. All equations are linearised at the homogeneous state and we study the evolution in time of a small perturbation.

We assume a clayey gouge at 7 Km depth. The values of the different parameters are retrieved from [6, 8]. In particular, we consider no dilatancy of the medium. The critical value of the plastic hardening modulus for which homogeneous shear becomes unstable is $H_{cr} = 2$ MPa considering THM couplings (Fig. 2b). Note that in absence of thermal effects, the critical hardening modulus is zero. Thermal pore fluid pressurization leads to instability in the hardening regime. Nevertheless if we consider a contractant medium (negative dilatancy), H_{cr} becomes positive with HM couplings due to the pressurization of the pore fluid. For example if dilatancy coefficient is 0.01, $H_{cr} = 133$ MPa.

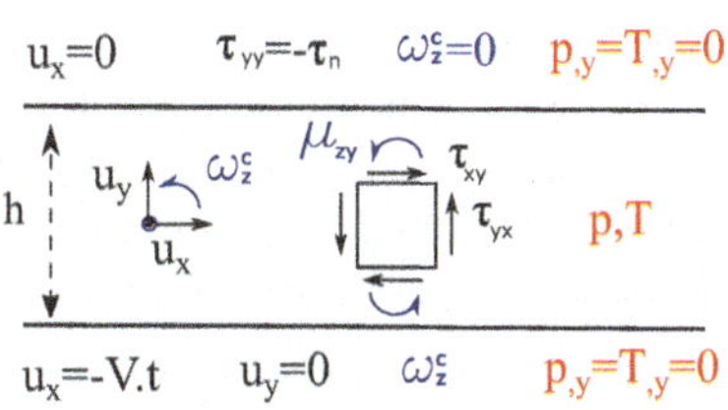

Fig. 1 Boundary conditions for the infinite sheared layer modelling a gouge. τ_n is the normal stress imposed and V is the shearing velocity

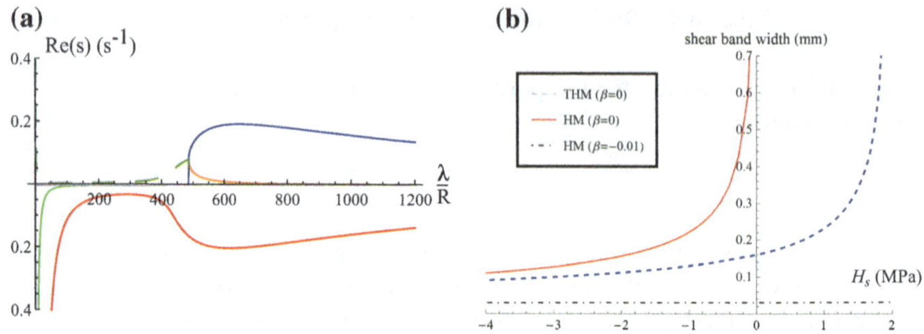

Fig. 2 **a** Real part of the exponential growth (s) of the perturbation as a function of the wavelength (λ) of the perturbation for a zero dilatancy coefficient ($\beta = 0$) and $H_s = 1.5$ MPa. The *dashed lines* represent the s that presents a nonzero imaginary part. **b** Thickness of the shear band as a function of the hardening modulus considering THM couplings or the Hydro-Mechanical (HM) couplings only

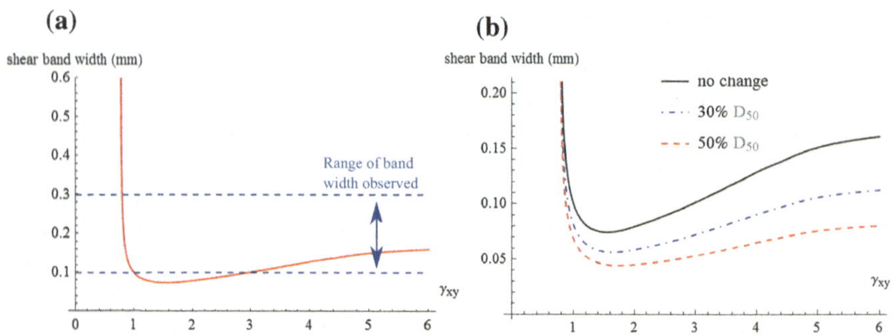

Fig. 3 Evolution of the shear band thickness with deformations compared with field observations (*left*) effect of grain cataclasis on the shear band size (*right*)

The onset of instability corresponds to an infinite wavelength (diffuse instability). Past this state, strain localizes rapidly in a band with a thickness controlled by a particular wavelength of the instability mode. This selected wavelength (λ) corresponds to the maximum growth coefficient (s) of the perturbation and is finite in the post-localisation regime when considering a Cosserat framework (Fig. 2a).

We plot in Fig. 3a, the evolution of the shear band thickness in terms of the hardening modulus. The values obtained with no dilatancy show a good agreement with field observations described in [6]. Note that for a contractant medium, the shear band thickness is much smaller (Fig. 2b).

At high mean stresses, like for faults at great depth, grain crushing is observed from exhumed samples and also in experiments. The introduction of a characteristic length related to the mean grain size in the Cosserat continuum enables to take into account the grain size evolution [1]. Assuming an empirical grain size reduction law during strain localisation $D(\gamma_{xy}) = (D_0 - D_f)e^{-\frac{\gamma_{xy}}{\gamma_c}} + D_f$; (where D_0, D_f are the

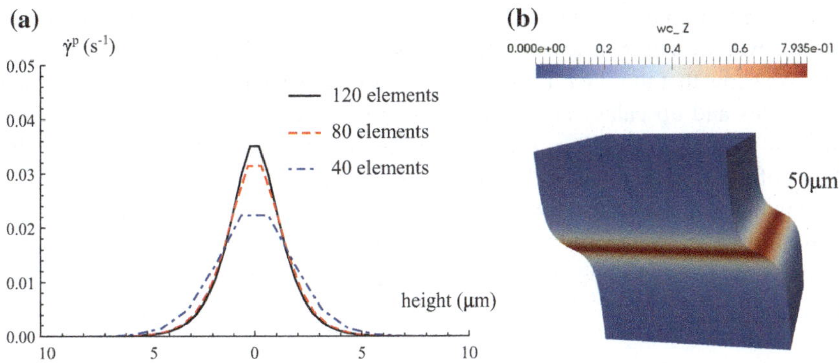

Fig. 4 Results of numerical analyses for HM couplings with $H_s = -4$ MPa and a dilatancy coefficient of −0.01 (contractant) for $h = 50\,\mu$m. **a** Rate of plastic strain increase ($\dot{\gamma}^p$) as a function of the height of the layer, centered in zero, for different meshes. **b** Cosserat rotation around the z-axis

initial and final mean grain size respectively and γ_c is a coefficient). The evolution of the shear band thickness is shown in Fig. 3b. Grain crushing leads to a shear band thickness reduction.

4 Numerical Simulation

The full system of equations is integrated numerically using a displacement-rotation incremental finite element formulation [2]. The integration in time is implicit using the solver provided by Redback [5] (Moose framework). In Fig. 4, the results of the simulation considering HM couplings are shown. The shear band thickness does not depend on the mesh size (Fig. 4a) and is in good agreement with the one predicted by the bifurcation analysis (Fig. 2b in black).

5 Conclusion

Energy dissipation and shear strain localisation in a fault gouge are of major importance. The shear band thickness is determined by various parameters and phenomena. Both rate dependent effects and the size of the microstructure (e.g. grain size) determine the shear band thickness [7]. Here rate effects are taken into account indirectly through the THM couplings, i.e. through the energy and mass balance equations. Cosserat continuum enables to upscale characteristic lengths to the macroscale. These internal lengths correspond here to the grain size of the gouge [3]. Bifurcation analysis shows the conditions for shear strain localisation that depend on different couplings.

The introduction of an internal length in Cosserat mechanics enables us to investigate the evolution of the shear band thickness in a fault and the effect of grain cataclasis. We show that the shear band size evolves with the hardening modulus, which encapsulates and upscales various microstructural phenomena. Moreover, consideration of the THM couplings seems of paramount importance as it modifies the stability of the system and also the shear band width.

The numerical analyses allow to verify analytical findings and permit to trace the post-peak regime. We found a good agreement with the size of the principal slip zone observed in experiments and from outcrops [6].

References

1. Bauer, E.: Simulation of the influence of grain damage on the evolution of shear strain localization. Continuous Media Microstruct. **2**, 231–244 (2016)
2. Godio, M., Stefanou, I., Sab, K., Sulem, J.: Multisurface plasticity for Cosserat materials: plate element implementation and validation. Int. J. Numer. Meth. Eng. (2016). doi:10.1002/nme.5219
3. Mühlhaus, H.B., Vardoulakis, I.: The thickness of shear bands in granular materials. Géotechnique **37**(3), 271–283 (1987)
4. Papanastasiou, P.C., Vardoulakis, I.: Numerical treatment of progressive localization in relation to borehole stability. Int. J. Numer. Anal. Meth. Geomech. **16**(6), 389–424 (1992)
5. Poulet, T., Paesold, M., Veveakis, M.: Multi-Physics Modelling of Fault Mechanics Using REDBACK: A Parallel Open-Source Simulator for Tightly Coupled Problems. Rock Mech. Rock Eng. 1–17 (2016)
6. Rice, J.R.: Heating and weakening of faults during earthquake slip. J. Geophys. Res. Sol. Earth **111**(5) (2006)
7. Sulem, J., Stefanou, I.: Thermal and chemical effects in shear and compaction bands. Geomech. Energy Environ. **6**, 4–21 (2016)
8. Sulem, J., Stefanou, I., Veveakis, E.: Stability analysis of undrained adiabatic shearing of a rock layer with Cosserat microstructure. Granular Matter **13**(3), 261–268 (2011)
9. Vardoulakis, I.: Dynamic thermo-poro-mechanical analysis of catastrophic landslides. Géotechnique **52**(3), 157–171 (2002)

Part III
Numerical Modeling of Failure, Localized Deformation and Gravity Driven Flows

Multiscale Poromechanics: Fluid Flow, Solid Deformation, and Anisotropic Plasticity

Ronaldo I. Borja, Shabnam J. Semnani and Jinhyun Choo

Abstract Natural geomaterials such as fissured rocks and aggregated soils often exhibit pore size distributions with two dominant porosity scales. In fractured rocks the dominant porosities are those of the fractures and rock matrix, whereas in aggregated soils the micropores and macropores comprise the two relevant porosity scales. When infiltrated with fluids this type of materials may also exhibit two permeability scales. In this paper we present a framework for so-called 'dual porosity-dual permeability' materials that covers both steady-state and transient fluid flow responses. The formulation revolves around a thermodynamically consistent *effective stress* previously developed for porous media exhibiting two porosity scales. Apart from the aspect of multiscale poromechanics, some geomaterials such as shale also exhibit pronounced anisotropy in their mechanical behavior due to the presence of distinct bedding planes. A transversely isotropic constitutive model is appropriate for this type of material behavior. Anisotropic plasticity models can easily be integrated into the aforementioned dual porosity-dual permeability framework.

1 Introduction

Many natural geomaterials such as fissured rocks and aggregated soils exhibit pore size distributions with two dominant scales of porosity. They are characterized as dual porosity-dual permeability materials, or materials with 'double porosity' for

R.I. Borja (✉) · S.J. Semnani
Department of Civil and Environmental Engineering, Stanford University,
Stanford, CA 94305, USA
e-mail: borja@stanford.edu

S.J. Semnani
e-mail: shabnamj@stanford.edu

J. Choo
Department of Civil Engineering and Engineering Mechanics, Columbia University,
New York, NY 10027, USA
e-mail: jinhyun.choo@columbia.edu

E. Papamichos et al. (eds.), *Bifurcation and Degradation of Geomaterials with Engineering Applications*, Springer Series in Geomechanics and Geoengineering, DOI 10.1007/978-3-319-56397-8_21

short. Some geomaterials also exhibit pronounced anisotropy in their hydromechanical behavior. An example is shale, a fine-grained sedimentary rock with a generally anisotropic fabric associated with the parallel alignment of clay particles preferentially oriented in the direction of the bedding plane. Anisotropy in fabric is generally attributed to deposition, compaction, and/or diagenetic processes.

In this paper we summarize the important features of a recently developed hydromechanical framework for materials exhibiting double porosity. The formulation revolves around the notion of *effective stress* introduced by Borja and Koliji [1] for porous materials with double porosity, which was subsequently integrated into a formulation for solving general boundary-value problems in multiscale poromechanics by Borja and Choo [2] and Choo et al. [3]. We note that the framework is sufficiently robust to accommodate anisotropy in the mechanical response. To this end, we recall the model of Semnani et al. [4] for transversely isotropic materials as a possible constitutive theory that can be integrated into the proposed dual porosity-dual permeability framework.

2 Two-Scale Poromechanics

For a solid matrix with double porosity and infiltrated with liquid and gas, Borja and Koliji [1] derived an expression for the *effective stress* of the form

$$\bar{\sigma}_{ij} = \sigma_{ij} + B\bar{p}\delta_{ij}\,, \tag{1}$$

where σ_{ij} is the total Cauchy stress tensor, B is the Biot coefficient, $\bar{p}$ is the mean fluid pressure weighted according to the local degrees of saturation and pore fractions, and δ_{ij} is the Kronecker delta. Choo et al. [3] used this expression for hydromechanical modeling of unsaturated flow in double porosity media, whereas Borja and Choo [2] presented a specialization of the theory for fully saturated aggregated soils to accommodate an evolving internal structure.

For fully saturated media, conservation of linear momentum takes the form

$$\sigma_{ij,i} + \rho g_j = c\left[(\widetilde{v}_m)_j - (\widetilde{v}_M)_j\right], \tag{2}$$

where g_j is the gravity acceleration component in the j direction, ρ is the total mass density of the mixture, $\widetilde{v}_m$ and $\widetilde{v}_M$ are the relative fluid velocities in the finer-scale and coarser-scale porosities, respectively, and c is a mass transfer term. Note that even in the quasi-static regime the right-hand side of the above equation is not zero due to the transfer of fluid mass. The linear momentum balance explicitly couples the solid displacement field with the fluid pressures in the finer-scale and coarser-scale porosities. Constitutive laws are necessary to close the boundary-value problem, see [2, 3]. In the following section we focus on the constitutive law for the mechanical deformation response of the solid matrix.

3 Anisotropic Plasticity

The idea behind the anisotropic constitutive formulation is to project the real stress space into an alternative stress space through a projection operator that captures transverse isotropy. To this end, we let $\boldsymbol{\ell}_1$, $\boldsymbol{\ell}_2$, and $\boldsymbol{\ell}_3$ denote three mutually orthogonal unit vectors such that $\boldsymbol{\ell}_1$ coincides with the axis of symmetry and $\boldsymbol{\ell}_2$ and $\boldsymbol{\ell}_3$ lie on the plane of isotropy. The physical analog to shale would be that $\boldsymbol{\ell}_1$ is the unit vector normal to the bedding plane. The microstructure tensor is then defined as

$$\boldsymbol{\phi} = \boldsymbol{\ell}_1 \otimes \boldsymbol{\ell}_1 . \tag{3}$$

Consider now the projection of the effective real Cauchy stress tensor $\bar{\sigma}_{kl}$ to an alternative stress configuration $\bar{\sigma}^*_{ij}$ of the form

$$\bar{\sigma}^*_{ij} = \mathbb{P}_{ijkl}\bar{\sigma}_{kl} , \tag{4}$$

where

$$\begin{aligned} \mathbb{P}_{ijkl} = & \frac{c_1}{2}\left(\delta_{ik}\delta_{jl} + \delta_{il}\delta_{jk}\right) \\ & + \frac{c_2}{2}\left(\phi_{ik}\phi_{jl} + \phi_{il}\phi_{jk}\right) \\ & + \frac{c_3}{4}\left(\delta_{ik}\phi_{jl} + \delta_{il}\phi_{jk} + \phi_{ik}\delta_{jl} + \phi_{il}\delta_{jk}\right) \end{aligned} \tag{5}$$

is the projector operator and δ_{ij} is the Kronecker delta.

When the coordinate system is aligned with the bedding planes, the projection operation is given by the matrix equation

$$\begin{Bmatrix} \bar{\sigma}^*_{11} \\ \bar{\sigma}^*_{22} \\ \bar{\sigma}^*_{33} \\ \bar{\sigma}^*_{23} \\ \bar{\sigma}^*_{13} \\ \bar{\sigma}^*_{12} \end{Bmatrix} = \begin{bmatrix} \alpha & & & & & \\ & \beta & & & & \\ & & \beta & & & \\ & & & \beta & & \\ & & & & \gamma & \\ & & & & & \gamma \end{bmatrix} \begin{Bmatrix} \bar{\sigma}_{11} \\ \bar{\sigma}_{22} \\ \bar{\sigma}_{33} \\ \bar{\sigma}_{23} \\ \bar{\sigma}_{13} \\ \bar{\sigma}_{12} \end{Bmatrix} , \tag{6}$$

where $\alpha = (c_1 + c_2 + c_3)$, $\beta = c_1$, and $\gamma = (c_1 + c_3/2)$. Note that the stress components are simply scaled by a diagonal matrix whose coefficients are chosen to preserve material symmetry requirements.

One can use any isotropic plasticity model in the alternative stress space while maintaining anisotropy in the real stress space. As an example, Fig. 1 shows the ellipsoidal modified Cam-Clay plasticity model in real and alternative stress configurations for various combinations of anisotropy parameters α and β, with $\gamma = 1$.

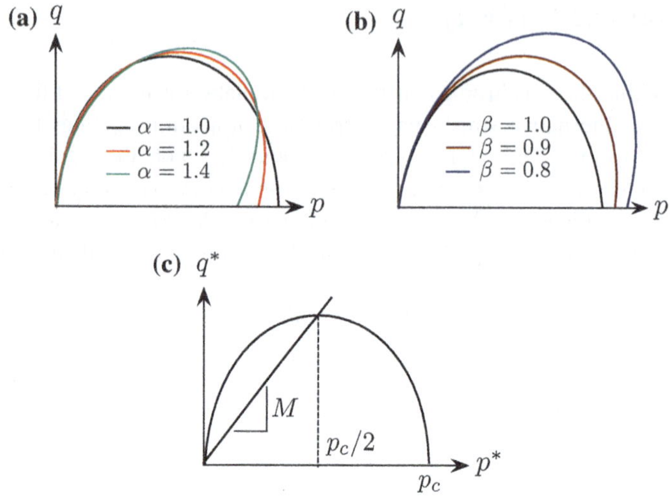

Fig. 1 Transversely isotropic modified Cam-Clay model in: **a** real stress space with $\beta = 1$; **b** real stress space with $\alpha = 1$; and **c** alternative stress space

Semnani et al. [4] employed the plasticity model shown in Fig. 1 to capture the stress-strain responses of highly anisotropic Tournemire shale subjected to triaxial compression.

4 Closing Remarks

Anisotropy in the material fabric could impact not only the mechanical response but also the fluid and transport properties. Preferentially aligned microfractures could result not only in anisotropy in permeability but also in directional fluid mass transfer between the pores of a dual porosity-dual permeability medium. Work is currently underway to investigate these complex mechanisms within the multiscale poromechanics framework presented in this paper.

Acknowledgements This material is based upon work supported by the U.S. Department of Energy, Office of Science, Office of Basic Energy Sciences, Geosciences Research Program, under Award Number DE-FG02-03ER15454, and by the National Science Foundation under Award Number CMMI-1462231. The second author acknowledges the National Science Foundation for a Graduate Research Fellowship under Grant No. DGE-114747.

References

1. Borja, R.I., Koliji, A.: On the effective stress in unsaturated porous continua with double porosity. J. Mech. Phys. Solids **57**, 1182–1193 (2009)
2. Borja, R.I., Choo, J.: Cam-Clay plasticity. Part VIII: A constitutive framework for porous materials with evolving internal structure. Comput. Methods Appl. Mech. Eng. **309**, 653–679 (2016)
3. Choo, J., White, J.A., Borja, R.I.: Hydromechanical modeling of unsaturated flow in double porosity media. Int. J. Geomech. D4016002 (2016). doi:10.1061/(ASCE)GM.1943-5622.0000558
4. Semnani, S.J., White, J.A., Borja, R.I.: Thermoplasticity and strain localization in transversely isotropic materials based on anisotropic critical state plasticity. Int. J. Numer. Anal. Methods Geomech. **40**, 2423–2449 (2016)

Micromechanical Modelling of Suffusion: Towards the Stability Analysis of Hydraulic Structures

Antoine Wautier, Stéphane Bonelli and François Nicot

Abstract Suffusion is the most controversial of the process of internal erosion and piping as there is a lack of microscopic understanding of the degradation processes at stake in saturated soils subjected to an internal flow. Thanks to the use of a DEM-PFV (Discrete Element Method—Pore-scale Finite Volume) approach, this study addresses the fully coupled problem of the suffusion of a non cohesive granular material. Eligible particles to grain detachment and grain transport are characterized from a micromechanical point of view thanks to the use of an enhanced force chain definition and the definition of a pore network. A Representative Elementary Volume with respect to the grain detachment and grain transport is then quantified.

1 Introduction

In various circumstances, permeable soils are subjected to internal flows which may modify their microstructure and by consequence their overall hydraulic and mechanical properties. At the microscale, this process consists in a rearrangement of particles driven by three elementary mechanisms, namely the detachment of grains from the granular skeleton, their transport through the pore network and possibly their reattachment to the granular skeleton farther away. At a larger scale, this process results in the selective erosion of the smallest particles of a soil and is referred to as suffusion within the geomechanics community [1].

A. Wautier (✉)
AgroParisTech-ENGREF, 19 Avenue du Maine, 75732 Paris, France
e-mail: antoine.wautier@irstea.fr

A. Wautier · S. Bonelli
Irstea UR RECOVER, 3275 Rte Cézanne, CS 40061, 13182
Aix-en-Provence Cedex 5, France

A. Wautier · F. Nicot
Université Grenoble Alpes, Irstea, UR ETGR, 2 Rue de la Papeterie-BP 76, F-38402
St-Martin-d'Hères, France

E. Papamichos et al. (eds.), *Bifurcation and Degradation of Geomaterials with Engineering Applications*, Springer Series in Geomechanics and Geoengineering, DOI 10.1007/978-3-319-56397-8_22

Many criteria where proposed in order to assess the internal stability of a particular soil with respect to suffusion. However, there is still a lack of global approaches able to assess suffusion susceptibility by considering in the meantime the soil microstructure, its stress state and the hydraulic loading. To this respect, this paper investigates the ability of a coupled DEM-PFV approach [2] to model the effect of an internal flow on cohesionless soils modelled as a poly-dispersed assembly of spheres. Thanks to the use of micromechanical tools based on the identification of chained particles [6] combined with the computation of autocorrelation functions [4] and the description of the pore space in terms of an oriented graph composed of pores and constrictions [7], the occurrence of the elementary mechanisms responsible for suffusion is explored and the length scales associated with grain detachment and grain transport processes are quantified.

The micromechanical analysis performed in this paper uses a discrete element method [3] implemented in YADE software [10]. Within this framework, a soil sample is modeled as a poly-disperse assembly of rigid spheres in interaction through the classical elasto-frictional contact law implemented in YADE [10] with a Young modulus of 325 MPa, a stiffness ratio of 0.42 and a friction angle of 35°.

Bearing in mind that suffusion involves the transport of the smallest particles of a soil, cubic assemblies of spheres are generated randomly with a uniform radius distribution between $r_{\min} = 3.6 \times 10^{-5}$ m and $r_{\max} = 10\, r_{\min}$ such that the smallest particles are likely to be transported through the voids created by the largest ones according to the Therzaghi filter rule: $r_{\min} \ll 4\, r_{\max}$.

Based on this grading, a dense cubic assembly of 10,000 particles with a void index of 0.6 is generated thanks to the radius expansion technique [5]. Then this initially isotropic sample is subjected to a drained triaxial test consisting under a confining pressure of 100 kPa up to a vertical strain $|\varepsilon_{zz}| = 20\%$.

2 Grain Detachment Analysis

The microscale analysis of the grain detachment mechanism is underpinned by the idea that the most detachable grains should also be the least stressed. As a result, the grains of the sample presented in the previous section are divided into two groups based on the force chain definition introduced in [6] according to on the following three characteristics:

- Chained particles have a higher principal stress than the mean particle principal stress;
- Their principal stress direction is aligned with the geometrical contact direction (less than 45° deviation);
- Force chains are composed of at least three contacting particles.

Based on this definition, force chains can be identified for several strain values. A typical visualization of the force chains can be seen in Fig. 1 in the initial and critical states. At the beginning of the triaxial test, the force chains are distributed in an

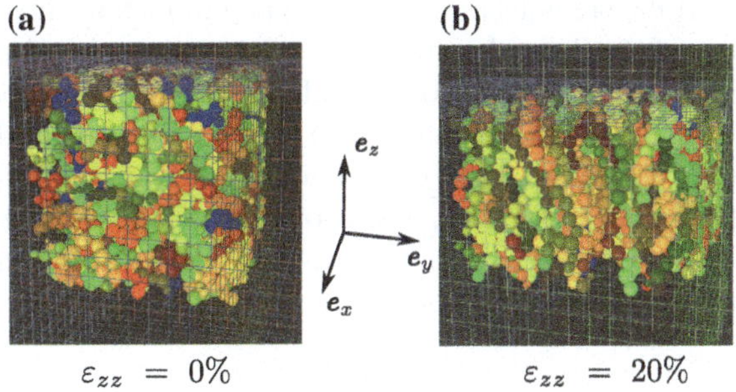

Fig. 1 Force chains visualisation under an isotropic confining pressure of 100 kPa (**a**) and at the end of the triaxial test (**b**)

isotropic way which is consistent with the fact that no principal direction of loading exists. Once the deviatoric loading is applied, the force chains tend to align in the vertical direction of the macroscopic principal stress.

In the literature, suffusion susceptibility is often considered based on a partition between fine and coarse particles according to a radius threshold. However, the ability of an internal flow to detach some particles should be preferentially linked to the loading state of the grains. This can be assessed by considering the probability density that a particle of a given radius belongs to a force chain which is illustrated in Fig. 2 in the initial configuration of Fig. 1.

Figure 2 shows that force chains are mainly composed of large particles. The dual comment is that the finest particles are the least loaded ones and therefore the most likely to be detached, provided that fluid forces are large enough. Fine particles are also good candidates for being transported over large distances, which will be estimated in Sect. 3. However, no sharp transition is visible in the probability densities in Fig. 2. As a result, no clear radius threshold exists between the loose particles of a granular assembly and the primary fabric responsible for stress transmission which is consistent with recent findings [8].

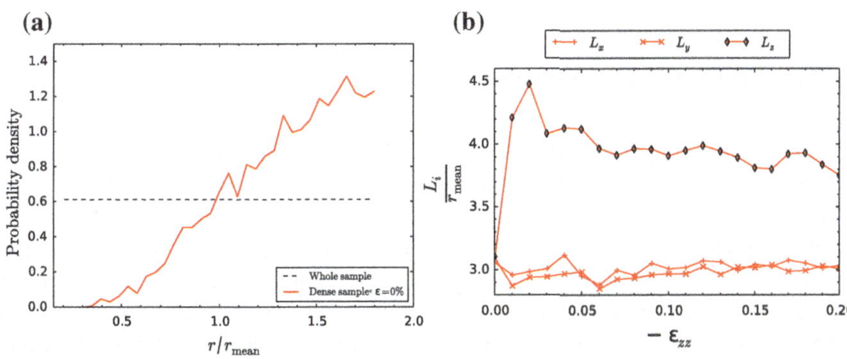

Fig. 2 Probability density for a grain of a given radius to be part of a force chain at $\varepsilon_{zz} = 0\%$ (**a**). Strain evolution of the autocorrelation lengths during the triaxial test (**b**)

If the previous result gives some information about the geometrical characteristics of the chained particles, it does not give any information on the spatial distribution of force chains. By considering the two-point spatial autocorrelation function between the chained particles, a length scale associated with the detachment process can be introduced as a typical distance between grains transmitting stresses through the granular assembly. The two-point spatial autocorrelation function C is defined for any vector $\boldsymbol{h} = (h_x, h_y, h_z)$ as the joint probability that a point $\boldsymbol{x}$ and the translated point $\boldsymbol{x} + \boldsymbol{h}$ simultaneously belong to the domain Ω occupied by chained particles:

$$C : \begin{cases} \mathbb{R}^3 \mapsto \mathbb{R} \\ \boldsymbol{h} \quad \mapsto \mathbb{P}\{\boldsymbol{x} \in \Omega \cap \boldsymbol{x} + \boldsymbol{h} \in \Omega\} \end{cases} \tag{1}$$

This autocorrelation function can be written in a dimensionless form as $\tilde{C}(\boldsymbol{h}) = \dfrac{C(\boldsymbol{h}) - C(0)^2}{C(0) - C(0)^2}$ and approximated as

$$\tilde{C}_{\text{fit}}(\boldsymbol{h}) = \exp\left(-2\pi^{1/3}\sqrt{\left(\frac{h_x}{L_x}\right)^2 + \left(\frac{h_y}{L_y}\right)^2 + \left(\frac{h_z}{L_z}\right)^2}\right). \tag{2}$$

In Eq. 2, three length scales L_x, L_y and L_z are introduced to quantify the decrease rate of $\tilde{C}$ and define a virtual unit cell in which the microstructure is highly correlated with respect to stress transmission. The strain evolution of these three autocorrelation lengths is plotted in Fig. 2 during the triaxial loading introduced in Sect. 1. As the horizontal autocorrelation lengths remains constant around $3\,r_{\text{mean}}$, the vertical autocorrelation length increases up to a peak value of $4.5\,r_{\text{mean}}$ before decreasing and stabilizing around $4\,r_{\text{mean}}$. The increase in the vertical autocorrelation corresponds to a lengthening of the force chains along the vertical direction which accounts for the initial hardening of our dense sample. The following decrease in the vertical autocorrelation accounts for the destruction of force chains and thus to the stress softening observed in the end of the triaxial test.

3 Grain Transport Analysis

In order to understand the grain transport mechanism from a micromechanical point of view, a description of the void phase based on the previous work of [7, 9] is proposed in this section. The description of the void space is reduced to an oriented graph composed of pores (the nodes of the graph) and constrictions (the edges of the graph) based on the tessellation of the granular assembly. Pore centers are defined at the center of each tetrahedron of a regular Delaunay triangulation and pore radii as the radius of the largest interior sphere in the associated tetrahedra. The constrictions are modeled as cylinders joining two adjacent pores and their radii are defined as the radius of the largest interior circle on the common face of the two tetrahedra defining the constriction.

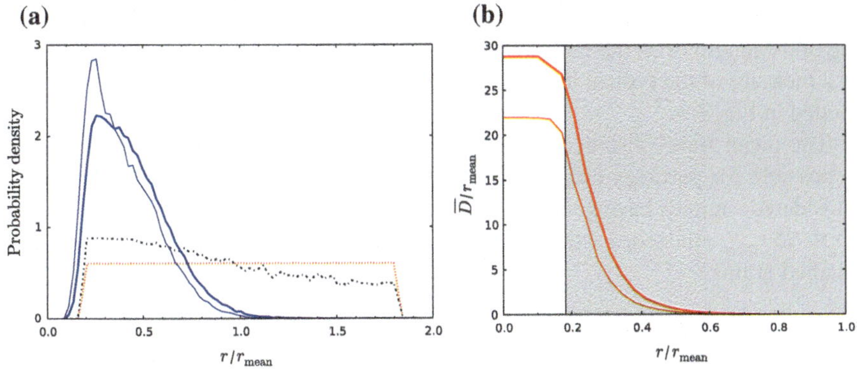

Fig. 3 Probability density for a constriction (*solid*), a grain (*dotted*) or a non-chained grain (*dash-dotted*) to be of a given radius (**a**). Mean travel distance for different radius thresholds (**b**). The thick line corresponds to $|\varepsilon_{zz}| = 20\%$, while the thin line to $|\varepsilon_{zz}| = 0\%$. The grey domain in the background corresponds to the radius values for the grains composing the sample

Based on this description of the pore space, the probability density for a free particle of a given radius to find a constriction large enough to travel from one pore to an other can be computed. This probability density is shown in Fig. 3 together with the probability density corresponding to the radius distribution of the non-chained grains identified in the previous section.

The comparison between the constriction probability density and the grain probability density in Fig. 3 highlights that a large number of grains are small enough to be transported through the pore network. An even larger fraction of the non-chained grains is concerned as 33–34% of these grains have a radius smaller than the mean constriction radius plus one standard deviation.

Even though many particles are identified in Fig. 3 as potentially transportable by analyzing the statistical properties of the pore network, determining the distance that a particle of a given radius can cover requires knowing the spatial distribution of pores and constrictions together with the fluid flow direction. This can be achieved thanks to the combined used of the DEM-PFV model implemented in YADE [2] together with the definition of transport paths based on a propagation criterion:

- First, the DEM-PFV scheme is used to generate a pressure map in the pore network under a horizontal fluid flow imposed by a small pressure drop of 10 Pa;
- Then, based on the Hagen-Poiseuille flow velocity profile for a circular tube and the expression of the drag force of a uniform flow acting on a sphere for low Reynolds numbers, a flow intensity indicator is defined in each constriction as $\frac{\Delta p}{\ell} \times R^2$ with ℓ being the length of the constriction, Δp the pressure drop between the two adjoining pores and R the constriction radius;
- Finally, a transport path is defined for a given particle of radius r and for a given starting pore as the path of maximum flow intensity on the restriction of the pore network to pores and constrictions of radii larger than r.

For a given radius r, a mean travel distance $\bar{D}(r)$ is computed from statistical averaging of the length of all the existing transport path defined above. This quantity stands as a measure of the typical length associated with grain transport mechanism and is plotted in Fig. 3.

The mean travel distances decreases from 22–29 r_{mean} to zero. If no transport is possible for particles larger than roughly 0.5 r_{mean} the smallest particles of the considered samples have good chances to be transported over large distances going up to 25 r_{mean}. Indeed, particles smaller than 0.5 r_{mean} represents 25% of the non-chained grains.

4 Conclusion

In this paper, specific micromechanical tools are developed to investigate the susceptibility of a polydisperse assembly of spherical particles to grain detachment and grain transport. By simultaneously considering the size distributions of the non-chained particles and the constrictions of the pore network, the ability of a fluid flow to modify the microstructure of the samples is analyzed with respect to the fraction of both potentially detachable and transportable particles. For the particular grading used in this study the typical length scale associated with grain transport can be up to ten times larger than that associated with grain detachment. As a result, a scale separation should exist between these two processes for the smallest particles.

Eventually, some fully coupled DEM-PFV numerical simulations were run. The numerical results were used to demonstrate the relevance of the developed micromechanical tools and validate the introduced length scales associated with grain detachment and grain transport. These numerical simulations also enable to track the microstructure evolution resulting from force chains rearrangements and grain transport which will pave the way for the micromechanical analysis of the internal fluid flow impact on the bifurcation domain of the considered sample.

References

1. Bonelli, S.: Erosion of geomaterials. Wiley (2012)
2. Chareyre, B., Cortis, A., Catalano, E., Barthélemy, E.: Pore-scale modeling of viscous flow and induced forces in dense sphere packings. Transp. Porous Media **94**(2), 595–615 (2012)
3. Cundall, P.A., Strack, O.D.: A discrete numerical model for granular assemblies. Geotechnique **29**(1), 47–65 (1979)
4. Kanit, T., Forest, S., Galliet, I., Mounoury, V., Jeulin, D.: Determination of the size of the representative volume element for random composites: statistical and numerical approach. Int. J. Sol. Struct. **40**(13), 3647–3679 (2003)
5. Lubachevsky, B.D., Stillinger, F.H.: Geometric properties of random disk packings. J. Stati. Phys. **60**(5–6), 561–583 (1990)
6. Peters, J.F., Muthuswamy, M., Wibowo, J., Tordesillas, A.: Characterization of force chains in granular material. Phys. Rev. E **72**(4), 041307 (2005)

7. Reboul, N., Vincens, E., Cambou, B.: A statistical analysis of void size distribution in a simulated narrowly graded packing of spheres. Granular Matter **10**(6), 457–468 (2008)
8. To, H.D., Torres, S.A.G., Scheuermann, A.: Primary fabric fraction analysis of granular soils. Acta Geotech. **10**(3), 375–387 (2015)
9. Vincens, E., Witt, K.J., Homberg, U.: Approaches to determine the constriction size distribution for understanding filtration phenomena in granular materials. Acta Geotech. **10**(3), 291–303 (2015)
10. Šmilauer, V., et al.: Yade Documentation, 2nd edn. The Yade Project (2015). http://yade-dem.org/doc/

Rigorous Comparison of the Rudnicki-Rice and Vermeer Bifurcation Criteria

Marte Gutierrez

Abstract This paper presents a rigorous comparison of the bifurcation criteria formulated by Rudnicki and Rice [1], and Vermeer [2, 3] for strain localization in geomaterials. It is shown that the simpler Vermeer's criterion is equivalent to the more complicated Rudnicki-Rice criterion for two-dimensional loading conditions. Differences between the two criteria exist for three-dimensional loading conditions.

1 Introduction

The Rudnicki-Rice [1] criterion has been widely used to detect bifurcation in geomaterials. Vermeer's [2, 3] bifurcation criterion has also been proposed and cited in numerous references, and is relatively simple making it easy to use and implement. The paper aims to improve the understanding bifurcation-induced strain localization in geomaterials. The specific objectives of this paper are to: (1) compare the bifurcation criteria of Rudnicki and Rice, and Vermeer; and (2) investigate the validity and applicability of Vermeer's criterion as an alternative to the more involved Rudnicki-Rice criterion.

2 Rudnicki and Rice Criterion

Localization of deformation into a planar band is an alternative to homogeneous deformation if a nontrivial solution exists to the following eigenvalue problem:

$$\{n_i D_{ijkl} n_l\} g_k = 0 \tag{1}$$

M. Gutierrez (✉)
Colorado School of Mines, Golden, CO 80401, USA
e-mail: mgutierr@mines.edu

E. Papamichos et al. (eds.), *Bifurcation and Degradation of Geomaterials with Engineering Applications*, Springer Series in Geomechanics and Geoengineering, DOI 10.1007/978-3-319-56397-8_23

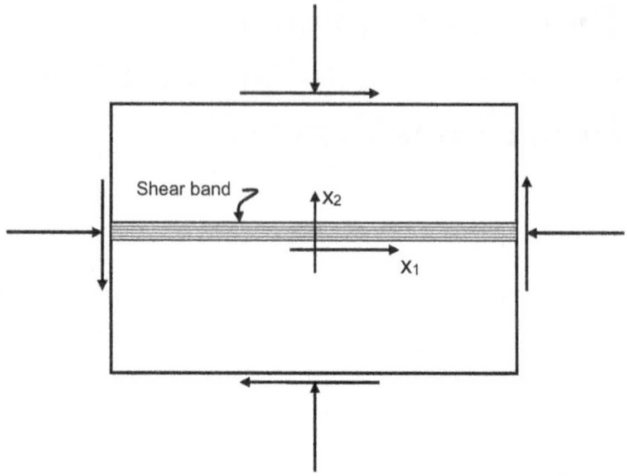

Fig. 1 Coordinate system for stain localization (Rudnicki and Rice [1])

where D_{ijkl} = constitutive stress-strain tensor, and n_i = normal to the localized band. A non-trivial solution for the discontinuity vector g_k in Eq. (1) is possible only when the acoustic tensor B_{ik}, defined below, has a zero or negative determinant, becomes non-positive definite, and produces negative eigenvalues. This condition may be stated for the case in which co-rotational terms are neglected as:

$$\det|B_{ik}| = \det|n_j D_{ijlk} n_l| \leq 0 \tag{2}$$

To facilitate the localization analysis, Rudnicki and Rice [1] formulated the bifurcation criterion by using a coordinate system with the x_1 and x_2 axes that are parallel and normal, respectively, to the yet undetermined shear band orientation (Fig. 1)

$$\det|D_{2jk2}| = 0 \tag{3}$$

Equation (3) implies kinematic conditions of uniform strain rates inside and outside the band $\Delta\dot{\varepsilon}_{ij} = 0$, except for the jump in the normal strain rate perpendicular to the band $\Delta\dot{\varepsilon}_{22} \neq 0$. Equilibrium conditions allow for jumps in the stresses inside and outside the band $\Delta\dot{\sigma}_{ij} \neq 0$, except for uniform stresses normal to the band $\Delta\dot{\sigma}_{2j} = 0$.

3 Vermeer Criterion

An easier way to formulate the bifurcation problem is via the compliance approach of Vermeer [2, 3]. Consider a soil element subjected to the principal stresses $\sigma_{\mathrm{I}}, \sigma_{\mathrm{II}}$ and σ_{III} (Fig. 2). Let the yet to be determined shear band plane be oriented along the x_1, x_2 and x_3 axes, with the x_1-axis oriented at angle θ from σ_{III}

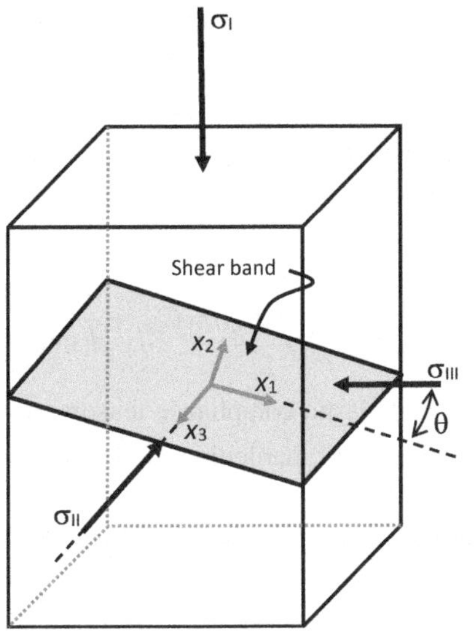

Axis x_1 is parallel to the shear band and θ is the shear band orientation measured from the σ_{III}-axis. Note that according to Rudnicki and Rice [1], and Vermeer [2], axis x_3 coincides with the σ_{II}-axis.

Fig. 2 Soil sample under triaxial loading conditions with a shear band

the-direction. According to [1], if the principal stresses are distinct, the normal to the critical plane of localization is perpendicular to σ_{II} for the expected range of friction and dilations angles for geomaterials, and the localization plane is perpendicular to σ_{III} only in exceptional cases. The comparison of the Rudnicki-Rice and Vermeer criteria is, therefore, carried out using the most expected shear band geometry as shown in Fig. 2 where σ_{II} is parallel to the shear band plane.

In Vermeer's criterion, kinematic conditions allow for jumps in strain rates inside and outside the band $\Delta\dot{\varepsilon}_{ij} \neq 0$, except for the uniform normal strain rates parallel to the band $\Delta\dot{\varepsilon}_{11} = 0$ and $\Delta\dot{\varepsilon}_{33} = 0$, and equilibrium conditions with the same stresses inside the outside the band $\Delta\dot{\sigma}_{ij} = 0$, except for the jump in normal stresses parallel to the band $\Delta\dot{\sigma}_{11} \neq 0$ and $\Delta\dot{\sigma}_{33} \neq 0$. Applying these conditions to the compliance relation $\Delta\varepsilon_{ij} = C_{ijkl}\Delta\dot{\sigma}_{kl}$, where C_{ijkl} is the elasto-plastic compliance matrix, yields Vermeer's criterion:

$$C_{1111} = 0 \tag{4}$$

Vermeer's approach simply requires that the compliance tensor C_{ijkl} be formulated using the axis coinciding with the yet unknown shear band direction, and then setting the component C_{1111} parallel to the shear band to zero. This gives the condition for the onset of instability and shear banding. It is noted that C_{ijkl} is undefined for perfect plasticity ($h = 0$) and strain softening ($h < 0$), however, both stages are beyond bifurcation, and are not the scope of the current study.

4 Comparison of Rudnicki-Rice and Vermeer Bifurcation Criteria

For elasto-plastic materials, the constitutive relations D_{ijkl} and C_{ijkl} can be expressed in terms of a yield criterion $f = f(\sigma_{ij}, \alpha)$, a plastic potential function $g = g(\sigma_{ij})$, and a plastic hardening parameter α:

$$D_{ijkl} = D^e_{ijkl} - \frac{1}{(h+h')}\left(D^e_{ijpq}\frac{\partial g}{\partial \sigma_{pq}}\frac{\partial f}{\partial \sigma_{rs}}D^e_{rskl}\right), C_{ijkl} = C^e_{ijkl} + \frac{1}{h}\frac{\partial f}{\partial \sigma_{kl}}\frac{\partial g}{\partial \sigma_{kl}} \quad (5)$$

where D^e_{ijkl} and C^e_{ijkl} are the elasticity and elastic compliance tensors, respectively, $h' = (\partial f/\partial \sigma_{ij})D^e_{ijkl}(\partial g/\partial \sigma_{kl})$, and h is the plastic hardening modulus which can be obtained from the consistency condition.

For two-dimensional loading conditions, it can be shown that the Rudnicki and Rice criterion (Eq. 3) can be written as:

$$\det\begin{bmatrix} D_{2222} & D_{2212} \\ D_{1222} & D_{1212} \end{bmatrix} = 0 \quad (6)$$

Substituting the elastoplastic relations given in the first equation in Eq. (5) to Eq. (6) will show that the determinant in Eq. (6) is exactly equal to C_{1111}, hence leading to Vermeer's criterion of Eq. (4). For a non-associated Mohr-Coulomb failure criterion with friction angles ϕ nd dilation angle ψ, the predicted critical strain hardening modulus h_{cr} at bifurcation and shear band orientations θ are given in Table 1. As can be seen, the predictions are identical for both criteria for two-dimensional loading.

For three-dimensional loading, the Rudnicki and Rice's elasto-plastic constitutive tensor D_{ijkl} takes the form:

$$D_{ijkl} = G(\delta_{ki}\delta_{lj} + \delta_{kj}\delta_{il}) + \left(K - \tfrac{2}{3}G\right)\delta_{ij}\delta_{kl} - \frac{\left(\frac{G}{\tau}s_{ij} - \beta K\delta_{ij}\right)\left(\frac{G}{\tau}s_{kl} - \mu K\delta_{kl}\right)}{h + G + \mu\beta K} \quad (7)$$

where s_{ij} = deviatoric part of σ_{ij}, $\tau = ((s_{ij}\ s_{ij})/2)^{(1/2)}$ = shear stress, K and G = elastic bulk and shear moduli, respectively, and μ and β are defined as the *internal friction coefficient* and *dilatancy factor*, respectively. Applying Rudnicki-Rice criterion (Eq. 3), and solving for h and minimizing its value yield the critical hardening modulus at bifurcation h_{cr} and the corresponding shear band orientation in Table 1.

Expressed with the same variables, the equation for elasto-plastic compliance component C_{1111} takes the form:

Table 1 Comparisons of the predicted shear band orientations θ and hardening moduli at bifurcation h_{cr} from Rudnicki and Rice [1], and Vermeer [2] for different loading conditions (ϕ = friction angle, ψ = dilation angle, G = shear modulus, υ = Poisson's ratio, μ = internal friction coefficient, β = dilatancy factor, τ = shear stress, ϑ = Lode angle, s_{ij} = deviatoric part of the stress tensor, and s_1, s_2 and s_3 = principal values of the deviatoric stress tensor)

Loading condition	Rudnicki and Rice (1975)	Vermeer (1982)
Biaxial loading	$\cos 2\theta = \frac{1}{2}(\sin\phi + \sin\psi)$ $\frac{h_{cr}}{G} = \frac{(1+\upsilon)}{8(1-\upsilon^2)}(\sin\phi - \sin\psi)^2$	
Three-dimensional loading	$\tan\theta = \left(\frac{\xi - s_3/\tau}{s_1/\tau - \xi}\right)^{1/2}, \xi = \frac{1}{3}(1+\upsilon)(\beta+\mu) - \frac{s_2}{\tau}(1-\upsilon)$ $\frac{h_{cr}}{G} = \frac{(1+\upsilon)}{9(1-\upsilon)}(\beta-\mu)^2 - \frac{1+\upsilon}{2}\left(\frac{s_2}{\tau} + \frac{\beta+\mu}{3}\right)$	$\cos 2\theta = \frac{\beta+\mu}{3\cos\vartheta} - \frac{1}{2}\tan\upsilon$ $\frac{h_{cr}}{G} = \frac{(1+\upsilon)}{18}(\beta-\mu)^2$
"Pure shear" ($s_2/\tau = -(\beta+\mu)/3$ for Rudnicki and Rice, and $\vartheta = 0°$ for Vermeer)	$\cos 2\theta = \frac{\beta+\mu}{3}$ $\frac{h_{cr}}{G} = \frac{1+\upsilon}{9(1-\upsilon)}(\beta-\mu)^2$	$\cos 2\theta = \frac{\beta+\mu}{3}$ $\frac{h_{cr}}{G} = \frac{(1+\upsilon)}{18}(\beta-\mu)^2$

$$C_{1111} = \frac{(G+3K)}{9GK} + \frac{1}{h}\left(\frac{s_{11}}{2\tau} - \frac{\mu}{3}\right)\left(\frac{s_{11}}{2\tau} - \frac{\beta}{3}\right) \tag{8}$$

Equating C_{1111} to 0 and solving for h and minimizing its value yield the critical hardening modulus at bifurcation h_{cr} and the corresponding shear band orientation in Table 1. As can be seen, significant differences exist in the predictions from the two criteria for three-dimensional loading conditions.

5 Conclusions

Comparisons of the predictions of the shear band orientation and hardening modulus at bifurcation from the Rudnicki-Rice and Vermeer criteria for two and three-dimensional loading conditions are summarized in Table 1. As can be seen, identical results are obtained for two-dimensional biaxial loading conditions. However, the results are completely different for three-dimensional loading except for the shear band orientation when $\vartheta = 0°$ for Vermeer's criterion, and $s_{II} = (\beta +$

Table 2 Comparisons of the Rudnicki and Rice [1], and Vermeer [2] bifurcation criteria

Rudnicki and Rice (1975)	Vermeer (1982)
(1) Applicable to 2D and 3D loading conditions	(1) Limited to 2D loading conditions
(2) Complicated requiring eigenvalue analysis	(2) Simple and easy to understand
(3) Difficult to use for complicated constitutive models	(3) Easy to implement even for complicated constitutive models
(4) Kinematic conditions require uniform strain rates inside and outside the band ($\Delta\dot{\varepsilon}_{ij} = 0$), except for the jump in the normal strain rate perpendicular to the band ($\Delta\dot{\varepsilon}_{22} \neq 0$)	(4) Kinematic conditions allow for jumps in strain rates inside and outside the band ($\Delta\dot{\varepsilon}_{ij} \neq 0$) except for the uniform normal strain rates parallel to the band ($\Delta\dot{\varepsilon}_{11} = 0$ and $\Delta\dot{\varepsilon}_{33} = 0$)
(5) Equilibrium conditions allow for jumps in the stresses inside and outside the band ($\Delta\dot{\sigma}_{ij} \neq 0$), except for uniform stresses normal to the band ($\Delta\dot{\sigma}_{2j} = 0$)	(5) Equilibrium conditions require the same stresses inside the outside the band ($\Delta\dot{\sigma}_{ij} = 0$), except for the jump in normal stresses parallel to the band ($\Delta\dot{\sigma}_{11} \neq 0$ and $\Delta\dot{\sigma}_{33} \neq 0$)
(6) Localization criterion is a function of the shear stress components s_{22}, s_{21} and s_{23}	(6) Localization criterion is a function of only the shear stress component s_{11}
(7) For 3D loading, shear band orientation and critical hardening modulus are functions of the intermediate principal stress	(7) For 3D loading conditions, predicts a much earlier bifurcation point than Rudnicki and Rice (1975), and a hardening modulus independent of the intermediate principal stress
(8) Agrees with experimental data	(8) Does not agree with experimental data for 3D loading

μ)/3 for Rudnicki and Rice's. For "simple shear" conditions, the critical hardening modulus from Vermeer's criterion is at most half of that of Rudnicki and Rice's.

Table 2 summarizes the comparison of the two criteria. The discrepancies between the two criteria stem from: (1) The differences in the kinematic and equilibrium conditions used to derive the criteria (Table 2), with these conditions being more restrictive in Vermeer's criterion, and (2) The fact that hardening modulus in Vermeer's criterion is a function of only a single stress component s_{11} (Eq. 8), while that of Rudnicki and Rice depend on three components s_{22}, s_{21} and s_{23} (Eq. 7). Thus, the latter can fully represent the state of stress in the shear band while the former cannot. The validity of the predictions from Rudnicki and Rice's criterion has been extensively validated experimentally and theoretically in the literature. Given the discrepancies, it appears that Vermeer's criterion is valid only for two-dimensional loading and not for three-dimensional loading.

References

1. Rudnicki, J.W., Rice, J.R.: Conditions for the localization of deformation in pressure-sensitive dilatant materials. J. Mech. Phys. Solids **23**, 371–394 (1975)
2. Vermeer, P.A.: A simple shear-band analysis using compliances. In: Proceedings IUTAM Conference on Deformation and Failure Granular Materials, Delft, Netherlands, pp. 493–499 (1982)
3. Vermeer, P.A., de Borst, R.: Non-associated plasticity for soils, concrete and rock. Heron **29**, 3–64 (1984)

Second Gradient Models and Concrete Structures

G. Jouan, M. Soufflet, P. Kotronis and F. Collin

Abstract Damage induces strain localization in quasi-brittle materials such as concrete. In order to correctly simulate this behaviour, it is necessary to introduce a length scale parameter in the continuum model. The second gradient model, a special case of kinematically enriched continua, uses an internal length parameter by taking into account the second order derivatives of the displacements in the virtual power principle. A penalty term is added in the original second gradient finite element formulation in order to improve convergence and to avoid erroneous damage distribution. A three points bending test of a reinforced concrete beam show the performance of the improved second gradient finite element formulation.

1 Introduction

As a material exhibiting softening, concrete is subject to strain localization. In order to correctly simulate this behavior a model with a length scale parameter is needed. Chambon, Caillerie and Hassan gave a closed form solution for the one dimensional problem of strain localization in a bar with a bilinear type constitutive law in a second gradient continuum [2]. This was soon followed with several applications for plane strain shear banding [4, 14, 17, 22], mostly in soils. Similar approaches were also recently used in the framework of damage mechanics combined with homogenization techniques by Li [15, 16], see also [10, 11, 13, 23] for application in concrete structures.

G. Jouan · M. Soufflet · P. Kotronis (✉)
Ecole Centrale de Nantes, Université de Nantes, CNRS UMR 6183,
GeM (Institut de Recherche en Génie Civil et Mécanique),
1 rue de la Noë, BP 92101, 44321 Nantes, France
e-mail: Panagiotis.Kotronis@ec-nantes.fr

F. Collin
Département Argenco, Institut de mécanique et Génie Civil,
Université de Liège, Bât. 52, 9 Allée de la Découverte, 4000 Liège, Belgium
e-mail: f.collin@ulg.ac.be

E. Papamichos et al. (eds.), *Bifurcation and Degradation of Geomaterials with Engineering Applications*, Springer Series in Geomechanics and Geoengineering, DOI 10.1007/978-3-319-56397-8_24

In this paper, it is shown that the original finite element formulation adopted in a previous work of the authors [11] presents some numerical difficulties to correctly simulate damage localization problems. The original finite element of Matsushima et al. [17] is thus improved by adding a penalty term, following previous works by Fernandes [6] for plasticity problems. The ability of the improved model to better simulate strain localization in concrete is studied with a three points bending test of a reinforced concrete beam.

2 Second Gradient Model

Considering a body of domain Ω and of boundary Γ, the general formulation of a second gradient model using the virtual power principle can be written as follows [2, 17]:

$$\int_{\Omega} (\sigma_{ij} u^{\star}_{i,j} + \Sigma_{ijk} u^{\star}_{i,jk})\, d\Omega = \int_{\Gamma} (p_i u^{\star}_i + P_i D u^{\star}_i) d\Gamma, \qquad (1)$$

where σ_{ij} is the macro stress (the classical second order stress tensor), Σ_{ijk} the double stress (a third order tensor) and p_i and P_i respectively the first (classical) and second order surface traction (first order tensors).

In linear elasticity, the constitutive law for σ_{ij} depends as usual on the first gradient of the displacements, while Σ_{ijk} is a function of the second gradient of the displacements [19]. Coupling between the first and second gradient parts is possible, see for example [3, 7, 12]. Chambon and co-workers considered mainly cases with full decoupling of the first and second gradient parts: nonlinear laws, independent of Σ_{ijk} and $u_{i,jk}$ for the first gradient part and an elastic linear isotropic law for the second gradient part [5, 14, 17, 20]. It should be highlighted however that the introduction of damage on the second gradient part has been studied by different authors [12, 15, 16, 23]. In [12] and for a 1D case, it is shown that a carefully formulated coupling law leads to an a priori control of the width evolution (constant, increase or decrease) of the localisation zone.

3 Finite Element Formulation

3.1 The Original Mixed Formulation

One way to circumvent the difficulties of the necessary C^1 continuity is to adopt a mixed formulation and to interpolate the gradient independently from the displacement field [17, 21]. More specifically, the displacement field u_i and its gradient noted v_{ij} are both interpolated with C^0 functions since only first order derivatives appear in the weak formulation provided either by the virtual power principle of a general micromorphic media with a kinematic constrain [3] or by integration of the strong form of the equilibrium equations [21]. In the former case, we can directly write:

$$\int_{\Omega} (\sigma_{ij} u^{\star}_{i,j} + \lambda_{ij}(v^{\star}_{ij} - u^{\star}_{i,j}) + \Sigma_{ijk} v^{\star}_{i,jk})\, d\Omega = \int_{\Gamma} (p_i u^{\star}_i + P_i v^{\star}_{ij} n_j) d\Gamma \tag{2}$$

with the added kinematic constraint (in weak form) :

$$\int_{\Omega} \lambda^{\star}_{ij}(v_{ij} - u_{i,j})\, d\Omega \tag{3}$$

where $\lambda^{\star}_{ij}$ is the field of Lagrange multipliers associated with the kinematic constraint. For the second gradient model, Matsushima et al. [1, 17] originally developed a 9-nodded finite element with the displacement field interpolated by biquadratic functions of the serendipity type and the gradient field by bilinear functions. Lagrange multipliers are constant on the element while numerical integration is performed using a classical Gauss scheme [1]. The numerical performance of this formulation is tested hereafter using a three points bending test of a reinforced concrete beam.

3.2 A Three Points Bending Test of a Reinforced Concrete Beam

The concrete beam has the following geometrical characteristics: thickness $b = 200$ mm, height $h = 500$ mm and span 5000 mm. The tested beam and the steel reinforcement are shown in Fig. 1.

The finite element mesh consists of 5180 elements, 4148 of which are second gradient elements and 1032 truss elements representing the horizontal reinforcement. The average size of the concrete elements is of 0.02 m × 0.035 m. The two supports (the two points at the bottom of the specimen, see Fig. 1) are blocked vertically while the right support is also blocked horizontally. For the finite element calculations, monotonically increased displacements are applied at the upper part of the beam through an elastic plate, which is very stiff compared to the other materials [11].

At both supports at the bottom of the beam and on the upper part, where the displacements are applied, an elastic linear law is introduced to prevent any artificial numerical damage. A classical damage mechanics law is used for the first gradient constitutive law [18]. An isotropic linear elastic constitutive law is adopted for the second gradient part depending on a single material parameter. No coupling is assumed between the first and the second gradient constitutive laws. An elastic perfectly plastic law is used for the reinforcement and a perfect bond is assumed.

Figure 2 shows the damage distribution in the entire beam as well as two close ups. It can be observed that in the localization bands and inside certain finite elements some integration points present an increasing damage variable while others are unloading leading to serious convergence problems.

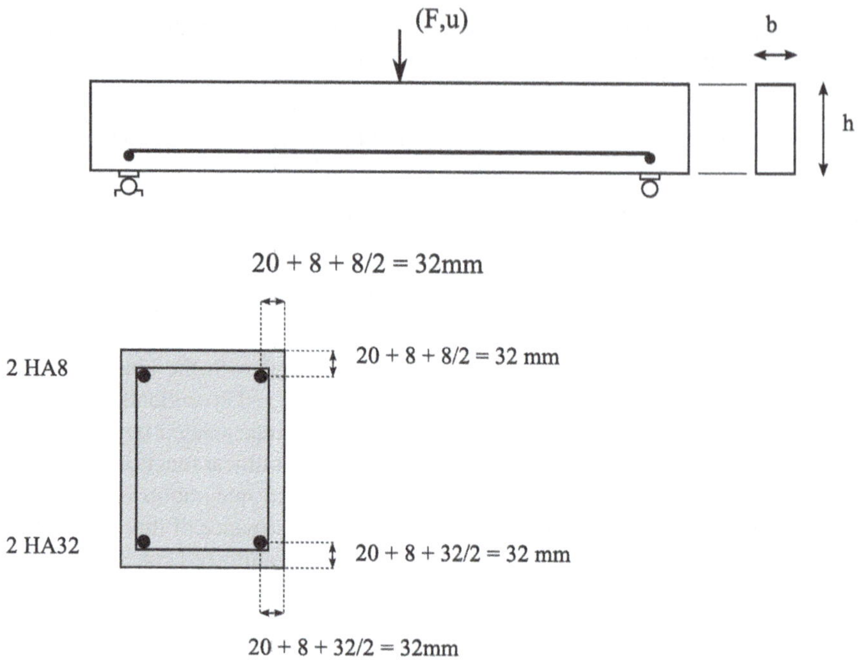

Fig. 1 3 point bending test: beam dimensions and steel reinforcement [8, 9]

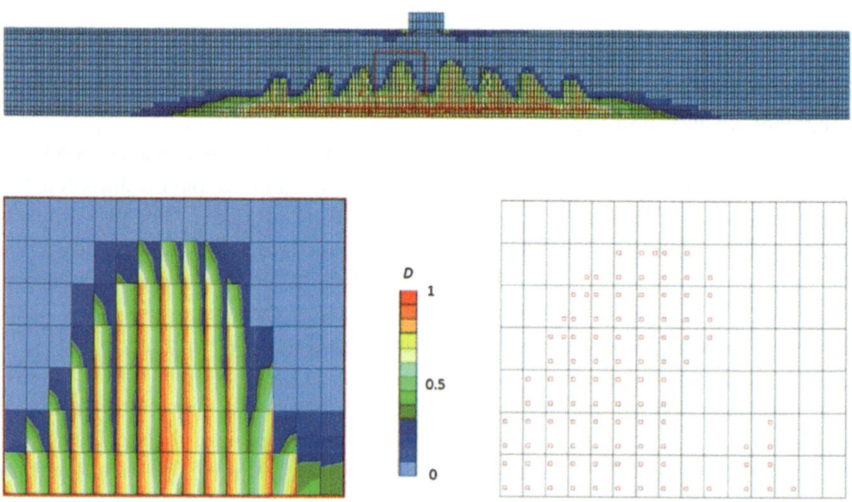

Fig. 2 Damage distribution in a three points beam test (*top figure*). Zoom on damage oscillation problems (*left figure*) and loading integration points (*red rectangles* correspond to integration points with an increasing damage variable, *right figure*)

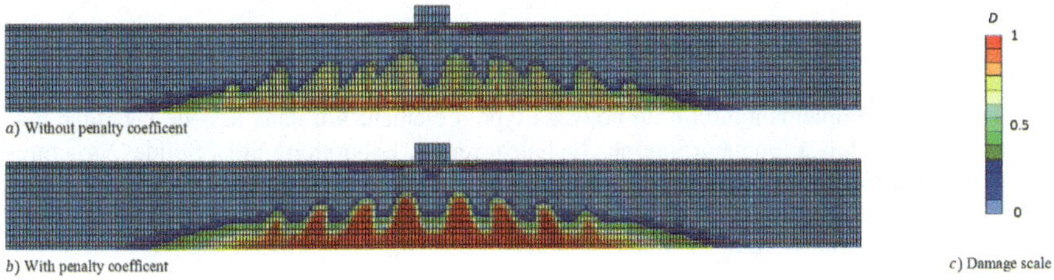

Fig. 3 3 point bending test: damage distribution for an imposed displacement of 6 mm **a** without and **b** with a penalty factor

3.3 Addition of a Penalty Term

A way to improve the numerical performance of the original mixed second gradient finite element formulation is to add a penalty term to enforce the kinematic constraint (while keeping the Lagrange multipliers) [10]. A similar approach was proposed by Fernandes et al. [6] in the context of plasticity. Noting *C* the penalty factor, the weak formulation of the problem becomes:

$$\int_{\Omega}(\sigma_{ij}u^{\star}_{i,j} + \lambda_{ij}(v^{\star}_{ij} - u^{\star}_{i,j}) + C(v^{\star}_{ij} - u^{\star}_{i,j})(v_{ij} - u_{i,j}) + \Sigma_{ijk}u^{\star}_{i,jk})\, d\Omega = \int_{\Gamma}(p_i u^{\star}_i + P_i v^{\star}_{ij,k} n_k) d\Gamma \tag{4}$$

$$\int_{\Omega} \lambda^{\star}_{ij}(v_{ij} - u_{i,j})\, d\Omega \tag{5}$$

The three points bending test of section is modelled here using the second gradient finite element with the added penalty term. Figure 3 gives the damage distribution corresponding to an imposed displacement $U = 6$ mm without and with a penalty factor. It is clear that the introduction of the penalty factor improves the numerical performance of the model and results to a more smooth and continuous damage distribution.

4 Conclusions

Following previous works [11, 13], the second gradient model is used as a regularization method for concrete structures in the context of damage induced strain localization. The performance of the 2*D* second gradient finite element is improved by introducing an additional penalty term in the Lagrangian mixed formulation. An example is provided considering a reinforced concrete beam.

References

1. Bésuelle, P.: Implémentation d'un nouveau type d'élément fini dans le code lagamine pour une classe de lois á longueur interne. Technical report, Laboratoire Sols, Solides Structures, Grenoble, France (2003). Rapport d'activité
2. Chambon, R., Caillerie, D., Hassan, N.E.: One-dimensional localisation studied with a second grade model. Eur. J. Mech. A/Solids **17**(4), 637–656 (1998). doi:10.1016/S0997-7538(99)80026-6
3. Chambon, R., Caillerie, D., Matsuchima, T.: Plastic continuum with microstructure, local second gradient theories for geomaterials: localization studies. Int. J. Solids Struct. **38**(4647), 8503–8527 (2001). doi:10.1016/S0020-7683(01)00057-9
4. Collin, F., Chambon, R., Charlier, R.: A finite element method for poro mechanical modelling of geotechnical problems using local second gradient models. Int. J. Numer. Methods Eng. **65**(11), 1749–1772 (2006)
5. Collin, F., Caillerie, D., Chambon, R.: Analytical solutions for the thick-walled cylinder problem modeled with an isotropic elastic second gradient constitutive equation. Int. J. Solids Struct. **46**(2223), 3927–3937 (2009). doi:10.1016/j.ijsolstr.2009.05.017
6. Fernandes, R., Chavant, C., Chambon, R.: A simplified second gradient model for dilatant materials: theory and numerical implementation. Int. J. Solids Struct. **45**(20), 5289–5307 (2008). doi:10.1016/j.ijsolstr.2008.05.032
7. Fleck, N., Hutchinson, J.: Strain gradient plasticity. Adv. Appl. Mech. **33**, 295–361 (1997)
8. Ghavamian, S., Carol, I., Delaplace, A.: Discussions over meca project results. Revue Franaise de Génie Civil **7**(5), 543–581 (2003)
9. Jason, L.: Réponse au benchmark statique monotone du projet national ceos.fr (2008). Rapport CEA SACLAY, DM2S/SEMT/LM2S, Rapport ANR MEFISTO
10. Jouan, G.: Modélisation numérique de la localisation des déformations dans le béton avec un modèle de second gradient. Ph.d., Ecole Centrale de Nantes, Université de Liège (cotutelle) (2014). http://bictel.ulg.ac.be/ETD-db/collection/available/ULgetd-11042014-112001/ (in French)
11. Jouan, G., Kotronis, P., Collin, F.: Using a second gradient model to simulate the behaviour of concrete structural elements. Finite Elem Anal Des **90**, 50–60 (2014). doi:10.1016/j.finel.2014.06.002
12. Kotronis, P.: Stratégies de modélisation de structures en béton soumises à des chargements sévères. H.D.R. (Habilitation Diriger des Recherches), Université Joseph-Fourier-Grenoble I (2008). http://tel.archives-ouvertes.fr/tel-00350461
13. Kotronis, P., Chambon, R., Mazars, J., Collin, F.: Local second gradient models and damage mechanics: application to concrete. In: 11th International Conference on Fracture, Turin, Italy, Org. ICF, cd, paper, vol. 5712, pp. 20–25 (2005)
14. Kotronis, P., Al Holo, S., Bésuelle, P., Chambon, R.: Shear softening and localization: modelling the evolution of the width of the shear zone. Acta Geotech. **3**(2), 85–97 (2008)
15. Li, J.: A micromechanics-based strain gradient damage model for fracture prediction of brittle materials-part i: Homogenization methodology and constitutive relations. Int. J. Solids Struct. **48**(24), 3336–3345 (2011)
16. Li, J., Pham, T., Abdelmoula, R., Song, S., Jiang, C.: A micromechanics-based strain gradient damage model for fracture prediction of brittle materials—part ii: Damage modeling and numerical simulations. Int. J. Solids Struct. **48**(24), 3346–3358 (2011)
17. Matsushima, T., Chambon, R., Caillerie, D.: Large strain finite element analysis of a local second gradient model: application to localization. Int. J. Numer. Methods Eng. **54**(4), 499–521 (2002)
18. Mazars, J.: Application de la mécanique de l'endommagement au comportement non linéaire et à la rupture du béton de structure. Université Pierre et Marie Curie, Paris VI, Thèse de doctorat d'état (1984)
19. Mindlin, R.: Second gradient of strain and surface-tension in linear elasticity. In. J. Solids Struct. **1**(4), 417–438 (1965)

20. Pardoen, B., Levasseur, S., Collin, F.: Using local second gradient model and shear strain localisation to model the excavation damaged zone in unsaturated claystone. Rock Mech. Rock Eng. **48**(2), 691–714 (2014)
21. Shu, J., King, W., Fleck, N.: Finite elements for materials with strain gradient effects. Int. J. Numer. Methods Eng. **44**(3), 373–391 (1999)
22. Sieffert, Y., Al Holo, S., Chambon, R.: Loss of uniqueness of numerical solutions of the borehole problem modelled with enhanced media. Int. J. Solids Struct. **46**(17), 3173–3197 (2009)
23. Yang, Y., Misra, A.: Micromechanics based second gradient continuum theory for shear band modeling in cohesive granular materials following damage elasticity. Int. J. Solids Struct. **49**(18), 2500–2514 (2012)

Fast Landslide Propagation: Alternative Modelling Techniques

M. Pastor, A. Yagüe, M. Martín Stickle, S. Moussavi, Chuan Lin, A. Furlanetto, D. Manzanal, P. Mira and J.A. Fernandez Merodo

Abstract We model debris flows using two sets of nodes, describing the water and the solid phases, which can move relative to each other. We present first the mathematical model which will be used, deriving it from Zienkiewicz-Shiomi model, and arriving to the depth integrated model proposed by Pitman and Le. Then, we present the rheological models describing solid, fluid and their interaction. Next, the SPH model for two phases will be described. Finally, we present some application cases where we will compare the results provided by the proposed model against those obtained using more simplified models.

Keywords Landslide propagation • Rheological models • SPH model • Debris flow

1 Introduction

The term landslide embraces a large variety of phenomena. Regarding flow type movements, we find rock avalanches, debris flows, mudflows and flow slides. In the case of rock avalanches the flowing material consists on solid blocks which are disintegrating into smaller particles as they propagate downhill.

Mudflows consist of mixtures of fine soil particles and water, and they can be modelled with continuum models using a single phase viscous fluid.

Flow slides are mixtures of soil particles and water, with relative displacements between both phases. The tendency of the solid skeleton to dilate or contract results

M. Pastor (✉) · A. Yagüe · M. Martín Stickle · S. Moussavi · C. Lin · A. Furlanetto · D. Manzanal
ETS de Ingenieros de Caminos, Universidad Politécnica de Madrid, Madrid, Spain
e-mail: manuel.pastor@upm.es

P. Mira
Laboratorio de Geotecnia, CEDEX, Madrid, Spain

J.A. Fernandez Merodo
IGME, Madrid, Spain

E. Papamichos et al. (eds.), *Bifurcation and Degradation of Geomaterials with Engineering Applications*, Springer Series in Geomechanics and Geoengineering, DOI 10.1007/978-3-319-56397-8_25

on pore pressure generation affecting the effective stresses. To describe them, mathematical models formulated in terms of velocities and pore pressures (Biot-Zienkiewicz of u-pw type) provide a good approximation.

Debris flows is the more complex phenomenon from the modelling point of view, as both solid particles and water can have different velocities. The models have to include velocities of both solid and fluid phases, and the stresses acting on them.

This purpose of this work is to present the alternative models (mathematical, constitutive and numerical) which can be applied to these phenomena. We will pay special attention to the case of debris flows, where both phases and their mutual interaction have to be modelled.

The paper is structured as follows:

(i) We will present first the mathematical model which will be used, deriving it from Zienkiewicz-Shiomi model, and arriving to the depth integrated model proposed by Pitman and Le.
(ii) Then, we will present the rheological models describing solid, fluid and their interaction.
(iii) Next, the SPH model for two phases will be described.
(iv) Finally, we will present some application cases where we will compare the results provided by the proposed model against those obtained using more simplified models.

2 Mathematical Models

While models for debris flows are relatively modern in engineering geology, their origin can be tracked back along two different paths:

(i) Zienkiewicz, following Biot [2], proposed a general formulation which can be applied to debris flows [14]. From here, simpler approximations such as the u-pw model were derived. This model is used today in most of coupled problems of geotechnical engineering.
(ii) Within the area of physics of granular media, where important contributions have been made to study industrial problems such as the fluidized beds, we can mention the work done by Anderson and Jackson [1]. Regarding applications to debris flows, Pitman and Le [11], and Pudasaini [13] have proposed two phase models for debris flows, arriving to most interesting two phase's simplified models of depth integrated type.

In the same manner than the general two phase model of Zienkiewicz and Shiomi [14] can be particularized to the u-pw model, depth integrated models based on this approach have been proposed by Iverson and Denlinger [5] and Pastor et al. [6, 8].

Once integration along depth is performed, we obtain the balance of mass and momentum equations for both phases solid (s) and fluid (w) as:

$$\frac{\bar{d}^{(\alpha)}}{dt}(h_\alpha) + h_\alpha \operatorname{div} \bar{\boldsymbol{v}}_\alpha = \bar{n}_\alpha e_R$$

where

$$\alpha = \{s, w\} \quad h_\alpha = \bar{n}_\alpha h$$
$$\bar{n}_s = (1 - \bar{n}) \quad \bar{n}_w = \bar{n}$$

In above equations, α refers to the phase, $\frac{\bar{d}^{(\alpha)}}{dt}$ is the derivative following phase α, $\bar{n}_\alpha$ the volume fraction, h the depth of the flow, $\bar{\boldsymbol{v}}_\alpha$ the depth averaged velocity and e_R the erosion rate.

Regarding balance of momentum, we have:

$$\rho_\alpha h_\alpha \frac{d^{(\alpha)} \bar{\boldsymbol{v}}_\alpha}{dt} = \operatorname{grad}\left\{\frac{1}{2}\rho_\alpha h h_\alpha b_3\right\} - \frac{1}{2}\rho_\alpha h^2 b_3 \operatorname{grad} \bar{n}_\alpha + \tau_b^{(\alpha)} + h_\alpha \overline{\boldsymbol{R}}_\alpha + \rho_\alpha h_\alpha \boldsymbol{b} - \bar{n}_\alpha \rho_\alpha \bar{\boldsymbol{v}}_\alpha e_R$$

where ρ_α is the density, b the body forces, $\tau_b^{(\alpha)}$ the basal friction, and $\overline{\boldsymbol{R}}_\alpha$ the interaction between phases.

The general two-fluid depth integrated equations described above can be simplified if we assume that $\frac{d^{(w)}}{dt} \approx \frac{d^{(s)}}{dt} = \frac{d}{dt}$ and neglect the acceleration of the fluid relative to the solid. Equations results on:

$$\frac{\bar{d}h}{dt} + h \operatorname{div} \boldsymbol{v} = 0$$
$$\rho h \frac{\bar{d}\boldsymbol{v}}{dt} = \frac{1}{2} \operatorname{grad}\left(\rho h^2 b_3\right) + \tau_b + \rho h \boldsymbol{b} - \rho \boldsymbol{v} e_R$$

The basal shear stress depends on the rheological law used. In the case of a pure frictional material, it is given by

$$-\{(\rho_s - \rho_w)(1 - n) h b_3 + \Delta p_{wb}\} \frac{\boldsymbol{v}_s}{\|\boldsymbol{v}_s\|} \tan \phi_b$$

where ϕ_b is the friction angle and Δp_{wb} the increment of pore water pressure at the basal surface.

3 Numerical Modelling

Regarding numerical models, there exist today some interesting approaches which could be used, i.e., the Smooth particle Hydrodynamics (SPH) and the Material Point Model (MPM). The authors have explored the former approach, finding that features such as coupling with pore water and failure mechanisms can be accurately described [3, 4]. Both MPM and SPH are lagrangian approaches. They present, in comparison to the eulerian approaches the advantage of not needing special techniques such as level set to track interfaces.

In the case of depth integrated models, both approaches can be used. So far, most of models are of eulerian type, using finite volumes or elements Pitman and Lee [11, 12], Pudasaini [13], Pastor et al. [6], or of lagrangian type (SPH), as Pastor et al. [8] and McDougall and Hungr [7].

In a previous publication, the authors have addressed the problem of coupling SPH with a series of finite difference meshes associated to each SPH node, which provides better accuracy to reproduce pore pressure changes Pastor et al. [9, 10].

Here, we will model debris flows using two sets of nodes, describing the water and the solid phases, which can move relative to each other.

We will introduce:

(i) Two sets of nodes $\{x_{\alpha K}\}$ with $K = 1 \ldots N_\alpha$ where N_s and N_w are the number of SPH nodes in the solid and fluid phases,

(ii) The nodal variables: $h_{\alpha I}$ heights of phases at node I, $\bar{v}_{\alpha I}$ depth averaged, 2D velocities, and $\tau_{bI}^{(\alpha)}$ shear stress at the bottom.

The balance of mass equation is written at node I as:

$$\frac{\bar{d}}{dt} h_I + h_I \langle \operatorname{div} \bar{v} \rangle_I = \langle \bar{n}\, e_R \rangle_I$$

where the divergence term is approximated as:

$$\operatorname{div} \bar{v}_I = - \sum_J \Omega_J v_J \operatorname{grad} W_{IJ}$$

Note that for the sake of simplicity we have omitted the indexes α. Alternatively, the height can be obtained once the position of the nodes is known as:

$$h_I = \langle h(x_I) \rangle = \sum_J h_J \Omega_J W_{IJ} = \sum_J m_J W_{IJ}$$

The height can be normalized, which allows improving the approximation close to the boundary nodes:

$$h_I = \frac{\sum_J m_J W_{IJ}}{\sum_J \left(\frac{m_J}{h_J}\right) W_{IJ}}$$

Next, we will discretize the balance of linear momentum equation

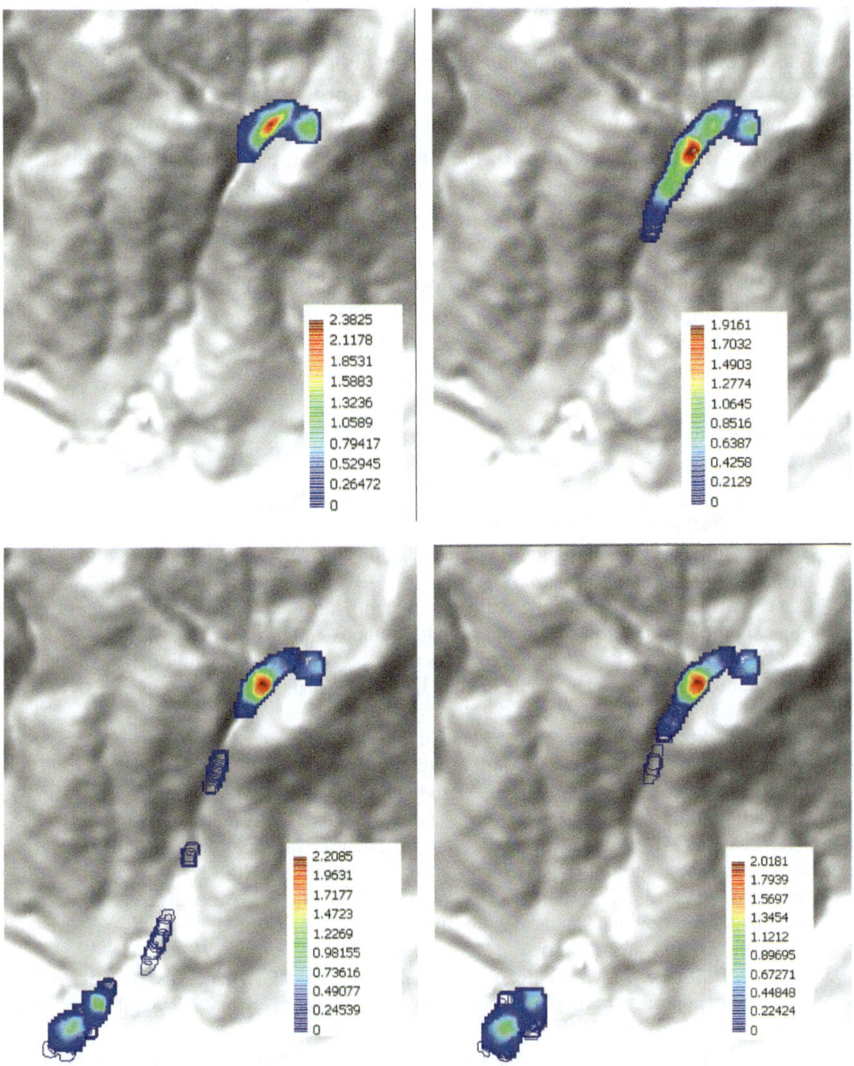

Fig. 1 Terrain topography and profiles of debris flow height at times 5, 10, 60 and 120 s

$$\frac{\bar{d}^{(\alpha)}}{dt}\bar{v}_{\alpha I} = \sum_J m_J\left(\frac{P_{\alpha I}}{h_I^2} + \frac{P_{\alpha J}}{h_J^2}\right)\text{grad}\, W_{IJ} + \sum_J m_J h_J^2 b_3\left(\frac{n_{\alpha I}}{h_I^2} + \frac{n_{\alpha J}}{h_J^2}\right)\text{grad}\, W_{IJ}$$
$$+ \frac{1}{\rho_\alpha h_{\alpha I}}\tau_{bI}^{(\alpha)} + \frac{1}{\rho_\alpha}\overline{\boldsymbol{R}}_\alpha + \boldsymbol{b} - \frac{1}{h_{\alpha I}}\bar{n}_\alpha \bar{\boldsymbol{v}}_\alpha e_R$$

where we have introduced the averaged pressure $P_\alpha = -\frac{1}{2}hh_\alpha b_3$.

In above equations, there is a term describing basal excess pore pressure at node I Δp_{wbI} which has to be obtained at each node and time step. One alternative is to use simple shape functions fulfilling boundary conditions at the surface and the basal surface. This approach presents the limitation of not being able to model changes of boundary conditions at the bottom. For instance, when a landslide runs over a very permeable basal layer- or a rack-pore pressure becomes zero there, while in the body of the landslide is not zero. If a single shape function is used, once the basal value is set to zero, the pressure becomes zero in the whole depth. The authors proposed in Pastor et al. [10] to use a set of finite difference meshes associated to each node in order to model pore water pressure dissipation or generation.

We have applied this model to reproduce the Sham Tseng San Tsuen Debris Flow, which happened in Hong Kong on the 23rd of August 1999 (Fig. 1).

References

1. Anderson, T.B., Jackson, R.: Fluid mechanical description of fluidized beds. Equations of motion. Ind. Eng. Chem. Fundam. **6**, 527–539 (1967)
2. Biot, M.A.: Theory of elasticity and consolidation for a porous anisotropic solid. J. Appl. Phys. **26**, 182 (1955)
3. Blanc, T., Pastor, M.: A stabilized fractional step, Runge-Kutta Taylor SPH algorithm for coupled problems in geomechanics. Comput. Methods Appl. Mech. Eng. **221**, 41–53 (2012)
4. Blanc, T., Pastor, M.: A stabilized smoothed particle hydrodynamics, Taylor-Galerkin algorithm for soil dynamics problems. Int. J. Numer. Anal. Methods Geomech. **37**(1), 1–30 (2013)
5. Iverson, R.M., Denlinger, R.P.: Flow of variably fluidized granular masses across three-dimensional terrain: 1. Coulomb mixture theory. J. Geophys. Res. **106**, 537 (2001)
6. Pastor, M., Quecedo, M., Fernández Merodo, J.A., Herreros, M.I., Gonzalez, E., Mira, P.: Modelling tailings dams and mine waste dumps failures. Géotechnique **52**, 579–591 (2002)
7. McDougall, S., Hungr, O: A model for the analysis of rapid landslide motion across three-dimensional terrain. Can. Geotech. J. **41**(6), 1084–1097 (2004)
8. Pastor, M., Haddad, B., Sorbino, G., Cuomo, S., Drempetic, V.: A depth-integrated, coupled SPH model for flow-like landslides and related phenomena. Int. J. Numer. Anal. Methods Geomech. **33**, 143–172 (2009)
9. Pastor, M., Stickle, M.M., Dutto, P., Mira, P., Merodo, J.F., Blanc, T., Benítez, A.S.: A viscoplastic approach to the behaviour of fluidized geomaterials with application to fast landslides. Contin. Mech. Thermodyn. **27**(1–2), 21–47 (2015a)
10. Pastor, M., Blanc, T., Haddad, B., Drempetic, V., Morles, M.S., Dutto, P., et al.: Depth averaged models for fast landslide propagation: mathematical, rheological and numerical aspects. Arch. Comput. Methods Eng. **22**(1), 67–104 (2015b)

11. Pitman, E.B., Le, L.: A two-fluid model for avalanche and debris flows. Philos. Trans. A. Math. Phys. Eng. Sci. **363**, 1573–1601 (2005)
12. Prime, N., Dufour, F., Darve, F.: Unified model for geomaterial solid/fluid states and the transition in between. J. Eng. Mech. **140**(6), 04014031 (2013)
13. Pudasaini, S.P.: A general two-phase debris flow model. J. Geophys. Res.: Earth Surf. (2003–2012)**117**(F3) (2012)
14. Zienkiewicz, O.C., Shiomi, T.: Dynamic behaviour of saturated porous media; the generalized Biot formulation and its numerical solution. Int. J. Numer. Anal. Methods Geomech. **8**, 71–96 (1984)

Bifurcation Criteria for Strain Localization in Multiphysical Systems

Manolis Veveakis, Thomas Poulet, Sotiris Alevizos and Martin Paesold

Abstract The study of bifurcation criteria for non-isothermal processes in geomaterials requires approaches that deviate from the classical material bifurcation approach. Indeed, in a quasi-static stress state of the medium, the admissible equilibria are of steady-creep-type and are governed also by the energy balance. Furthermore the steady-state temperature profiles are far from homogeneous as shown by [3], leading the analytical stability analysis to be rather complex. In this contribution, we adopt an overstress plasticity approach and present approximations for the criteria of the onset of localisation of deformation in a plane strain setting, using numerical continuation methods.

1 Modelling Considerations

In plane strain conditions, the general system of equations under consideration consists of the momentum and energy balance equations which can be written as,

$$\sigma_{ji,j} = 0, \tag{1}$$

$$\rho c \dot{T} = \alpha T_{,ii} + \Phi. \tag{2}$$

where index notation is applied ($i = 1, 2$), the over imposed dot denotes time differentiation and comma denotes spatial differentiation. In this system, α is the thermal conductivity, σ_{ij} is the stress tensor, T is the temperature, ρ the density, c the specific heat capacity and Φ the mechanical work dissipated into heat. The temperature diffusion equation (2), was derived by neglecting the higher order energy terms, like

M. Veveakis (✉) · T. Poulet · S. Alevizos
University of New South Wales, Sydney, NSW 2052, Australia
e-mail: e.veveakis@unsw.edu.au

M. Veveakis · M. Paesold
School of Mathematics and Statistics, University of Western Australia, Perth, Australia

T. Poulet · M. Paesold
CSIRO Mineral Resources, North Ryde, NSW 2113, Australia

E. Papamichos et al. (eds.), *Bifurcation and Degradation of Geomaterials with Engineering Applications*, Springer Series in Geomechanics and Geoengineering, DOI 10.1007/978-3-319-56397-8_26

the thermo-elastic heating and by combining the energy balance law with the second law of thermodynamics.

In addition, we adopt the stress decomposition $\sigma_{ij} = p\delta_{ij} + s_{ij}$, with p the volumetric mean stress and s_{ij} the deviatoric stress. For such a problem, we formulate the governing equations in the equivalent coordinate system where the stress tensor is rotated such that its elements are the mean stress $p = I_1/3$ and the von Mises stress $q = \sqrt{3J_2}$. In these expressions $I_1 = tr(\sigma_{ij})$ is the first invariant of the stress tensor and $J_2 = s_{ij}s_{ij}/2$. The corresponding coordinate transformation dates back to Levy [4] and has the form:

$$\sigma_{11} = p - q\sin 2\psi, \quad \sigma_{22} = p + q\sin 2\psi, \quad \sigma_{12} = \sigma_{21} = q\cos 2\psi, \tag{3}$$

where ψ is the rotation angle of the coordinate system.

The mechanical dissipation takes the form (see also [7, 11]) $\Phi = \beta\sigma_{ij}\dot{\varepsilon}^p_{ij}$, where β is the Taylor-Quinney coefficient [10]. In order to close the system of equations, we need to provide a mechanical constitutive law. To that end, we first split the strain rate into elastic (reversible) and plastic (irreversible) parts $\dot{\varepsilon}_{ij} = \dot{\varepsilon}^e_{ij} + \dot{\varepsilon}^p_{ij}$. For the elastic component we adopt a linear elastic law of the form $\dot{\varepsilon}^e_{ij} = C^{-1}_{ijkl}\dot{\sigma}_{kl}$.

For the irreversible part, we assume that the Helmholtz free energy is invertible, such that the evolution of the plastic strain depends on the stress and temperature and takes the following form,

$$\dot{\varepsilon}^p_{ij} = \dot{\varepsilon}_0\, e^{-T_0/T}\langle f(\sigma_{ij})\rangle \frac{\partial g}{\partial \sigma_{ij}}, \tag{4}$$

where $\dot{\varepsilon}_0$ is a reference strain rate, where T_0 is the activation temperature for the thermal hardening mechanism, f is the so called over-stress function [8], g is the plastic potential function and the Macaulay brackets $\langle\cdot\rangle$ ensure zero plastic strain before yield. This decomposition is supported by experimental data at elevated temperatures, below the phase transition temperature of the material [2] and the two most representative constitutive responses of temperature and rate dependent materials are an Arrhenius-type dependency on temperature, with either a power-law or an exponential dependency on stress. The function $f(\sigma_{ij})$ is an arbitrary flow stress function.

The exact form of the constitutive equation is not prescribed during the analysis of the bifurcation in order to emphasize the generic nature of the formulation, where the onset of plastic deformation is derived from the basic assumptions of the energetics. The only important aspect of the constitutive response of the material is that it must obey a visco-plastic relationship linking the plastic strain-rate with the stress. This is required so that in the steady-state limit of the equations ($\dot{\sigma}_{ij} = \dot{T} = 0$) the mechanical dissipation remains non-zero. Finally, as shown in earlier studies (see for example [6]), the choice of the form of the temperature dependence of the plastic flow law is not central for the results of the present study and the Arrhenius type exponential dependency is the one that allows for the most convenient mathematical treatment.

2 Energy Bifurcation

We start by studying the steady-state limit of the system, in which $\dot{T} = \dot{\sigma}_{ij} = 0$. Therefore, in this regime the elastic contribution is neglected and the problem reduces to that of the study of the response of a rigid (visco-)plastic material. This setting can therefore be considered to be a direct extension of the slip line field theory to thermo-visco-plastic materials. We note that in the present formulation the temperature equation (2) is active only when dissipation is non-zero. Since this is achieved in the post-yield regime, we expect that the orientation of possible localization planes would arise from the characteristics of the stress equilibrium, in accordance with the theory of plasticity [4].

At the point of initial yield, where the temperature equation is inactive, the response of the system is governed by the stress equilibrium equations. We assume a generalized yield surface at a reference temperature of the form $q = q_Y(p)$. In order to find the characteristics of the hyperbolic differential stress equilibrium equations (slip lines) we substitute the Levy stress transformations Eq. (3), into the stress equilibrium, Eq. (1). The slip lines are parametrized in terms of the arc-length s and the slopes of the characteristics along the x_1- and x_2-axes read

$$\frac{\partial x_1}{\partial s_k} = \frac{\mu^{(k)}}{\sqrt{1+\mu^{(k)2}}}, \quad \frac{\partial x_2}{\partial s_k} = \frac{1}{\sqrt{1+\mu^{(k)2}}}, \quad (k = 1, 2) \tag{5}$$

where $\mu^{(k)}$ is the kth left eigenvalue of the stress equilibrium equations. For a general yield surface $q(p)$

$$\mu^{(1,2)} = \frac{\mp\sqrt{1-h^2} + \cos(2\psi)}{h + \sin(2\psi)} \tag{6}$$

and the quantity $h = q'_Y(p)$ is a generalized pressure modulus. Note that in this expression p must be critical, i.e. equal to its yield value. For incompressible materials $h = 0$.

Rescaling Eq. (1), along the characteristics Eq. (5) provide the generalized Hencky's equations

$$0 = \sqrt{1-h^2}\, p_s \pm 2q\psi_s. \tag{7}$$

In the case of a von Mises material, where $q_Y = k = const.$, Eq. (6) simplifies (as $h = 0$) to the familiar form $\mu^{(1)} = -\tan\psi,\ \mu^{(2)} = \cot\psi$, and the corresponding traces are commonly known as α/β-slip lines. Along the slip lines Eq. (7) reduce to the classical Hencky's equations $p \pm 2k\psi = C_{\alpha,\beta}$.

The energy balance can be brought into dimensionless form by setting

$$\theta = \frac{T - T_b}{T_b}, \quad \hat{x}_i = \frac{x_i}{L_i}\ (i = 1, 2), \quad Ar = \frac{T_0}{T_b}, \tag{8}$$

where T_b is the boundary temperature and L_i is an appropriate length scale. Since we are interested in deformation taking place under isothermal boundary conditions the final dimensionless equation is rescaled along the characteristics to obtain (the superimposed hats are dropped for convenience)

$$\frac{\partial^2 \theta}{\partial s^2} + Gr^{1D} \exp\left(\frac{Ar\,\theta}{1+\theta}\right) = 0, \tag{9}$$

where the Gruntfest number admits the following form

$$Gr^{1D} = \frac{\beta \dot{\varepsilon}_0 L^2}{\alpha T_b} \frac{\mu^2}{\left(1+\mu^2\right)^2} e^{-Ar} \sigma_{ij} \langle f(\sigma_{ij}) \rangle \frac{\partial g}{\partial \sigma_{ij}}, \tag{10}$$

where L is a length scale. This Gruntfest number is spatially dependent through μ, and also incorporates the dimensionality of the system at hand, through L.

In Fig. 1, we present the steady state response of the system with respect to the Gruntfest number, in a bifurcation diagram expressed for the dissipation function ϕ. We find three distinct steady states of alternating stability. The unstable branch BC corresponds to a localization instability across the characteristic trace for the normalized dissipation function ϕ. Since in Fig. 1a the branches AB and CD are stable in the dissipation the unstable branch BC cannot be admitted and the solution jumps from stable near isothermal homogeneous plastic deformation on AB to localized near adiabatic deformation on CD. The saddle point B therefore corresponds to the necessary condition for the loss of ellipticity of the thermomechanical system, after which localisation is progressively achieved. It is a similar condition to the tradi-

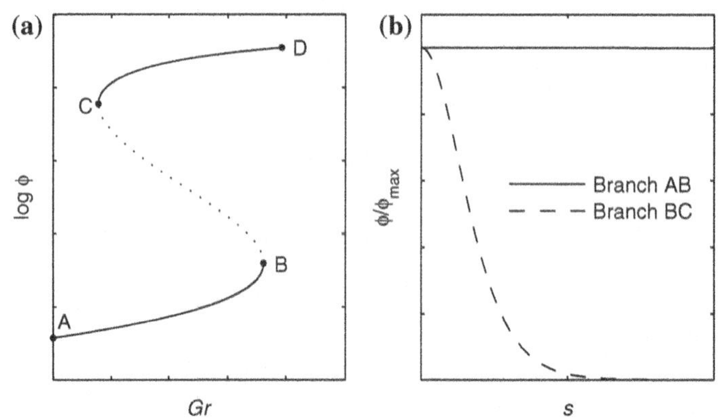

Fig. 1 **a** Folded S-curve. Along the branch AB the solution of Eq. (9) corresponds to an isothermal temperature profile whereas along the section between the turning points B and C the solutions localizes. **b** Examples of the one-dimensional spatial pattern of the dissipation profile for $Ar = 10$. The profiles are normalized with respect to the maximum value of dissipation

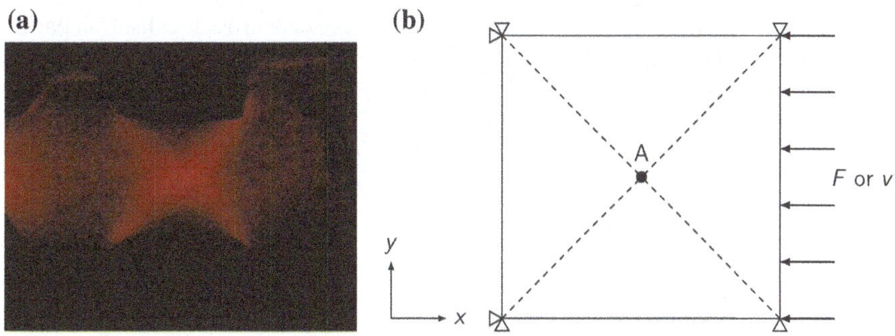

Fig. 2 **a** The thermal heat cross observed in warm metals when forged [5]. **b** Geometry of the problem configuration. A *square* is deformed by applying either a constant force F or constant velocity v at the *right hand side* of the sample. The sample is pinned on the left and in the y-direction. At the centre 'A' the sample is probed for various quantities. The theoretical analysis presented in the previous sections for a rate-dependent von Mises material suggests that the slip lines of this model run diagonally across the specimen and are represented by *dashed lines* [5]

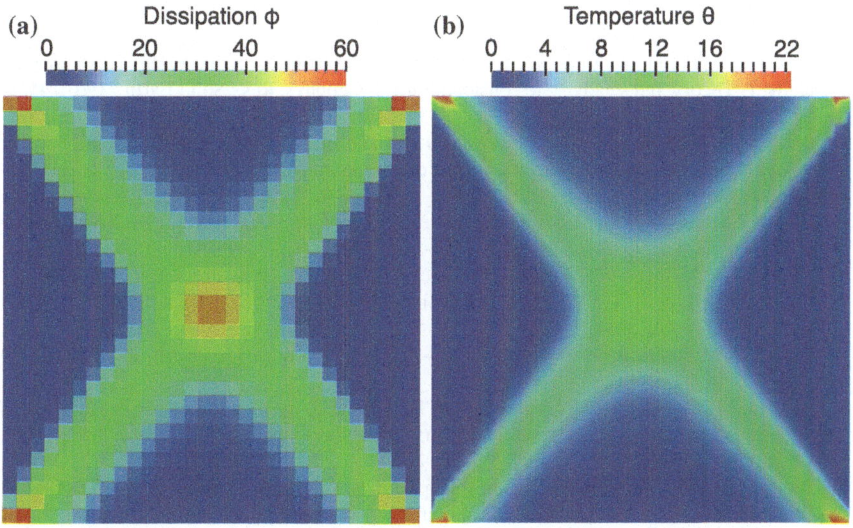

Fig. 3 Heat lines under fast constant velocity loading. The heat lines emerge from an initially homogeneous temperature state. The heat lines follow the slip lines and can be observed as a localisation of **a** mechanical dissipation and **b** temperature

tional jump conditions for stress and velocity discontinuities in the classical slip line field solution but is expressed here in terms of their product, the dissipation.

This transition from homogeneous to localized deformation is shown in Fig. 1b, where the normalized dissipation is shown to localize towards the center of the domain as it crosses the unstable BC branch of the S-curve (see also the results of [6, 11]). The critical threshold for the transition from the stable branch AB to the new steady state CD is characterized by the turning point B.

To showcase this condition, we consider the simple heat cross localisation pattern of Fig. 2a. The mathematical problem can be idealised following Fig. 2b, to obtain localisation of temperature (dissipation) along the slip lines of the mechanical deformation when $Gr > Gr_B$ (Fig. 3). No localisation is observed before the turning point of the S-curve.

3 Conclusion

In this work we have summarised an energy based localisation theory for temperature sensitive viscoplastic materials. The generalisation of this criterion for more complicated physics is straightforward as shown recently in the literature [1, 9, 12].

References

1. Alevizos, S., Poulet, T., Veveakis, E.: Thermo-poro-mechanics of chemically active creeping faults. 1: Theory and steady state considerations. J. Geophys. Res.: Solid Earth **119**(6), 4558–4582 (2014)
2. Bauwens-Crowet, C., Ots, J.-M., Bauwens, J.-C.: The strain-rate and temperature dependence of yield of polycarbonate in tension, tensile creep and impact tests. J. Mater. Sci. **9**(7), 1197–1201 (1974)
3. Chen, H., Douglas, A.S., Malek-Madani, R.: An asymptotic stability condition for inhomogeneous simple shear. Q. Appl. Math. **47**, 247–262 (1989)
4. Hill, R.: The Mathematical Theory of Plasticity. Oxford University Press, London (1950)
5. Johnson, W., Baraya, G., Slater, R.: On heat lines or lines of thermal discontinuity. Int. J. Mech. Sci. **6**(6), 409–414 (1964)
6. Leroy, Y., Molinari, A.: Stability of steady-states in shear zones. J. Mech. Phys. Solids **40**, 181–212 (1992)
7. Paesold, M., Bassom, A., Regenauer-Lieb, K., Veveakis, E.: Conditions for the localisation of plastic deformation in temperature sensitive viscoplastic materials. J. Mech. Mater. Struct. **11**(2), 113–136 (2016)
8. Perzyna, P.: Fundamental problems in viscoplasticity. Adv. Appl. Mech. **9**, 243–377 (1966)
9. Peters, M., Herwegh, M., Paesold, M.K., Poulet, T., Regenauer-Lieb, K., Veveakis, M.: Boudinage and folding as an energy instability in ductile deformation. J. Geophys. Res.: Solid Earth, **121**(5):3996–4013 (2016)
10. Taylor, G., Quinney, H.: The latent energy remaining in a metal after cold working. Proc. R. Soc., Ser. A. **143**, 307–326 (1934)
11. Veveakis, E., Alevizos, S., Vardoulakis, I.: Chemical reaction capping of thermal instabilities during shear of frictional faults. J. Mech. Phys. Solids **58**, 1175–1194 (2010)
12. Veveakis, E., Poulet, T., Alevizos, S.: Thermo-poro-mechanics of chemically active creeping faults: 2. Transient considerations. J. Geophys. Res.: Solid Earth **119**(6), 4583–4605 (2014)

Modelling Creeping and Catastrophic Failure of Thermomechanically Driven Landslides

A. Zervos, E. Alonso and N. Pinyol

Abstract Frictional heating of the slip zone is a possible explanation for why some catastrophic landslides moved so fast and so far. The transition from a creeping phase to catastrophic collapse governed by frictional heating is also a pertinent problem: what governs it and why does it occur in some landslides but not others? Here we consider the possibility that frictional heating combined with non-linear velocity-dependent friction hardening suffices to explain both phenomena as well as the transition. The paper outlines the relevant theoretical considerations and a numerical treatment. It discusses the integration of the governing equations with adequate temporal resolution for both regimes and the assumed location of the far boundaries, which needs to strike a balance between accuracy and cost.

1 Introduction

Large catastrophic landslides are a constant threat to human communities and our infrastructure. The morphology and dynamics of these events is quite diverse, and a classification can be found in Hungr et al. [8]. Here we consider landslides that can be described, at least in the timeframe of interest, as a coherent mass sliding on a well-defined slip surface. This surface is assumed fully developed; we are thus concerned with the post-failure stage of first-time failures, and with reactivated slides where movement restarts along a pre-existing failure surface. The slide's motion and its evolution in time are governed by processes occurring at this shearing surface, which is orders of magnitude thinner than the sliding mass. The difference of scale makes it very difficult to numerically deal with both in a common framework. The approach followed here is to simplify the dynamics of the

A. Zervos (✉)
Engineering and the Environment, University of Southampton, Southampton, UK
e-mail: az@soton.ac.uk

E. Alonso · N. Pinyol
Department of Geotechnical Engineering and Geosciences, Universitat Politecnica di Catalunya (UPC), Barcelona, Spain

E. Papamichos et al. (eds.), *Bifurcation and Degradation of Geomaterials with Engineering Applications*, Springer Series in Geomechanics and Geoengineering, DOI 10.1007/978-3-319-56397-8_27

moving mass by assuming it is rigid in comparison to the shearband material, and model in detail the latter; this practice is well established, e.g. Vardoulakis [13], Veveakis et al. [14], Pinyol and Alonso [9, 10], Goren and Aharonov [5], Cecinato et al. [3], Cecinato and Zervos [2], Alonso et al. [1]. In all these cases frictional heating resulting to pressurization were fundamental parts of the analysis; their relevance was first highlighted in thc pioneering contributions of Habib [6], Uriel and Molina [12] and Voigt and Faust [15].

Frictional heating is a destabilising process that leads to immediate catastrophic failure. Slow movement at near-constant velocity is only possible if a stabilising mechanism is present to counter-balance its effect. Rate-dependent friction is a physically meaningful such mechanism [1]. We consider:

$$\tan \phi = \tan \phi_0 \left(1 + A \ \ln \frac{v}{v_0}\right) \quad (1)$$

where v is the sliding velocity, v_0 is a reference velocity and the rate parameter A can be taken to vary between extremes of 10^{-5} and 10^{-2} [1].

2 Governing Equations

As in Alonso et al. [1], the governing equations are written using the dimensionless variables: $\hat{z} = \frac{z}{D}$, $\hat{t} = \frac{t\sqrt{gD}}{D}$, $\theta(\hat{z},\hat{t}) = \frac{\theta(z,t)}{\theta_0}$, $\hat{u}_w(\hat{z},\hat{t}) = \frac{u_w(z,t)}{\rho g D}$, $\hat{p}_w = \frac{p_w}{\rho g D}$ and $\hat{v}(\hat{t}) = \frac{v(t)}{\sqrt{gD}}$, where t is the time, θ the temperature, u_w the excess pore water pressure, p_w the initial pore water pressure, v the landslide velocity, ρ the density of saturated soil, D the slide thickness, g the gravitational acceleration and θ_0 a reference initial temperature, assumed equal to 10 °C. The governing equations and initial and boundary conditions are expressed as follows:

$$\frac{d\hat{v}(\hat{t})}{d\hat{t}} = \sin(\beta) - \left(\cos(\beta) - \frac{\hat{p}_{wh}}{\cos(\beta)} - \frac{\hat{u}_w^{\max}(\hat{t})}{\cos(\beta)}\right) \tan \varphi' \quad (2)$$

$$\frac{\partial \theta(\hat{z},\hat{t})}{\partial \hat{t}} - \frac{\gamma}{(\rho c)_m D\sqrt{gD}} \frac{\partial^2 \theta(\hat{z},\hat{t})}{\partial \hat{z}^2} = \hat{H}(\hat{z},\hat{t}) \quad (3)$$

$$-\frac{\beta_{soil}\theta_0}{m_{soil}\rho g D} \frac{d\theta(\hat{z},\hat{t})}{d\hat{t}} + \frac{d\hat{u}_w(\hat{z},\hat{t})}{d\hat{t}} - \frac{k}{\gamma_w m_{soil} D\sqrt{gD}} \frac{d^2 \hat{u}_w(\hat{z},\hat{t})}{d\hat{z}^2} = 0 \quad (4)$$

and describe the slide dynamics (Eq. 2), heat generation and transfer (Eq. 3) and excess pore pressure generation and diffusion (Eq. 4). $(\rho c)_m$ is the specific heat of the saturated soil and Γ its thermal conductivity. The source term in Eq. 3 represents heat production within the shearband due to mechanical energy dissipation:

$$\hat{H}(\hat{z},\hat{t}) = \frac{\rho g D^2}{(\rho c)_m \theta_0 2e}\left(\cos^2(\beta) - \hat{p}_{wh} - \hat{u}_w^{\max}(\hat{t})\right)\tan\varphi' \hat{v}(\hat{t}) \quad (5)$$

where e is the half-thickness of the shearband and β the slope angle.

The initial conditions are zero excess pore pressure and ambient temperature in the whole domain, which consists of the upper half of the shearband and the overlying sliding mass. The source terms of Eqs. 3 and 4 are only active within the shearband. The mid-plane of the shearband is a symmetry boundary. The far boundary is at ambient temperature and zero excess pore pressure.

3 Numerical Treatment

Equations 1–5 form a system of coupled, non-linear partial differential equations. Past work has relied on centred-space forward-time (explicit) finite differences (e.g. Vardoulakis [13]; Alonso et al. [1]; Cecinato et al. [2, 3]). However explicit integration is unstable and requires very small timesteps, here typically of the order of 10^{-5} s, to mitigate. This is tolerable when modelling the catastrophic phase over the final few tens of seconds, but makes modelling of the long-term creeping behaviour that may precede failure impractical. An alternative is to use an iterative, unconditionally stable implicit scheme that allows larger timesteps.

Two different classes of implicit methods were considered: Runge-Kutta and linear multistep methods. Runge-Kutta methods generally require calculations at intermediate points; these are only useful for the timestep in question. Linear multistep methods calculate the solution at the next timestep using results of previous timesteps only, so were preferred as more efficient. Their accuracy depends on the number n of previous steps used; higher n leads to a more expensive, more accurate calculation. The general form of a linear multistep method is:

$$y_{i+n} = h(b_n f(t_{i+n}, y_{i+n}) + b_{n-1} f(t_{i+n-1}, y_{i+n-1}) + \cdots + b_0 f(t_1, y_1)) \quad (6)$$

where y the unknown function, f its first derivative that is being integrated, and h the integration step. We specifically considered *implicit* multistep methods, i.e. ones where $b_n \neq 0$, also known as Adams-Moulton methods.

In principle we could employ an Adams-Moulton method of as high accuracy as we wish; an n-step method is $O(h^{n+1})$ accurate. However, Eq. 6 implies a constant step equal to h. Therefore using an Adams-Moulton method of high order will improve accuracy but deprive us of the ability to vary the size of the integration step (timestep) as the calculation progresses. The latter is essential for modelling with the same numerical scheme periods of prolonged creeping motion, where the solution changes very little, as well as the last few seconds of catastrophic collapse, during which the solution varies dramatically. For this reason we resort to the two-step Adams-Moulton method, which is the (implicit) trapezoidal rule: It is $O(h^2)$ accurate and uses the results of the current and the previous timesteps only, naturally allowing for a varying timestep. In principle, it is possible to produce higher order Adams-Moulton methods that accommodate varying timestep, however they will be significantly more complex and computationally costly, e.g. Skeel [11].

Using superscripts to denote the timestep and subscripts the spatial location, Eq. 3 can be discretised in terms of original, rather than normalised, quantities thus:

$$\theta_i^j = \theta_i^{j-1} + \frac{\Delta t}{2(\rho c)_m}\left[H^{j-1} + H^j + \frac{\Gamma}{(\Delta z)^2}\left(\theta_{i+1}^{j-1} - 2\theta_i^{j-1} + \theta_{i-1}^{j-1} + \theta_{i+1}^{j-1} - 2\theta_i^{j-1} + \theta_{i-1}^{j-1}\right)\right] \tag{7}$$

and similarly Eq. 4. Discretising Eq. 2 yields an equation of the general form:

$$v^j = v^{j-1} + \frac{\Delta t}{2}\left\{c_0 + c_1 \tan\phi^j + c_2 u_w^{max,j} \tan\phi^j\right\} \tag{8}$$

where c_0, c_1 and c_2 depend on quantities already known from the previous timestep. Equation 8 is strongly nonlinear as it involves the velocity v^j, $\ln(v^j)$, as well as a cross term of $\ln(v^j)$ and the unknown excess pore pressure at the shearband centre. When solving the system of discretised equations using Newton-Raphson, the strong nonlinearity of Eq. 8 impedes convergence unless $\Delta t \sim 10^{-3}$ s. This makes modelling of long-term creeping motions problematic.

To mitigate a staggered solution scheme was devised: the system corresponding to the discretised forms of Eqs. 3 and 4 was solved by Newton-Raphson using the current approximation of the velocity; the dynamics equation was subsequently solved for the velocity by Pegasus [4], a method of the *regula falsi* family that is fast and efficient in finding roots of nonlinear equations. For the timestepping used here it was sufficient to seek a solution in [v/2, 2v]. Using the new approximation of the velocity Eqs. 3 and 4 were solved again, etc. After each iteration the residual norms were calculated for the dynamic equation, the pore pressure and the temperature equations; iterations continued till all were below 10^{-6}.

The initial timestep was $\Delta t = 1$ s and was varied based on the rate of convergence. It was increased by 10% following convergence within 6 iterations, while decreased by 33% if no convergence was achieved after 20 iterations. The maximum timestep was 24 h and was possible to use for the most part of an analysis, making the efficient modelling of years of creeping motion possible.

An additional challenge was the location of the far boundary. Previous work considered that ambient values of temperature and excess pore pressure a small multiple of the shearband thickness away. Modelling the creep phase should require a larger domain, as heat and pore water will potentially have the time to travel longer distances. Here, the far boundary was initially considered at a distance 10 times the shearband thickness. If, during the calculation, the value of either the excess pore pressure or the temperature next to the boundary was found to differ by more than 10^{-10} kPa or °C respectively from the corresponding boundary condition, the domain was expanded by 100 times the shearband thickness before the calculation continued. Therefore all models progressively became more expensive; typical calculations started with about 300 unknowns but eventually involved in excess of 300,000. Nevertheless, due to the direct, sparse multi-frontal solver [7] used, all models could run on a desktop computer within hours.

4 Some Results

To investigate the effect of the rate parameter *A* of Eq. 1, a set of analyses is run where the slide is triggered by increasing the initial pore pressure to a safety factor of 0.99. Calculations end when a runout of twice the slide thickness is reached. Figure 1 summarises the results. For $A \leq 10^{-4}$, catastrophic failure occurs almost immediately. For $A = 5 * 10^{-4}$ the slide creeps with near-constant velocity followed by catastrophic failure, while for $A > 5 * 10^{-4}$ the slide simply creeps. A threshold value $6 * 10^{-4} > A_{crit} > 5 * 10^{-4}$ determines whether catastrophic failure will occur. In this range, even a 2% difference in the rate parameter can have order-of magnitude impact on the duration of the creep phase, although the predicted creep velocity is less sensitive. For comparison, Fig. 1 also presents the (baseline) constant creep velocity each slide would attain if thermal effects were ignored: it is evident that, even where thermal effects do not eventually lead to catastrophic collapse, they still increase slide velocity by at least one order of magnitude.

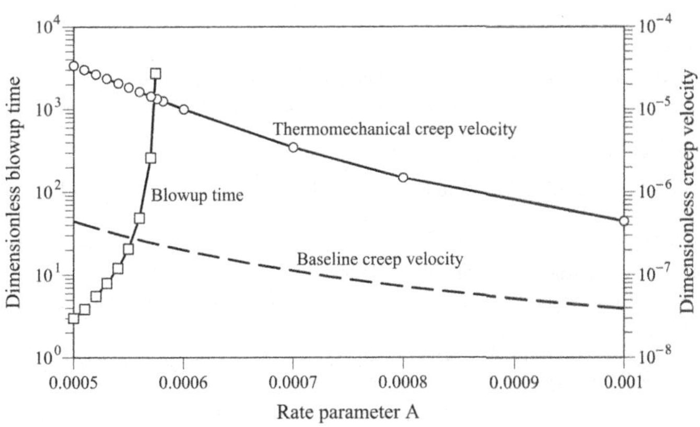

Fig. 1 Effect of rate parameter on creep velocity and blowup time

5 Conclusions

A combination of rate dependent friction hardening and frictional heating suffices to describe the prolonged creeping phase followed by sudden catastrophic collapse that is seen in some landslides. Both regimes of movement can be modelled numerically using the same implicit finite difference scheme with adaptive time stepping. Rate dependent friction leads to a strongly nonlinear dynamics equation which, when combined with heat and pore pressure generation and diffusion, necessitates a staggered solution scheme. In addition, the location of the "far" boundary needs to be continually adjusted to avoid numerical artefacts.

Finally, the predicted behaviour of a slide is very sensitive to the value of the rate parameter; for a given runout distance, the latter determines whether catastrophic collapse will eventually occur or not.

References

1. Alonso, E.E., Zervos, A., Pinyol, N.M.: Thermo-poro-mechanical analysis of landslides: from creeping behaviour to catastrophic failure. Geotechnique **66**(3), 202–219 (2016)
2. Cecinato, F., Zervos, A.: Influence of thermomechanics in the catastrophic collapse of planar landslides. Can. Geotech. J. **49**(2), 207–225 (2012)
3. Cecinato, F., Zervos, A., Veveakis, E.: A thermo-mechanical model for the catastrophic collapse of large landslides. Int. J. Numer. Anal. Methods Geomech. **35**, 1507–1535 (2011)
4. Dowell, M., Jarratt, P.: The "Pegasus" method for computing the root of an equation. BIT **12**, 503–508 (1972)
5. Goren, L., Aharonov, E.: Long runout landslides: the role of frictional heating and hydraulic diffusivity. Geophys. Res. Lett. **34**(L07301), 1–7 (2007)
6. Habib, P.: Production of gaseous pore pressure during rock slides. Rock. Mech. Rock. Eng. **7**, 193–197 (1975)
7. HSL: A collection of Fortran codes for large scale scientific computation. STFC Rutherford Applieton Laboratory, Harwell, Didcot, Oxfordshire, UK. http://www.hsl.rl.ac.uk/ (2015)
8. Hungr, O., Leroueil, S., Picarelli, L.: The Varnes classification of landslide types, an update. Landslides **11**, 167–194 (2014)
9. Pinyol, N.M., Alonso, E.E.: Criteria for rapid sliding II. Eng. Geol. **114**(3–4), 211–227 (2010a)
10. Pinyol, N.M., Alonso, E.E.: Fast planar slides. A closed-form thermo-hydro-mechanical solution. Int. J. Numer. Anal. Methods Geomech. **34**, 27–52 (2010)
11. Skeel, R.D.: Construction of variable-stepsize multistep formulas. Math. Comput. **47**(176), 503–510 (1986)
12. Uriel Romero, S., Molina, R.: Kinematic aspects of Vaiont slide. Proceedings of the 3rd International Conference of the ISRMR, Denver, USA. Natl. Acad. Sci. **2B**, 865–870 (1977)
13. Vardoulakis, I.: Dynamic thermo-poro-mechanical analysis of catastrofic landslides. Géotechnique **52**(3), 157–171 (2002)
14. Veveakis, E., Vardoulakis, I., Di Toro, G.: Thermoporomechanics of creeping landslides: the 1963 Vaiont slide, northern Italy. J. Geoph. Res. **112**(F3), F03026 (2007)
15. Voigt, B., Faust, C.: Frictional heat and strength loss in some rapid landslides. Géotechnique **32**(1), 43–54 (1982)

Material Point Method Modelling of Landslides with Coupled Segregation

Benjy Marks and Itai Einav

Abstract Landslides, debris flows and avalanches often exhibit strong segregation during flow and deposition. The largest particles are usually found at the nose of the avalanche, with moderate sized particles at the free surface, and smaller particles at the base of the flow. At the same time, we know that the constitutive behaviour of such a system is strongly influenced by the local average grainsize. In numerical modelling of these flows, the coupling of the spatial heterogeneity and constitutive behaviour has heretofore only been weakly coupled, if addressed at all. Here, we will present a unified framework for coupling the feedback between these two phenomena using the material point method. Several examples of landslide propagation will be investigated. The effect of flow lubrication via segregation will be highlighted.

1 Introduction

Granular flows in nature, such as landslides, debris flows and avalanches, often are composed of particles ranging in size from clay platelets to boulders. These particles are constructed from a variety of materials, with significantly varying properties. Unifying the description of these flows has proven to be a considerable challenge, not least because of the spatial variability of the material [1].

One significant issue controlling the spatial variability of these flows is grainsize segregation [2]. In the context of a gravity current, this phenomenon causes larger particles to rise to the surface of the flow, and smaller particles to sink to the base. A typical phenomenological description of this mechanism is that after a collision between particles, a new void space is formed, which is preferentially filled by a smaller particle falling into, as it is less likely that a large particle will fit into the void than a small one.

B. Marks (✉) · I. Einav
The University of Sydney, Sydney, NSW, Australia
e-mail: benjy.marks@sydney.edu.au

E. Papamichos et al. (eds.), *Bifurcation and Degradation of Geomaterials with Engineering Applications*, Springer Series in Geomechanics and Geoengineering, DOI 10.1007/978-3-319-56397-8_28

Whilst the physical mechanisms responsible for this phenomenon are still under investigation [3], many analytic [4–6] and numerical [7–9] models exist. These models, however, generally treat separately the bulk flow and the segregation induced flow, with some recent notable exceptions [10, 11]. This lack of coupling between the bulk flow and the segregation prevents a systematic study of the bulk rheology of the material as it segregates. Here, we develop a comprehensive framework for studying grainsize evolution that is weakly coupled to the bulk rheology. This is then implemented in a large deformation continuum solver, using the material point method (MPM), which is distributed under as open source code (under the GPL3.0 license) and freely available at http://www.benjymarks.com.

2 Grainsize Dynamics

Representing a polydisperse granular material as a five dimensional continuum has been the subject of several previous studies [8, 12], and for brevity only the salient details will be discussed here. The novelty of this method is the inclusion of an additional coordinate, s, into the continuum fields, which is used to represent the grainsize distribution. For example, the concentration of each grainsize $\phi(s)$, can easily be stated. Following [13], and assuming that particles do not expend or shrink, mass and momentum conservation in the entire domain can be written in a local form as:

$$\frac{\partial \rho}{\partial t} + \nabla \cdot (\rho \mathbf{u}) = 0, \tag{1}$$

$$\frac{\partial}{\partial t}(\rho \mathbf{u}) + \nabla \cdot (\rho \mathbf{u} \otimes \mathbf{u}) = \mathbf{F}. \tag{2}$$

where ρ is the partial density of the material, $\mathbf{F}$ the total force per unit volume and $\mathbf{u}$ the velocity field. In a recently submitted manuscript [12], a significant development was the formulation of a relationship between the segregation velocity, $\hat{\mathbf{u}}$ and the bulk kinetic stress field, $\bar{\boldsymbol{\sigma}}_k$, as

$$\hat{\mathbf{u}} = \frac{1 - \bar{s}/s}{\bar{\rho} c_0} \nabla \cdot \bar{\boldsymbol{\sigma}}_k. \tag{3}$$

where s is the grainsize, $\bar{s}$ is the local mean grainsize, $\bar{\rho}$ is the bulk density and c_0 is a rate which controls the rate of segregation.

This relationship allows us to describe the segregation phenomenon, once the kinetic stress field is known. This field, which is closely related to the granular temperature T_g, has recently been described using its own set of evolution equations [14]. Here, we merely assume that the granular temperature scales to first order as the rate-of-shear tensor, $\mathbf{S}$, following [15], such that $\bar{\boldsymbol{\sigma}}_k \approx c_1 \bar{\rho} \mathbf{S}$, where c_1 is a free parameter (Fig. 1).

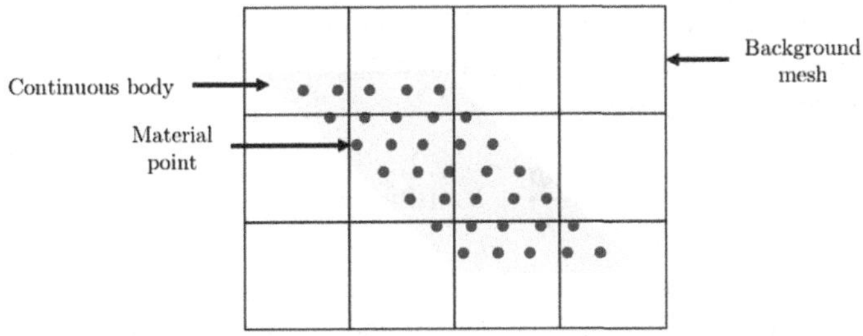

Fig. 1 Schematic diagram of the material point method. A continuous body, represented by the *gray* region, is discretised onto a set of material points (in *blue*). At each time step, information is projected from the material points to the background mesh (shown in *black*). Continuum equations are then solved on the background mesh, and the updated quantities projected back to the material points

3 Material Point Method Implementation

A material point method code is implemented following [16], with additional parameters included to describe the segregation process. MPM has successfully been used to model a variety of large deformation problems, from the macro scale (whole landslides) [17] to the micro scale (modelling granular materials) [18]. There are at least two options available for impregnating such a continuum solver with a grainsize distribution. One option would be to add material points for each phase of material, such that if a bidisperse material was modelled, there would be two distinct material points, each carrying a proportion of the total mass. This method suffers from a significant increase in computational time (for arbitrarily large polydispersity, the computational time also scales arbitrarily large). This method also has difficulty dealing with particle breakage (which is not explicitly modelled here), as material points must lose mass, and new particles must be added to represent fragments. A second option involves describing the grainsize distribution within each material point as a discrete histogram of grainsize (or alternatively an analytic function, if required). This is the method pursued here.

The grainsize distribution is discretised into N_s components, such that the solid fraction in a particular grainsize bin $s_a \leq s < s_b$ can be expressed as $\Phi(s_i) = \int_{s_a}^{s_b} \phi(s)\, ds$. This solid fraction is stored as an additional vector for each material point, and at each time step is mapped from the material points to the background mesh, along with the mass and momentum as:

$$m_i^n = \sum_p w_{ip}^n m_p, \tag{4}$$

$$m_i^n \mathbf{u}_i^n = \sum_p w_{ip}^n m_p \mathbf{u}_i^n, \tag{5}$$

$$m_i^n \Phi_i^n = \sum_p w_{ip}^n m_p \Phi_p^n, \tag{6}$$

where n is the current time step, i the grid points, p the material points and w_{ip} the interpolation weights. During the update stage of the MPM, a further evolution equation is solved for $\boldsymbol{\Phi}$, as

$$\frac{\partial \boldsymbol{\Phi}}{\partial t} + \nabla \cdot (\boldsymbol{\Phi} \hat{\mathbf{u}}) = 0, \tag{7}$$

which is the typical statement of conservation of grainsize [4], but extended from bidisperse to polydisperse materials. Due to the non-linear nature of this equation [5], and the requirement that for perfect segregation there must exist discontinuities in the solution, we require a sophisticated numerical solution scheme that is capable of properly accounting for this behaviour. The advective fluxes are solved for using a high-resolution central scheme that is total variation diminishing, as described in [19]. Putting together (7) and (3), we recover the governing equation for segregation in this system:

$$\frac{\partial \boldsymbol{\Phi}}{\partial t} + \nabla \cdot \left(\boldsymbol{\Phi} c \left(1 - \frac{\bar{s}}{s}\right) \nabla \cdot \mathbf{S} \right) = 0, \tag{8}$$

where $c = c_1/c_0$. Using the formulation SD2 from [19], incremental changes of $\boldsymbol{\Phi}$, termed $\Delta\boldsymbol{\Phi}_i^{n+1}$, can be computed. The final step required to complete the simulations is a mapping back from the nodal grainsize distribution $\boldsymbol{\Phi}_i^n$ to the new values at the material points. This is done in the same manner as for position, $\mathbf{x}$ and velocity, $\mathbf{u}$, by solving

$$\mathbf{u}_p^{n+1} = \mathbf{u}_p^n + \sum_i w_{ip} \mathbf{a}_i^n \Delta t, \tag{9}$$

$$\mathbf{x}_p^{n+1} = \mathbf{x}_p^n + \sum_i w_{ip} \mathbf{u}_i^{n+1} \Delta t, \tag{10}$$

$$\boldsymbol{\Phi}_p^{n+1} = \boldsymbol{\Phi}_p^n + \sum_i w_{ip} \Delta\boldsymbol{\Phi}_i^{n+1}, \tag{11}$$

where $\mathbf{a}_i^n$ is the acceleration calculated from conservation of momentum at each grid point.

4 1D Flow

An initial test was conducted for the case of one dimensional laminar flow down an inclined plane, with a rough basal plane that maintains $\mathbf{u} = 0$ at the base (but not $\hat{\mathbf{u}} = 0$). Initially, the material is at rest, and time $t = 0$, gravity of 9.81 m/s is applied at an angle of 18 degrees. A two dimensional system is modelled, with width of 2 grid cells in the x-direction (parallel to the flow) and 51 grid cells in the y-direction (perpendicular to the slope), with the grid spacing in both directions equal to 0.02 m, such that the flow is 1 m deep. In each grid cell, we initially seed 9 regu-

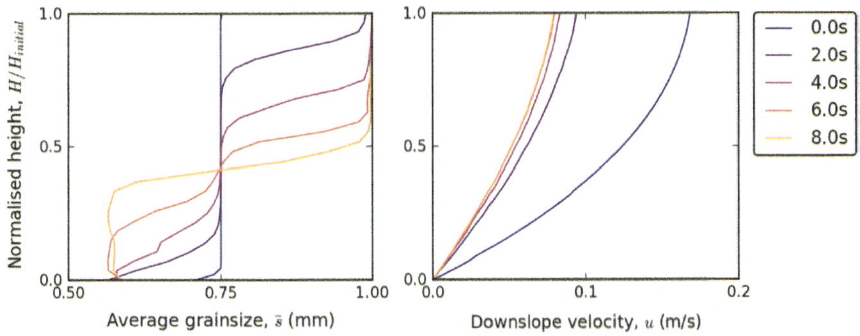

Fig. 2 Changing average grainsize in a one dimensional laminar flow down an inclined plane

larly spaced material points, each with equal mass. The initial grainsize distribution of each material point is chosen to be bidisperse, with 50% each of small and large particles (nominally 0.5 and 1 mm, but note that only their relative sizes affect the segregation velocity). A material density of 1000 kg/m^3 is used. This system models a bidisperse fluid, initially homogeneous in space and in grainsize distribution.

We introduce rheological coupling by allowing the material properties to depend on the grainsize distribution. For illustrative purposes, this work has been conducted using a newtonian rheology ($\tau = \mu\dot{\gamma}$), where the newtonian shear stress is proportional to the average grainsize, as $\mu_s = 10^3\bar{s}$, such that material points composed of primarily small particles will flow faster than larger ones. This is not designed to be a rigorous test of the rheology of landslides.

The results of this test are shown in Fig. 2. It is observed that over time, large particles segregate towards the free surface, and small particles collect at the base of the flow. This is accompanied by a slight change in rheology.

5 Conclusions

An initial investigation has shown that it is possible to couple bulk flow and segregation dynamics using the material point method. By solving the segregation equations directly on the background mesh, it is relatively straightforward to include these effects. Once segregation is included in the description of the flow, rheological coupling is trivial to model, given that there is a known relationship between the grainsize distribution and the constitutive behaviour of the material. In the future, this method will be used to study unsteady, two and three dimensional problems, such as the formation of granular bores, and levee deposits, where in natural systems significant segregation is observed [20, 21].

References

1. Iverson, R.M.: The debris-flow rheology myth. Debris-Flow Hazards Mitigation: Mechanics, Prediction, and Assessment, vol. 1, pp. 303–314 (2003)
2. Bridgwater, J.: Fundamental powder mixing mechanisms. Powder Technol. **15**(2), 215–236 (1976)
3. Staron, L., Phillips, J.C.: Stress partition and microstructure in size-segregating granular flows. Phys. Rev. E **92**(2), 022210 (2015)
4. Dolgunin, V.N., Ukolov, A.A.: Segregation modeling of particle rapid gravity flow. Powder Technol. **83**(2), 95–104 (1995)
5. Gray, J.M.N.T., Thornton, A.R.: A theory for particle size segregation in shallow granular free-surface flows. Proc. Royal Soc. A: Math. Phys. Eng. Sci. **461**(2057), 1447–1473 (2005)
6. Savage, S.B., Lun, C.K.K.: Particle size segregation in inclined chute flow of dry cohesionless granular solids. J. Fluid Mech. **189**, 311–335 (1988)
7. Marks, Benjy, Einav, Itai: A mixture of crushing and segregation: the complexity of grainsize in natural granular flows. Geophys. Res. Lett. **42**(2), 274–281 (2015)
8. Marks, Benjy, Rognon, Pierre, Einav, Itai: Grainsize dynamics of polydisperse granular segregation down inclined planes. J. Fluid Mech. **690**, 499–511 (2012)
9. Thornton, Anthony, Weinhart, Thomas, Luding, Stefan, Bokhove, Onno: Modeling of particle size segregation: calibration using the discrete particle method. Int. J. Mod. Physics C **23**(08), 1240014 (2012)
10. Baker, J.L., Barker, T., Gray, J.M.N.T.: A two-dimensional depth-averaged $\mu(I)$-rheology for dense granular avalanches. J. Fluid Mech. **787**, 367–395 (2016)
11. Gray, J.M.N.T., Ancey, Christophe: Segregation, recirculation and deposition of coarse particles near two-dimensional avalanche fronts. J. Fluid Mech. **629**, 387–423 (2009)
12. Marks, B., Eriksen, J.A., Dumazer, G., Måxløy, K.J., Sandnes, B.: Segregation of Intruders in Perpetual Granular Avalanches. Unpublished Manuscript (2016)
13. Ramkrishna, D.: Population Balances: Theory and Applications to Particulate Systems in Engineering. Academic press (2000)
14. Jiang, Yimin, Liu, Mario: Granular solid hydrodynamics. Granular Matter **11**(3), 139–156 (2009)
15. Jiang, Yimin, Liu, Mario: Granular solid hydrodynamics (gsh): a broad-ranged macroscopic theory of granular media. Acta Mech. **225**(8), 2363–2384 (2014)
16. Sulsky, D., Zhou, S.-J., Schreyer, H.L.: Application of a particle-in-cell method to solid mechanics. Comput. Phys. Commun. **87**(1), 236–252 (1995)
17. Andersen, S., Andersen, L.: Modelling of landslides with the material-point method. Comput. Geosci. **14**(1), 137–147 (2010)
18. Bardenhagen, S.G., Brackbill, J.U., Sulsky, D.: The material-point method for granular materials. Comput. Methods Appl. Mech. Eng. **187**(3), 529–541 (2000)
19. Kurganov, Alexander, Tadmor, Eitan: New high-resolution central schemes for nonlinear conservation laws and convection-diffusion equations. J. Comput. Phys. **160**(1), 241–282 (2000)
20. Möbius, M.E., Lauderdale, B.E., Nagel, S.R., Jaeger, H.M.: Brazil-nut effect: size separation of granular particles. Nature **414**(6861), 270–270 (2001)
21. Schröter, M., Ulrich, S., Kreft, J., Swift, J.B., Swinney, H.L.: Mechanisms in the size segregation of a binary granular mixture. Phys. Rev. E **74**(1), 011307 (2006)

Microstructural Effects on Strain Localization in a Multiscale Model for Hydro-Mechanical Coupling

A.P. van den Eijnden, P. Bésuelle, F. Collin and R. Chambon

Abstract The formulation and implementation of a double-scale finite element model for hydro-mechanical coupling in the framework of the finite element squared method has allowed studying macroscale boundary value problems in a poromechanical continuum. The macroscale constitutive relations are directly derived from the micromechanical interaction between fluid and solid microstructure, captured in representative elementary volumes. The application of this model in the simulation of a biaxial test and a gallery excavation problem is presented here to give examples of the model in strain localization problems. While using simple micromechanical models, the results demonstrate the ability of the model to provide complex macroscale material behaviour, that controls the initiation and development of the strain localization.

1 Introduction

The modelling of micromechanical behaviour of geomaterials has provided means to describe material behaviour based on its microstructural constituents. This has allowed phenomenological constitutive laws for continuous media to be replaced by a direct simulation of the interaction between its microstructural constituents and different phases. However, the application of these microscale models in direct

A.P. van den Eijnden (✉)
Section of Geo-Engineering, Faculty of Civil Engineering and Geosciences, Delft University of Technology, Delft, The Netherlands
e-mail: A.P.vandenEijnden@tudelft.nl

P. Bésuelle
Université Grenoble Alpes, CNRS, 3SR, 38000 Grenoble, France
e-mail: pierre.besuelle@3sr-grenoble.fr

F. Collin
Université de Liège, ArGEnCo, 4000 Liège, Belgium
e-mail: f.collin@ulg.ac.be

R. Chambon
Université Grenoble Alpes, 3SR, 38000 Grenoble, France

E. Papamichos et al. (eds.), *Bifurcation and Degradation of Geomaterials with Engineering Applications*, Springer Series in Geomechanics and Geoengineering, DOI 10.1007/978-3-319-56397-8_29

numerical simulations of macroscale problems can lead to excessive computational loads, as the length scale of common macroscale problems can be several orders of magnitude larger than the length scale of the microstructural constituents. To overcome this issue, the different scales can be treated in separate computations that are coupled in a homogenization scheme, using the homogenized material behaviour of so-called representative elementary volumes (REV) as the local material behaviour of a continuum description of the macroscale problem. As such, coupling between the scales of observation is used to take account of the microstructure in an averaged sense.

This paper presents some of the recent advances in the doublescale modelling of hydro-mechanical coupling in the framework of computational homogenization. A finite element squared (FE^2) formulation is used to derive the poromechanical continuum behaviour from a microscale model for microstructural solid-fluid interaction. After a general introduction of the modelling concept, examples of numerical simulations of laboratory tests and engineering structures are given. The examples are used to demonstrate the results obtained with the model and highlight the interplay between microstructural characteristics and macroscale initiation and propagation of strain localization.

2 The Doublescale Model

On the macroscale, a poromechanical continuum is formulated, of which the mechanical part of the solution of the macroscale boundary value problem is regularized by a local second gradient paradigm [3, 8]. This implicitly prescribes the internal length scale, needed for computations of softening behaviour without mesh dependency. For the second gradient part of the material behaviour, a linear-elastic constitutive relation is used [2, 3, 11], whereas the classical HM-coupled part is derived by means of computational homogenization from the computed equilibrium state of the REV [5]. For each integration point in the FE computation, an equilibrium state of the corresponding REV is computed with boundary conditions dictated by the local kinematics of the iterative macroscale test solution. From this equilibrium state, a consistent tangent operators (with respect to the rate of change of the test solution kinematics) and the homogenized response (stress state, fluid content, fluid mass flux) is derived by means of computational homogenization [5]. In this way, the REV boundary conditions and the computational homogenization for the coupling from macro to micro and from micro to macro respectively.

A microscale model was formulated in continuation of the model developments in Grenoble [1, 7, 10], in which the microstructure is modeled as an assembly of solid grains using the finite element method. Grains are considered elastic and the interfaces between grains are modelled by triple-noded interface elements to account for relative displacements between grains and fluid flow in the resulting interface channels (Fig. 1). Cohesive normal and tangential traction components T_n and T_t, acting over the interfaces between the grains, are described independently as functions of

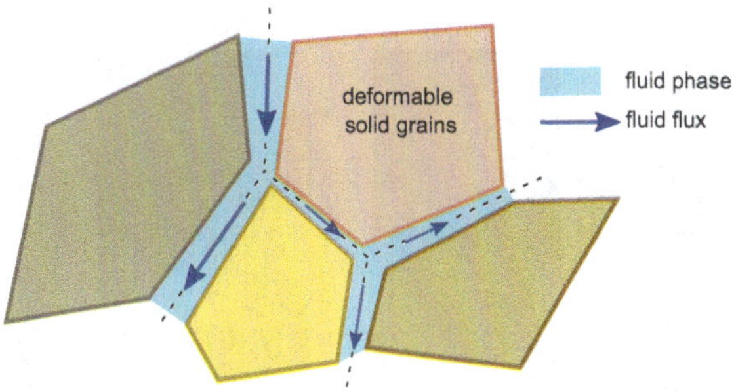

Fig. 1 Microscale model concept of elastic grains separated by cohesive interfaces [4]

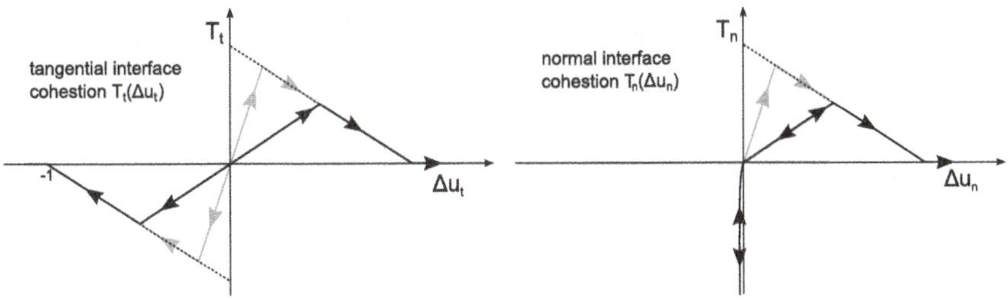

Fig. 2 Interface damage model for normal and tangential cohesion between grains over the interface. Penalization is used to account for contact between grains under local compression [5]

components Δu_n and Δu_t of the relative displacement between sides of the interface. A simple damage formulation is used for interface cohesion, introducing softening in the computed material behaviour (Fig. 2). Although coupling between the two components of cohesion are not used in the microscale formulation, confining stress-dependency of the macroscale behaviour is found as an effect of interlocking of the grains at the microscale. Nevertheless, interface mechanical behaviour can easily be modified within the same framework of homogenization.

A pore channel network is formed by the interfaces between the grains, allowing pore fluid to percolate. Fluid flow is modelled by means of one-dimensional channel elements with equivalent hydraulic conductivity, based on Poiseuille flow between smooth parallel plates. In addition, the water content depends on the relative volume of the pore space that is formed by the opened interfaces.

Upscaling from the equilibrated microscale model to the local macroscale response terms and constitutive behaviour is obtained through computational homogenization [9, 12]. This framework was extended to the case of hydro-mechanical coupling in the steady-state microscale model presented above, and provides the homogenized response of the REV as well as the tangent operators consistent to the current loading direction [5]. The classical, first order part of the material behaviour is thereby derived completely from the micromechanical model, without the need for determining derivatives of state by means of numerical perturbations.

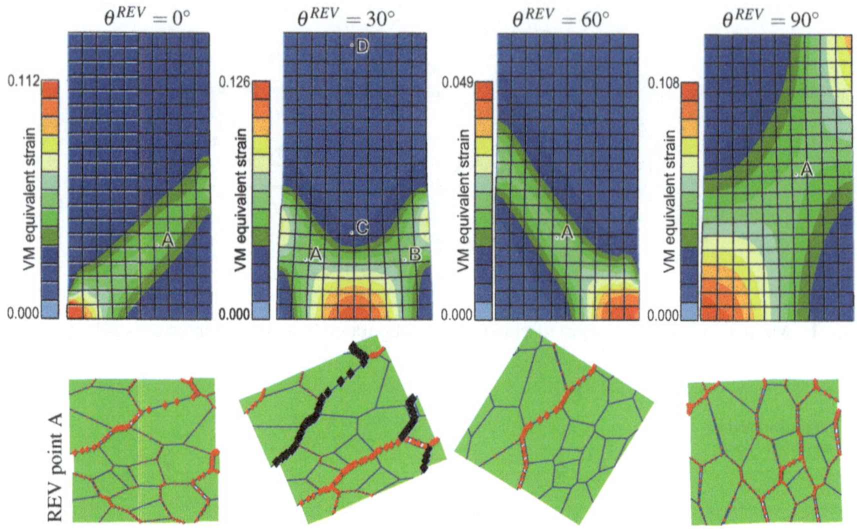

Fig. 3 Deformed macroscale samples and corresponding microstructures after loading at axial strain rate $\dot{\varepsilon}_a = 1 \times 10^{-8}$. Symbols ♦ and ♦ represent the interface state in softening and decohesion respectively [5]

3 Simulations of a Biaxial Compression Test

A biaxial compression test of a fluid-saturated material was modeled under transient conditions. The REV characterizing the material microstructure was varied in orientation with respect to the sample orientation to have different orientations of anisotropy. No confining pressure was applied and samples were drained at the bottom and top; a weak element was introduces at both lower corners to attract the initiation of strain localization. The specific microstructure, including an average elongation to represent the effect of a bedding plane, is rotated between the different samples to obtain different orientations of the anisotropic behaviour with respect to the loading direction.

Figure 3 shows the deviatoric strain fields of four of these tests in post-peak conditions. A deformed microstructure is given for each test, corresponding to a characteristic point *A* in the shear band that has developed. Mesh-independent results for shear bands are obtained through the regularization of the macroscale solution.

4 Simulations of the Excavation of a Tunnel

The excavation damaged zone around a deep tunnel was modeled. The excavation process in a 2D plane strain model is simulated by gradually reducing the initial stress state ($\sigma_v = -12.7$ MPa, $\sigma_h = -16.1$ MPa, $p = 4.7$ MPa) at the future tunnel

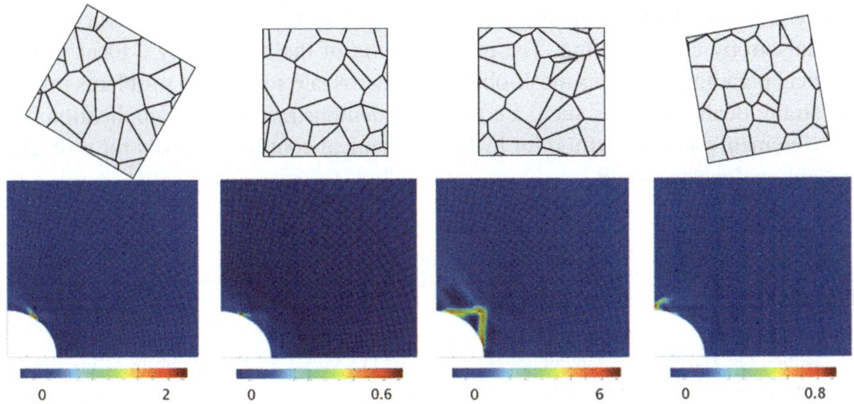

Fig. 4 Initiation of strain localization for material behaviour derived from different microstructures. *Colorscale* represents deviatoric strain rate, normalized against relative stage of unloading on tunnel wall [6]

wall, until zero-stress state is reached at the tunnel wall after 28 days. Four different microstructures are used to characterize the material microstructure, generated using a Voronoï-based algorithm [6]. Although not large enough to give statistically representative elementary volumes, the specific realizations of microstructures represent macroscale material behaviour related to their specific type of granular assembly and grain geometry and the resulting difference in macroscale behaviour is studied in relation with initiation of strain localization.

Figure 4 shows four REVs with microstructure, rotated to align the peak strength with the reference axes. The macroscale strain rate fields at the end of the excavation are given for simulations with each of these microstructures. The strain rate fields show the localized activity at the final stage of excavation, controlled directly by the microstructure in the REV. Building further on these observations, the influence of the microstructure on the macroscale initiation of strain localization can be explored.

In addition to the examples of strain localization given above, the presented simulations can be used to study the hydro-mechanical coupling in the doublescale model. As part of the hydro-mechanical coupled behaviour, the macroscale fluid transport phenomena can be derived from the micromechanical model of solid-fluid interaction. Both controlled by the deformation of the microstructure, the evolution of the permeability tensor and the zones of strain localization can be related.

5 Conclusions

The development and implementation of a finite element squared method for hydro-mechanical coupling has provided a doublescale approach for modelling strain localization in fluid-saturated rock-like materials. The given examples demonstrate the ability of the model to account for complex macroscale material behaviour, including anisotropy and softening. The regularization of the solution by a local second gradient paradigm provides mesh-objective macroscale solution to problems involving strain localization. As such, a conceptual demonstration of the application of a finite element squared method for hydro-mechanical coupling is given for problems at semi-engineering scale.

References

1. Bilbie, G., Dascalu, C., Chambon, R., Caillerie, D.: Micro-fracture instabilities in granular solids. Int. J. Solids Struct. **3**(1), 25–35 (2008)
2. Chambon, R., Caillerie, D., Matsuchima, T.: Plastic continuum with microstructure, local second gradient theories for geomaterials: localization studies. Int. J. Solids Struct. **38**(46), 8503–8527 (2001)
3. Collin, F., Chambon, R., Charlier, R.: A finite element method for poro mechanical modelling of geotechnical problems using local second gradient models. Int. J. Numer. Methods Eng. **65**(11), 1749–1772 (2006)
4. van den Eijnden, B.: Multiscale modelling of the hydromechanical coupling in argillaceous rocks. Ph.D. Thesis, Université Grenoble Alpes (2015)
5. van den Eijnden, A.P., Bésuelle, P., Chambon, R., Collin, F.: A FE^2 modelling approach to hydromechanical coupling in cracking-induced localization problems. Int. J. Solids Struct. **97**, 475–488 (2016)
6. van den Eijnden, A.P. Bésuelle, P., Collin, F., Chambon, R., Desrues, J.: Modeling the strain localization around an underground gallery with a hydro-mechanical double scale model; effect of anisotropy. Comput. Geotech. (2016)
7. Frey, J., Chambon, R., Dascalu, C.: A two-scale poromechanical model for cohesive rocks. Acta Geotech. **8**(2), 107–124 (2013)
8. Germain, P.: The method of virtual power in continuum mechanics. Part 2: Microstructure. SIAM J. Appl. Math. **25**(3), 556–575 (1973)
9. Kouznetsova, V., Brekelmans, W.A.M., Baaijens, F.P.T.: An approach to micro-macro modeling of heterogeneous materials. Comput. Mech. **27**(1), 37–48 (2001)
10. Marinelli, F., van den Eijnden, A.P., Sieffert, Y., Chambon, R., Collin, F.: Modeling of granular solids with computational homogenization: comparison with Biot's theory. Finite Elem. Anal. Des. **119**, 45–62 (2016)
11. Mindlin, R.D.: Micro-structure in linear elasticity. Arch. Ration. Mech. Anal. **16**, 51–78 (1964)
12. Özdemir, I., Brekelmans, W.A.M., Geers, M.G.D.: Computational homogenization for heat conduction in heterogeneous solids. Int. J. Numer. Methods Eng. **73**(2), 185–204 (2008)

Review and Comparison of Numerical Implementations for Cosserat Plasticity

Fahad Gulib and Stefanos-Aldo Papanicolopulos

Abstract The seminal paper of Mühlhaus and Vardoulakis (1987) led to an extensive study of generalised continuum plasticity models for numerical modelling of localisation of deformation. While different models and numerical implementations have been proposed in the literature, all capable of capturing the initial development of localisation, there is a lack of comparative studies to highlight the differences between models and their numerical implementations. In this work we present a review of existing numerical implementations of Cosserat plasticity models. We then implement a number of these models and finite elements within a commercial code to provide a comparison of the localisation behaviour based on numerical results of a simple test.

1 Introduction

The seminal paper of [13] demonstrated the ability of Cosserat plasticity to model the localisation of deformation in geomaterials. Based on this, many authors have presented numerical simulations of localisation in geomaterials using finite element implementations of Cosserat plasticity models.

All these numerical simulations successfully model localisation, however we are not aware of any detailed comparison of the different constitutive models proposed. While [10, 11] made a comparative study of integral-type and gradient-type plasticity models (for one-dimensional problems), no such studies were found for Cosserat continua.

F. Gulib · S.-A. Papanicolopulos (✉)
School of Engineering, Institute for Infrastructure and Environment,
The University of Edinburgh, The King's Buildings, Edinburgh EH9 3FG, UK
e-mail: S.Papanicolopulos@ed.ac.uk

F. Gulib
e-mail: F.Gulib@ed.ac.uk

E. Papamichos et al. (eds.), *Bifurcation and Degradation of Geomaterials with Engineering Applications*, Springer Series in Geomechanics and Geoengineering, DOI 10.1007/978-3-319-56397-8_30

Finite elements for Cosserat continua must separately discretise and interpolate the displacement and the rotation fields. For a given shape (e.g. triangle or quadrilateral in two dimensions) it is therefore necessary to select the interpolation used, and therefore the type and location of the degrees of freedom, for the displacements and the rotations; clearly these two interpolations can be different. Additionally, it is necessary to select the spatial integration scheme to be used. While this can be selected based solely on the displacement field, using a scheme that avoids locking and hourglassing, the way integration applies to the two different fields should also be taken into account in determining the appropriate scheme.

We briefly review here numerical implementations of Cosserat plasticity, highlighting their main aspects in terms of the constitutive models and the finite element formulation used. We then implement some of these models and elements and compare their numerical results for three different simulations presented in the literature. The results obtained provide insight into which elements perform better.

2 An Overview of Proposed Constitutive Models

Table 1 shows the main publications surveyed here, indicating the type of constitutive model employed. Von Mises and Drucker-Prager models are extended into a Cosserat formulation by modifying the quantity J_2, which in the classical formulation is the second deviatoric invariant of the stress, to include the couple stresses (scaled by an appropriate internal length parameter) and the antisymmetric part of the Cosserat stress. The combination of the symmetric and antisymmetric parts of the stress in J_2 is however not unique, with two main choices (static and kinematic) being proposed. Alsaleh [3] also uses a generalised second invariant.

The generalisation of Mohr-Coulomb models is less obvious. Manzari [12] modifies the J_2 invariant, but computes the Lode angle θ based only on stresses, while [16] computes θ using the modified J_2 invariant. Adhikary et al. [1, 2] use a Mohr-Coulomb yield surface that does not involve the couple stresses. More details on two-dimensional Cosserat Mohr-Coulomb models are given by [18].

3 A Review of Proposed Cosserat Finite Elements

Table 1 lists the elements proposed for Cosserat elasticity and plasticity, and the spatial integration scheme employed. The interpolation of displacements/rotations can be linear/linear, quadratic/linear or quadratic/quadratic, resulting in the triangular elements COS3, COS6(3) and COS6 and in the rectangular elements COS4, COS8(4) and COS8. A "serendipity" (bi-)quadratic interpolation is used, though a Lagrangian interpolation is also possible to obtain COS9(4) and COS9 elements.

Table 1 Summary of surveyed literature on Cosserat finite element analysis, indicating constitutive model, element type and integration scheme

References	Constitutive model	Element	Integration scheme
Nakamura et al. [15]	Elasticity	COS3	*Not stated*
Nakamura and Lakes [14]	Elasticity	COS3/4	*Not stated*
Providas and Kattis [20]	Elasticity	COS3/6/6(3)	Analytical
Zhang et al. [24]	Elasticity	COS4/8/8(4)	*Not stated*
de Borst [5]	Von Mises	COS6	Cubic
de Borst and Sluys [7]	Von Mises	COS6	Quartic
Tejchman and Wu [23]	Von Mises	COS3	Quadratic
Sharbati and Naghdabadi [21]	Von Mises	COS4	*Not stated*
Papanastasiou and Vardoulakis [17]	Drucker-Prager	COS8	Quintic
de Borst [6]	Drucker-Prager	COS6	*Not stated*
Sulem and Cerrolaza [22]	Drucker-Prager	COS4	*Not stated*
Iordache and Willam [9]	Drucker-Prager	COS3	*Unclear*
Arslan and Sture [4]	Drucker-Prager	COS4	cubic (SR)
Peng et al. [19]	Drucker-Prager	COS8	Quintic
Adhikary et al. [2]	Mohr-Coulomb	COS8	Quintic
Manzari [12]	Mohr-Coulomb	COS4	*Not stated*
Adhikary and Dyskin [1]	Mohr-Coulomb	COS8?	*Not stated*
Papamichos [16]	Mohr-Coulomb	COS8	*Not stated*
Alsaleh [3]	Lade-Kim	COS4	Cubic (SR)

For Cosserat elasticity, [20] for triangles and [24] for quadrilaterals both show that all elements pass the patch test, with COS6(3) slightly more accurate than COS6 and COS8(4) slightly less accurate than COS8, while COS3 and COS4 in each case have the lowest accuracy. For Cosserat plasticity, however, we are not aware of similar comparisons.

Indeed, we are not aware of any Cosserat plasticity results where quadratic/linear interpolation is used. Sulem and Cerrolaza [22] suggests that a quadratic/linear interpolation should be used (but uses linear/linear interpolation), while [2] suggest that the same order of interpolation should be used for both kinematic fields. de Borst [5] mentions numerical experimentation for a COS6(3) element, but only shows results for COS6, while [9] present results only for COS3, though they also mention COS6(3) and COS6 elements.

Papanastasiou and Vardoulakis [17] mention that a comparison of results shows that the COS9 element is slightly better than the COS8 element, but the difference is negligible and thus they use the computationally cheaper COS8 element.

From Table 1 it is also evident that not enough information is generally given concerning the spatial integration scheme used, and no detailed comparisons between different schemes are presented. de Borst [5] mentions that COS6 requires a quartic integration scheme to eliminate the possibility of zero-energy modes, but uses a cubic scheme, and also mentions that a COS6(3) element only needs quadratic integration without showing detailed results. Selective reduced (SR) integration is used by some authors for COS4 elements, but no reduced integration is mentioned for COS8 elements (a discussion of reduced integration for quadratic/quadratic Cosserat elements is given by [8] for elastic Cosserat plates).

4 Implementation and Comparison

We have implemented in the finite element code Abaqus the three Cosserat quadrilaterals in Table 1, with both full and reduced integration. We therefore obtain elements COS4F and COS4R with cubic (2×2) and linear (1×1) Gauss integration, as well as elements COS8F/COS8(4)F and COS8R/COS8(4)R with quintic (3×3) and cubic (2×2) integration respectively. We have additionally implemented a COS4SR element with selective reduced integration [3, 4].

The widely used benchmark of biaxial (plane strain) compression is used to compare the different finite element formulations for Von Mises and Drucker-Pragel models. The first benchmark test is the Von Mises model described in [5], for the case $l/H = 1/12$ (where l is the internal length and H is the height of the specimen). Figure 1 shows a detail of the load-displacement curve during the softening stage. Elements COS8F, COS8R, COS8(4)F and COS8(4)R give practically indistinguishable results, all very slightly stiffer than the ones reported by [5], so only COS8F is shown. Element COS4SR gives a slightly stiffer but still very similar result, while COS4F gives a much stiffer behaviour. COS4R gives results similar to the quadratic displacement elements, but suffers from hourglassing.

A similar set of results is obtained for the Von Mises model of [21] (biaxial test with $l/H = 1/60$), as shown in Figs. 1 and 2. Once more COS4F is overly stiff, while COS4R gives a reasonable load-displacement graph but suffers from hourglassing; COS8(4)R gives an almost identical load-displacement curve (not shown) but with no hourglassing. Elements COS8(4)F and COS8F give slightly different, but still very close results, while COS4SR is again stiffer, more visibly so than in the previous case. Except for COS4F, all other elements give the correct localised result.

The third benchmark (not shown here) uses the Drucker-Prager model and the biaxial simulation presented by [6]. We initially simulate a completely homogeneous material. Even in this case, however, only element COS4F gives the homogeneous solution, while the other tested elements show localisation of deformation. Elements COS8F, COS8R and COS8(4)R all give very similar results, similar to but stiffer than the localised behaviour given by [6], with COS4SR giving an even stiffer localised behaviour. Introducing an imperfection as described in [6] leads to a different behaviour, with all tested elements showing a less stiff behaviour and a more diffuse failure mode. By reducing the internal length it was however again possible to obtain localised results.

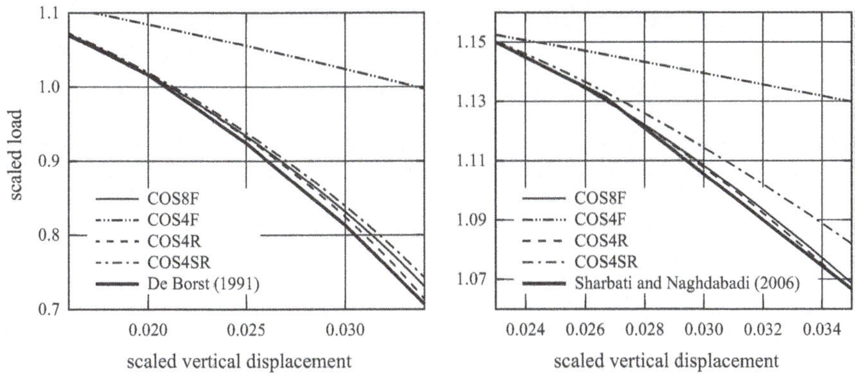

Fig. 1 Detail of the load-displacement curves for the two Von Mises models (load scaled by initial cross section area and yield stress, displacement scaled by initial specimen height)

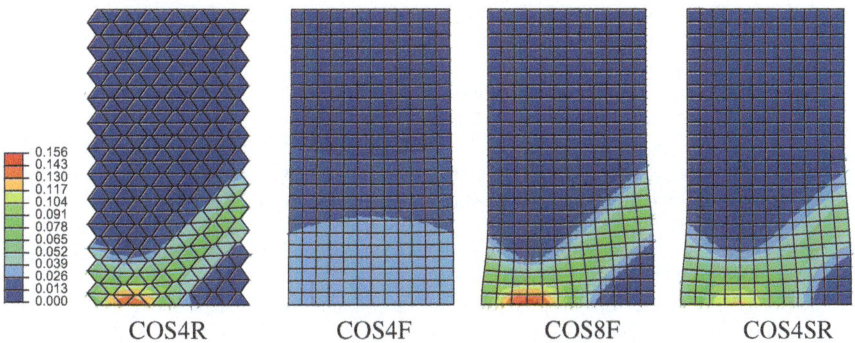

Fig. 2 Deformed meshes and contours of equivalent plastic strain for the model by [21] (deformation is scaled, with non-uniform scaling for COS4R in the x direction)

5 Conclusions

Elements COS4F and COS4R are not recommended as they give stiff (locking) behaviour and hourglassing, respectively. All elements with quadratic interpolation of displacements give comparable results with no indication of numerical issues; in this case the quadratic/linear reduced integration element COS8(4)R would be preferable as being the least expensive computationally. The even less expensive linear/linear selective reduced integration element COS4SR shows good numerical behaviour, however it's deviation from the other results necessitates further investigation. For the case of the Drucker-Prager model, it is seen that introduction of an imperfection changes the results significantly, though the relative performance of different elements does not seem to be greatly affected.

This work identifies the need for comparative studies of Cosserat finite element implementations for plasticity models and forms part of a wider research effort to provide guidance on the appropriate use of generalised continuum models.

Acknowledgements This research effort is funded from the People Programme (Marie Curie Actions) of the European Union's Seventh Framework Programme (FP7/2007–2013) under REA grant agreement n° 618096 (Generalised Continuum Models and Plasticity–GECOMPL). F. Gulib also acknowledges the support received from the School of Engineering of the University of Edinburgh.

References

1. Adhikary, D.P., Dyskin, A.V.: Modelling of progressive and instantaneous failures of foliated rock slopes. Rock. Mech. Rock. Eng. **40**(4), 349–362 (2007)
2. Adhikary, D.P., Mühlhaus, H.B., Dyskin, A.V.: Modelling the large deformations in stratified media–the Cosserat continuum approach. Mech Cohes-Frict Mater **4**(3), 195–213 (1999)
3. Alsaleh, M.I.: Numerical modeling of strain localization in granular materials using Cosserat theory enhanced with microfabric properties. Ph.D. Thesis, Louisiana State University and Agricultural and Mechanical College (2004)
4. Arslan, H., Sture, S.: Finite element analysis of localization and micro-macro structure relation in granular materials. Part ii: Implementation and simulations. Acta Mech. **197**(3), 153 (2008)
5. de Borst, R.: Simulation of strain localization: a reappraisal of the Cosserat continuum. Engl. Compos. **8**(4), 317–332 (1991)
6. de Borst, R.: A generalisation of J_2-flow theory for polar continua. Comput. Methods Appl. Mech. Eng. **103**(3), 347–362 (1993)
7. de Borst, R., Sluys, L.: Localisation in a Cosserat continuum under static and dynamic loading conditions. Comput. Methods Appl. Mech. Eng. **90**(1), 805–827 (1991)
8. Godio, M., Stefanou, I., Sab, K., Sulem, J.: Dynamic finite element formulation for Cosserat elastic plates. Int. J. Numer. Methods Eng. **101**(13), 992–1018 (2015)
9. Iordache, M.M., Willam, K.: Localized failure analysis in elastoplastic Cosserat continua. Comput. Methods Appl. Mech. Eng. **151**(3–4), 559–586 (1998)
10. Jirásek, M., Rolshoven, S.: Comparison of integral-type nonlocal plasticity models for strain-softening materials. Int. J. Eng. Sci. **41**(13–14), 1553–1602 (2003)
11. Jirásek, M., Rolshoven, S.: Localization properties of strain-softening gradient plasticity models. Part I: strain-gradient theories. Int. J. Solids Struct. **46**(11–12), 2225–2238 (2009)
12. Manzari, M.T.: Application of micropolar plasticity to post failure analysis in geomechanics. Int. J. Numer. Anal. Methods Geomech. **28**(10), 1011–1032 (2004)
13. Mühlhaus, H.B., Vardoulakis, I.: The thickness of shear bands in granular materials. Géotechnique **37**(3), 271–283 (1987)
14. Nakamura, S., Lakes, R.: Finite element analysis of stress concentration around a blunt crack in a cosserat elastic solid. Comput. Methods Appl. Mech. Eng. **66**(3), 257–266 (1988)
15. Nakamura, S., Benedict, R., Lakes, R.: Finite element method for orthotropic micropolar elasticity. Int. J. Eng. Sci. **22**(3), 319–330 (1984)
16. Papamichos, E.: Borehole failure analysis in a sandstone under anisotropic stresses. Int. J. Numer. Anal. Methods Geomech. **34**(6), 581–603 (2010)
17. Papanastasiou, P.C., Vardoulakis, I.G.: Numerical treatment of progressive localization in relation to borehole stability. Int. J. Numer. Anal. Methods Geomech. **16**(6), 389–424 (1992)
18. Papanicolopulos, S.A., Veveakis, E.: Sliding and rolling dissipation in Cosserat plasticity. Granular Matter **13**, 197–204 (2011)

19. Peng, C., Kong, X., Xu, C.: Numerical implementation of pressure-dependent elasto-plastic cosserat continuum model in abaqus. In: 2011 Second International Conference on Mechanic Automation and Control Engineering (MACE), pp. 6621–6624 (2011)
20. Providas, E., Kattis, M.: Finite element method in plane cosserat elasticity. Comput. Struct. **80**(27–30), 2059–2069 (2002)
21. Sharbati, E., Naghdabadi, R.: Computational aspects of the cosserat finite element analysis of localization phenomena. Comput. Mater. Sci. **38**(2), 303–315 (2006)
22. Sulem, J., Cerrolaza, M.: Finite element analysis of the indentation test on rocks with microstructure. Comput. Geotech. **29**(2), 95–117 (2002)
23. Tejchman, J., Wu, W.: Numerical study on patterning of shear bands in a Cosserat continuum. Acta Mech. **99**(1), 61–74 (1993)
24. Zhang, H., Wang, H., Liu, G.: Quadrilateral isoparametric finite elements for plane elastic Cosserat bodies. Acta Mech. Sin. **21**(4), 388–394 (2005)

Part IV
Soil-Tool/Machine Interaction

Sand to Mud to Fording: Modeling and Simulation for Off-Road Ground Vehicle Mobility Analysis

Dan Negrut and Hammad Mazhar

Abstract This contribution highlights modeling and numerical solution techniques that enable the simulation of ground vehicles operating in off-road conditions. We briefly outline the equations of motion governing the time evolution of complex systems such as, for instance, large collections of granular material or full vehicles. Herein, we demonstrate how these equations of motion have been used in an open source multi-physics simulation software called Chrono to investigate the dynamics of wheeled and tracked vehicles operating on granular material. Next, we briefly discuss the fluid-solid interaction problem, which comes into play, for instance, when simulating fording operations in which a vehicle negotiates a body of water. The two salient attributes of the approach proposed are reliance on differential algebraic inequalities to model the dynamics of solid and fluid phases; and, leverage of parallel computing to handle systems with millions of degrees of freedom.

Keywords Multi-physics modeling and simulation · Rigid and flexible multi-body dynamics · Friction and contact · Fluid-solid interaction · Granular dynamics · Vehicle dynamics · Parallel computing

1 Large Scale Displacements/Rotations via Many-Body Dynamics

Herein, the set of generalized coordinates used to position and orient a rigid body j in the 3D Euclidean space are $\mathbf{r}_j \in \mathbb{R}^3$ and $\epsilon_j \in \mathbb{R}^4$ [8]. The former provides the absolute position of the center of mass of body j, while the latter represents a set of Euler parameters (quaternions) that characterize body orientation in a global reference frame. The set of generalized coordinates for a system of n_b bodies works out to be $\mathbf{q} = \left[\mathbf{r}_1^T, \epsilon_1^T, \ldots, \mathbf{r}_{n_b}^T, \epsilon_{n_b}^T\right]^T \in \mathbb{R}^{7n_b}$ and their time derivatives $\dot{\mathbf{q}} =$

D. Negrut (✉) · H. Mazhar
University of Wisconsin–Madison, Madison, USA
e-mail: negrut@engr.wisc.edu

E. Papamichos et al. (eds.), *Bifurcation and Degradation of Geomaterials with Engineering Applications*, Springer Series in Geomechanics and Geoengineering, DOI 10.1007/978-3-319-56397-8_31

$\left[\dot{\mathbf{r}}_1^T, \dot{\epsilon}_1^T, \ldots, \dot{\mathbf{r}}_{n_b}^T, \dot{\epsilon}_{n_b}^T\right]^T \in \mathbb{R}^{7n_b}$. Rather than using $\dot{\mathbf{q}}$ to pose the Newton-Euler equations of motion, the array $\mathbf{v} = \left[\dot{\mathbf{r}}_1^T, \bar{\omega}_1^T, \ldots, \dot{\mathbf{r}}_{n_b}^T, \bar{\omega}_{n_b}^T\right]^T \in \mathbb{R}^{6n_b}$ is used since it leads to: (i) a smaller problem; and (ii) a constant, symmetric and positive definite mass matrix. There is a simple linear transformation that for each body B relates its angular velocity expressed in the body-fixed reference frame, $\bar{\omega}_B$, to the time derivatives of the Euler parameters $\dot{\epsilon}_B$. Specifically, $\bar{\omega}_B = 2\mathbf{G}(\epsilon_B)\dot{\epsilon}_B$, where the entries in the matrix $\mathbf{G} \in \mathbb{R}^{3\times 4}$ depend linearly on the Euler parameters ϵ_B [8]. Defining the block diagonal matrix $\mathbf{L}(\mathbf{q}) \equiv \text{diag}\left[\mathbf{I}_{3\times 3}, \frac{1}{2}\mathbf{G}^T(\epsilon_1), \ldots, \mathbf{I}_{3\times 3}, \frac{1}{2}\mathbf{G}^T(\epsilon_{n_b})\right] \in \mathbb{R}^{7n_b \times 6n_b}$, where $\mathbf{I}_{3\times 3}$ is the identity matrix, yields $\dot{\mathbf{q}} = \mathbf{L}(\mathbf{q})\mathbf{v}$. For the fluid phase in a system of n_f objects each fluid marker or rigid body k consists of a position $\mathbf{r} \in \mathbb{R}^3$ and a velocity $\dot{\mathbf{r}} \in \mathbb{R}^3$. The fluid phase only has translational degrees of freedom and does not rotate, in this document objects that have six degrees of freedom will be denoted as 6 degrees of freedom (dof) and objects with three degrees of freedom as 3dof.

Consider the contact between two bodies A and B represented in Fig. 1. Assuming that the body geometries are regular at the contact point, the contact point along with the shared tangent plane are used to define two local reference frames, one for each body. For body A, the normal $\mathbf{n}_{i,A}$ at contact point i is chosen to be perpendicular on the tangent plane at the shared contact point and to point towards the exterior of body A. Two mutually perpendicular unit vectors $\mathbf{u}_{i,A}$ and $\mathbf{w}_{i,A}$ are chosen to define a right-hand local reference frame associated with contact i on body A. A similar sequence of steps is followed to define a local reference frame for body B based on $\mathbf{n}_{i,B}, \mathbf{u}_{i,B}, \mathbf{w}_{i,B} \in \mathbb{R}^3$. The Lagrange multiplier $\widehat{\gamma}$ associated with contact i is used to

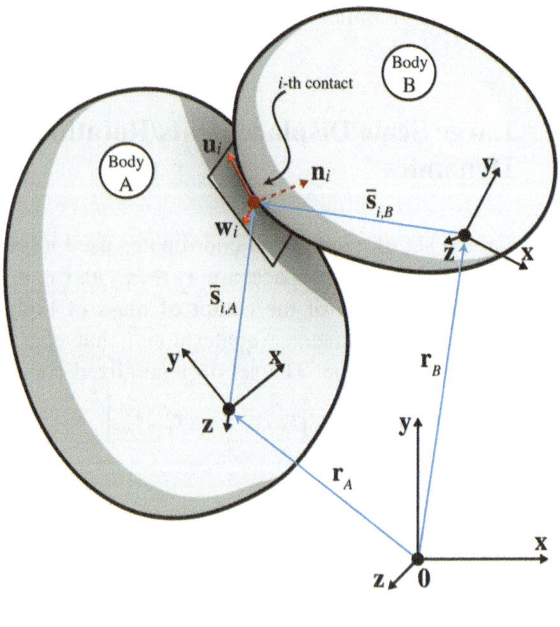

Fig. 1 Contact i between two bodies $A, B \in \{1, 2, \ldots, n_b\}$

pose a complementarity condition in relation to the gap (distance) Φ between bodies A and B: $0 \leq \widehat{\gamma}_{i,n} \perp \Phi_i(\mathbf{q}) \geq 0$.

In what follows, by convention, $\alpha_i \equiv \alpha_{i,A}$, for $\alpha \in \{\mathbf{n}, \mathbf{u}, \mathbf{w}\}$. The force associated with contact i can then be decomposed into the normal component, $\mathbf{F}_{i,N} = \widehat{\gamma}_{i,n}\mathbf{n}_i$, and the tangential component, $\mathbf{F}_{i,T} = \widehat{\gamma}_{i,u}\mathbf{u}_i + \widehat{\gamma}_{i,w}\mathbf{w}_i$, where the multipliers $\widehat{\gamma}_{i,n} > 0$, $\widehat{\gamma}_{i,u}$, and $\widehat{\gamma}_{i,w}$ represent the magnitude of the force in each direction. The friction forces are assumed to satisfy the Coulomb dry-friction model, which can be expressed as a set of three conditions: $\sqrt{\widehat{\gamma}_{i,u}^2 + \widehat{\gamma}_{i,w}^2} \leq \mu_i \widehat{\gamma}_{i,n}$, $\quad ||\mathbf{v}_{i,T}|| \left(\sqrt{\widehat{\gamma}_{i,u}^2 + \widehat{\gamma}_{i,w}^2} - \mu_i \widehat{\gamma}_{i,n} \right) = 0$, and $\langle \mathbf{F}_{i,T}, \mathbf{v}_{i,T} \rangle = -||\mathbf{F}_{i,T}|| \, ||\mathbf{v}_{i,T}||$, where $\mathbf{v}_{i,T}$ represents the relative tangential velocity between bodies A and B at the point of contact. These equations represent the first order Karush-Kuhn-Tucker optimality condition for the following optimization problem in two dummy variables $y, z \in \mathbb{R}$:

$$\left(\widehat{\gamma}_{i,u}, \widehat{\gamma}_{i,w}\right) = \underset{\sqrt{y^2+z^2} \leq \mu_i \widehat{\gamma}_{i,n}}{\operatorname{argmin}} \ \mathbf{v}_{i,T}^T \left(y\mathbf{u}_i + z\mathbf{w}_i\right).$$

The force at the ith contact point can be expressed as $\mathbf{F}_i = \mathbf{F}_{i,N} + \mathbf{F}_{i,T} = \widehat{\gamma}_{i,n}\mathbf{n}_i + \widehat{\gamma}_{i,u}\mathbf{u}_i + \widehat{\gamma}_{i,w}\mathbf{w}_i \in \Upsilon_i$, where Υ_i is a 3D cone of slope $\tan^{-1}\mu_i$, i.e., $\Upsilon_i = \{\left[x, y, z\right]^T \in \mathbb{R}^3 | \sqrt{y^2 + z^2} \leq \mu_i x\}$, oriented along $\mathbf{n}_i$ and with its tip at the contact point.

The Newton-Euler equations of motion [2, 24] then assume the following expression [23]:

$$\overbrace{\dot{\mathbf{q}}}^{\text{Generalized positions}} = \underbrace{\mathbf{L}(\mathbf{q})}_{\text{Velocity transformation matrix}} \mathbf{v} \tag{1a}$$

$$\mathbf{M}\dot{\mathbf{v}} = \underbrace{\mathbf{f}(t^{(l)}, \mathbf{q}^{(l)}, \mathbf{v}^{(l)})}_{\text{Applied \& Coriolis force}} \overbrace{- \mathbf{G}(\mathbf{q}, t)\hat{\lambda}}^{\text{Reaction force}} + \underbrace{\sum_{i \in \mathcal{A}(q^{(l)}, \delta)} \left(\widehat{\gamma}_{i,n}\mathbf{D}_{i,n} + \widehat{\gamma}_{i,u}\mathbf{D}_{i,u} + \widehat{\gamma}_{i,w}\mathbf{D}_{i,w}\right)}_{\text{Frictional contact forces}} \tag{1b}$$

$$0 = \underbrace{\mathbf{g}(\mathbf{q}, t)}_{\text{Bilateral constraints}} \tag{1c}$$

$$i \in \mathcal{A}(q^{(l)}, \delta): \quad \overbrace{0 \leq \Phi_i(\mathbf{q})}^{\text{Unilateral constraint}} \perp \widehat{\gamma}_{i,n} \geq 0 \tag{1d}$$

$$\left(\widehat{\gamma}_{i,u}, \widehat{\gamma}_{i,w}\right) = \underset{\sqrt{\gamma_{i,u}{}^2 + \gamma_{i,w}{}^2} \leq \mu_i \gamma_{i,n}}{\operatorname{argmin}} \ \mathbf{v}^{\mathrm{T}} \left(\gamma_{i,u}\mathbf{D}_{i,u} + \gamma_{i,w}\mathbf{D}_{i,w}\right), \tag{1e}$$

where the tangent space generators $\mathbf{D}_i = [\mathbf{D}_{i,n}, \mathbf{D}_{i,u}, \mathbf{D}_{i,w}] \in \mathbb{R}^{6n_b \times 3}$ are defined as

$$\mathbf{D}_i^T = \left[\mathbf{0} \ \ldots \ -\mathbf{A}_{i,p}^T \ \ \mathbf{A}_{i,p}^T\mathbf{A}_A\tilde{\bar{\mathbf{s}}}_{i,A} \ \ \mathbf{0} \ \ldots \ \mathbf{0} \ \ \mathbf{A}_{i,p}^T \ -\mathbf{A}_{i,p}^T\mathbf{A}_B\tilde{\bar{\mathbf{s}}}_{i,B} \ \ldots \ \mathbf{0}\right], \tag{2}$$

where $\mathbf{A}_{i,p} = [\mathbf{n}_i, \mathbf{u}_i, \mathbf{w}_i] \in \mathbb{R}^{3\times3}$ is the orientation matrix associated with contact i; $\mathbf{A}_A = \mathbf{A}\left(\epsilon_A\right)$ and $\mathbf{A}_B = \mathbf{A}\left(\epsilon_B\right)$ are the rotation matrices of bodies A and B respectively; and the vectors $\bar{\mathbf{s}}_{i,A}$ and $\bar{\mathbf{s}}_{i,B}$ represent the contact point positions in body-relative coordinates as illustrated in Fig. 1. Finally, the set of active and potential unilateral constraints is denoted by $\mathcal{A}(\mathbf{q}, \delta)$ and is defined based on the bodies that are mutually less than a distance $\delta \geq 0$ apart.

For interaction between a 6dof rigid body and a frictionless 3dof rigid body the equations can be greatly simplified. Tangential friction forces do not need to be computed and as a result the size of the problem is reduced from three constraints per contact to one constraint. A 3dof does not experience torque and the contact position only needs to be computed in body-relative coordinates for the 6dof rigid body. This means that for a contact between a 6dof object and a 3dof frictionless object $\mathbf{D}_i$ simplified to $\mathbf{D}_i = [\,\mathbf{D}_{i,n}, \mathbf{D}_{i,u}, \mathbf{D}_{i,w}] \in \mathbb{R}^{6n_b\times1}$ for the 6dof body and $\mathbf{D}_i = [\,\mathbf{D}_{i,n}, \mathbf{D}_{i,u}, \mathbf{D}_{i,w}] \in \mathbb{R}^{3n_b\times1}$ for the 3dof body. For a contact between two 3dof bodies each body contributes a Jacobian of size $\mathbf{D}_i \in \mathbb{R}^{3n_b\times1}$ for a contact.

2 Fluid-Solid Interaction Modeling via Smoothed Particle Hydrodynamics

The continuity and momentum equations associated with the fluid dynamics problem assume the form [6]

$$\frac{D\rho}{dt} = -\rho\nabla\cdot\mathbf{v} \tag{3a}$$

$$\frac{D\mathbf{v}}{dt} = -\frac{1}{\rho}\nabla p + \frac{\mu}{\rho}\nabla^2\mathbf{v} + \mathbf{f}\ , \tag{3b}$$

where μ is the fluid viscosity, ρ is fluid density, $\mathbf{v}$ and p are the flow velocity and pressure, respectively, and $\mathbf{f}$ is the body force. Assuming a Newtonian and incompressible flow, the first equation translates either into $\dfrac{D\rho}{dt} = 0$, or equivalently, imposing a divergence-free flow condition, into $\nabla \cdot \mathbf{v} = 0$.

The approach embraced in Chrono for the spatial discretization of the Navier-Stokes equations draws on the Smoothed Particle Hydrodynamics (SPH) methodology [5, 10], a meshless method that dovetails well with the Lagrangian modeling perspective adopted for the dynamics of the solid phase. The term *smoothed* in SPH refers to the approximation of point properties via a smoothing kernel function W, defined over a support domain S. This approximation reproduces functions with up to second order accuracy, provided the kernel function: (i) approaches the Dirac delta function as the size of the support domain tends to zero, that is $\lim_{h\to0} W(\mathbf{r}, h) = \delta(\mathbf{r})$, where $\mathbf{r}$ is the spatial distance and h is a characteristic length that defines the kernel smoothness; (ii) is symmetric, i.e., $W(\mathbf{r}, h) = W(-\mathbf{r}, h)$; and (iii) is normal, i.e., $\int_S W(\mathbf{r}, h)dV = 1$, where $d\,V$ denotes the differential volume.

The term *particle* in SPH terminology indicates the discretization of the domain by a set of Lagrangian particles. To remove ambiguities caused by the use of the term *rigid particles* in the context of FSI problems, the term *marker* is also used herein to refer to the SPH discretization process. Each marker a has mass m_a associated with the representative volume dV and carries all of the essential field properties. As a result, any field property at a certain location is shared and represented by the markers in the vicinity of that location [25]. Within this framework, Eqs. (3a) and (3b) are discretized at an arbitrary location $\mathbf{x}_a$ within the fluid domain as [16]:

$$\frac{d\rho_a}{dt} = \rho_a \sum_b \frac{m_b}{\rho_b} \left(\mathbf{v}_a - \mathbf{v}_b\right) \cdot \nabla_a W_{ab} \tag{4a}$$

$$\frac{d\mathbf{v}_a}{dt} = -\sum_b m_b \left[\left(\frac{p_a}{\rho_a{}^2} + \frac{p_b}{\rho_b{}^2} \right) \nabla_a W_{ab} - \frac{(\mu_a + \mu_b)\mathbf{x}_{ab} \cdot \nabla_a W_{ab}}{\bar{\rho}_{ab}^2 (x_{ab}^2 + \varepsilon h^2)} \mathbf{v}_{ab} \right] + \mathbf{f}_a, \tag{4b}$$

which are augmented by the kinematic differential equation

$$\frac{d\mathbf{x}_a}{dt} = \mathbf{v}_a, \tag{5}$$

which is to update the location of the discretization markers. In the above equations, quantities with subscripts a and b are associated with markers a and b, respectively; $\mathbf{x}_{ab} = \mathbf{x}_a - \mathbf{x}_b$, $\mathbf{v}_{ab} = \mathbf{v}_a - \mathbf{v}_b$, $W_{ab} = W(\mathbf{x}_{ab}, h)$, $\bar{\rho}_{ab}$ is the average density of markers a and b, ∇_a is the gradient with respect to $\mathbf{x}_a$, i.e., $\partial/\partial \mathbf{x}_a$, and ε is a regularization coefficient to prevent infinite reaction between markers sharing the same location.

We have experimented with two approaches for solving the SPH-based solution of the FSI problem. The first one relies on a equation of state that ties the expression of the pressure p to the value of the density ρ. Specifically, under the assumption of handling an incompressible flow, we expect that $\rho = \rho_0 =$ const. Then, the pressure is introduced in the problem as a forcing term that corrects any deviation in density from the nominal value ρ_0. This is the so called weakly compressible model, namely finding ρ via Eq. (4a) followed by an equation of state to update pressure p [15]:

$$p = \frac{c_s^2 \rho_0}{\gamma} \left\{ \left(\frac{\rho}{\rho_0} \right)^{\gamma} - 1 \right\}, \tag{6}$$

where γ tunes the stiffness of the pressure-density relationship, and c_s is the speed of sound. The value c_s is adjusted depending on the maximum speed of the flow, $V_{\max}$, to keep the flow compressibility below some arbitrary value. Typically, $\gamma = 7$ and $c_s = 10V_{\max}$, which allows 1% flow compressibility [15].

In a second approach, we experimented with constraint fluids [4], which is an approach that leverages SPH and enforces incompressibility as *a many-body density constraint* between a fluid particle and its neighbors. Depending on the numerical discretization, a corrective impulse [4] or a position correction [11] is applied to each marker to enforce the density constraint. This holonomic constraint tries to match the current density of a fluid marker ρ_i with the target density ρ_0, which results in the following indicator function

$$d_i = \frac{\rho_i - \rho_0}{\rho_0}. \tag{7}$$

These kinematic constraint equations are the analog of the conditions in Eq. (1c). Fortunately, this analogy with the dynamics of the solid phase goes one step further. Indeed, for the fluid-solid interaction problems, a non-penetration; i.e., unilateral, constraint is imposed between a fluid particle and a solid thus enforcing a non-penetration condition between the fluid and solid. These non-penetration unilateral constraints are the analog of the conditions in Eq. (1d). Coupling in this manner is straightforward and allows interaction between fluid and rigid bodies in a physically accurate manner using *one modeling framework*; i.e., that provided by differential algebraic inclusion problems.

3 Overall Numerical Solution Methodology

In what follows we will compound the many-body dynamics equations of motion of Eq. (1) with the equations of motion that capture the fluid dynamics. The dynamics of the fluid phase will be described using the constrained-fluid approach; i.e., using Eq. (7), such that the overall set of equations of motion is posed as a differential algebraic inclusion problem. The coupled fluid-solid approach that falls back on the equation of state of Eq. (6) is not discussed here but is described at length in [18–20].

The numerical solution methodology for the aforementioned differential algebraic inclusion problem, which captures both the dynamics of the fluid and solid phases, is built around the following two decisions: ($D1$) following the approach proposed in [24], a symplectic half implicit Euler methods is used to discretize the dynamics; and ($D2$) a zero gap non-penetration condition between bodies in mutual contact is enforced at the new time step $t^{(l+1)}$. Given a consistent position $\mathbf{q}^{(l)}$ and velocity $\mathbf{v}^{(l)}$ at time $t^{(l)}$, the numerical solution at $t^{(l+1)} = t^{(l)} + h$ is obtained by solving the following mathematical programming problem with complementarity and equilibrium constraints [12]:

$$\overbrace{\mathbf{q}^{(l+1)}}^{\text{Generalized positions}} = \mathbf{q}^{(l)} + \overbrace{\Delta t}^{\text{Step size}} \underbrace{\mathbf{L}}_{\text{Velocity transformation matrix}} (\mathbf{q}^{(l)})\mathbf{v}^{(l+1)} \tag{8a}$$

$$\mathbf{M}(\overbrace{\mathbf{v}^{(l+1)}}^{\text{Generalized speeds}} - \mathbf{v}^{(l)}) = \underbrace{\Delta t\mathbf{f}(t^{(l)}, \mathbf{q}^{(l)}, \mathbf{v}^{(l)})}_{\text{Applied impulse}} - \overbrace{\mathbf{G}(\mathbf{q}^{(l)}, t)\lambda}^{\text{Joint impulse}} - \overbrace{\mathbf{D}_d(\mathbf{q}^{(l)}, t)\gamma_d}^{\text{Density Impulse}} + \overbrace{\mathbf{D}_c(\mathbf{q}^{(l)}, t)\gamma_c}^{\text{Contact Impulse}} - \overbrace{\mathbf{D}_b(\mathbf{q}^{(l)}, t)\gamma_b}^{\text{Boundary Impulse}} \tag{8b}$$

$$0 = \underbrace{\frac{1}{\Delta t}\mathbf{g}(\mathbf{q}^{(l)}, t)}_{\text{Stabilization term}} + \mathbf{G}^{\mathrm{T}}\mathbf{v}^{(l+1)} + \frac{1}{\Delta t}\mathbf{g}_t \tag{8c}$$

$$-\mathbf{D}_d{}^T\mathbf{v}^{(l+1)} = \underbrace{\mathbf{g_d}(\mathbf{q}^{(l)}, t)}_{\text{Stabilization term}} \tag{8d}$$

$$j \in \mathcal{A}(q^{(l)}, \delta) : 0 \leq \overbrace{\frac{1}{\Delta t}\Phi_{b,j}(\mathbf{q}^{(l)})}^{\text{Stabilization term}} + \mathbf{D}_{b,j,n}^T \mathbf{v}^{(l+1)} \perp \gamma_{b,j,n} \geq 0 \tag{8e}$$

$$i \in \mathcal{A}(q^{(l)}, \delta) : 0 \leq \overbrace{\frac{1}{\Delta t}\Phi_{c,i}(\mathbf{q}^{(l)})}^{\text{Stabilization term}} + \mathbf{D}_{c,i,n}^T \mathbf{v}^{(l+1)} \perp \gamma_{c,i,n} \geq 0 \tag{8f}$$

$$\left(\gamma_{c,i,u}, \gamma_{c,i,w}\right) = \underset{\sqrt{\gamma_{c,i,u}{}^2 + \gamma_{c,i,w}{}^2} \leq \mu_i \gamma_{c,i,n}}{\operatorname{argmin}} \quad \mathbf{v}^{\mathrm{T}}\left(\gamma_{c,i,u}\mathbf{D}_{c,i,u} + \gamma_{c,i,w}\mathbf{D}_{c,i,w}\right) \tag{8g}$$

Here, λ represents a constraint impulse associated with a bilateral constraint, $\gamma_c, \gamma_d, \gamma_b$ are constraint impulses associated with contacts, density constraints and boundary contacts respectively. The superscript $(l+1)$ on γ_c was dropped for brevity. All forces acting on the system except the frictional contact, density and boundary constraint forces are evaluated at time $t^{(l)}$ and denoted by $\mathbf{f} \equiv \mathbf{f}(t^{(l)}, \mathbf{q}^{(l)}, \mathbf{v}^{(l)})$. The term $\frac{1}{h}\Phi_i(\mathbf{q}^{(l)})$ achieves constraint stabilization by eliminating any penetration within one time-step.

For large models with millions of contacts, no effective methods are available for solving the numerical problem in Eqs. (8b)–(8g), which assumes the form of a nonlinear complementarity problem (NCP). This observation motivated a third decision ($D3$), which was to recast the aforementioned numerical problem into a more amenable one by a convexification of the NCP; i.e., by relaxing the complementarity condition in Eq. (8f) [3] to

$$i \in \mathcal{A}(q^{(l)}, \delta) : 0 \leq \tfrac{1}{h}\Phi_i(\mathbf{q}^{(l)}) + \mathbf{D}_{i,n}^T\mathbf{v}^{(l+1)} - \mu_i\sqrt{(\mathbf{v}^T\mathbf{D}_{i,u})^2 + (\mathbf{v}^T\mathbf{D}_{i,w})^2} \perp \gamma_n^i \geq 0\,. \tag{9}$$

Owing to this relaxation, the resulting set of equations become a cone complementarity problem (CCP), see, for instance, [17]. Specifically, solving for $\mathbf{v}^{(l+1)}$ from Eq. (8b) and plugging its expression in Eq. (9) yields a mathematical programming problem with complementarity constraints formulated exclusively in the set of Lagrange multipliers γ [1]. To pose the CCP, the following notation is used: the number of contacts in $\mathcal{A}(\mathbf{q}, \delta)$ is n_c; $\mathbf{D} \equiv [\mathbf{D}_1, \ldots, \mathbf{D}_{n_c}] \in \mathbb{R}^{6n_b \times 3n_c}$ is the generalized contact transformation matrix; $\mathbf{D}_i \equiv [\mathbf{D}_{i,n}, \mathbf{D}_{i,v}, \mathbf{D}_{i,w}] \in \mathbb{R}^{6n_b \times 3}$ is the contact transformation matrix associated with contact $i \in \mathcal{A}(q^{(l)}, \delta)$; $\mathbf{r}_i \equiv \mathbf{b}_i + \mathbf{D}_i^T\mathbf{M}^{-1}\mathbf{f} \in \mathbb{R}^3$ is the generalized contact velocity for contact i; $\mathbf{b}_i \equiv \left[\frac{1}{h}\Phi_i(\mathbf{q}^{(l)}), 0, 0\right]^T \in \mathbb{R}^3$ is the unilateral constraint stabilization term; and $\mathbf{N} \equiv \mathbf{D}^T\mathbf{M}^{-1}\mathbf{D} \in \mathbb{R}^{3n_c \times 3n_c}$ is the contact associated symmetric positive-semidefinite Schur complement matrix, which is typically very sparse. The new quantities introduced – n_c, $\mathbf{D}$, $\mathbf{D}_i$, $\mathbf{r}_i$, $\mathbf{b}_i$, and $\mathbf{N}$—should be further qualified by a superscript (l) to indicate that they are evaluated in the system configuration corresponding to $t^{(l)}$. For brevity, the superscript was omitted. Then, the CCP then assumes the form

$$\begin{aligned} \text{Find:} \quad & \gamma_i^{(l+1)}, \text{ for } i = 1, \dots, n_c \\ \text{such that} \quad & \Upsilon_i \ni \gamma_i^{(l+1)} \quad \perp \quad -\left(\mathbf{N}\gamma^{(l+1)} + \mathbf{r}\right)_i \in \Upsilon_i^{\circ} , \\ \text{where} \quad & \Upsilon_i = \{ \left[x, y, z\right]^T \in \mathbb{R}^3 | \sqrt{y^2 + z^2} \leq \mu_i x \} , \\ \text{and} \quad & \Upsilon_i^{\circ} = \{ \left[x, y, z\right]^T \in \mathbb{R}^3 | x \leq -\mu_i \sqrt{y^2 + z^2} \} . \end{aligned} \tag{10}$$

This CCP represents the first order optimality condition of a quadratic optimization problem with conic constraints whose solution provides the set of normal and friction forces associated with the set of contacts in $\mathcal{A}(\mathbf{q}, \delta)$:

$$\begin{aligned} & \min f(\gamma) = \frac{1}{2}\gamma^T \mathbf{N} \gamma + \mathbf{r}^T \gamma \\ & \text{subject to } \gamma_i \in \Upsilon_i \text{ for } i = 1, 2, \dots, n_c . \end{aligned} \tag{11}$$

The solution of the optimization problem in Eq. (11) is found using a Nesterov method [14].

The overall approach is summarized as follows: an application that leads to a solid-body dynamics, fluid dynamics, or fluid-solid interaction problem is posed as a set of differential algebraic inclusion equations. These equations are discretized to lead to a NCP via $(D1)$ and $(D2)$. The NCP is convexified based on $(D3)$ to become a CCP. The CCP is solved by considering an equivalent quadratic optimization problem with conic constraints, whose solution is the desired $\gamma \in \mathbb{R}^{3n_c}$. Equation (8b) is then used to expeditiously compute the velocity $\mathbf{v}^{(l+1)}$. The generalized coordinates $\mathbf{q}^{(l+1)}$ are recovered using Eq. (8a) and the simulation is advanced to $t^{(l+1)}$.

4 Numerical Experiments

The solution framework described in Sect. 3 has been implemented in an freely-available open source simulation engine called Chrono [13, 21, 27]. In this paper we focus on a fording maneuver that is posed as a fluid-solid interaction problem. To this end, we use a nine body HMMWV vehicle model that consists of a 2086.52 kg chassis connected to four rigid lugged wheels via a double wishbone suspension. For this four-wheel drive vehicle, a driver model was used to move in a straight line while providing throttle and brake input to the vehicle to maintain a vehicle speed of 2 m/s during the entire fording maneuver.

Fig. 2 Fording setup. No fluid shown

(a) Fording simulation, fluid is represented as a collection of 1 426 663 rigid frictionless spheres. The simulation leads to problem as stated in Eq. (1).

(b) Fording scenario simulated with constraint fluids via Navier-Stokes and SPH. The problem is posed as a set of differential algebraic inclusion equations, see Eq. (8).

Fig. 3 A fording scenario, simulated via two representations of the fluid

The goal of this numerical experiment was to simulate a river fording operation where two different techniques would be compared. The first technique represents the fluid as a very large number of rigid frictionless spheres, more precisely 1 426 663 of them, and the equations of motion are exclusively governed by the differential algebraic inclusion problem of Eq. (1). The second technique represents the fluid via the Navier-Stokes equations and uses a constraint fluid formulation as detailed in Sect. 2, see Eq. (7). Using the results from these simulations we can understand how well they match each other and assess the benefits and pitfalls associated with each of them.

The fording setup, shown in Fig. 2 consists of two end platforms approximately 4.87 m (16 ft) long followed by a slope of length 5.18 m (17 ft) and a bottom slope of length 4.57 m (15 ft). The distance from the end platform to the bottom was 2.43 m (8 ft) and was filled with fluid such that half of the chassis would be under water during the fording operation, see Fig. 3b. Each of the 1 426 663 SPH fluid particles has a mass of 3.96×10^{-3} kg, a kernel radius of 0.032 m and a rest density of 1000 kg/m^3. The simulation is 12 s long and carried out with a time step of 0.001 s. The Barzilai Borwein solver was used with 1000 iterations for the bilateral constraints and an additional 50 iterations for the entire problem [9]. Simulations were performed using 40 parallel threads on a workstation that had two 10 physical core Intel(R) Xeon® CPU E5-2650 v3 @ 2.30 GHz processor with 128 GB of DDR4 2133 MHz memory.

Two simulation snapshots are shown in Fig. 3a, where the fluid is modeled as a collection of rigid spheres, and Fig. 3b, where the fluid motion is captured via Navier-Stokes and SPH. As reported in detail in [7, 12], over the entire fording operation the engine used approximately 278 530 J of energy to drive through the fluid and 372 112 J to drive through the rigid spheres, a difference of 25%. Indeed, rigid

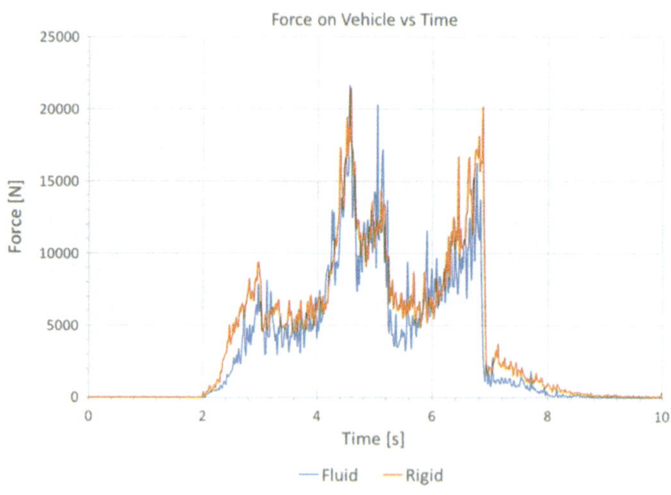

Fig. 4 Magnitude of the overall force impressed on the vehicle in a fording operation as it moves through a fluid (*blue*) and a large collection of frictionless spheres that mimic a fluid (*red*)

spheres are more difficult to move through even when frictionless as spheres interlock when settled and more energy is required to make them flow. This difference in flowability causes the increase in energy required to perform the fording operation. Moreover, although not shown in these plots, the controller required more throttle while attempting to maintain a 2 m/s speed through the rigid spheres when compared to the fluid due to the larger forces exerted upon the chassis, see Fig. 4.

(a) Chrono simulation of vehicle operating on plastic material that is meant to capture clay/silt-type soils. Performed using parallel computing via multiple CPU cores and GPU computing.

(b) Chrono simulation of vehicle operating on granular/gravel-type material. The motion of every particle in the terrain is monitored by leveraging parallel computing on GPU cards.

Fig. 5 Two other simulations that draw on the same differential algebraic inclusion analytical foundation, and same open-source Chrono simulation infrastructure

5 Conclusions

We outlined a modeling approach that relies on differential algebraic inclusion problems to pose the equations of motion for complex systems in the presence of friction and contact. This approach was used to demonstrate the dynamics of a vehicle performing a fording operation when in one case the fluid was represented by means of approximately 1.5 million frictionless spheres, and in the second case using the Navier-Stokes equations of motion. Both cases lead to problems with bilateral and unilateral constraints. In this context, insofar the "differential algebraic inclusion" concept is concerned, the "differential" attribute of the approach is associated with the momentum balance equations. The "algebraic" attribute is tied to the presence of holonomic and nonholonomic kinematic constraints that relate the generalized positions, and potentially velocities, used to describe the state of the mechanical system. Finally, the "inclusion" attribute goes back to the unilateral constraints and their connection to complementarity conditions—they appear, for instance, in handling the Coulomb friction, in non-penetration conditions, and in plasticity constitutive equations [26]. Although we have not touched upon the handling of plasticity, a Chrono simulation that uses the same differential algebraic inclusion approach is shown in Fig. 5a. For completeness, in Fig. 5b we provide a snapshot of a simulation in which a vehicle operates on granular/gravel-type material. Looking ahead, work is under way to implement solutions for the analysis of tracked vehicles operating in off-road conditions [22].

Acknowledgements This project was made possible through US National Science Foundation grant GOALI-CMMI 1362583.

References

1. Anitescu, M.: Optimization-based simulation of nonsmooth rigid multibody dynamics. Math. Program. **105**, 113–143 (2006)
2. Anitescu, M., Cremer, J.F., Potra, F.A.: Formulating 3D contact dynamics problems. Mech. Struct. Mach. **24**(4), 405–437 (1996)
3. Anitescu, M., Hart, G.D.: A constraint-stabilized time-stepping approach for rigid multibody dynamics with joints, contact and friction. Int. J. Numer. Methods Eng. **60**(14), 2335–2371 (2004)
4. Bodin, K., Lacoursiere, C., Servin, M.: Constraint fluids. IEEE Trans. Visual. Comput. Graph. **18**, 516–526 (2012)
5. Gingold, R.A., Monaghan, J.J.: Smoothed particle hydrodynamics-theory and application to non-spherical stars. Mon. Not. R. Astron. Soc. **181**, 375–389 (1977)
6. Gurtin, M.E., Fried, E., Anand, L.: The Mechanics and Thermodynamics Of Continua. Cambridge University Press (2010)
7. Hammad, M., Negrut, D.: Representing fluid dynamics as a many-body dynamics problem: a vehicle fording analysis test case. Technical Report TR-2016-01, Simulation-Based Engineering Laboratory, University of Wisconsin-Madison (2016)
8. Haug, E.J.: Computer-Aided Kinematics and Dynamics of Mechanical Systems Volume-I. Prentice-Hall, Englewood Cliffs (1989)

9. Heyn, T., Anitescu, M., Tasora, A., Negrut, D.: Using Krylov subspace and spectral methods for solving complementarity problems in many-body contact dynamics simulation. Int. J. Numer. Methods Eng. **95**, 541–561 (2013)
10. Lucy, L.B.: A numerical approach to the testing of the fission hypothesis. Astron. J. **82**, 1013–1024 (1977)
11. Macklin, M., Müller, M.: Position based fluids. ACM Trans. Graph. **32** 104:1–104:12 (2013)
12. Mazhar, H.: Physics Based Simulation Using Complementarity and Hybrid Lagrangian-Eulerian Methods. Ph.D. Thesis, University of Wisconsin–Madison (2016)
13. Mazhar, H., Heyn, T., Pazouki, A., Melanz, D., Seidl, A., Bartholomew, A., Tasora, A., Negrut, D.: Chrono: a parallel multi-physics library for rigid-body, flexible-body, and fluid dynamics. Mech. Sci. **4**, 49–64 (2013)
14. Mazhar, H., Heyn, T., Tasora, A., Negrut, D.: Using Nesterov's method to accelerate multibody dynamics with friction and contact. ACM Trans. Graph. **34** (2015)
15. Monaghan, J.J.: Smoothed particle hydrodynamics. Rep. Prog. Phys. **68**, 1703–1759 (2005)
16. Morris, J.P., Fox, P.J., Zhu, Y.: Modeling low Reynolds number incompressible flows using SPH. J. Comput. Phys. **136**, 214–226 (1997)
17. Negrut, D., Serban, R.: Posing multibody dynamics with friction and contact as a differential algebraic inclusion problem. Technical Report TR-2016-12, Simulation-Based Engineering Laboratory, University of Wisconsin-Madison. http://sbel.wisc.edu/documents/TR-2016-12.pdf (2016)
18. Pazouki, A., Negrut, D.: Numerical investigation of microfluidic sorting of microtissues. Comput. Math. Appl. (2015)
19. Pazouki, A., Negrut, D.: A numerical study of the effect of particle properties on the radial distribution of suspensions in pipe flow. Comput. Fluids **108**, 1–12 (2015)
20. Pazouki, A., Serban, R., Negrut, D.: A Lagrangian-Lagrangian framework for the simulation of rigid and deformable bodies in fluid. In: Terze, Z. (ed.) Multibody Dynamics, vol. 35 of Computational Methods in Applied Sciences, pp. 33–52. Springer (2014)
21. Project Chrono: Chrono: an open source framework for the physics-based simulation of dynamic systems. https://github.com/projectchrono/chrono. Accessed 18 Sep 2016
22. Serban, R., Taylor, M., Melanz, D., Negrut, D.: NG-NRMM phase I benchmarking: Chrono tracked vehicle simulation results summary. Technical Report TR-2016-08. Simulation-Based Engineering Laboratory, University of Wisconsin-Madison. http://sbel.wisc.edu/documents/TR-2016-08.pdf (2016)
23. Stewart, D.E.: Rigid-body dynamics with friction and impact. SIAM Rev. **42**(1), 3–39 (2000)
24. Stewart, D.E., Trinkle, J.C.: An implicit time-stepping scheme for rigid-body dynamics with inelastic collisions and Coulomb friction. Int. J. Numer. Methods Eng. **39**, 2673–2691 (1996)
25. Swegle, J., Hicks, D., Attaway, S.: Smoothed particle hydrodynamics stability analysis. J. Comput. Phys. **116**, 123–134 (1995)
26. Tasora, A., Anitescu, M., Negrini, S., Negrut, D.: A compliant visco-plastic particle contact model based on differential variational inequalities. Int. J. Non-Linear Mech. **53**, 2–12 (2013)
27. Tasora, A., Serban, R., Mazhar, H., Pazouki, A., Melanz, D., Fleischmann, J., Taylor, M., Sugiyama, H., Negrut, D.: Chrono: an open source multi-physics dynamics engine. In: Kozubek, T. (ed.) High Performance Computing in Science and Engineering, Lecture Notes in Computer Science, pp. 19–49. Springer (2016)

Large Deformation Modeling of Soil-Machine Interaction in Clay

Chong Peng, Mozhen Zhou and Wei Wu

Abstract The evaluation of soil reaction force on tillage tool is important for design and management optimization. The large deformation, dynamic nature and complex soil-tool contact in the problem make it a challenge in numerical modeling. The popular finite element method (FEM) has distorted mesh in large deformation, and complex adaptive remeshing techniques need to be used in soil-tool interaction simulation. On the other hand, numerical approaches based on computational fluid mechanics (CFD) and discrete element method (DEM), although avoid mesh distortion, have difficulties in capturing the true mechanical properties of soils. In this study, we develop a total Lagrangian SPH (TL-SPH) approach for simulating the large deformation soil-machine interaction in clay. The TL-SPH is simple in formulation and computationally efficient. The accuracy and stability of the method is improved by employing an hourglass control technique. Soil-tool contact is modeled using a node-to-segment (NTS) contact algorithm. Preliminary numerical studies are carried out, it is demonstrated that the presented approach is capable of capturing salient soil-tool interaction properties.

1 Introduction

The study of of soil-tool/machine interaction focus on soil failure and deformation patterns and force prediction for design and management optimization. Previously, the reaction forces are evaluated based on theoretical models with assumed failure and deformation mode [1]. These models have difficulties when complex geometries and loading conditions present. Furthermore, soils are modeled

C. Peng (✉) · W. Wu
Institut Für Geotechnik, Universität Für Bodenkultur, Feistmantelstr. 4, 1180 Vienna, Austria
e-mail: pengchong@boku.ac.at

M. Zhou
CAEP Software Center for High Performance Numerical Simulation, Institute of Applied Physics and Computational Mathematics, Beijing 100088, China

E. Papamichos et al. (eds.), *Bifurcation and Degradation of Geomaterials with Engineering Applications*, Springer Series in Geomechanics and Geoengineering, DOI 10.1007/978-3-319-56397-8_32

as rigid-perfectly-plastic media. Some important soil behaviors, such as rate-dependency, plastic hardening/softening and hydro-mechanical coupling are not considered. Numerical methods can deal with complex geometry and loading conditions, as well as arbitrarily complex soil constitutive responses. However, the large deformation, dynamic nature and complex soil-tool contact in soil-tool interaction make it a challenge for numerical methods. The finite element method (FEM) (so as other grid-based method) has distorted mesh in large deformation; thus, complex adaptive remeshing techniques need to be adopted. On the other hand, numerical approaches based on computational fluid mechanics (CFD), although avoiding mesh distortion, have difficulties in capturing the true mechanical properties of soils [2].

Meshfree methods have gained popularity in the last two decades, and have advantages in special fields like large deformation and crack propagation [3, 4]. Smoothed particle hydrodynamics (SPH) is a well-established meshfree method mainly applied in astrophysics and computational fluid mechanics. SPH has also been employed in large deformation analysis and granular flow modeling [3]. However, the conventional SPH has two deficiencies: tensile instability and low accuracy in stress computation caused by rank-deficiency, which severely restrict its application in solid mechanics. Several techniques were developed to mitigate the deficiencies. The artificial stress technique proposed by Gray et al. [5] is usually employed to alleviate the tensile instability. However, the artificial stress inevitably introduces additional dissipation which is unphysical. Later, it was revealed that the use of Eulerian kernel function is the cause of the tensile instability, and the alternative total Lagrangian SPH (TL-SPH) is completely free of tensile instability. However, the hourglass mode caused by rank-deficiency is still problematic in the TL-SPH method, leading to low accuracy and unphysical particle distribution. Researchers introduced additional stress points to cure the rank-deficiency [6]. However, where to place the stress points and how to move them seem ambiguous. Also, the employment of stress points introduces additional computational cost. As a consequence, the stress points method does not see wide application.

Recently, an hourglass control technique analogous to that used for reduced integration finite element was developed for TL-SPH [7]. It is demonstrated that this hourglass control technique dramatically improves the accuracy and stability of SPH. The TL-SPH thus becomes a reliable numerical method for large deformation analysis in solid mechanics. In this paper, we present a TL-SPH method for the large deformation soil-machine interaction in clay. As a preliminary study, the clay is modeled using a rate-independent elastoplastic model with von-Mises yield criterion. A node-to-segment (NTS) contact algorithm is employed to correctly simulate the soil-tool contact. The stiffness-based hourglass control technique is used to stabilize the simulations. A simple soil cutting problem is simulated to validate the presented method.

2 Total Lagrangian SPH for Clay

Conventionally, SPH employs an updated Lagrangian strategy. The node connectivity used for approximation is based on the current configuration, and it constantly changes during the computation. The kernel functions are also evaluated based on the current configuration, giving the name Eulerian kernel. It is found that the tensile instability in SPH is caused by the use of Eulerian kernel. Consequently, total Lagrangian formulations were developed, which define the support domain using the undeformed reference configuration. In TL-SPH, the particles in a support domain is fixed so the domain deforms with the material. This TL-SPH does not have tensile instability; thus, it is more suitable for solid mechanics where material strength is encountered.

2.1 Total Lagrangian Formulation

Let $\boldsymbol{X}$ be the coordinates of the initial undeformed reference configuration. The kinematics of the deformed medium are as follows. A mapping between the current configuration and $\boldsymbol{X}$ is used to describe the motion at time t

$$\boldsymbol{x} = \varphi(\boldsymbol{X}, t) \tag{1}$$

The displacement $\boldsymbol{u}$ is given by

$$\boldsymbol{u} = \boldsymbol{x} - \boldsymbol{X} \tag{2}$$

The deformation of the medium is measured by the deformation gradient $\boldsymbol{F}$

$$\boldsymbol{F} = \frac{\mathrm{d}\boldsymbol{x}}{\mathrm{d}\boldsymbol{X}} = \frac{\mathrm{d}\boldsymbol{u}}{\mathrm{d}\boldsymbol{X}} + \boldsymbol{I} \tag{3}$$

where $\boldsymbol{I}$ is the identity matrix. The rate of deformation gradient $\dot{\boldsymbol{F}}$ is given by

$$\dot{\boldsymbol{F}} = \frac{\mathrm{d}\boldsymbol{v}}{\mathrm{d}\boldsymbol{X}} \tag{4}$$

where $\boldsymbol{v}$ is the velocity defined as $\mathrm{d}\boldsymbol{x}/\mathrm{d}t$. Note that $\dot{\boldsymbol{F}}$ is evaluated based on the reference configuration. The instantaneous velocity gradient $\boldsymbol{l}$ in the current configuration is written as

$$\boldsymbol{l} = \frac{\mathrm{d}\boldsymbol{v}}{\mathrm{d}\boldsymbol{x}} = \frac{\mathrm{d}\boldsymbol{v}}{\mathrm{d}\boldsymbol{X}}\frac{\mathrm{d}\boldsymbol{X}}{\mathrm{d}\boldsymbol{x}} = \dot{\boldsymbol{F}}\boldsymbol{F}^{-1} \tag{5}$$

Consequently, the rate of deformation tensor $\boldsymbol{d}$ and spin tensor $\boldsymbol{w}$ are obtained by $\boldsymbol{d} = (\boldsymbol{l} + \boldsymbol{l}^{\mathrm{T}})/2$ and $\boldsymbol{w} = (\boldsymbol{l} - \boldsymbol{l}^{\mathrm{T}})/2$.

The conservation of mass and momentum is based on the reference configuration

$$\rho J = \rho_0 \tag{6}$$

$$\frac{\mathrm{d}\boldsymbol{v}}{\mathrm{d}t} = \frac{1}{\rho_0}\nabla_0 \cdot \boldsymbol{P} + \frac{\boldsymbol{f}}{\rho_0} \tag{7}$$

where J is the determinant of the deformation gradient $\boldsymbol{F}$, ρ is the density, and $\boldsymbol{P}$ is the first Piola Kirchhoff stress tensor. $\boldsymbol{f}$ includes all external forces such as body force, contact force and surface traction. The subscript 0 indicates that a quantity is evaluated in the reference configuration.

2.2 Kernel Corrected Total Lagrangian SPH Approximation

The Lagrangian kernel is based on the reference coordinates $\boldsymbol{X}$. At the beginning of simulations, the nodal connectivity is determined and unchanged later on, unless the reference configuration $\boldsymbol{X}$ is updated in extremely large distortion. Therefore, the nodal connectivity is only required to calculate once, which saves much computational cost. SPH approximation of a scalar function f and its derivatives can be written as

$$f(\boldsymbol{X}_i) = \sum_{j=1}^{n} f(\boldsymbol{X}_j) W_i(X_{ij}) V_j^0 \tag{8}$$

$$\nabla_0 f(\boldsymbol{X}_i) = \sum_{j=1}^{n} f(\boldsymbol{X}_j) \nabla_0 W_i(X_{ij}) V_j^0 \tag{9}$$

where n is the number of particles in the influence domain of $\boldsymbol{X}_i$. $W_i(X_{ij})$ is the kernel function of particle $\boldsymbol{X}_i$ dependent on the distance between particle $\boldsymbol{X}_i$ and particle $\boldsymbol{X}_j$, with $X_{ij} = \| \boldsymbol{X}_j - \boldsymbol{X}_i \|$.

The conventional SPH approximation lacks zeroth and first order consistency. The kernel gradient correction is used to restore the consistency. At the beginning of each step, a shape matrix is computed as

$$\boldsymbol{L}_i = \sum_{j=1}^{n} (\boldsymbol{X}_j - \boldsymbol{X}_i) \otimes \nabla_0 W_i(X_{ij}) V_j^0 \tag{10}$$

The kernel gradient $\nabla W_i(X_{ij})$ is replaced by a corrected gradient $\tilde{\nabla} W_i(X_{ij}) = \boldsymbol{L}_i^{-1} \nabla W_i (X_{ij})$. Together with the symmetric discretization of the momentum equation, the zeroth and first order consistency is restored.

As total Lagrangian formulation is used, the mass conservation is automatically satisfied. For the momentum equation, the nodal forces caused by stress are given by the following equation in TL-SPH [7]

$$f_i = \sum_{j=1}^{n} (P_i L_i^{-1} + P_j L_j^{-1}) \nabla_0 W_i(X_{ij}) V_i^0 V_j^0 \tag{11}$$

In the TL-SPH, boundary conditions can be enforced as either velocity constraints or external nodal forces. Therefore, if a constitutive model is provided to obtain stress tensor $\boldsymbol{P}$, the large deformation problem can be solved using explicit integration.

2.3 Constitutive Law

Large plastic deformation occurs during soil-tool interaction. The implementation of plastic models in large deformation analysis is non-trivial. In finite strain theory, the treatment of plasticity is based on multiplicative decomposition of deformation gradient. Another popular but less theoretically solid approach is the hypoelastic-plastic method making use of objective stress rate in the current configuration. It is found that the hypoelastic-plastic approach is accurate enough if: (1) the elastic strain is small compared to plastic strain, (2) the time step is sufficiently small. In this paper, a hypoelastic-plastic approach is employed due to its convenience in implementation.

The rate of deformation tensor has the following additive form

$$\boldsymbol{d} = \boldsymbol{d}^e + \boldsymbol{d}^p \tag{12}$$

where $\boldsymbol{d}^e$ and $\boldsymbol{d}^p$ are the elastic and plastic parts of the rate of deformation tensor. The constitutive equation is written as

$$L_v(\boldsymbol{\tau}) = \boldsymbol{D} : (\boldsymbol{d} - \boldsymbol{d}^p) \tag{13}$$

where L_v denotes the Lie objective stress rate defined as $L_v(\boldsymbol{\tau}) = \dot{\boldsymbol{\tau}} - \boldsymbol{l} \cdot \boldsymbol{\tau} - \boldsymbol{\tau} \cdot \boldsymbol{l}^{\mathrm{T}}$, $\boldsymbol{\tau} = J\boldsymbol{\sigma}$ is the Kirchhoff stress. $\boldsymbol{D}$ is the fourth-order elastic tensor.

The von-Mises yield criterion is employed

$$F(\boldsymbol{\tau}) = \sqrt{2J_2} - \sqrt{2/3}\sigma_Y = 0 \tag{14}$$

where J_2 is the second invariant of the Kirchhoff stress tensor, σ_Y is the yield stress of clay.

At each time step, the velocity gradient $\boldsymbol{l}$ and rate of deformation tensor $\boldsymbol{d}$ can be obtained from $\boldsymbol{F}$ and $\dot{\boldsymbol{F}}$ through Eq. (4). The Kirchhoff stress tensor is calculated by time integration of the rate of Kirchhoff stress. The first Piola-Kirchhoff stress $\boldsymbol{P}$ is finally obtained by $\boldsymbol{P} = \boldsymbol{\tau}\boldsymbol{F}^{-\mathrm{T}}$.

2.4 Hourglass Control for Total Lagrangian SPH

The TL-SPH method outlined above has rank-deficiency which leads to hourglass mode distribution of particles. The deformation gradient of a particle in SPH is the average of individual particle pair contribution. This average process may lead to information loss. If more than one particle configurations result in the same deformation gradient hence the same potential energy, the solution may oscillates between these configurations, leading to the so-called hourglass mode.

An hourglass control algorithm was proposed recently by Ganzenmüller [7]. The deformation gradient operator is a linear operator, which means that the displacement in the support domain is a linear field. Nodal displacements incompatible with the linear field described by the average deformation gradient are identified as hourglass displacement. For a particle pair $\boldsymbol{X}_i$ and $\boldsymbol{X}_j$ in a domain centered at $\boldsymbol{X}_i$ with deformation gradient $\boldsymbol{F}_i$, the exact linear separation is

$$< \boldsymbol{x}_{ij} >^i = \boldsymbol{F}_i \boldsymbol{X}_{ij} \tag{15}$$

However, the actual separation $\boldsymbol{x}_{ij}$ is not identical to $< \boldsymbol{x}_{ij} >^i$ and the error is $\boldsymbol{e}^i_{ij} = < \boldsymbol{x}_{ij} >^i - \boldsymbol{x}_{ij}$. For symmetry we apply the same operation to particle $\boldsymbol{X}_j$. Consequently, the error vector is obtained as

$$\boldsymbol{e}_{ij} = \frac{1}{2}(\boldsymbol{e}^i_{ij} + \boldsymbol{e}^j_{ij}) = \frac{1}{2}(\boldsymbol{F}_i + \boldsymbol{F}_j)\boldsymbol{X}_{ij} - \boldsymbol{x}_{ij} \tag{16}$$

To suppress the hourglass mode the error vector should be minimized. Therefore, a force in line with $\boldsymbol{X}_{ij}$ is used to control the hourglass mode

$$\hat{\boldsymbol{f}}^{HG}_{ij} = -\frac{1}{2}\alpha E \frac{\boldsymbol{e}_{ij} \cdot \boldsymbol{x}_{ij}}{X^2_{ij}} \frac{\boldsymbol{x}_{ij}}{x^2_{ij}} \tag{17}$$

where E is the Young's modulus of the material, and α is a constant to control the magnitude of hourglass control force. Note that $\hat{\boldsymbol{f}}^{HG}_{ij}$ is the force per unit volume, and it is only the force between the particle pair $\boldsymbol{X}_{ij}$. In the SPH context, the total hourglass control force on particle i is the SPH average of all the contributions from the neighboring particles

$$\boldsymbol{f}^{HG}_i = V^0_i \sum_{j=1}^{n} \hat{\boldsymbol{f}}^{HG}_{ij} W_i(X_{ij}) V^0_j = -\frac{1}{2} \sum_{j=1}^{n} \alpha E \frac{\boldsymbol{e}_{ij} \cdot \boldsymbol{x}_{ij}}{X^2_{ij}} \frac{\boldsymbol{x}_{ij}}{x^2_{ij}} W_i(X_{ij}) V^0_i V^0_j \tag{18}$$

The adopted hourglass control algorithm works perfectly for elastic problems. But the stiffness-based hourglass control forces prohibit the development of plastic deformations, which are intrinsically nonlinear. In this paper, if a particle is identified undergoing plastic deformation, the constant α in the hourglass control force is divided by a factor of 100, to allow the normal plastic deforming.

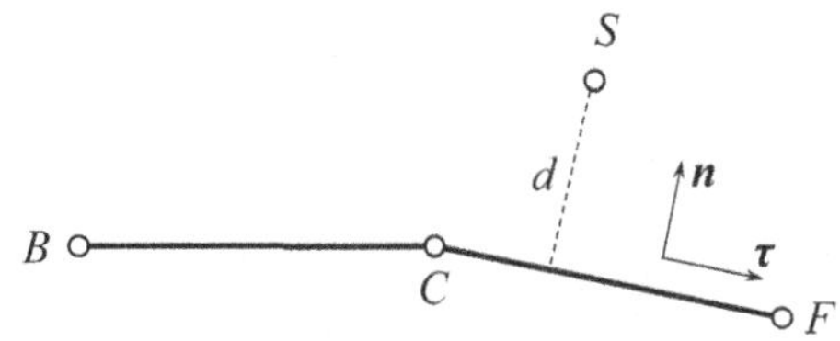

Fig. 1 The contact between rigid tool surface and SPH soil particle

3 Node-to-Segment Contact for TL-SPH

The contact between soil and tool is modeled using a node-to-segment algorithm. The tool is discretized with line segments with length of Δp, the initial average spacing of SPH particles, as shown in Fig. 1. The commonly used penalty method is used to compute the contact force

$$\boldsymbol{f}_i^n = (1 - \chi)\frac{2m_i}{(\Delta t)^2}(d_0 - d)\boldsymbol{n} \tag{19}$$

where $\boldsymbol{n}$ is the unit normal vector of the master contact face. The penalty coefficient $2m_i/(\Delta t)^2$ is applied to make sure that the penetration $d_0 - d$ become zero in the time duration Δt, where $d_0 = \Delta p/2$ is the contact threshold between deformable soil and rigid tool face and d is the distance between the master surface and the contact slave node. χ defines the allowed residual penetration, which usually ranges from 0.9 to 0.99. In this work, the contact is modeled as frictionless. Therefore, the contact force in the tangential direction is zero.

A important aspect in contact analysis is the search for potential contact and determination of the slave node-master surface pair. For a soil particle, we first search for the closest surface node C. Then the next and previous surface nodes F and B are determined. Naturally, the particle node is the slave node, and the two surfaces $\overrightarrow{BC}$ and $\overrightarrow{CF}$ are the potential master surfaces. However, the determination of master surface in NTS algorithm is complicated by many special cases. The procedure described in [8] is used in this work to decide the master surface. Once the master surface is determined, the distance d can be obtained by

$$d = (\boldsymbol{x}_S - \boldsymbol{x}_C) \cdot \boldsymbol{n} \tag{20}$$

4 Numerical Example

A case of rigid tool cutting into a clay is modeled, as shown in Fig. 2. The width and angle of the blade is 0.12 m and 45°, respectively. The material parameter for the clay is: Young's modulus $E = 1.0$ MPa, Poisson's ratio $\nu = 0.25$ and yield stress $\sigma_Y = 5.0$ kPa. As a preliminary study, the rate-dependent and frictional effects are not considered. The blade moves horizontally, accelerating linearly to $v_x = -4$ cm/s in

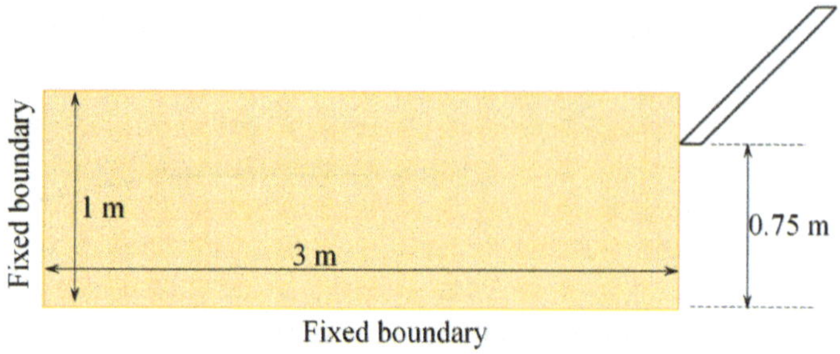

Fig. 2 The geometry and boundary conditions of the soil cutting problem

Fig. 3 The horizontal and vertical reaction force on the tool surface

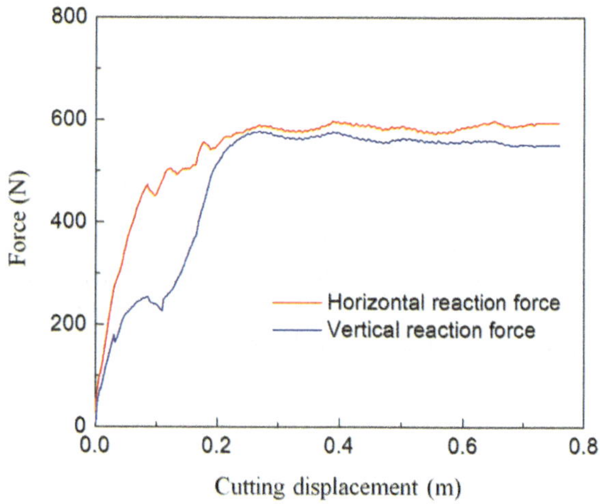

2 s, ensuring a quasi-static loading condition. After reaching −0.04 cm/s, the velocity is then kept constant. The spatial resolution of the numerical discretization is $\Delta p =$ 0.02 m. Gravitation is not considered. The coefficient α for the hourglass control is taken as 50.

The horizontal and vertical reaction forces are shown in Fig. 3. As the sharp tool cutting into the soil body, the reaction forces increase gradually, and eventually become stable. The stable horizontal and vertical reaction forces are approximately 595 and 550 N, respectively. The deformation pattern of the cutting is given in Fig. 4, together with the displacement and equivalent plastic strain. It is found that the deformation of the soil-tool cutting is reasonably modeled.

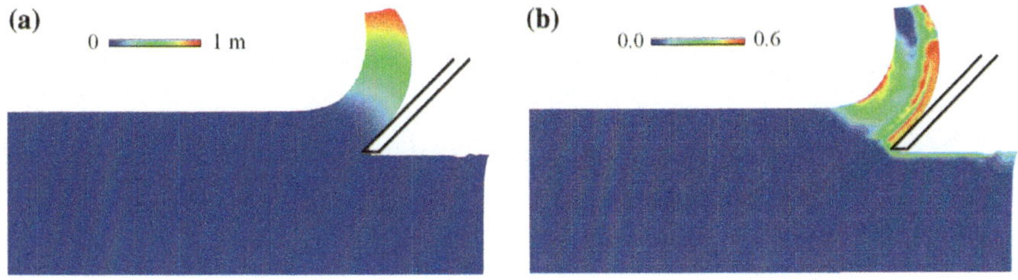

Fig. 4 The shape of the soil after cutting: **a** displacement; **b** plastic strain

5 Conclusion

Large deformation modeling of soil-tool interaction is performed using a total Lagrangian smoothed particle hydrodynamics. The TL-SPH method is free of tensile instability. Its accuracy and robustness is further improved using a stiffness-based hourglass control algorithm. With a hypoelastic-plastic approach, any constitutive model for infinitesimal strain can be used in large deformation analysis. The node-to-segment contact formulation is employed to model the soil-tool contact. The simulation of a blade cutting into a clay demonstrates that the method reasonably captures the forces and the deformation pattern, which are the two main concerns in soil-machine interactions. The preliminary results show that the presented method is promising and can be applied to model various soil-tool interaction problems.

Acknowledgements C. Peng gratefully acknowledges the financial support from the Otto Pregl Foundation for Fundamental Geotechnical Research in Vienna.

References

1. Onwualu, A.P., Watts, K.C.: Draught and vertical forces obtained from dynamic soil cutting by plane tillage tools. Soil Tillage Res. **48**, 239–253 (1998)
2. Subrata, K., Lal Kushwah, R.: Dynamic modelling of soil-tool interaction: An overview from a fluid flow perspective. J. Terramech. **43**, 411–425 (2006)
3. Peng, C., Wu, W., Yu, H.S., Wang, C.: A SPH approach for large deformation analysis with hypoplastic constitutive model. Acta Geotech. **10**, 703–717 (2015)
4. Peng, C., Wu, W., Zhang, B.Y.: Three-dimensional simulations of tensile cracks in geomaterials by coupling meshless and finite element method. Int. J. Numer. Anal. Methods Geomech. **39**, 135–154 (2015)
5. Gray, J.P., Monaghan, J.J., Swift, R.P.: SPH elastic dynamics. Comput. Methods Appl. Mech. Eng. **190**, 6641–6662 (2001)
6. Randles, P.W., Libersky, L.D.: Normalized SPH with stress points. Int. J. Numer. Methods Eng. **48**, 1445–1462 (2000)
7. Ganzenmüller, G.C.: An hourglass control algorithm for Lagrangian smoothed particle hydrodynamics. Comput. Methods Appl. Mech. Eng. **286**, 87–106 (2015)
8. Zavarise, G., De Lorenzis, L.: The node-to-segment algorithm for 2D frictionless contact: Classical formulation and special cases. Comput. Methods Appl. Mech. Eng. **198**, 3428–3451 (2009)

Tracing the Salt Crystallization Front in Limestone Using the DRMS

Sevasti Modestou and Ioannis Ioannou

Abstract Salt crystallization is widely recognised as one of the most damaging factors affecting stone monuments and buildings. The conservation of stone heritage suffering from weathering due to the presence of salts can be very expensive and laborious. Furthermore, the assessment of salt-laden buildings alone is often challenging and costly in its own right. This paper focuses on the use of a micro-destructive technique, the Drilling Resistance Measurement System (DRMS), to trace the salt crystallization front in natural building stone subjected to continuous partial immersion in a salt solution. The results provide strong evidence of the efficacy of micro-drilling to "map" the crystallization front. This is facilitated by increased resistances recorded during drilling in areas where pore clogging due to salt crystallization occurs. The experimental results are important for future research aiming to validate models relating rising damp to salt crystallization. The combination of DRMS and continuous partial immersion in salt solution laboratory tests may also be used to investigate the effect of water repellents and coatings/finishes on salt transport and crystallization in porous materials. Last but not least, the potential of the portable DRMS to trace the salt crystallization front in limestone may be used to detect cryptoflorescence in situ, on stone monuments, before it becomes damaging.

S. Modestou · I. Ioannou (✉)
Department of Civil and Environmental Engineering, University of Cyprus, Nicosia, Cyprus
e-mail: ioannis@ucy.ac.cy

S. Modestou
Scottish Universities Environmental Research Centre, University of Glasgow, Glasgow, UK

E. Papamichos et al. (eds.), *Bifurcation and Degradation of Geomaterials with Engineering Applications*, Springer Series in Geomechanics and Geoengineering, DOI 10.1007/978-3-319-56397-8_33

1 Introduction

The crystallization of soluble salts within the pores of building materials, such as stone, is widely accepted as an important cause of decay [1, 3, 5, 10, 12, 20]. Conservation and repair of stone structures and monuments suffering from salt crystallization can be a costly and time-consuming practice. At the same time, the assessment of salt-laden structures may be a challenging task itself. Thus, tools and methods that could be used in preventive conservation of monuments to detect salt crystallization within the pores of stone before it becomes damaging, could provide an early warning system to conservators and may be of utmost importance in the protection of such structures.

In previous work [16], we have successfully used the micro-destructive Drilling Resistance Measurement System (DRMS) to detect the salt front in a natural limestone, both in the laboratory and in situ. Here, we focus on the use of the DRMS to trace salt crystallization in a building limestone subjected to continuous partial immersion in a salt solution in the laboratory.

2 Materials and Methodology

A limestone sample measuring 5 × 5 × 30 cm was subjected to continuous partial immersion in the laboratory using a 1.6 M magnesium sulphate salt solution. The stone under investigation was quarried in the village of Agios Theodoros in Cyprus. It is a packstone or poorly washed biosparite with rather complex mineralogy, including a large proportion of siliciclastic grains with a significant fraction of basaltic lithoclasts. The intergrain cement is microsparry calcite and micrite [15].

The stone sample was positioned upright, with the long axis vertical, on two thin supports inside a glass container containing the salt solution and a layer of diabase aggregates (Fig. 1). Paraffin wax was used to provide a barrier to evaporation from the free solution surface. The test specimen was left partially immersed in the salt solution for six months. During this period, the salt solution was replenished as necessary using a rubber tube inserted in the set-up.

Following the continuous partial immersion test, the stone sample was drilled along its entire length using the DRMS in order to map the location of the salt crystallization front. The DRMS is a cordless and portable micro-destructive tool that was primarily developed for laboratory and in situ studies in the field of cultural heritage (e.g. [4, 8, 23]). More recently, the system has also been used to characterize various limestones from across Europe [22]. During drilling tests, the penetration rate and rotational speed of the drill bit are kept constant, while the force required to drill is continuously recorded in relation to the drilling depth. This force is sensitive to pore clogging and can thus be used to map the location of the salt front in salt-laden stones fairly accurately [16].

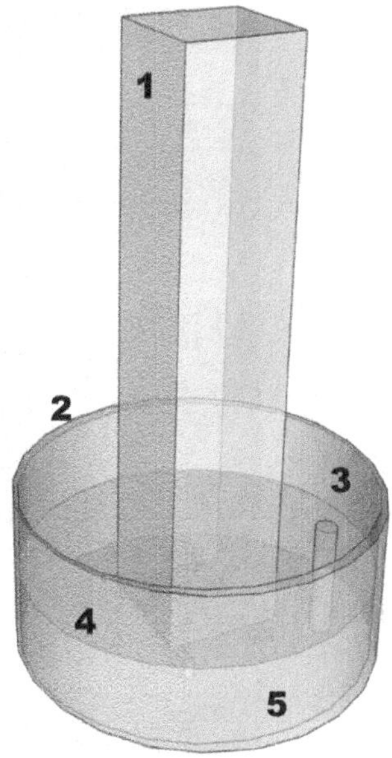

Fig. 1 Continuous partial immersion test set-up (*1* Stone specimen, *2* Glass container, *3* Rubber tube to facilitate solution replenishment, *4* Paraffin wax layer, *5* Salt solution and diabase aggregates)

In the present research, the stone sample was drilled to a depth of 4 cm. The diameter of the diamond drill bit used was 5 mm, whilst the operating conditions were 600 rpm for the rotational speed and 10 mm/min for the penetration rate.

3 Results and Discussion

Photographic evidence of the results of the continuous partial immersion test is presented in Fig. 2. Figure 3 presents the DRMS mapping patterns for the first 2.5–3.0 cm of drilling for each successful test.

From the results, it appears that the lowest section of the test specimen, below the red dotted line (Fig. 2), remains damp throughout the test. This is attributed to the fact that the rate of salt solution rise in that area is faster, compared to the rate of evaporation. The section immediately above that area and below the upper limit of capillary rise produces harmless efflorescence, since evaporation and salt solution supply are approximately equal, thus allowing the drying/crystallization front to locate itself on the surface of the material. The uppermost section is where cryptoflorescence persists. Here, evaporation is greater than the rate of supply of salt solution by capillarity; thus the evaporative surface or drying front moves into the material, preventing salts from being transported to the surface [19]. In that area, close to the centre of the test specimen, cracks appear (see Figs. 2 and 3) as a result of crystallization pressures exceeding the tensile strength of the sample under investigation.

Fig. 2 Continuous partial immersion test results at various stages (*left* 1 day; *middle* 3 months; *right* 6 months). Dotted red line on 1-day sample emphasizes limit of continuously damp area. *Arrows* indicate crack positions

The DRMS patterns (Fig. 3) vividly show increased resistances (peaks) at varying positions along the central vertical axis of the sample. The positions of the peaks in the DRMS profiles allow the prediction of the position of the crystallization front and consequently the generation of a theoretical crystallization profile (see Fig. 3, right). The parabolic shape of the predicted crystallization front is in line with idealized diagrams produced by Scherer [19] and Ruiz-Agudo et al. [18] for salt crystallization incidents occurring due to the upward movement of moisture in unsaturated porous materials. This shape is attributed to the fact that evaporation is more efficient near the surface of a material, rather than within its body. As a result, fluid flow deeper within the test specimen is more efficient than near its surface, allowing the salt solution to move further up [14].

The parabolic shape of the predicted crystallization front also confirms the macroscopic analysis of the results shown in Fig. 2. Whilst there is little (if any) resistance to drilling in the visibly damp region of the sample, as well as in the region with efflorescence (see DRMS mapping patterns for holes nr. 9–12, Fig. 3), above these regions, where cracking due to the action of cryptoflorescence appears, there is increased resistance to drilling (see DRMS mapping patterns for holes nr. 3–8, Fig. 3). The peaks in this upper region gradually move towards the centre of

Fig. 3 DRMS mapping patterns (*left* sample face subjected to DRMS tests; *middle* side profile of sample; *right* corresponding DRMS patterns overlaid on sample profile with predicted crystallization front marked in *red*)

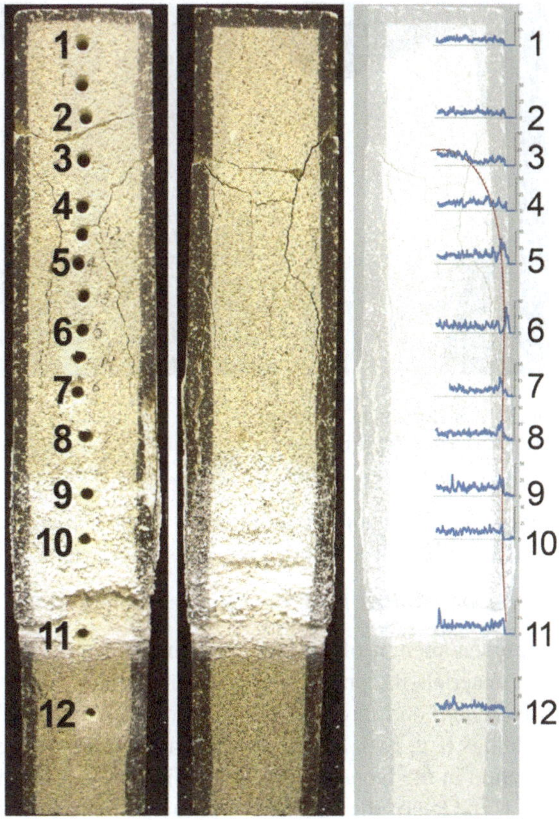

the test specimen, likely due to pore clogging and cryptoflorescence occurring progressively further inside the body of the material.

Figure 4 confirms that the DRMS patterns shown in in Fig. 3 provide a reliable estimate of the salt crystallization front. In this Fig., which shows a horizontal section through the material at the position of drill hole nr. 3, the crystallization front can clearly be observed (white residue). The DRMS pattern follows this front almost perfectly; this is reflected in the increased drilling resistances observed across the white residue area (Fig. 4, right).

It is interesting to note that, at least for the particular sample under investigation here, moisture does not seem to affect the overall resistance to drilling. Comparing the DRMS patterns for holes nr. 1 and 2, where the material appears dry, to the pattern for hole nr. 12, where the material is clearly quite moist, there does not appear to be any significant difference. The average drilling resistances at these drill positions were 9.1 N (drill hole nr. 1), 8.5 N (drill hole nr. 2) and 10.7 N (drill hole nr. 12); these are more or less equal, bearing in mind the sensitivity of the drilling methodology to the mineralogy and microstructure of a test specimen. This

Fig. 4 *Left* Horizontal section through the test specimen at the position of drill hole nr. 3. *Right* DRMS pattern from drill hole nr. 3 superimposed on photograph at location of drill hole

observation is important with respect to the application of the DRMS in situ, where it is very difficult to control the moisture content of the material.

A Note on the Practical Significance of the Results

The phenomenon of upward movement of moisture (also referred to as *wicking*) into materials in contact with a water reservoir, such as saturated soil, is known as rising damp and has long been noted to be related to salt weathering problems in buildings (e.g. [6]). Many scientists have approximated the rising damp weathering situation in the laboratory using continuous partial immersion tests (e.g. [2, 13, 17, 19, 21]). One of the advantages of the partial immersion test, particularly when the environmental conditions are controlled or when many tests are run in parallel, is that the results between different stone varieties and/or salt solution combinations can readily be compared [7]. Furthermore, specific elements of salt crystallization, such as the location/distribution of salt crystals, can be examined at the same time as the durability of a specific stone, provided a non- or micro-destructive mapping methodology can be applied. This renders the results of the present research particularly useful, since they may be used in future work aiming to validate models relating rising damp to salt crystallization. Combining results from continuous partial immersion tests and DRMS salt mapping, with physical models of rising damp (see, for example, [11]), provides a means to estimate the effect of salts on water transport within masonry walls, since salts can affect parameters such as the viscosity of the transporting liquid, evaporation rate, and permeability.

Moreover, the combination of DRMS and continuous partial immersion in salt solution laboratory tests can be adapted to investigate the effect of water repellents and coatings/finishes on salt transport and crystallization in porous materials. Such treatments can inhibit evaporation, fully or partially, and thus may induce an increase in the height of capillary rise of moisture and subsequently to the position of the salt crystallization front (e.g. [9, 24]).

4 Conclusions

This research provides strong evidence of the efficacy of micro-drilling to "map" the crystallization front in natural building stone subjected to continuous partial immersion in a salt solution. This is facilitated due to the increased resistances recorded during drilling in areas where pore clogging due to salt crystallization had occurred. The aforementioned potential of the DRMS may be used to detect cryptoflorescence in situ, on stone monuments, before it becomes damaging.

The experimental results may also be utilized in future research work aiming to validate models relating rising damp to salt crystallization. The combination of DRMS and continuous partial immersion in salt solution laboratory tests may further be used to investigate the effect of water repellents and coatings/finishes on salt transport and crystallization in porous materials.

References

1. Amoroso, G., Fassina, V.: Stone Decay and Conservation—Atmospheric Pollution, Cleaning, Consolidation and Protection. Elsevier, Amsterdam (1983)
2. Benavente, D., García del Cura, M.A., Bernabéu, A., Ordóñez, S.: Quantification of salt weathering in porous stones using an experimental continuous partial immersion method. Eng. Geol. **59**, 313–325 (2001)
3. Benavente, D., García del Cura, M.A., García-Guinea, J., Sánchez-Moral, S., Ordóñez, S.: Role of pore structure in salt crystallization in unsaturated porous stone. J. Cryst. Growth **260**, 532–544 (2004)
4. Delgado, R.J., Ferreira, P.A., Rodrigues da Costa, D.: Tracing of decay profiles and evaluation of stone treatments by means of microdrilling techniques. J. Cult. Heritage **3**, 117–125 (2002)
5. Desarnaud, J., Bonn, D., Shahidzadeh, N.: The pressure induced by salt crystallization in confinement. Sci. Rep. **6**, 30856 (2016)
6. Doehne, E.: Salt weathering: a selective review. In: Siegesmund et al. (ed.) Natural stone, weathering phenomena, conservation strategies and case studies, vol. 205, pp. 51–6. Geological Society, London (2002)
7. Espinosa-Marzal, R.M., Scherer, G.W.: Mechanisms of damage by salt. In: Smith, B.J. et al. (eds) Limestone in the Built Environment: Present-Day Challenges for the Preservation of the Past, vol. 331, pp. 61–77. Geological Society, London (2010)
8. Exadaktylos, G., Tiano, P., Filareto, C.: Validation of a model of rotary drilling of rocks with the drilling force measurement. Int. J. Restor. Build. Monuments **3**, 307–340 (2000)
9. Gonçalves, T.D., Pel, L., Rodrigues, J.D.: Influence of paints on drying and salt distribution processes in porous building materials. Constr. Build. Mater. **23**, 1751–1759 (2009)
10. Goudie, A.S.: Further experimental investigation of rock weathering by salt and other mechanical processes. Zeitschift fur Geomorphologie, Supplementband **21**, 1–12 (1974)
11. Hall, C., Hoff, W.D.: Rising damp: capillary rise dynamics in walls. Proc. R. Soc. A **463**, 1871–1884 (2007)
12. Ioannou, I., Hoff, W.D.: Water repellent influence on salt crystallisation in masonry. Proc. ICE Constr. Mater. **161**, 17–23 (2008)

13. La Iglesia, A., Gonzalez, V., López-Acevedo, V., Viedma, C.: Salt crystallization in porous construction materials I. Estimation of crystallization pressure. J. Cryst. Growth **177**, 111–118 (1997)
14. López-Acevedo, V., Viedma, C., Gonzalez, V., La Iglesia, A.: Salt crystallization in porous construction materials II: mass transport and crystallization processes. J. Cryst. Growth **182**, 103–110 (1997)
15. Modestou, S., Theodoridou, M., Fournari, R., Ioannou, I.: Physico-mechanical properties and durability performance of building and decorative carbonate stones from Cyprus. In: Přikryl et al. (eds.) Sustainable Use of Traditional Geomaterials in Construction Practice, 416, pp. 145–162. Geological Society, London (2016)
16. Modestou, S., Theodoridou, M., Ioannou, I.: Micro-destructive mapping of the salt crystallization front in limestone. Eng. Geol. **193**, 337–347 (2015)
17. Rivas, T., Alvarez, E., Mosquera, M.J., Alejano, L., Taboada, J.: Crystallization modifiers applied in granite desalination: the role of the stone pore structure. Constr. Build. Mater. **24**, 766–776 (2010)
18. Ruiz-Agudo, E., Mees, F., Jacobs, P., Rodriguez-Navarro, C.: The role of saline solution properties on porous limestone salt weathering by magnesium and sodium sulfates. Environ. Geol. **52**, 269–281 (2007)
19. Scherer, G.W.: Stress from crystallization of salt. Cem. Concr. Res. **34**, 1613–1624 (2004)
20. Scherer, G.W.: Crystallization in pores. Cem. Concr. Res. **29**, 1347–1358 (1999)
21. Sperling, C.H.B., Cooke, R.U.: Laboratory Simulation of rock weathering by salt crystallization and hydration processes in hot, arid environments. Earth Surf. Proc. Land. **10**, 541–555 (1985)
22. Theodoridou, M., Dagrain, F., Ioannou, I.: Micro-destructive cutting techniques for the characterization of natural limestone. Int. J. Rock Mech. Min. Sci. **76**, 98–103 (2015)
23. Tiano, P.: The use of microdrilling techniques for the characterisation of stone materials. In: PRO 26: International RILEM Workshop on On-Site Control and Non-Destructive Evaluation of Masonry Structures, 12–14 November 2001, Mantova, Italy (2001)
24. van Hees, R.P.J., Brocken, H.J.P.: Damage development to treated brick masonry in a long-term salt crystallisation test. Constr. Build. Mater. **18**, 331–338 (2004)

Bearing Capacity in Sand Under Eccentric and Inclined Loading Using a Bounding Surface Plasticity Model

Dimitrios Loukidis and Nektaria Ygeionomaki

Abstract The bearing capacity of strip footings on sand under inclined or eccentric loading is investigated using finite element analysis. The sand behavior is simulated using a bounding surface plasticity constitutive model based on critical state theory that accounts for strain softening, pre-failure non-linearity and material anisotropy. The parametric study focuses on the rate of decrease of the bearing capacity with increasing load inclination or load eccentricity for various values of sand relative density. The numerical predictions are compared with experimental data from centrifuge tests, as well as predictions from existing design equations.

Keywords Sand · Bearing capacity · Finite elements

1 Introduction

The bearing capacity Q_{bL} of shallow foundations is traditionally calculated using the bearing capacity equation. For the specific case of a strip footing resting on the free surface of an uncemented (cohesionless) sand deposit and loaded with eccentric and inclined loading, the bearing capacity equation reduces to the following:

$$Q_{\mathrm{bL}} = \frac{1}{2}\gamma B'^{2} N_{\gamma} i_{\gamma} = \frac{1}{2}\gamma B^{2} N_{\gamma,\mathrm{eq}} \tag{1}$$

where γ is the soil unit weight and N_{γ} is the bearing capacity factor. The factor i_{γ} introduces the effect of load inclination and B' is the effective width, which incorporates the effect of load eccentricity. According to Meyerhof [8], the effective footing width is calculated as

D. Loukidis (✉) · N. Ygeionomaki
Department of Civil and Environmental Engineering, University of Cyprus, Kallipoleos 75, 1678 Nicosia, Cyprus
e-mail: loukidis@ucy.ac.cy

E. Papamichos et al. (eds.), *Bifurcation and Degradation of Geomaterials with Engineering Applications*, Springer Series in Geomechanics and Geoengineering, DOI 10.1007/978-3-319-56397-8_34

$$B' = B - 2e \tag{2}$$

where e is the load eccentricity (i.e. the distance of the point of application of the load from the footing central axis) and B is the actual footing width. The i_γ is commonly calculated using the Vesic [9] formula:

$$i_\gamma = (1 - \tan\alpha)^3 \tag{3}$$

where α is the load inclination angle measured from the vertical axis. Both Eqs. (2) and (3) have been adopted by Eurocode 7. The $N_{\gamma,\mathrm{eq}}$ $(= N_\gamma i_\gamma (1 - 2e/B)^2)$ appearing in Eq. (1) is an equivalent form of the bearing capacity factor that includes the effects of load inclination and eccentricity.

The current state of practice (Eq. 1) for the estimation of the bearing capacity of shallow foundations relies on the assumption that the soil behaves as an elastic-perfectly plastic Mohr-Coulomb material with a constant friction angle ϕ. However, the ϕ in reality is a function of many variables, such the mean effective stress, the magnitude of shear straining (strain softening), and the direction of loading relative to the soil fabric (anisotropy). All these change both spatially and temporally inside the bearing capacity failure mechanism and proper analysis of the problem requires consideration of advanced constitutive models. Loukidis and Salgado [7] investigated the bearing capacity of centrally and vertically loaded footings on sand, i.e. the N_γ factor, using finite element analysis combined with advanced constitutive modeling. The present study extends that work to the case of strip footings loaded by eccentric or inclined loads.

2 Numerical Methodology

The finite element (FE) simulations are performed using the code SNAC of Abbo and Sloan [1]. The FE mesh consists of 15-noded plane-strain triangular elements. A typical mesh and boundary conditions are shown in Fig. 1. The mechanical behavior of the sand is a simulated using a bounding surface plasticity constitutive model that is based on critical state theory [3]. The model captures realistically the pre-failure non-linearity and contractiveness, as well as the post-failure softening to critical state, while taking into account the effects of mean effective stress and fabric-related anisotropy on the sand dilatancy. FE analyses are performed for sets of constitutive model parameter values for two different sands [5], Toyoura sand (subangular to angular grains with mean diameter 0.2 mm) and Ottawa C778 sand (rounded to subrounded grains with mean diameter 0.4 mm).

The footing is modeled as a very stiff weightless elastic material, so that it moves practically as a rigid body. No interface elements are placed between footing and soil (perfectly rough footing). The unit weight γ of the sand is set equal to 20 kN/m^3 in all analyses. The choice of the γ value is practically immaterial since the problem

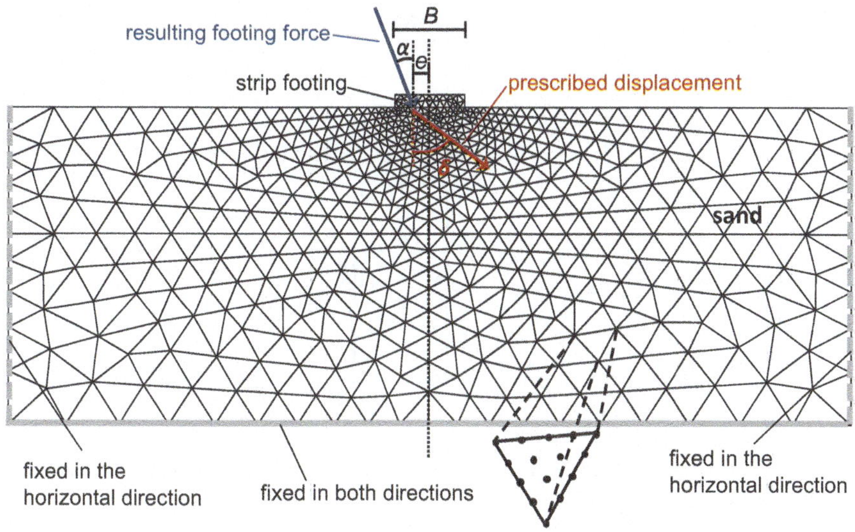

Fig. 1 Typical finite element mesh and boundary conditions

results can be normalized with respect to the product $\gamma B'$. Loading is applied in the form of a prescribed incremental displacement at a node lying at the footing-soil interface and, thus, the footing force results from the analyses as a reaction force (Fig. 1). For a footing loaded by an eccentric vertical force ($e > 0$, $\alpha = 0$), the prescribed displacement is vertical and applied at a distance e from the footing center. For a footing loaded by an inclined central load ($e = 0$, $\alpha > 0$), the displacement is applied at the footing center with a certain inclination δ with respect to the vertical axis. It should be noted that the inclination angle α of the resulting reaction load turns to be different (smaller) than δ and varies during the course of a simulation. Analyses were performed for B ranging from 1 to 3 m, relative density D_R from 45 to 90%, eccentricity ratio e/B from 0 to ¼, and δ from 0° to 79° (yielding values of load inclination α at limit state in the 0°–34° range). The distance between adjacent nodes inside the region where the failure mechanism develops ranges from 37 to 450 times the mean particle diameter of the sand. Hence, the shear band thickness in the simulations is roughly between 2.5 and 30 times the thickness of the shear bands that would develop in reality.

3 Simulation Results

Figure 2 shows examples of vertical load versus vertical prescribed displacement for various values of initial relative density D_R and for two loading cases: 1 m wide footing loaded by an eccentric vertical force (Fig. 2a) and by a central inclined force (Fig. 2b). It can be observed that for dense (D_R = 75, 90%) and medium

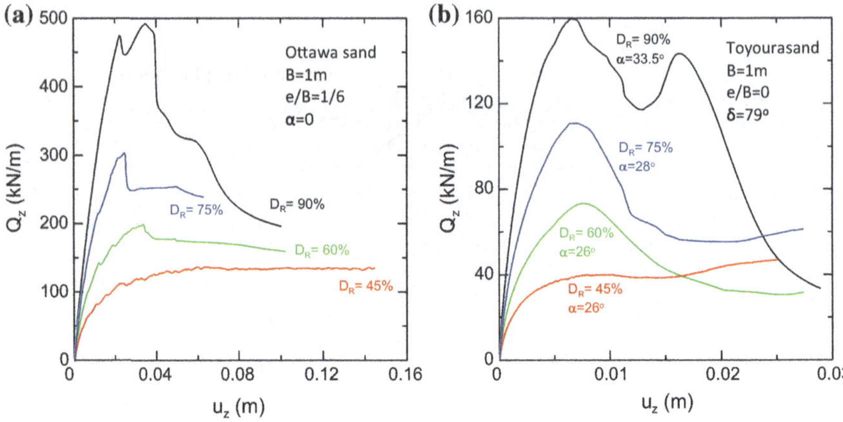

Fig. 2 Load displacement curves from analyses of 1 m wide strip footing with **a** eccentric load and **b** inclined load

dense sand (D_R = 60%), the load-displacement curve reaches a well defined peak (failure) load (i.e. Q_{bL}), while for loose sand (D_R = 45%) the response is much more ductile. Moreover, the curves seem to approach the same post-peak ultimate (residual) capacity (Fig. 2a) regardless of the value of initial D_R. This observation is consistent with critical state theory, according to which the operative soil friction angle at very large shear strains attains its minimum value, which is independent of the soil density and mean effective stress. The prescribed displacement vector inclination δ is 79° in all simulations of Fig. 2b. The load inclination α values reported in Fig. 2b (ranging from 26° to 33.5°) correspond to the peak load state.

Figure 3 shows typical examples of the plastic failure mechanism developing under the strip footings, as depicted by contour plots of the plastic maximum shear strain increment $\Delta\gamma_{max,pl}$ (= $\Delta\varepsilon_{1,pl} - \Delta\varepsilon_{3,pl}$). In the eccentrically loaded footing case (Fig. 3a), the failure mechanism consist of two distinct parts. On the right hand side, we have a "scoop" that the footing sits on and rotates counterclockwise, while a comparatively smaller in size combination of a "fan" zone and passive wedge

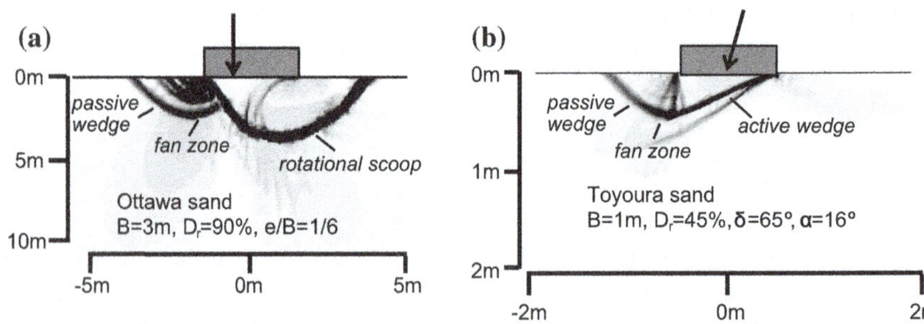

Fig. 3 Examples of plastic failure mechanism: **a** eccentric loading, **b** inclined loading

develops on the opposite side. In the case of inclined loading, the failure mechanism is one-sided and consists of an active (rigid) wedge under the footing, a narrow fan zone and a comparatively small passive wedge. The geometries of the mechanisms of Fig. 3 resemble closely those observed in sandbox (1 g) model experiments by Jumikis [4], as well as in FE analyses in which the soil was modeled as a Mohr-Coulomb elastic-perfectly plastic material [6].

The equivalent bearing capacity factor $N_{\gamma,\mathrm{eq}}$ increases with increasing D_R, for both centrally and eccentrically loaded footings, while it decreases with increasing B (Fig. 4). The latter is due to the reduction of the operative friction angle with increasing mean effective stress inside the failure mechanism. Nevertheless, the relative reduction of Q_bL with increasing e/B appears to be independent of D_R and B (Fig. 5). On the other hand, in the case of inclined loading, the rate of relative

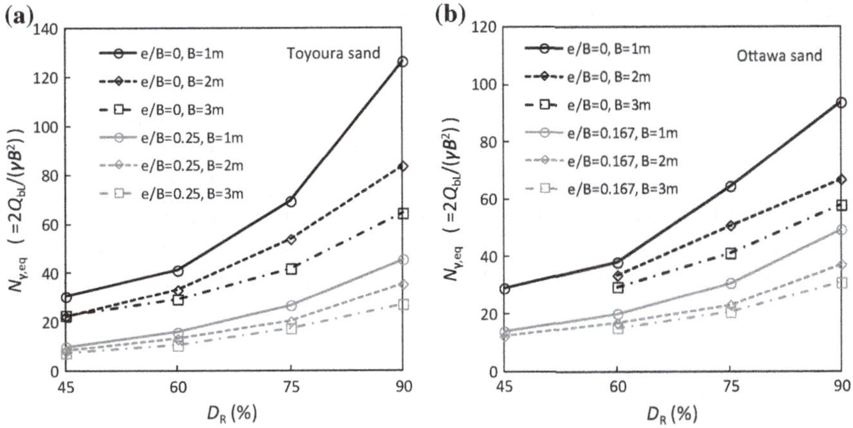

Fig. 4 Dependence of bearing capacity on relative density and footing size

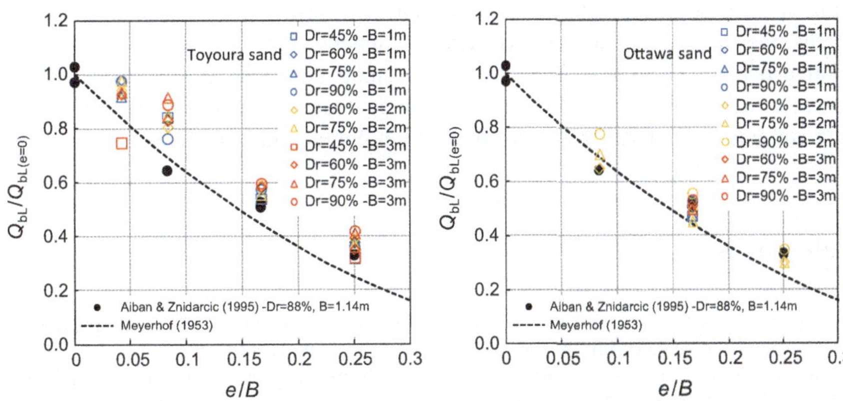

Fig. 5 Effect of load eccentricity on bearing capacity: comparison between numerical predictions and experimental data

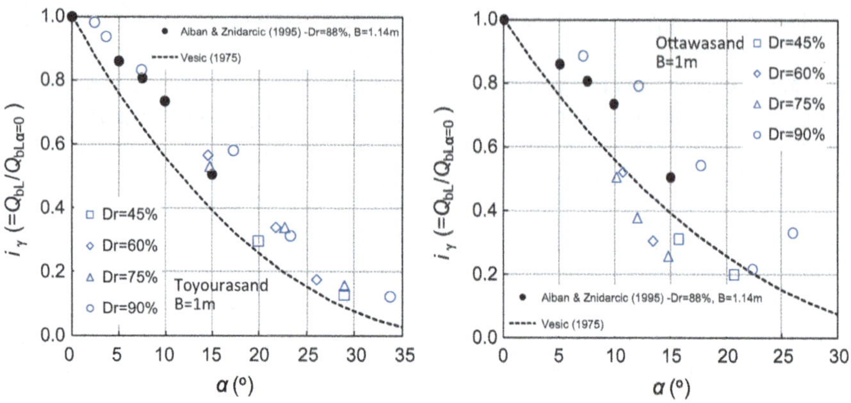

Fig. 6 Effect of load inclination on bearing capacity: comparison between numerical predictions and experimental data

reduction of Q_{bL} with α appears to be smaller for dense sand than for loose sand (Fig. 6). Aiban and Znidarcic [2] performed centrifuge tests of eccentrically or obliquely loaded strip footing (B = 1.14 m) on dense sand. The FE results are in general agreement with the centrifuge data (Figs. 5 and 6). Finally, both FEA and experimental data suggest that the Meyerhof [8] effective width approach (Eq. 2) and the Vesic [9] i_γ factor may be slightly on the conservative side (Figs. 5 and 6).

4 Conclusions

Finite element analyses were performed to investigate the effects of load inclination angle α and eccentricity e on the bearing capacity Q_{bL} of strip footings on sand, with the soil modeled using bounding surface plasticity. According to the FE results, the rate of decrease of Q_{bL} with e is independent of footing size and sand relative density D_R, while there may be a small dependence of the rate of Q_{bL} decrease with α on D_R. The FE results are in general agreement with centrifuge test data and equations used in current foundation engineering practice.

References

1. Abbo, A.J., Sloan, S.W.: SNAC User Manual, Version 2.0. Department of Civil, Surveying and Environmental Engineering, University of Newcastle, Australia (2000)
2. Aiban, S.A., Znidarcic, D.: Centrifugal modeling of bearing capacity of shallow foundations on sands. J. Geotech. Eng. **121**(10), 704–712 (1995)
3. Dafalias, Y.F., Papadimitriou, A.G., Li, X.S.: Sand plasticity model accounting for inherent fabric anisotropy. J. Eng. Mech. ASCE **130**(11), 1319–1333 (2004)

4. Jumikis, A.A.: General discussion: rupture of soil. In: Vesic, A.S. (ed.) Proceedings of the Symposium on Bearing Capacity and Settlement of Foundations, Duke University, pp. 103–107 (1965)
5. Loukidis, D., Salgado, R.: Modeling sand response using two-surface plasticity. Comput. Geotech. **36**(1–2), 166–186 (2009)
6. Loukidis, D., Chakraborty, T., Salgado, R.: Bearing capacity of strip footings on purely frictional soil under eccentric and inclined loads. Can. Geotech. J. **45**(6), 768–787 (2008)
7. Loukidis, D., Salgado, R.: Effect of relative density and stress level on the bearing capacity of footings on sand. Géotechnique **61**(2), 107–119 (2011)
8. Meyerhof, G.G.: The bearing capacity of foundations under eccentric and inclined loads. In: Proceedings of the 3rd ICSMFE, Zürich, vol. I, pp. 440–445 (1953)
9. Vesic, A.S.: Bearing capacity of shallow foundations. In: Fang, H.-Y. (ed.) Winterkorn HF, pp. 121–147. Van Nostrand Reinhold Foundation Engineering Handbook, New York (1975)

Part V
Geomechanics for Energy, Environment and Geophysics

An X-FEM Algorithm for Modeling of Multi-zone Hydraulic Fracturing in Saturated Porous Media

M. Vahab and N. Khalili

Abstract In this paper, a fully coupled hydro-mechanical model is presented for the study of multi-zone hydraulic fracturing within saturated porous media. The well-known $(\mathbf{u}-p)$ formulation is employed in order to simulate the saturated porous domain. An X-FEM Penalty algorithm is adopted to prevent the overlap of the fracture edges. Meanwhile, the large time increment method (LATIN) is utilized to remove the leak-off flow in the case of crack closure. Multiple crack growth pattern is determined by means of energy based stress functions for cohesive fractures. A computational algorithm is proposed for distribution of the fracturing fluid flow across existing discontinuities. Finally, several examples are provided to demonstrate the robustness of the proposed formulation.

Keywords Saturated porous media · $(\mathbf{u}-p)$ formulation · Multizone hydraulic fracturing · X-FEM

1 Introduction

Hydraulic fracturing is a well-known technique in which well bores are stimulated by injection of a highly pressurized liquid to induce permeable cracks. Recent advances have made multizone multistage hydraulic fracturing a common practice [1]. General issues in this regard emanate from: (i) the extent of the propagation of competing fractures, (ii) coalescence of the initiated fractures, (iii) uncontrolled fracture growth towards shallower or deeper layers, and (iv) unequal distribution of fracturing fluid at different strata [2] (see Fig. 1).

M. Vahab · N. Khalili (✉)
School of Civil and Environmental Engineering, UNSW, Sydney 2052, Australia
e-mail: n.khalili@unsw.edu.au

E. Papamichos et al. (eds.), *Bifurcation and Degradation of Geomaterials with Engineering Applications*, Springer Series in Geomechanics and Geoengineering, DOI 10.1007/978-3-319-56397-8_35

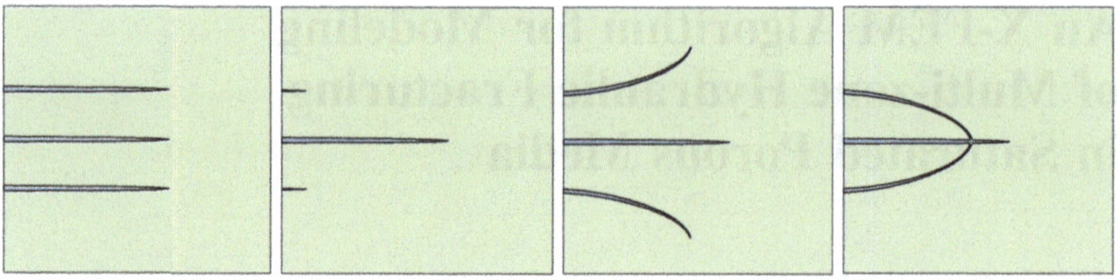

(i) Simulatenous (ii) Asynchronous (iii) Diverted (iv) Converging

Fig. 1 Possible scenarios encountered in multizone hydraulic fracturing treatments

The growth and interaction of cracks has remained a challenging problem due to the diverse kinematic and constitutive complexities. A key simplification adopted in the literature is the application of the isolated crack growth using either the maximum circumferential tensile stress (see Zhang and Ghassemi [3]) or maximum energy release rate criterion to multiple cracks (see Ooi and Yang [4]; Dong and de Pater [5]). However, it is well-established that the real evolution pattern of multiple competitive cracks can only be recognized by means of stability analysis (Budyn et al. [6]). Li and Ghosh [7, 8] developed an innovative technique for multiple cohesive crack growth in brittle materials based on multi-resolution stress functions. Accordingly, the precise crack growth increment and direction were determined through the application of a cohesive fracture energy based criterion so that the fracture energy is maximized.

An innovative numerical framework for the simulation of propagating cracks has been the X-FEM which was first used for the implementation of stationary fractures in saturated porous media by de Borst et al. [9] and Réthoré et al. [10]. Khoei et al. [11] presented a model for hydro-mechanical fracture growth in saturated porous media, which was extended to multiphase porous media by Mohammadnejad and Khoei [12] and Salimzadeh and Khalili [13]. Taleghani [1], and Wang and Taleghani [2] developed an X-FEM model for mode I multizone multistage hydraulic fracturing. More contributions on the simulation of hydro-fractures interacting with natural discontinuities can be found in Khoei et al. [14, 15], Olsen and Taleghani [16], and Salimzadeh and Khalili [17].

In this paper, a fully coupled hydro-mechanical model is presented for the study of multizone hydraulic fracturing in saturated deformable porous media. The hydro-fracture inflow is modeled based on the Darcy law, where the fracture permeability is calculated via the well-known cubic law. Multiple cohesive crack growth is handled by taking advantages from the cohesive potential functions introduced by Li and Ghosh [7, 8]. Using the X-FEM, the strong discontinuity in the displacement field due to fracture opening as well as the weak discontinuity within the pressure field due to leakage is incorporated. Finally, through several numerical simulations the robustness and versatility of the developed framework are demonstrated.

2 Governing Equations of Fractured Porous Media

The mixture theory which is based on the Biot theory [18] is employed in order to describe the mechanical behaviour of the saturated porous media. The motion of the solid-fluid mixture and the relative motion of the pore fluid respective to the mixture are denoted by $\mathbf{u}(\boldsymbol{x}, t)$ and $\mathbf{w}(\boldsymbol{x}, t)$, respectively. The momentum balance equation of the total mixture is written as

$$\nabla \cdot \sigma + \rho \, \mathbf{b} = \mathbf{0}, \tag{1}$$

where $\mathbf{b}$ is the body force vector. The linear momentum balance equation for the pore fluid known as the generalized Darcy relation is represented as

$$-\nabla p - \mathbf{R} + \rho_f \mathbf{b} = \mathbf{0}, \tag{2}$$

where $\mathbf{R}$ denotes the viscous drag force defined by using the Darcy seepage law as, $\boldsymbol{k}_f$ is the intrinsic permeability matrix of the porous medium, and μ_f is the fluid dynamic viscosity. The generalized Darcy law can be rewritten as

$$\dot{\mathbf{w}} = \frac{\boldsymbol{k}_f}{\mu_f}(-\nabla p + \rho_f \mathbf{b}). \tag{3}$$

The continuity equation of the pore fluid is in turn expressed as

$$\nabla \cdot \dot{\mathbf{w}} + \alpha \nabla \cdot \dot{\mathbf{u}} + \frac{1}{Q}\dot{p} = \mathbf{0}, \tag{4}$$

in which Q is the compressibility coefficient. Substituting Eq. (2) into Eq. (4) it follows that

$$\nabla . \left[\frac{\boldsymbol{k}_f}{\mu_f}(-\nabla p + \rho_f \, \mathbf{b})\right] + \alpha \nabla . \dot{\mathbf{u}} + \frac{1}{Q}\dot{p} = 0. \tag{5}$$

Equations (1) and (5) form the governing equations of the saturated porous media under quasi-static conditions. Consider a two dimensional domain Ω bounded by external boundary Γ as depicted in Fig. 2, which contains internal discontinuities $\Gamma_d = \bigcup_{i=1}^{NOD} \Gamma_d^i$ (i = 1: *NOD*) with *NOD* denoting the number of discontinuities. The boundary conditions on the internal discontinuity Γ_d is defined as

$$\begin{array}{l} \sigma \cdot \mathbf{n}_{\Gamma_d} = \mathbf{t}_{cont} + \mathbf{t}_{cohs} - p \, \mathbf{n}_{\Gamma_d} \\ [[\mathbf{w}]] \cdot \mathbf{n}_{\Gamma_d} = \bar{q}_d \end{array} \qquad \text{on } \Gamma_d, \tag{6}$$

where $\mathbf{t}_{cont}$ is contact traction, $\mathbf{t}_{cohs}$ is cohesive stress, p is fluid pressure, and $\bar{q}_d$ is the leak-off flow.

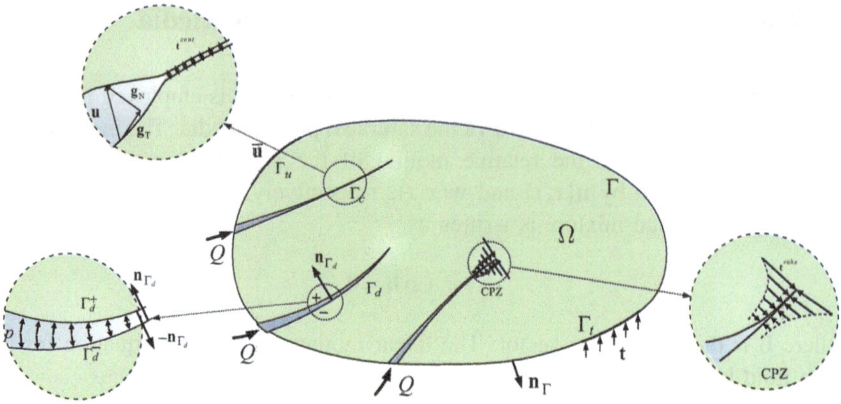

Fig. 2 Problem definition and the boundary conditions

2.1 Model for Hydro-Fracture Inflow

The hydro-fracture inflow is modeled by applying the mass balance Eq. (4) within the discontinuity, where the Biot parameter α is omitted, and the compressibility coefficient Q is simplified to K_f since the fracture porosity n is equal to unity. It follows that,

$$\nabla \cdot \dot{\mathbf{w}} + \nabla \cdot \dot{\mathbf{u}} + \frac{1}{K_f}\dot{p} = 0. \tag{7}$$

The generalized Darcy law given by Eq. (3) is adapted for description of the flow velocity within the fracture as

$$\dot{\mathbf{w}} = \frac{k_{f_d}}{\mu_f}(-\nabla p + \rho_f \mathbf{b}). \tag{8}$$

The fracture intrinsic permeability is obtained through the assumption of a Newtonian viscous laminar flow which is given by the well-known cubic law as $k_{f_d} = w^2/12\kappa$ (see Zimmerman and Bodvarsson [19]), where $w = 2h$ is the fracture opening, and κ is a modification coefficient introduced by Witherspoon et al. [20] in range of 1.04–1.65.

2.2 Model for Contact Along Discontinuities

The contact constrains are imposed on the solid phase in order to avoid the penetration of contacting surfaces along the discontinuities. The contact constrain is introduced along the faces of the discontinuities using the first *Kuhn-Tucker* inequality as (Hirmand et al. [21])

$$g_{\mathrm{N}} \geq 0, \qquad t_{cont_{\mathrm{N}}} \leq 0, \qquad g_{\mathrm{N}}\ t_{cont_{\mathrm{N}}} = 0 \qquad \forall \boldsymbol{x} \in \Gamma_d \tag{9}$$

where $t_{cont_{\mathrm{N}}} = \mathbf{t}_{cont} \cdot \mathbf{n}_{\Gamma_d}$ is the normal contact stress, and $g_{\mathrm{N}} = [[\mathbf{u}]] \cdot \mathbf{n}_{\Gamma_d}$ is the normal opening of the crack. In tangential direction, the "stick" and "slip" conditions are recognized by using the standard Coulomb's friction law as $F_f = ||\mathbf{t}_{cont_{\mathrm{T}}}|| - \mu_f ||\mathbf{t}_{cont_{\mathrm{N}}}||$ with μ_f and c_f being the Coulomb's friction coefficient and the tangential cohesion, respectively. In addition, $t_{cont_{\mathrm{T}}} = \mathbf{t}_{cont} \cdot \mathbf{m}_{\Gamma_d}$ is the tangential component of the contact traction with $\mathbf{m}_{\Gamma_d}$ being the unit tangential vector of the discontinuity. The contact constitutive relation is defined in form of $\mathrm{d}\mathbf{t}_{cont} = \mathbf{D}^{cont}\,\mathrm{d}[[\mathbf{u}]]$ as (Liu and Borja [22])

$$\mathbf{D}^{cont} = \begin{cases} \epsilon_{\mathrm{N}}(\mathbf{n}_{\Gamma_d} \otimes \mathbf{n}_{\Gamma_d}) + \epsilon_{\mathrm{T}}(\mathbf{m}_{\Gamma_d} \otimes \mathbf{m}_{\Gamma_d}) & \text{for stick} \\ \epsilon_{\mathrm{N}}(\mathbf{n}_{\Gamma_d} \otimes \mathbf{n}_{\Gamma_d}) + \mu\epsilon_{\mathrm{N}}(\mathbf{m}_{\Gamma_d} \otimes \mathbf{n}_{\Gamma_d}) & \text{for slip} \end{cases} \tag{10}$$

where $\mathbf{D}^{cont}$ denotes the contact constitutive matrix defined, and ϵ_{N} and ϵ_{T} are the normal and tangential Penalty coefficients, respectively. On the other hand, for the fluid phase the flow exchange between the discontinuity and surrounding porous medium must vanish in case of crack closure ([23]) as

$$\bar{q}_d = \dot{\mathbf{w}} \cdot \mathbf{n}_{\Gamma_c} = 0 \qquad \text{on} \quad \Gamma_c. \tag{11}$$

Following Khoei and Vahab [23], the LATIN method is adopted in order to impose the contact constrains onto the fluid phase. To this end, a reverse computational algorithm is utilized to eliminate leakage from the discontinuity into the domain. In this approach, using an arbitrary penalization parameter, $\bar{\boldsymbol{k}}_f$, a reverse leakage $\bar{\bar{q}}_d$ is imposed as $\dot{\mathbf{w}} = -\bar{\boldsymbol{k}}_f/\mu_f[[\nabla p]]$ till $\bar{q}_d$ is vanished.

2.3 *Model for Multiple Cohesive Crack Growth*

Employing the linear cohesive zone model on the basis of Camacho and Ortiz [24] the cohesive traction is expressed in terms cohesive fracture energy potential ϕ as

$$\mathbf{t}_{cohs} = \frac{\partial\phi}{\partial g_{\mathrm{N}}}\mathbf{n}_{\Gamma_d} + \frac{\partial\phi}{\partial g_{\mathrm{T}}}\mathbf{m}_{\Gamma_d} = \frac{t_{cohs}}{g}(g_{\mathrm{N}}\mathbf{n}_{\Gamma_d} + \beta^2\mathbf{m}_{\Gamma_d}). \tag{12}$$

in which g is the effective crack opening defined as $g = \sqrt{g_{\mathrm{N}}^2 + \beta^2 g_{\mathrm{T}}^2}$, and t is the cohesive traction magnitude evaluated as $t_{cohs} = \partial\phi/\partial g = \sqrt{(t_{cohs}^{\mathrm{N}})^2 + \beta^{-2}(t_{cohs}^{\mathrm{T}})^2}$ with t_{cohs}^{N} and t_{cohs}^{T} being the normal and tangential components of the cohesive traction, respectively. Moreover, β is the coupling coefficient of the normal and tangential opening displacement. Based on Li and Ghosh [7, 8] the crack growth direction is determined by imposing the equivalence of the cohesive fracture energy ϕ at the stage of complete decohesion to the fracture energy release rate, G_c. The cohesive fracture energy at crack tip T_i in direction α is expressed as

$$\phi_{\mathrm{T}_i} = \int_0^{g(\alpha)} t_{cohs}(\alpha)\,\mathrm{dg} = \frac{g_e}{2\sigma_{\max}}(\sigma_{\max}^2 - t_{cohs}(\alpha)^2), \tag{13}$$

where $t_{cohs}(\alpha)$ is the effective cohesive traction associated with the direction α , g_e is the separation value at which the cohesive traction is vanished. Accordingly, the crack growth direction is predicted as the direction along which the cohesive fracture energy is maximized. The cohesive fracture energy on the other hand is used to evaluate the crack growth increment associated with each crack tip. In this manner, each crack tip propagates an incremental length such that the cohesive fracture energy is vanished. To this end, an arbitrary point, $\bar{\mathrm{T}}_i$, located along the extension of crack tip T_i in direction α is chosen and the crack growth increment, $\Delta\ell_i$, corresponding to the crack tip T_i is computed as

$$\Delta\ell_i = \frac{\phi_{\mathrm{T}_i}}{\phi_{\mathrm{T}_i} - \phi_{\bar{\mathrm{T}}_i}} \left|\ell_{\mathrm{T}_i\bar{\mathrm{T}}_i}\right|, \tag{14}$$

in which $\ell_{\mathrm{T}_i\bar{\mathrm{T}}_i}$ is the length of line segment T_i to $\bar{\mathrm{T}}_i$.

3 Discretization of the Governing Equations

The strong discontinuity in displacement field due to the existence of fractures necessitates that the displacement field to be enhanced by employing the Heaviside enrichment function. Meanwhile, the weak discontinuity in the pressure field due to discontinuous pressure gradient is incorporated by adopting the modified level-set enrichment function. Accordingly, the enriched approximation for the displacement field $\mathbf{u}(\boldsymbol{x}, t)$ is written as

$$\mathbf{u}^h(\boldsymbol{x}, t) = \mathbf{N}_u^{std}(\boldsymbol{x})\,\mathbf{U}(t) + \mathbf{N}_u^{Hev}(\boldsymbol{x})\tilde{\mathbf{U}}(t), \tag{15}$$

where $\mathbf{N}_u^{std}$ is the standard shape function matrix associated with the standard DOFs $\mathbf{U}(t)$, and $\mathbf{N}_u^{Hev} = \sum_{I=1}^{NOD} N_{uI}(\boldsymbol{x})\,(\mathcal{H}(\varphi(\boldsymbol{x})) - \mathcal{H}(\varphi(\boldsymbol{x}_I)))$ is the shifted Heaviside enrichment shape function matrix corresponding to the enriched DOFs $\tilde{\mathbf{U}}(t)$. The X-FEM approximation of the pressure field given by

$$p^h(\boldsymbol{x}, t) = \mathbf{N}_p^{std}(\boldsymbol{x})\,\mathbf{P}(t) + \mathbf{N}_p^{\text{ridge}}(\boldsymbol{x})\,\tilde{\mathbf{P}}(t), \tag{16}$$

where $\mathbf{N}_p^{std}(\boldsymbol{x})$ is the standard shape function matrix associated with the standard DOFs $\mathbf{P}(t)$, and $\mathbf{N}_p^{\text{ridge}} = \sum_{I=1}^{NOD} N_{pI}(\boldsymbol{x})(\psi(\boldsymbol{x}) - \psi(\boldsymbol{x}_I))$ is the ridge enrichment shape function matrix associated with the enriched DOFs $\tilde{\mathbf{P}}(t)$, where $\psi^J(x) = \sum_{I=1}^{NOD} N_{pI}(x)|\varphi_I| - |\varphi(x)|$.

The time domain discretization is carried out using the well-known generalized finite difference theta-scheme. The successive value of the unknowns at time t_{n+1} is expressed versus known variables at time t_n with time increment of $\Delta t = t_{n+1} - t_n$ as [25]

$$\dot{\mathbb{U}}_{n+1} = \frac{\theta}{\Delta t}(\mathbb{U}_{n+1} - \mathbb{U}_n) + (1-\theta)\dot{\mathbb{U}}_n, \tag{17}$$

for the displacement field and,

$$\dot{\mathbb{P}}_{n+1} = \frac{\overline{\theta}}{\Delta t}(\mathbb{P}_{n+1} - \mathbb{P}_n) + \left(1 - \overline{\theta}\right)\dot{\mathbb{P}}_n, \tag{18}$$

for the pressure field where $\mathbb{U} = \langle \mathbf{U}, \tilde{\mathbf{U}} \rangle$ and $\mathbb{P} = \langle \mathbf{P}, \tilde{\mathbf{P}} \rangle$ are the vector of DOFs of the displacement and pressure fields, respectively. Note that θ and $\overline{\theta}$ are the method parameters in range of 0–1.

4 Computational Algorithm for Multizone HF

Hydraulic fracturing treatments are generally performed by injection of the fracturing fluid into the bore hole at a prescribed injection rate. For multizone hydraulic fracturing treatments a new challenge arises due to the fact that the fracturing fluid is injected into a group of perforations instead of a single one. In this case, the fracturing fluid inflow to various perforations will be different, but the fluid pressure at the opening of the perforations will be identical. Here, a straightforward iterative algorithm is proposed to fulfil the constant fracturing fluid pressure constrain at all perforations whilst the total injection rate is maintained. To this end, an initial distribution in accordance to the previous time step is supposed for the fracturing fluid flow across all active perforations. The cumulative fluid pressure $\mathcal{P}_{Total}$, the total flow rate $\mathcal{Q}_{Total}$ and the average fluid pressure $\mathcal{P}_{Total}$ at the ith iteration are calculated as

$$\mathcal{P}_{Total} = \sum_{j=1}^{2N_{perf}} \mathcal{P}_j^i, \qquad \mathcal{Q}_{Total} = \sum_{j=1}^{2N_{perf}} \mathcal{Q}_j^i, \qquad \mathcal{P}_{Average} = \mathcal{P}_{Total}/2N_{perf} \tag{19}$$

where $\mathcal{P}_j^i$ and $\mathcal{Q}_j^i$ denote the nodal values of the fluid pressure and flow rate corresponding to the *j*th node of the perforated element edge, with N_{perf} being the total number of perforations. So to obtain a similar value for the fracturing fluid pressure at all perforations, the flow rate at each perforation is modified by

$$\mathcal{Q}_j^{i+1} = \mathcal{Q}_j^i \left(1 + \left| \frac{\mathcal{P}_j^i - \mathcal{P}_{Average}}{\mathcal{P}_{Total}} \right|^{\beta} * \text{Sign}(\mathcal{P}_j^i - \mathcal{P}_{Average}) \right), \tag{20}$$

where β is a modification factor in range of 0–1. Here, $\beta = 0.8$ is assumed to deliver the optimal convergence rate.

5 Numerical Simulations

In order to illustrate the robustness of the proposed framework in the study of multizone hydraulic fracturing treatments, several examples are investigated numerically. The first example deals with a group of perforations subjected to similar injection rates in order to highlight difficulties encountered in modeling of multizone hydraulic fracturing. In the second example, the algorithm proposed for the multizone hydraulic fracturing is elaborated to show the extent of its success in resolving the modeling issues as well as to further imply the practical complexities of the problem.

5.1 *Multiple Hydro-Fractures Subjected to Identical Injection Rates*

Consider a 2.5 × 5 m block containing three perforations with initial length of $\ell_{\mathcal{HF}} = 0.15$ m at varying vertical spacings of $d = 0.25$, 0.5 m (see Fig. 3). The fracturing fluid is injected at a total rate of $Q = 0.001$ m^2/s, and is distributed uniformly across all three existing perforations. Meanwhile, the results for an isolated hydro-fracture subjected to the same injection rate are also included for reference. In this manner, the primary idea behind multizone hydraulic fracturing is investigated and compared with that of the isolated case presented in the previous example. The material properties of poroelastic medium are also given in Table 1.

The final crack growth paths associated with the perforations are represented in Fig. 4. As can be seen, in case of $d = 0.25$ m the upper and lower branches are diverted towards the neighboring layers (outside of the pay zone). However, for $d = 0.5$ m all branches have evolved in a nearly parallel pattern, yet for considerably different lengths. For the same total injection rate, the analysis shows that the cumulative hydro-fracture lengths are increased by a ratio of 47% and 36%,

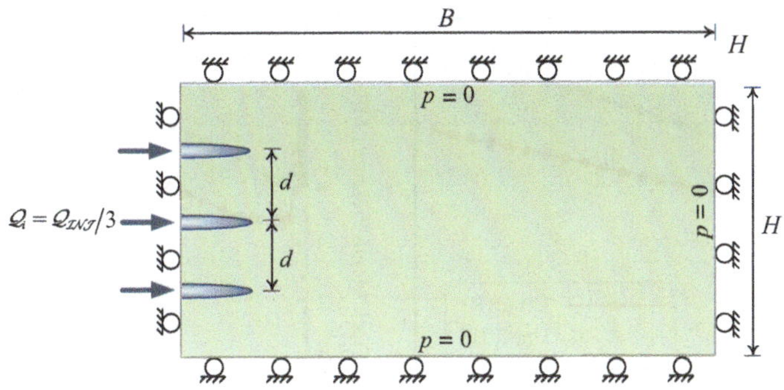

Fig. 3 Multiple hydro-fractures subjected to similar injection rate; problem definition and boundary conditions

Table 1 Material properties of the saturated porous block

Elasticity modulus	$E = 15.96\,\text{GPa}$
Poisson's ratio	$\nu = 0.33$
Biot's constant	$\alpha = 1$
Porosity	$n = 0.19$
Density	$\rho_s = 2000\ \text{kg/m}^3$
Water density	$\rho_w = 1000\ \text{kg/m}^3$
Bulk modulus of water	$K_w = 3.0\,\text{GPa}$
Bulk modulus of solid grains	$K_s = 36.0\,\text{GPa}$
Permeability	$k_f = 1.0 \times 10^{-12}\,\text{m}^2$
Viscosity of water	$\mu_w = 1 \times 10^{-3}\,\text{Pa s}$
Penalty number in normal direction	$K_N = 1 \times 10^8\ \text{MPa/m}$
Friction coefficient	$\mu_f = 0.0$
Tensile strength	$\sigma_t = 1.0\,\text{MPa}$
Fracture energy	$G_f = 100\,\text{N/m}$

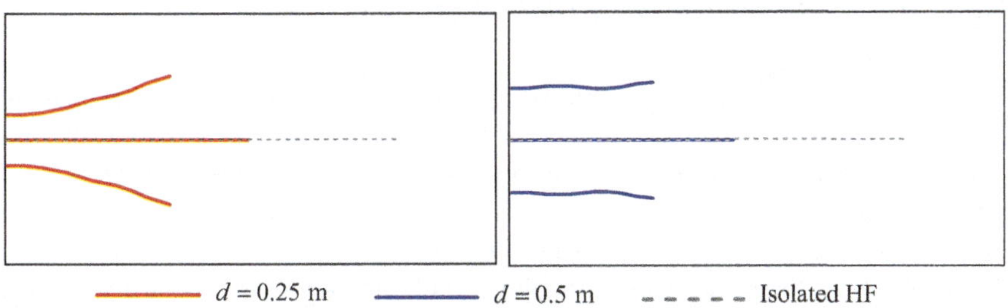

Fig. 4 Final crack growth paths for multiple hydro-fractures subjected to similar injection rate

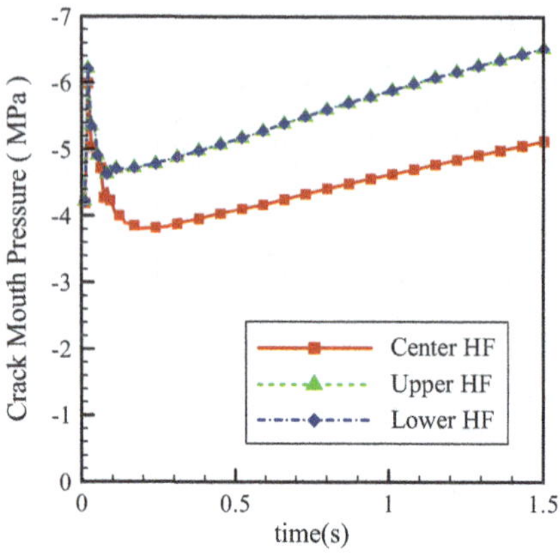

(a) $d = 0.25\,\text{m}$

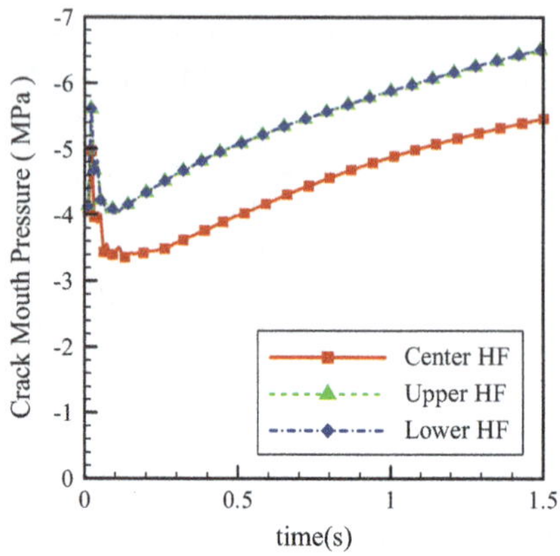

(b) $d = 0.5\,\text{m}$

Fig. 5 The variations with time of the crack mouth pressure (CMP) multiple hydro-fractures subjected to similar injection rate

respectively, for $d = 0.25$ and $0.5\,\text{m}$ in comparison to the isolated hydro-fracture, which is a substantial improvement.

The variations with time of the crack mouth pressure (CMP) for all perforation spacings in Fig. 5. As can be seen, the fracturing fluid pressure at the upper and lower perforation openings is considerably greater than the central branch. Provided that all hydro-fractures are connected to the same well bore, the values obtained for the fracturing fluid pressures clearly violate the physics of the problem as reported by Wu and Olson [26]. Thus, the assumption of a uniform distribution for the fracturing fluid flow across all perforations assumed in the literature (e.g. see Yao [27] and Samuelson et al. [28]) is unrealistic. Thus the results suggested in this example cannot be relied upon.

5.2 *Multizone Hydraulic Fracturing*

The final example explores the multizone hydraulic fracturing treatment and the associated practical issues. Using the configurations of the previous example, the hydraulic fracturing treatment is performed through the imposition of a total injection rate of $Q = 0.001\ \text{m}^2/\text{s}$, which is distributed across existing perforations by incorporating the multizone HF algorithm introduced in Sect. 4. In this example, the capability of the proposed methodology in elimination of the inconsistencies in the fracturing fluid pressure is evaluated.

The final crack growth trajectory is presented for all perforation spacings in Fig. 6. As can be seen, the majority of crack propagation occurs along the central branch, while the upper and lower perforations only show a negligible growth. Our

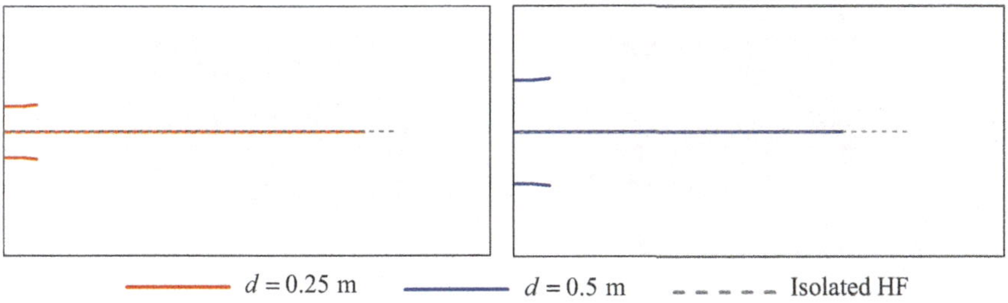

Fig. 6 Final crack growth paths for multi-zone hydro-fractures growth in a saturated porous block

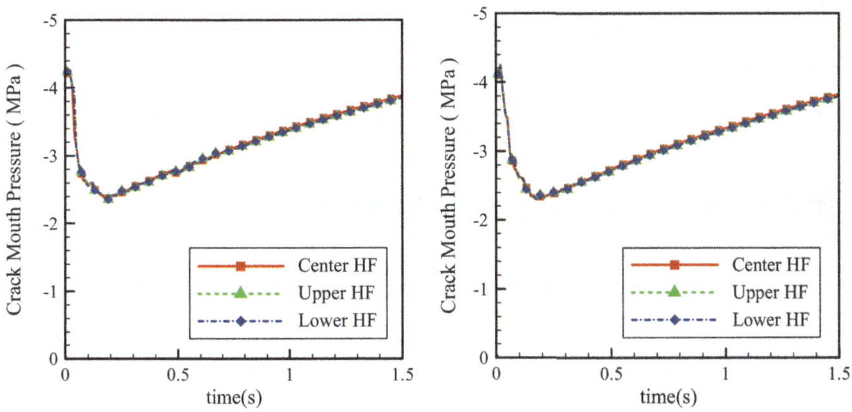

(a) $d = 0.25\,\text{m}$ (b) $d = 0.5\,\text{m}$

Fig. 7 The variations with time of the crack mouth pressure (CMP) for multi-zone hydraulic fracturing

analysis shows that the cumulative crack lengths in comparison to the isolated perforation are slightly increased by a ratio of 8% and 1%, respectively, for $d = 0.25$ and 0.5 m. Taking into account that the upper and lower perforations are nearly remained at their initial configuration, the multizone hydraulic fracturing executed can be interpreted as an unsuccessful treatment. This appears to accord with various reported cases of failed multiple fracturing treatments in the field (Ahmad et al. [29]).

The variations with time of the crack mouth pressure (CMP) is presented in Fig. 7. Unlike the previous example, no inconsistency is observed for the fracturing fluid pressure at the perforation openings. This indicates the success of the proposed computational algorithm in modelling of multizone hydraulic fracturing treatments. Contours of the pore fluid pressure are presented in Fig. 8, with identical fluid pressures obtained at the perforation openings. Note that the fracturing fluid pressure is fully developed along the central branch, while for the upper and lower branches it is remained within the initial segments of the perforations.

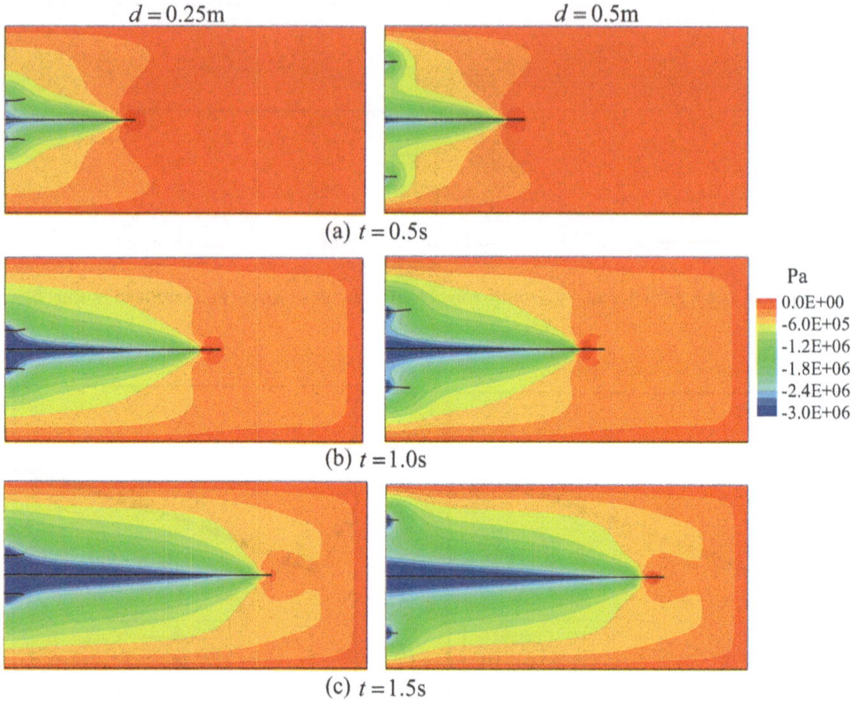

Fig. 8 Contours of fluid pressure for multi-zone hydraulic fracturing at different time steps

6 Conclusions

In this paper, a computational framework is presented for the study of multizone hydraulic fracturing treatments within saturated two-phase porous media. The hydro-fracture inflow is modeled based on the Darcy law, where the fracture permeability is evaluated by using the cubic law. The cohesive fracture energy functions are invoked to determine the multiple crack growth pattern. Provisions are made for the plausible crack closure in the solid phase by means of Kuhn-tucker inequalities embedded in an X-FEM Penalty method. In addition, for the fluid phase the zero leak-off constrain is imposed through the application of a LATIN based contact algorithm. X-FEM is utilized in order to incorporate the strong discontinuities in the displacement field as well as the weak discontinuities within the pressure field. Time domain discretization is performed by means of the finite difference theta-scheme. A computational algorithm is introduced for dealing with the fracturing fluid flow distribution across the existing perforations. Several numerical examples are presented to demonstrate the robustness of the proposed X-FEM framework in the study of multizone hydraulic fracturing treatments in saturated porous media. It is shown that the computational algorithm successfully can resolve all the inconsistencies encountered in the process of multizone hydraulic fracturing.

Acknowledgement Submitted to: 11th International Workshop on Bifurcation and Degradation in Geomaterials

References

1. Taleghani, A.D.: Modeling simultaneous growth of multi-branch hydraulic fractures. In: 45th US Rock Mechanics/Geomechanics Symposium, 2011. American Rock Mechanics Association
2. Wang, W., Taleghani, A.D.: Simulating multizone fracturing in vertical wells. J. Energ. Res. Technol. **136**(4), 042902 (2014)
3. Zhang, Z., Ghassemi, A.: Simulation of hydraulic fracture propagation near a natural fracture using virtual multidimensional internal bonds. Int. J. Numer. Anal. Meth. Geomech. **35**(4), 480–495 (2011)
4. Ooi, E., Yang, Z.: Modelling multiple cohesive crack propagation using a finite element–scaled boundary finite element coupled method. Eng. Anal. Boundary Elem. **33**(7), 915–929 (2009)
5. Dong, C., De Pater, C.: Numerical implementation of displacement discontinuity method and its application in hydraulic fracturing. Comput. Methods Appl. Mech. Eng. **191**(8), 745–760 (2001)
6. Budyn, E., Zi, G., Moës, N., Belytschko, T.: A method for multiple crack growth in brittle materials without remeshing. Int. J. Numer. Meth. Eng. **61**(10), 1741–1770 (2004)
7. Li, S., Ghosh, S.: Extended Voronoi cell finite element model for multiple cohesive crack propagation in brittle materials. Int. J. Numer. Meth. Eng. **65**(7), 1028–1067 (2006)
8. Li, S., Ghosh, S.: Multiple cohesive crack growth in brittle materials by the extended Voronoi cell finite element model. Int. J. Fract. **141**(3–4), 373–393 (2006)
9. De Borst, R., Réthoré, J., Abellan, M.-A.: A numerical approach for arbitrary cracks in a fluid-saturated medium. Arch. Appl. Mech. **75**(10–12), 595–606 (2006)
10. Rethore, J., Borst, Rd, Abellan, M.A.: A two-scale approach for fluid flow in fractured porous media. Int. J. Numer. Meth. Eng. **71**(7), 780–800 (2007)
11. Khoei, A., Vahab, M., Haghighat, E., Moallemi, S.: A mesh-independent finite element formulation for modeling crack growth in saturated porous media based on an enriched-FEM technique. Int. J. Fract. **188**(1), 79–108 (2014)
12. Mohammadnejad, T., Khoei, A.: Hydro-mechanical modeling of cohesive crack propagation in multiphase porous media using the extended finite element method. Int. J. Numer. Anal. Meth. Geomech. **37**(10), 1247–1279 (2013)
13. Salimzadeh, S., Khalili, N.: A three-phase XFEM model for hydraulic fracturing with cohesive crack propagation. Comput. Geotech. **69**, 82–92 (2015)
14. Khoei, A., Hirmand, M., Vahab, M., Bazargan, M.: An enriched FEM technique for modeling hydraulically-driven cohesive fracture propagation in impermeable media with frictional natural faults; Numerical and experimental investigations. Int. J. Numer. Meth. Eng. **104**, 439–468 (2015)
15. Khoei, A., Vahab, M., Hirmand, M.: Modeling the interaction between fluid-driven fracture and natural fault using an enriched-FEM technique. Int. J. Fract. **197**(1), 1–24 (2016)
16. Olson, J.E., Taleghani, A.D.: Modeling simultaneous growth of multiple hydraulic fractures and their interaction with natural fractures. In: SPE Hydraulic Fracturing Technology Conference. Society of Petroleum Engineers (2009)

17. Salimzadeh, S., Khalili, N.: (2015) Fully Coupled XFEM model for flow and deformation in fractured porous media with explicit fracture flow. Int. J. Geomech. 04015091
18. Biot, M.: General solutions of the equations of elasticity and consolidation for a porous material. J. Appl. Mech. **23**(1), 91–96 (1956)
19. Zimmerman, R.W., Bodvarsson, G.S.: Hydraulic conductivity of rock fractures. Transp. Porous Media **23**(1), 1–30 (1996)
20. Witherspoon, P.A., Wang, J., Iwai, K., Gale, J.: Validity of cubic law for fluid flow in a deformable rock fracture. Water Resour. Res. **16**(6), 1016–1024 (1980)
21. Hirmand, M., Vahab, M., Khoei, A.: An augmented Lagrangian contact formulation for frictional discontinuities with the extended finite element method. Finite Elem. Anal. Des. **107**, 28–43 (2015)
22. Liu, F., Borja, R.I.: A contact algorithm for frictional crack propagation with the extended finite element method. Int. J. Numer. Meth. Eng. **76**(10), 1489–1512 (2008)
23. Khoei, A., Vahab, M.: A numerical contact algorithm in saturated porous media with the extended finite element method. Comput. Mech. **54**(5), 1089–1110 (2014)
24. Camacho, G.T., Ortiz, M.: Computational modelling of impact damage in brittle materials. Int. J. Solids Struct. **33**(20), 2899–2938 (1996)
25. Zienkiewicz, O.C., Chan, A., Pastor, M., Schrefler, B., Shiomi, T.: Computational Geomechanics. Wiley, Chichester (1999)
26. Wu, K., Olson, J.E.: Mechanisms of simultaneous hydraulic-fracture propagation from multiple perforation clusters in horizontal wells. SPE J. (2015)
27. Yao, Y.: Linear elastic and cohesive fracture analysis to model hydraulic fracture in brittle and ductile rocks. Rock Mech. Rock Eng. **45**(3), 375–387 (2012)
28. Samuelson, M., Stefanski, J., Downie, R.: Field development study: channel fracturing achieves both operational and productivity goals in the barnett shale. In: SPE Americas Unconventional Resources Conference, pp. 5–7, Pittsburgh, Pennsylvania, USA (2012)
29. Ahmed, U., Thompson, T., Kelkar, S., Veghte, R., Hathaway, S.: Perforation placement optimization: a modified hydraulic fracturing technique. In: SPE Unconventional Gas Recovery Symposium, Society of Petroleum Engineers (1984)

Seismic Wave Field Generation in Heterogeneous Geological Media Containing Multiple Cavities

Ioanna-Kleoniki Fontara, Petia S. Dineva, Frank Wuttke and George D. Manolis

Abstract We develop a boundary integral equation (BIE) method for the numerical simulation of seismic motions in geological media containing multiple cavities under anti-plane strain conditions. We consider a half-plane of heterogeneous structure subjected to either time-harmonic incident shear waves or to body waves radiating from a seismic point source. Three different types of material heterogeneity are considered: (a) The density and shear modulus vary proportionally as quadratic functions of depth, but the wave speed remains constant; (b) the material is viscoelastic, with a shear modulus and density that vary with respect to the spatial coordinates in an arbitrary fashion, with a wave velocity is frequency and position–dependent; (c) the material has a depth-dependent shear modulus and constant density, yielding a linear wave velocity profile. This necessitates the development of three frequency-dependent integral equation schemes based on: (a) A Green's function for a quadratically-graded elastic half-plane; (b) a fundamental solution for a viscoelastic full-plane with position–dependent wave speeds; and (c) a fundamental solution for an elastic full-plane with a linearly varying wave speed. Numerical examples are presented for inhomogeneous geological media containing any number of cavities of arbitrary geometry and position.

Keywords Wave propagation · Site effects · Seismic waves · Graded media

I.-K. Fontara
Department of Civil Engineering, TU Berlin, Berlin, Germany

P.S. Dineva
Bulgarian Academy of Sciences, Sofia, Bulgaria

F. Wuttke
Christian-Albrechts University, Kiel, Germany

G.D. Manolis (✉)
Aristotle University, Thessaloniki, Greece
e-mail: gdm@civil.auth.gr

E. Papamichos et al. (eds.), *Bifurcation and Degradation of Geomaterials with Engineering Applications*, Springer Series in Geomechanics and Geoengineering, DOI 10.1007/978-3-319-56397-8_36

1 Introduction

The geological depth scale reveals that the upper surface of the Earth has a strongly heterogeneous structure due to the existence of both free-relief, non-parallel soil layers, pronounced material variation with distance, plus the presence of cavities, inclusions, cracks and faults, all in addition to man-made buried infrastructure. One consequence of this complex geology of the Earth is that it causes significant spatial variations in the seismically-induced ground motions in earthquake-prone regions of the world, which may result in large amplification effects during earthquakes. From a mathematical point of view, the development of models describing the mechanical behavior of graded media, i.e., materials whose properties are spatially-dependent, involves the solution of partial differential equations with non-constant coefficients. The predominant numerical method used currently is the standard finite element method (FEM) along with its various extended forms. Boundary integral equation methods (BIEM) are an alternative, because of their efficient handling of infinite and semi-infinite domains and their high accuracy in evaluating stress fields [1]. In general, dynamic analyses of graded materials by BIEM are possible only if the appropriate fundamental solutions are available [2]. An overview of BIEM numerical solution of dynamic problems involving graded media containing multiple discontinuities can be found in the recent review by Manolis and Dineva [3].

2 Problem Statement

Consider a geological half-plane Ω with free surface S_f containing N cavities of arbitrary shape and subjected to the following loads: (a) an SH elastic wave propagating in the plane $x_3 = 0$, tracing an incident angle θ with respect to the coordinate axis Ox_1; (b) a point force at source $\boldsymbol{x_0} = (x_{01}, x_{02})$. In both cases, time harmonic conditions are assumed to hold and the loads are functions of frequency ω. Next, the deformation mode is anti-plane strain and the only non-zero variables are the out-of-plane displacement component $u_3(\boldsymbol{x}, \omega)$ and the stresses $\sigma_{i3}(\boldsymbol{x}, \omega)$, i = 1, 2. The material is described by a position-dependent shear module $\mu(x)$ and a density ρ. The governing equation and corresponding boundary conditions are

$$\nabla.\{\mu(\mathbf{x})\nabla u_3(\mathbf{x},\omega)\} = \rho(\mathbf{x})\omega^2 u_3(\mathbf{x},\omega) - \psi f_{03}\delta(\mathbf{x},\mathbf{x_0}), \quad \mathbf{x}\in\Omega \tag{1}$$

$$t_3(x_1,x_2,\omega) = \sigma_{3j}(x_1,x_2,\omega)\, n_j(x_1,x_2) = 0, \ \mathbf{x}\in S_f \cup \Gamma \tag{2}$$

In the above, position vector is $\boldsymbol{x} = (x_1, x_2)$, n_i are the components of the outward pointing normal vector along a surface, index ψ is either $= 0$ or $= 1$ when the load is respectively an incident wave or a point load at $\boldsymbol{x_0}$ with magnitude f_{03}. Also, ∇ is the gradient operator, (.) is the inner product and a summation convention over repeated indices is implied. For an infinitely extending boundary, the Sommerfeld radiation condition is satisfied. Three different types of inhomogeneity (i.e., material gradient) are now considered:

(1) Type A, where the material is isotropic, but both density and shear modulus vary proportionally in a quadratic way in respect to the depth of the half-plane, yielding a macroscopically constant wave speed:

$$\mu(x_1,x_2)=\mu^0 h(x_2);\quad \rho(x_1,x_2)=\rho^0 h(x_2);\quad h(x_2)=(bx_2+1)^2 \tag{3}$$

We define $h(x)$ as the inhomogeneity function, where $b \leq 0$ is the magnitude and the reference material constants are $\mu^0 = \mu(x_1, 0)$; $\rho^0 = \rho(x_1, 0)$.

(2) Type B, where the material is viscoelastic (Kelvin-Voigt model) and both shear modulus and density have a variation with respect to depth. This yields a wave speed that is both frequency and position–dependent. The shear modulus is now complex-valued and is defined as $\mu = \mu^* = Re(\mu^*) - i\omega\mu'$, where the real part is position-dependent. The imaginary part μ' indicates dissipation and is computed from the logarithmic decrement in the motion amplitude as $\delta = \pi/Q = 2\pi\sqrt{\omega\bar{\mu}/\mathrm{Re}(\mu^*)}$, with Q the dimensionless quality factor, see Manolis and Shaw [4].
(3) Type C, with position-dependent shear modulus and constant density, i.e., $\mu = \mu(x_2)$; $\rho = \rho^0(0)$, yielding linear profiles in the wave speed as $C_s(x_2) = C_s^0(1 + x_2/H)$. Parameter $H > 0$ is defined in the interval $10^{-7} < H/\alpha < 10^{+7}$, where α is the characteristic heterogeneity length [5], and $C_s(-H) = 0$. In all, the total wave field comprises free-field motion (u_3^{ff}, t_3^{ff}) plus the scattered waves (u_3^{sc}, t_3^{sc}).

3 BIEM Formulation and Solution

The above boundary-value problem (BVP) is reformulated as integral equations along the inhomogeneous domain boundaries $S_f \cup \Gamma$ through Betti's reciprocal theorem in conjunction with the appropriate fundamental solution (or Green's function). As discussed in Fontara et al. [6], two basic BIEM schemes are developed, one using a Green's function for a graded half-plane in the case of the material profile of *Type A* and another using a fundamental solution for the graded full-plane in the case of material profiles of *Type B* and *Type C*.

The system of BIE which results following discretization of all surfaces in question by single node, line boundary elements (BE) and subsequent collocation of the field variables (displacement and tractions) at all nodal points is numerically solved using matrix inversion. This way, all unknown field variables are expressed in terms of their prescribed values. For a well-posed BVP, this procedure results in a system of N equations obtained with respect to all N unknowns at the surfaces of the BVP in question. The numerical procedure was implemented using Matlab.

4 Numerical Simulation and Results

A large number of verification examples were conducted [6] to ascertain the robustness of the BIEM formulation, with comparisons with published work (e.g., [7]). Following that, a series of parametric studies were conducted involving buried cavities (circular, vertically elongated and horizontally elongated ellipses) in three

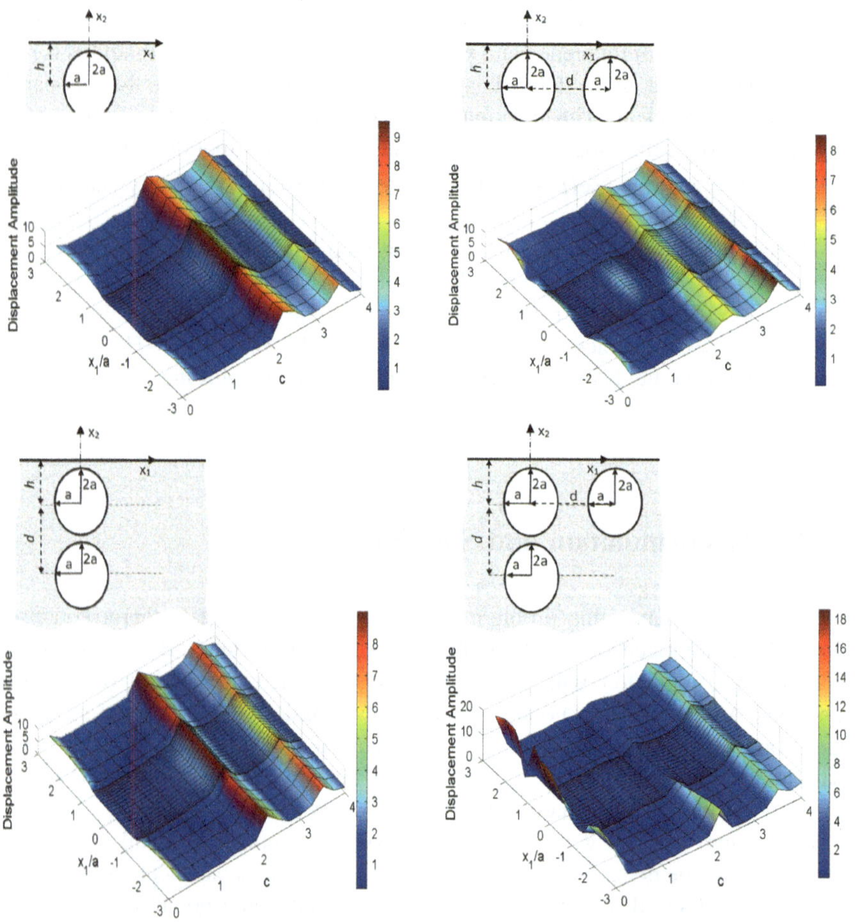

Fig. 1 Displacement amplitude along the free surface of *Type B* half-plane with inhomogeneity parameter c due to concentrated load at point $\mathbf{x_0}(0, -15\alpha)$ with dimensionless frequency $\eta = 0.5$: All embedded cavities are vertically elongated ellipses with semi-axes α and 2α

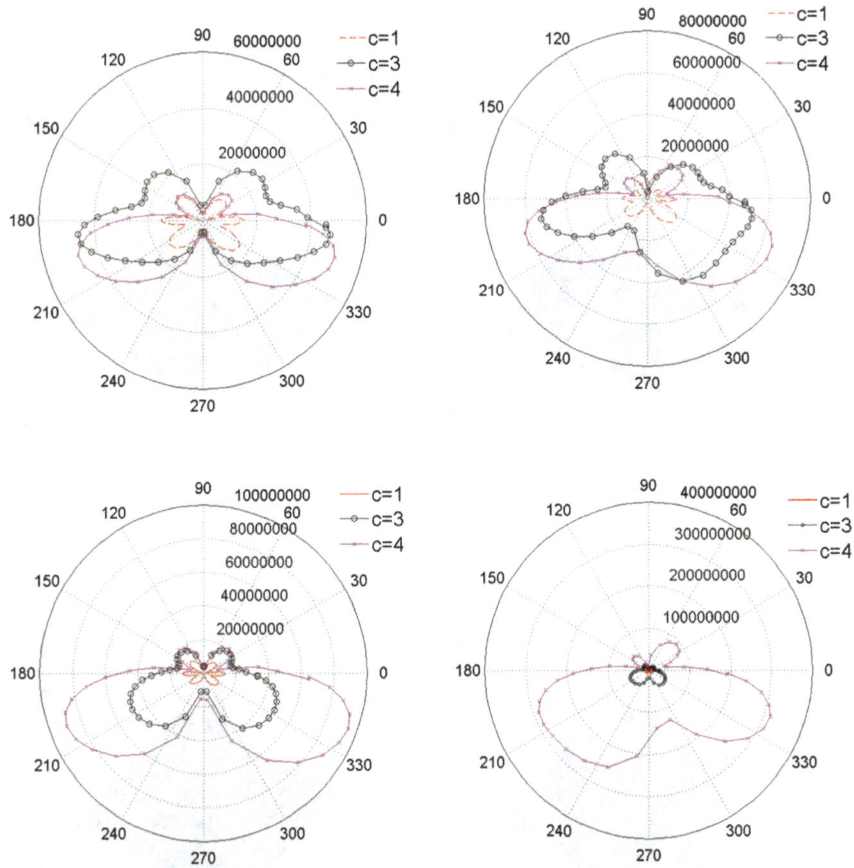

Fig. 2 Stress distribution at the perimeter of a reference cavity in a half plane with inhomogeneity magnitude c at dimensionless frequency $\eta = 0.5$ due to an SH-wave from an embedded source located at point $(0, -15\alpha)$: All cavities are vertically elongated ellipses with semi-axes α, 2α. (c−w from top *left*: 1 cavity, 2 horizontal cavities, 2 vertical cavities, 3 cavities)

inhomogeneous material profiles. These involve a centered point force at a depth of 15α from the surface with dimensionless frequency $\eta = 0.5$. A selective presentation of these results in given in Figs. 1, 2, and 3 for the stress field distribution and the displacement amplitude along the reference cavity perimeter, plus the displacement amplitude at the free surface as functions of the inhomogeneity parameter c.

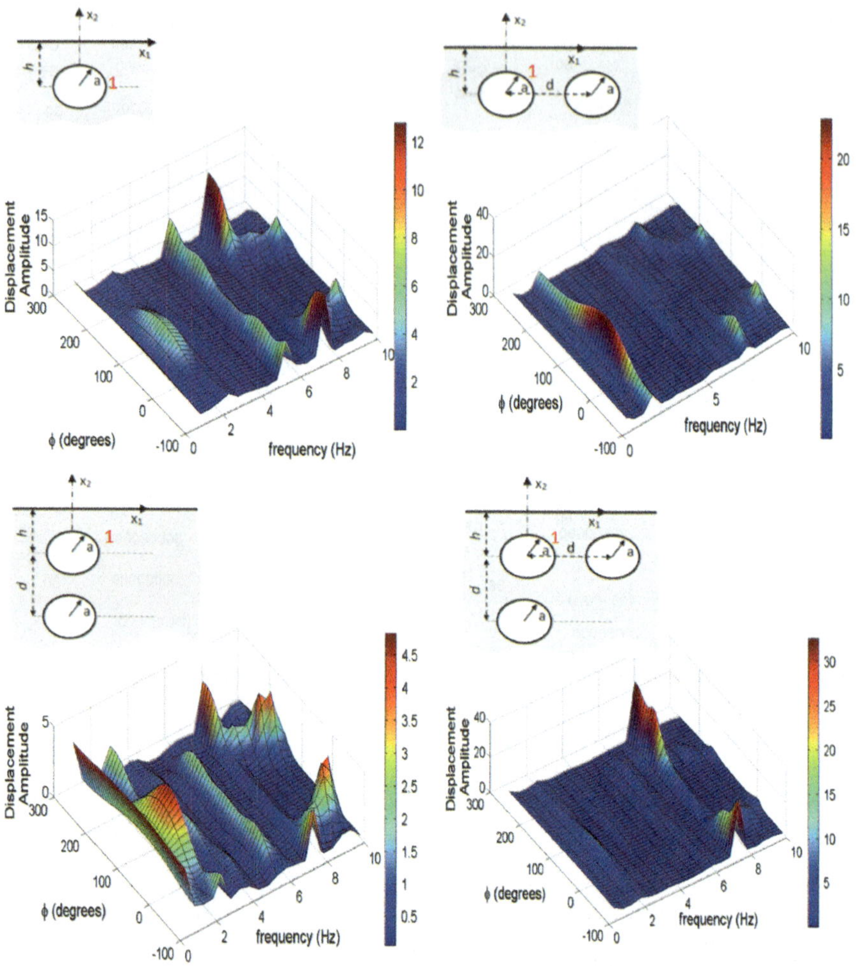

Fig. 3 Displacement amplitude along cavity 1 embedded in the half-plane with *Type B* inhomogeneity magnitude c = 2 versus frequency due to SH-wave emanating from a source located at $(0, -15\alpha)$: All embedded cavities are circular with radius α

5 Conclusions

In this work, we presented a new BIEM development for the numerical solution of SH-wave propagation in graded geological media containing any number of cavities of arbitrary shape and position. The numerical results reveal the dependency of the wave fields and of the zones of dynamic stress concentration to the following key factors: (a) material property variation within the soil mass; (b) characteristics of the applied seismic loads; (c) shape, position and number of cavities; and (d) interaction phenomena between the cavities and the free surface.

References

1. Manolis, G.D., Beskos, D.E.: Boundary Element Methods. Unwin, London (1988)
2. Kausel, E.: Fundamental Solutions in Elastodynamics: A Compendium. Cambridge University Press, Cambridge (2006)
3. Manolis, G.D., Dineva, P.S.: Elastic waves in continuous and discontinuous geological media by boundary integral equation methods: a review. Soil Dyn. Earthquake Eng. **70**, 11–29 (2015)
4. Manolis, G.D., Shaw, R.P.: Harmonic wave propagation through viscoelastic heterogeneous media exhibiting mild stochasticity—I. Fundamental solutions. Soil Dyn. Earthquake Eng. **15**, 119–127 (1996)
5. Luzon, F., Sanchez-Sesma, F.J., Perez-Ruiz, A., Ramirez, L., Pech, A.: In-plane seismic response of inhomogeneous alluvial valleys with vertical gradients of velocities and constant Poisson ratio. Soil Dyn. Earthquake Eng. **29**, 994–1004 (2009)
6. Fontara, I.K., Dineva, P.S., Manolis, G.D., Wuttke, F.: Numerical simulation of seismic wave fields in graded geological media containing multiple cavities. Geophys. J. Int. **206**, 921–940 (2016)
7. Dravinski, M., Yu, M.C.: Scattering of plane harmonic SH-waves by multiple inclusions. Geophys. J. Int. **186**(3), 1331–1346 (2011)

Multiscale Investigation of Microcrack-Induced Instability in Rocks

Mahdad Eghbalian and Richard Wan

Abstract The paper deals with a mathematical description of instability and localization in rocks through a multi-scale modeling strategy with coupled multiphysics. We adopt a homogenization approach to work out an effective description of microscopic phenomena at the continuum through defining a Representative Volume Element (RVE). The role of hydro-mechanical coupling in the transition from an otherwise initial diffuse state to a localized state can be investigated through the gradual degradation of the kinematics due to micro-fracture formation.

1 Introduction

Unconventional reservoirs such as tight gas and shales gas are not permeable enough for the fluid production to be economically viable. Thus, increasing the hydraulic conductivity of such porous media through induced material failure and instabilities is of prime interest. Pressurized fluid causes the dispersed micro-fractures to propagate so as to ultimately increase the permeability of the reservoir. Thus, the hydro-poromechanics of such heterogeneous media is central to the understanding of material failure processes during hydraulic fracturing.

Phenomenological approaches to formulate the hydro-poroelasticity of rocks lack micromechanical insight and involve several parameters to be calibrated, many of them are devoid of any physical meaning. This motivates the pursuit of micromechanical approaches where all the underlying multiphasic, micromechanical phenomena can be accounted for in an effective medium formulation. However, current micromechanical models are limited in the sense that they either consider only one scale of porosity, or neglect the hydraulic coupling on fracture behavior of the material (see e.g. [1, 2]).

M. Eghbalian (✉) · R. Wan
Civil Engineering Department, University of Calgary, Calgary, Canada
e-mail: meghbali@ucalgary.ca

R. Wan
e-mail: wan@ucalgary.ca

E. Papamichos et al. (eds.), *Bifurcation and Degradation of Geomaterials with Engineering Applications*, Springer Series in Geomechanics and Geoengineering, DOI 10.1007/978-3-319-56397-8_37

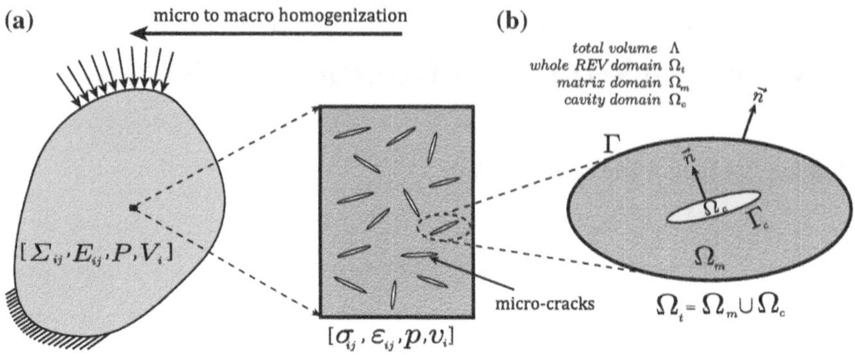

Fig. 1 Structure of fractured porous medium: **a** Homogenized medium in a macroscale problem, **b** Tiny cracks dispersed within the porous matrix

This paper aims at addressing the mentioned gaps and building a continuum framework that contains the key elements necessary for describing the hydro-mechanical behavior of a saturated porous medium with arbitrarily distributed strong kinematic discontinuities (Fig. 1) and their transition to a localized state.

2 Mathematical Framework

In this section, we address the question of how an arbitrary distribution of micro-fractures affects the hydro-poroelastic behavior of the porous matrix. In the homogenization method, we replace a microscopically heterogeneous material point of the body with a fictitious homogeneous one. The average-field theory defines the upscaled quantities as the volume average of the microscopic fields. In this manner, the material point behavioral equations are closely related to a boundary value problem defined on the microstructure.

2.1 Fracture Behavior of Rock

In what follows, a multiscale description of fractured rock is presented. The behavior of the material under compressive loading is addressed relevant to the occurrence of instability in a triaxial rock sample.

We start with the behavioral equations of solid and fluid phases in the microstructure of the rock (Fig. 1b). Darcy's law is invoked for fluid flow in both matrix and micro-cracks:

$$v_i = \begin{cases} -\Pi \; \partial p/\partial x_i & \Omega_m \\ -\Pi' \; \partial p/\partial x_i & \Omega_c \end{cases} \tag{1}$$

with Π and $\Pi^{'}$ are constants related to the permeability of the medium. Moreover, the following relation for the divergence of the fluid velocity is recalled:

$$\frac{\partial v_i}{\partial x_i} = \begin{cases} -\left(\dot{\Phi}_f - \dot{\Phi}_f^0\right)/\Phi_f & \Omega_m \\ 0 & \Omega_c \end{cases} \tag{2}$$

where Φ_f is the change in the fluid fraction. The momentum balance and constitutive equations of the poroelastic matrix together with the mass balance equation are:

$$\partial \sigma_{ij}/\partial x_j = 0 \tag{3a}$$

$$\varepsilon_{ij} = \frac{1}{2G}\sigma_{ij} + \frac{\kappa - 3}{8G}\sigma_{kk}\delta_{ij} + \frac{\alpha\kappa^{'}}{2G}p\delta_{ij} \tag{3b}$$

$$\Phi_f - \Phi_f^0 = \alpha\varepsilon_{kk} + \frac{p}{Z} \tag{4}$$

where (α) is the Biot's constant, (Z) is the storage coefficient and (G) is the shear modulus. Moreover, κ and $\kappa^{'}$ are constants for plane (stress/strain) conditions. The coupling between solid and fluid field equations arises from the mass balance equations. All the parameters of the system are derived micro-mechanically through a homogenization process based on the porosity of the matrix (n_f), bulk and shear modulus of the solid skeleton (K^s and G^s respectively) in [1, 3].

The macroscopic stress tensor (Σ_{ij}) and pressure (P) are prescribed through the uniform traction boundary condition:

$$t_i = \sigma_{ij}n_j = \begin{cases} \Sigma_{ij}n_j & \Gamma \\ p^c n_i & \Gamma_c \end{cases} \tag{5}$$

and uniform pressure $p = \nabla_i P x_i$ on Γ. The interface conditions necessary for solving the boundary value problem are $v_i^f = v_i^p$ and $\sigma_{ij}^f n_j = \sigma_{ij}^p n_j$ defined on Γ_c. The subscripts f and p refer to the flow in the crack and in the porous matrix respectively; while p^c refers to the pressure inside the micro-crack. The above uniform boundary conditions ensure that the volume average of stress and pressure gradient are equal to their prescribed macroscopic values, i.e. $\langle\sigma_{ij}\rangle^{\Omega_t} = \Sigma_{ij}$ and $\langle\nabla_i p\rangle^{\Omega_t} = \nabla_i P$. The macroscopic description is constructed by formulating the average strain (E_{ij}), average fluid velocity (V_i) and average change in fluid fraction $\left(\overline{\Phi}_f - \overline{\Phi}_f^0\right)$ from the solution of the field variables (p, u_i).

One-way coupling of the above system of equations can be advocated in the steady state case where the pressure field is not affected by changes in the strain field. The Darcy/Darcy coupling has been investigated in [4]. Also, closed-form solutions exist in the literature for deformation analysis of line-cracks subjected to internal traction (saturated crack) and far-field stress [5].

Furthermore, micro-fractures are considered as ellipses with high aspect ratio β. Also, n^θ cracks per unit area (volume) are considered in direction θ each having half-foci length r_θ with their surface density denoted by ω^θ. The Mori-Tanaka scheme assumes the effect of micro-cracks presence and their interaction can be approximated by considering each of them being embedded in an infinite matrix subjected to far-field average stress and pressure gradient of the matrix. Based on the above assumptions, the average law is written in the form of integrals over the micro-cracks and matrix domains. Solving for the defined system of equations, along with application of the divergence theorem and Mori-Tanaka scheme lead to:

$$\begin{aligned} E_{ij} = & \frac{1}{2G}\Sigma_{ij} + \frac{\kappa - 3}{8G}\Sigma_{kk}\delta_{ij} + \frac{(\kappa+1)\beta}{8G}\Big(\Sigma_{im}T\left(\xi_{jm}\right) + \Sigma_{jm}T\left(\xi_{im}\right) \\ & - \Sigma_{kk}T\left(f\left(r_\theta\right)\xi_{ij}\right) + 2\Sigma_{mn}T\left(f\left(r_\theta\right)\xi_{ij}\xi_{mn}\right)\Big) \\ & + \frac{1}{4G}\left(\left(\left(\kappa - 1 - 2\alpha\kappa'\right)T\left(\left(n^\theta - w^\theta\right)/n^\theta\right) + 2\alpha\kappa'\right)\delta_{ij} + (\kappa+1)\beta T\left(\xi_{ij}\right)\right)\overline{p} \end{aligned} \tag{6}$$

$$\begin{aligned} \overline{\Phi}_f - \overline{\Phi}_f^0 = & \frac{(\kappa+1)\beta}{4G}\left(\Sigma_{lm}T\left(\xi_{lm}\right) - \frac{1}{2}\Sigma_{kk}T\left(f\left(r_\theta\right)\xi_{ll}\right) + \Sigma_{mn}T\left(f\left(r_\theta\right)\xi_{ll}\xi_{mn}\right)\right) \\ & + \frac{\alpha(\kappa-1)}{4G}\Sigma_{kk} + \left(\frac{\alpha^2\kappa'}{G} + \frac{1}{Z}\right)\left(1 - T\left(\left(n^\theta - w^\theta\right)/n^\theta\right)\right)P \\ & + \frac{\alpha(\kappa-1)}{2G}T\left(\left(n^\theta - w^\theta\right)/n^\theta\right)P + \frac{(\kappa+1)\beta}{4G}T(1)\,P \end{aligned} \tag{7}$$

from which the mass balance equation is easily formulated. Also:

$$\overline{\Pi}_{ij} = \Pi\left(1 + (1+\beta)T(1)\right)\delta_{ij} + \frac{1-\beta^2}{\beta}\Pi\, T\left(\xi_{ij}\right) \tag{8}$$

$$\xi_{ij} = \begin{pmatrix} \sin^2\theta & -\frac{\sin 2\theta}{2} \\ -\frac{\sin 2\theta}{2} & \cos^2\theta \end{pmatrix}, \quad T(x^\theta) = \int_0^\pi \frac{w^\theta n^\theta}{n^\theta - w^\theta}x^\theta\, d\theta \tag{9}$$

The function $f(r)$ is used to take into account the effect of micro-crack lengths on the tensile field generated around them under compressive loading. This fictitious tensile force is proportional to the deviatoric stress $\Sigma_t = f(r)\langle S_n\rangle^+$ [6]; with S_n as the effective normal deviatoric stress acting on the micro-crack and added to the stresses that the micro-crack is subjected to. Furthermore, $\langle x\rangle^+$ implies $\langle x\rangle^+ = (x + |x|)/2$.

2.2 *Micro-Structure Evolution (Instability)*

A hydro-mechanical crack growth criterion based on Fracture Mechanics is used here $(K_I - K_{IC} = 0)$ [7]. Mode-I fractures are assumed to be dominant. The effects of pore pressure is considered through introducing the effective stress in the formulation; while the local tensile stress discussed in previous section is also considered:

$$K_I = \sqrt{r_\theta}\left[\Sigma_n^e\left(\frac{f_c}{f_c + \langle -\Sigma_n^e\rangle^+}\right)^m + 3g\left(r_\theta\right)\langle S_n^e\rangle^+\right] \tag{10a}$$

$$\Sigma_n^e = n_i^\theta \Sigma_{ij}^e n_j^\theta \;,\; S_n^e = n_i^\theta S_{ij}^e n_j^\theta \;,\; S_{ij}^e = \Sigma_{ij}^e - \left(\frac{\Sigma_{kk}^e}{3}\right)\delta_{ij} \tag{10b}$$

where K_I and K_{IC} are the mode-I stress intensity factor and fracture toughness respectively, and n_i^θ is the normal to the micro-crack face. Also, the following form has been suggested in [7] for $g\left(r_\theta\right)$, which ensures a stable micro-crack growth:

$$g\left(r_\theta\right) = \sqrt{\pi} f\left(r_\theta\right)/3 = \begin{cases} \eta r_f/r_\theta & r_\theta \le r_f \\ \eta & r_\theta > r_f \end{cases} \tag{11}$$

where $f_c = K_{IC}/\left(\eta\sqrt{r_f}\right)$ and η is a material constant and r_f is the threshold micro-crack length for the RVE to become unstable and fail.

2.3 Degradation Behavior of Lac Du Bonnet Granite

Experimental results on Lac Du Bonnet granite are chosen to compare with the output of the model set up here. The RVE includes $m = 500$ sets of microcracks uniformly distributed in the initial state and all having the same length. The parameters used in the calculations are $K_{IC} = 1.03\,\text{MPa}\,\text{m}^{\frac{1}{2}}$, $K^s = 328\,\text{GPa}$, $G^s = 54\,\text{GPa}$, $r_0 = 3\,\text{mm}$, $r_f = 9\,\text{mm}$, $n_f = 0.01$, $\Pi = 5.4 \times 10^{-22}\,\text{m}^2/\text{Pa s}$, $\eta = 0.06$, $n^\theta = 6.0 \times 10^4\,\text{m}^{-2}$ and $\beta = 2000$. The poroelastic parameters are calibrated from results reported in [7]; while parameters related to crack evolution are the same as the one used in the above-mentioned work.

We first let the RVE reach an initial equilibrium state by applying an equal lateral and axial compressive stresses. Then, an axial strain increment of $\Delta\overline{E}_y = -2 \times 10^{-5}$ is applied to the RVE in each loading step. The boundary conditions are such that the fluid pressure P is constant and equal to zero during the test.

In Fig. 2a, the deviatoric stress at failure is plotted against the confining pressure. A good agreement between the model predictions and experimental results is observed. The distributions of microcrack lengths at the initial configuration, including intermediate states towards failure are depicted in Fig. 2b, where as expected, the microcrack growth is indeed aligned with the direction of the applied axial stress.

It is worth mentioning that the permeability enhancement is not investigated herein and is the subject of a separate study. The reason is that the model developed in Sect. 2.1 is valid when we deal with a disconnected network of micro-cracks.

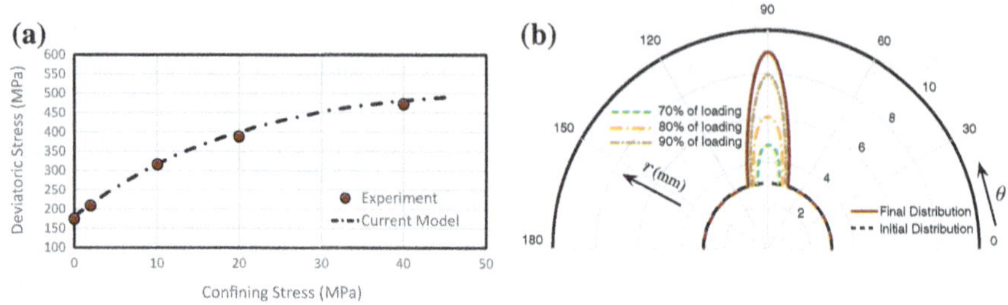

Fig. 2 **a** Deviatoric stress at failure versus the confining pressure. **b** Crack length distribution at different stages of loading

Numerical simulation results [8] show that the situation is different when some part of these micro-cracks are connected such that they form channels. Hence, in order to do a comparison with experimental results; forming channels should be included by considering the Navier-Stokes solution for laminar fluid flow between two parallel channel faces along the length of the channel.

3 Concluding Remarks

A comprehensive hydro-mechanical formulation is presented for a porous medium with embedded micro-cracks in the fully saturated case. The generally anisotropic behavior of the material is described by a set of closed-form relations through a homogenization process starting from the underlying coupled physics at the microscale. The upscaling process is carried out within the framework of mean-field theory to finally obtain the macroscopic properties solely from those of the micro-constituents. It is demonstrated in the case study of Lac Du Bonnet granite that the model predictions based on material instabilities through micro-crack evolution and coalescence are quite satisfactory.

Acknowledgements This work is jointly funded by the Natural Science and Engineering Council of Canada and the Foundation Computer Modeling Group (FCMG).

References

1. Dormieux, L., Kondo, D., Ulm, F.J.: Microporomechanics. Wiley, The Atrium, Southern Gate, Chichester, West Sussex PO19 8SQ, England (2006). doi:10.1002/0470032006
2. Nemat-Nasser, S., Hori, M.: Micromechanics: Overall Properties of Heterogeneous Materials, North-Holland (1993)
3. Wan, R.G., Eghbalian, M.: Int. J. Solids Struct. **96**, 274 (2016). doi:10.1016/j.ijsolstr.2016.06.002

4. Zhao, C., Hobbs, B.E., Ord, A., Hornby, P., Shenglin, P., Liu, L.: Geophys. J. Int. **166**(2), 970 (2006). doi:10.1111/j.1365-246X.2006.03049.x
5. Zehnder, A.T.: In Fracture Mechanics, pp. 7–32, Springer Netherlands (2012). doi:10.1007/978-94-007-2595-9_2
6. Golshani, A., Okui, Y., Oda, M., Takemura, T.: Mech. Mater. **38**(4), 287 (2006). doi:10.1016/j.mechmat.2005.07.003
7. Shao, J.F., Zhou, H., Chau, K.T.: Int. J. Numer. Anal. Meth. Geomech. **29**(12), 1231 (2005). doi:10.1002/nag.457
8. Huang, Z., Yao, J., Li, Y., Wang, C., Lv, X.: Commun. Comput. Phys. **9**(1), 180 (2011). doi:10.4208/cicp.150709.130410a

Borehole Strengthening in Gas Wells from Near Borehole Drying

Euripides Papamichos

Abstract The effect of flow-through drying in sand production is analyzed. In flow-through drying evaporation of the connate water of the reservoir sandstone occurs into the gas phase during natural gas production. The strength of sandstone is affected by the water saturation with dry rock being stronger than water saturated rock. Drying of the formation near the well due to gas flow will thus lead to an increase in the critical stresses for failure and sand production. Flow-through drying of a near wellbore formation is coupled with a constitutive model with water saturation dependent strength and stiffness to analyze the coupled problem of flow-through drying and mechanical behavior of a material. Results from finite element simulations are presented and compared with experimental results.

1 Introduction

Evaporation into the gas phase of the connate water of reservoir sandstones occurs during flow-through drying in natural gas production, as a result of the compressibility of the gas. Near the wellbore, the gas pressure drops and thus its relative humidity level drops as well. When an unsaturated gas flows through a medium, it can evaporate the connate water in the medium. Drying of the rock may lead to higher strength and stiffness as sandstones are often water sensitive. The result is higher borehole strength and delayed sand production in gas wells compared to oil wells in the same formation. Modeling of flow-through drying involves the solution of axisymmetric simultaneous, two-phase flow equations of two immiscible fluids, a single-component liquid (water) phase and a binary (air-vapor) gas phase. These comprise the conservation equations for the water and gas components, constitutive

E. Papamichos (✉)
Department of Civil Engineering, Aristotle University of Thessaloniki,
54124 Thessaloniki, Greece
e-mail: euripides.papamichos@sintef.no; epapamic@civil.auth.gr

E. Papamichos
SINTEF Petroleum Research, N-7465, Trondheim, Norway

E. Papamichos et al. (eds.), *Bifurcation and Degradation of Geomaterials with Engineering Applications*, Springer Series in Geomechanics and Geoengineering, DOI 10.1007/978-3-319-56397-8_38

equations for relative permeability and capillary pressure, and Raoul's law for phase equilibrium [1–3]. Liquid evaporation is governed by a diffusion–reaction equation, where compressibility-driven drying acts as the reaction and capillarity acts as diffusion. The reaction rate increases with factors that favor gas expansion, i.e. gas mobility, flow rate, radius of well influence. Results show the significance of drying in wellbore strengthening and sand production avoidance in gas wells, an effect that is verified by experimental results.

2 Flow-Through Drying in Gas Wells

The formulation is based on the overall compositional balance equations for each component. In the gas fluid flow and evaporation problem under consideration, there exists two components (*c*) and two phases (*p*). Phase $p = 1$ (or *g*) is the binary (air-vapor) gas phase and phase $p = 2$ (or *w*) is the single-component liquid (water) phase. Component $c = 1$ (or H_2O) is the connate water in the pores of the solid and component $c = 2$ (or N_2+) is the dry gas. Symbol N_2+ signifies that dry gas is a mixture of various constituents, among others nitrogen, oxygen, argon and carbon dioxide, which are all viewed here as comprising one component. The solid phase does not enter in the formulation since it is assumed that there is no adsorption of the H_2O or N_2+ in the solid. In the absence of mass generation or destruction e.g. by chemical reactions, the overall conservation of each component over the two phases is written as [4]

$$\frac{\partial W_{N_2+}}{\partial t} + N_{i,i}^{N_2+} = 0, \qquad \frac{\partial W_{H_2O}}{\partial t} + N_{i,i}^{H_2O} = 0 \tag{1}$$

where t denotes time, a subscript preceded by comma denotes differentiation with respect to that coordinate and repeated indices are summed. In Eq. (1), the first term in both equations corresponds to the rate of accumulation of each component in both phases in a volume V and the second term to the rate of transport of each component in minus out of V. Thus, W_{N_2+} is the sum of the concentrations of N_2+ in each phase and W_{H2O} is the sum of the concentrations of H_2O in each phase with units of mass/volume (kg/m^3). Similarly, $N_i^{N_2+}$ is the sum of the fluxes of N_2+ in each phase and $N_i^{H_2O}$ the sum of the fluxes of H_2O in each phase along the i-direction with units of mass/(surface × time) (kg/(m$^2 \cdot$ s)). In the absence of dispersion in the phase velocities and no dissolution of N_2+ in the water phase

$$\frac{\partial}{\partial t}\left[\phi(1-S_w)\frac{(1-y_{H_2O,g})\rho_g}{M_g}\right] + \left[\frac{(1-y_{H_2O,g})\rho_g}{M_g}q_i^g\right]_{,i} = 0 \tag{2}$$

$$\frac{\partial}{\partial t}\left[\phi\frac{\rho_w}{M_{H_2O}}S_w + \phi(1-S_w)\frac{y_{H_2O,g}\rho_g}{M_g}\right] + \left(\frac{\rho_w}{M_{H_2O}}q_i^w + \frac{y_{H_2O,g}\rho_g}{M_g}q_i^g\right)_{,i} = 0 \tag{3}$$

where ϕ is the porosity, ρ_g and ρ_w are the gas and water phase densities, S_w is the water phase saturation, q_i^g and q_i^w are the gas and water phase fluxes (m/s), $M_{H_2O} = 18.01528$ g/mol is the molecular weight of H_2O, $M_{N_2+} = 28.97$g/mol the molecular weight of N_2+, i.e. dry air, M_g the average mass of a mole in the gas phase given as $M_g = y_{H_2O,g}M_{H_2O} + y_{N_2+,g}M_{N_2+}$, and $y_{H_2O,g}$ and $y_{N_2+,g}$ the H_2O and N_2+ mole fractions in the gas phase, respectively. For isothermal processes and for constant with time porosity, using the ideal gas law, Raoult's law for partial vapor pressures and Dalton's empirical law for partial pressures of gas mixtures, Eqs. (2) and (3) result in

$$
\begin{aligned}
&\phi\frac{\partial}{\partial t}\left[(1-S_w)p_{gs}\right] + m_{i,i} = 0 \quad N_2+ \ \text{conservation}\\
&\phi\frac{\partial S_w}{\partial t} + q^w_{i,i} + \varepsilon q^g_{i,i} = 0 \qquad H_2O \ \text{conservation}
\end{aligned}
\tag{4}
$$

where $p_{gs} = p_g - p_{sat}$ with p_g the total gas phase pressure and p_{sat} the vapor saturation pressure which can be calculated using Antoine's equation [5] as a function of temperature. The mass flux parameter $m_i = p_{gs}q_i^g$, $\varepsilon = \beta_{H_2O,g}/(\beta_{H_2O,w} - \beta_{H_2O,g})$, where $\beta_{H_2O,w} = 55508$ mol/m^3 and $\beta_{H_2O,g} = p_{sat}/R_gT$ are the mole concentrations of H_2O in the water and gas phases, respectively, the universal gas constant is R_g = 8.3145 m^3 Pa/(mol · °K) and T is the temperature.

In Eq. (8), the gas flux q_i^g is given by Darcy's law for multiphase flow while the water flux q_i^w is modeled with corner film flow theory [6]

$$
q_i^g = -\kappa_{rg}(S_w)p_{gs,i}, \qquad q_r^w = -D(S_w)S_{w,i} \tag{5}
$$

where the relative mobility κ_{rg} of gas is defined as $\kappa_{rg}(S_w) = kk_{rg}(S_w)/\mu_g$, where k is the permeability of the medium, k_{rg} the relative permeability with respect to gas and μ_g the gas viscosity. Moreover, $D(S_w)$ is the capillary diffusivity function given as $D(S_w) = a_g\gamma r_c\sqrt{S_w}/\mu_w$ where a_g a dimensionless pore geometric constant that ranges between 0.05 × 10^{-3} and 2 × 10^{-3} for the rocks reported by Mahadevan et al. [1], γ the interfacial tension between gas and water and r_c the mean pore throat radius. The variational form of Eq. (4) within a volume V is written as

$$
\begin{aligned}
&\int_V \delta p_{gs}\phi\frac{\partial}{\partial t}\left[(1-S_w)p_{gs}\right]dV - \int_V \delta p_{gs,i}m_i dV + \int_{\partial V_m} \delta p_{gs}m_i n_i dS = 0\\
&\int_V \delta S_w\phi\frac{\partial S_w}{\partial t}dV - \int_V \delta S_{w,i}q_i^w dV - \varepsilon\int_V \delta S_{w,i}q_i^g dV + \int_{\partial V_{qw}} \delta S_w q_i^w n_i dS +\\
&\quad + \varepsilon\int_{\partial V_{qg}} \delta S_w q_i^g n_i dS = 0
\end{aligned}
\tag{6}
$$

The above system of nonlinear equations is discretized in space with a finite element scheme using linear time interpolation functions. The resulting system of nonlinear equations is solved with the Newton–Raphson method. For the hollow cylinder or borehole problem, an axisymmetric formulation is implemented. The equations are solved under the initial condition $S_w = S_{wi}$ at $t = 0$, zero water phase flux boundary conditions at the external r_e and internal r_i radii

$$q_r^w\big|_{r_e} = -D(S_w)S_{w,r}\big|_{r_e} = 0, \qquad q_r^w\big|_{r_i} = -D(S_w)S_{w,r}\big|_{r_i} = 0 \tag{10}$$

and prescribed gas pressures $p_g\big|_{r_e} = p_e$ at r_e and $p_g\big|_{r_i} = p_i$ at r_i.

Figure 1 presents saturation profiles for five different times for a hollow cylinder specimen with applied gas pressure $p_e = 0.3$ MPa and $p_i = 0.1$ MPa at the external and internal radii, $r_e = 0.1$ m and $r_i = 0.01$ m, respectively. The initial $S_w = S_{wi} = 0.4$, $\phi = 0.25$, $a_g = 0.001$, $k = 0.2$ D, $k_{rg} = 0.7$, $\mu_g = 0.018137$ cP, $T = 293.15$ °K, $\gamma = 0.072$ N/m, $\mu_w = 1$ cP and $r_c = \sqrt{8k/\phi}$. Two cases with and without capillary wicking are compared. In the second, the surface tension is set equal to zero. The effect of capillarity is to move the water in the water phase towards the hole smoothing out the saturation profile.

Moreover, it fuels evaporation by supplying water close to the hole where evaporation is stronger due to decompression. The effect of flow-through drying in borehole stability and sand production is investigated by comparing the cavity closure curves and the plastic shear strains at the cavity of a hollow cylinder for the case of zero drying, i.e. with $S_w = S_{wi}$ and the case after 5 h (18,000 s) of drying (with no capillary wicking). The material obeys the Mohr-Coulomb yield and plastic potential laws with a tension cutoff T_c (Fig. 2) that follows an exponential decay function with saturation $T_c(S_w) = T_{c0} + T_{c1}\exp(-T_{c2}S_w)$ where T_{c0}, T_{c1} and T_{c2} are calibration parameters (Fig. 2). Such a model has been used to model the water sensitivity of chalks [7].

Figure 3 compares the cavity closure curves versus the internal tangential strain $\varepsilon_{\theta i} = u_{ri}/r_i$ which expresses the radial hole closure normalized by the cavity radius.

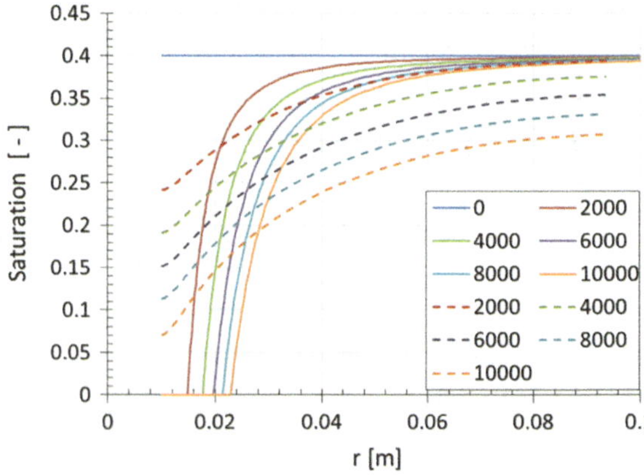

Fig. 1 Saturation radial profiles for constant relative permeability for various times in seconds. Results for zero surface tension (*solid lines*) and non-zero surface tension (*dashed lines*)

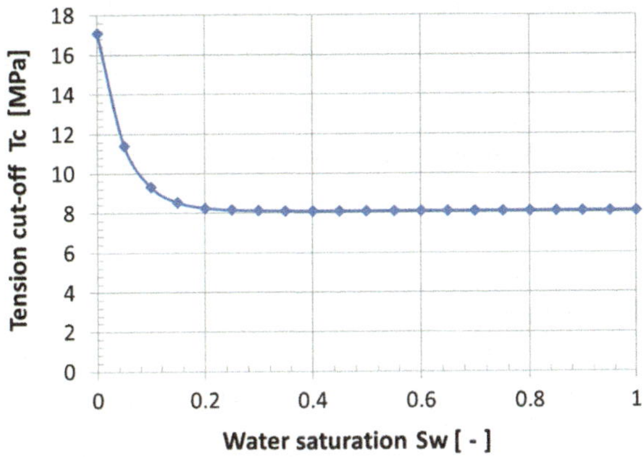

Fig. 2 Tension cutoff as a function of water saturation

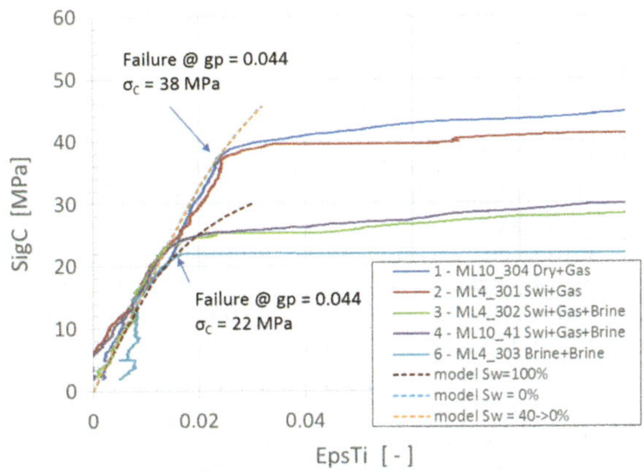

Fig. 3 External stress versus cavity closure. Experimental results on Red Wildmoor sandstone under different saturation and flow conditions (*solid lines*) and modeling results (*dashed lines*)

It can be seen that after drying the sandstone becomes both stiffer and stronger. The experimental results are for Red Wildmoor sandstone that is strongly water sensitive [8, 9]. Moreover, if a criterion for hole failure is set to a critical plastic strain $g^p = 0.044$, then the same model can predict failure in both dry and water saturated rock. It has been shown that the critical plastic strain criterion corresponds to a bifurcation type condition for hole failure [10].

3 Conclusions

Flow through drying is a strength enhancement mechanism in gas wells. Drying occurs due to gas decompression near well. A saturation dependent elastoplastic model was developed to study the mechanical effect of drying in gas wells. The flow through drying effect was verified with experimental findings. The results show the significance of drying in wellbore strengthening and sand production avoidance in gas wells.

Acknowledgements This work has been supported by the Sand production JIP at SINTEF Petroleum Research sponsored by DEA and ConocoPhillips.

References

1. Mahadevan, J., Sharma, M.M., Yortsos, Y.C.: Flow through drying of porous media. Am. Inst. Chem. Eng. J. **52**(7), 2367–2380 (2006)
2. Mahadevan, J., Sharma, M.M., Yortsos, Y.C.: Water removal from porous media by gas injection: experiments and simulation. Transp. Porous Media **66**, 287–309 (2007a)
3. Mahadevan, J., Sharma, M.M., Yortsos, Y.C.: Capillary wicking in gas wells. SPE 103229. Soc. Pet. Eng. J. **12**(4), 429–437 (2007b)
4. Lake, L.W.: Enhanced Oil Recovery. Prentice Hall, Englewood Cliffs, New Jersey (1989)
5. Antoine, C.: Tensions des vapeurs; nouvelle relation entre les tensions et les températures Comptes Rendus des Séances de l'Académie des Sciences 107, 681–684, 778–780, 836–837 (1888)
6. Yiotis, A.G., Boudouvis, A.G., Stubos, A.K., Tsimpanogiannis, I.N., Yortsos, Y.C.: The effect of liquid films on the drying of porous media. Am. Inst. Chem. Eng. J. **50**(11), 2721–2737 (2004)
7. Papamichos, E., Brignoli, M., Santarelli, F.J.: An experimental and theoretical study of a partially-saturated collapsible rock. Mech. Cohes.-Frict. Mater. **2**(3), 251–278 (1997)
8. Papamichos, E., Tronvoll, J., Vardoulakis, I., Labuz, J.F., Skjærstein, A., Unander, T.E., Sulem, J.: Constitutive testing of Red Wildmoor sandstone. Mech. Cohes.-Frict. Mater. **5**(1), 1–40 (2000)
9. Sulem, J., Vardoulakis, I., Papamichos, E., Oulahna, A., Tronvoll, J.: Elasto-plastic modelling of Red Wildmoor sandstone. Mech. Cohes.-Frict. Mater. **4**(3), 215–245 (1999)
10. Papamichos, E.: Borehole failure analysis in a sandstone under anisotropic stresses. Int. J. Num. Anal. Meth. Geomech. **34**, 581–603 (2010). doi:10.1002/nag.824

Wormholes: Chemically Damaged States in Carbonate Rocks During CO_2-Acidized Fluid Flow

A.P.S. Selvadurai and C.-B. Couture

Abstract Carbonate rocks and carbonate zones in sandstone formations can experience chemical reaction with CO_2 acidized water; this can degrade the fabric of the carbonate rock leading to the formation of damaged regions that are referred to as wormholes. The experimental research shows that once a wormhole is formed in a carbonate zone, the retention capabilities of the storage medium are lost, leading to unrestricted flow of the injected fluids in the defective zones. Collateral effects of such erosion processes is the accumulation of dissolved solids in remote locations that can promote permeability reduction causing potential zones for the initiation of hydraulic fracture and void-compaction-induced distress to the caprock barriers.

1 Introduction

The geologic sequestration of fluidized greenhouse gases is identified as a means for to mitigating the effects of climate change [1, 3, 6, 10, 11, 15–22]. Geologic sequestration relies on several trapping mechanisms to provide storage security and longevity of the activity; these include primary trapping mechanisms commonly identified include adsorption, structural and stratigraphic trapping and hydrodynamic trapping, which are activated at the time of sequestration for a period of a few hundred years or more while secondary trapping mechanisms, including dissolution and mineralization, require in excess of a thousand years. The latter mechanisms offer a degree of permanence to the sequestration process but the primary

A.P.S. Selvadurai (✉) · C.-B. Couture
Environmental Geomechanics Laboratory, Department of Civil Engineering and Applied Mechanics, McGill University, H3A 0C3, Montréal, QC, Canada
e-mail: patrick.selvadurai@mcgill.ca

C.-B. Couture
e-mail: cyrille.couture@mail.mcgill.ca

E. Papamichos et al. (eds.), *Bifurcation and Degradation of Geomaterials with Engineering Applications*, Springer Series in Geomechanics and Geoengineering, DOI 10.1007/978-3-319-56397-8_39

mechanisms are essential for the long term trapping actions to take place. Hydrodynamic trapping relies on a geological setting where the injected CO_2 can form a stable plume that displaces the resident groundwater. Many current storage settings rely on the chemical inertness of the storage formations provided by rocks that are predominantly of sandstone compositions. Storage rocks with low calcium carbonate content have also been selected as host storage rocks.

The basic objective of this research is to investigate whether the interactions of the injected fluidized CO_2 can be detrimental to the fabric of carbonate rocks that can be encountered as seams or lenses even in sandstone formations. The large scale erosion of surficial deposits and karst formations created by internal chemical and mechanical erosion of carbonate rocks is well documented [5] as are laboratory studies [7, 23]; Bauer et al., [2, 8, 9, 12–14, 24, 25] that investigate the development of wormhole features during acidized CO_2 migration in carbonate rocks.

2 Experimental Procedures

To investigate defect development in carbonate rocks during acidized CO_2 flow experiments conducted experiments using Indiana Limestone ad the candidate storage rock. In order to simulate the geostatic stress states that can occur in a typical storage setting we used a triaxial testing facility capable of accommodating cylindrical specimens 50 mm diameter and 100 mm length, and applied radial and axial stresses to specimens jacketed by an impervious CO_2-resistant neoprene membrane, which isolates the Indiana Limestone sample from the pressurizing fluid in the triaxial cell. Steady state water flow experiments were first performed to estimate the reference permeability of the Indiana Limestone samples. The permeabilities varied between 3.52×10^{-15} m^2 and 10.21×10^{-15} m^2. The CO_2-acidized water supply was prepared by mixing CO_2 and water in an autoclave maintained at a pressure of 10 MPa (approximately 98.7 atmospheres) and the pH of the CO_2-acidized water was measured using a colour graded litmus paper contained in an inspection chamber. In the six tests performed, the temperatures varied between 24.9 and 26.2 °C.

A cell pressure of 8 MPa was maintained to ensure that flow takes place through the sample thus eliminating leakage through the Indiana Limestone sample-neoprene membrane interface. The cell pressure and the axial stress applied to the test specimen are provided by a high pressure nitrogen source. Steady flow through the rock sample is developed using a precision pump, capable of maintaining flow rates as low as 0.0001 ml/min for periods up to 240 h. The inflow pressure is measured using an in-line transducer and the outflow pressure in the flow process is controlled using a back-pressure regulator. A schematic view of the

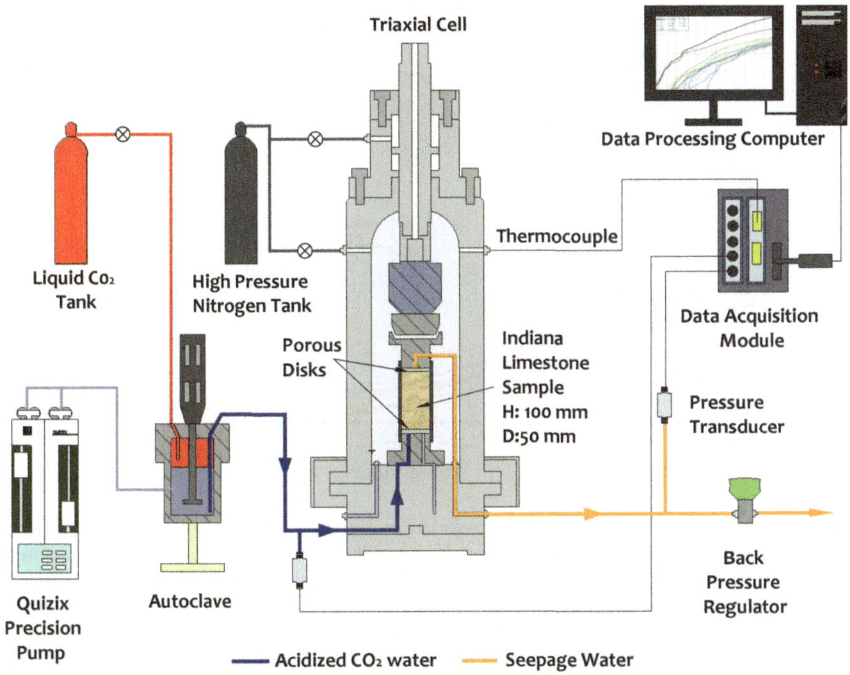

Fig. 1 Experimental facilities for initiating CO_2 acidized flow through samples of Indiana Limestone

experimental arrangements is shown in Fig. 1. Steady flow of CO_2-acidized water at a pH estimated to be between 3 and 4 is maintained through the Indiana Limestone sample.

As the fluid migrates through the sample the upstream pressure needed to maintain the flow rate decreases and the time history of the upstream pressure obtained for six separate experiments is shown in Fig. 2. The reacted Indiana Limestone samples showed visible signs of defect development at the entry surface and nominal evidence of defects at the exit surface. To ascertain detailed configuration of the defects created within the Indiana Limestone, CT scans were performed on all the samples. Since sandstones constitute the predominant geologic material associated with storage formations, we also conducted acidized CO_2 flow experiments on Rudna Sandstone. Figure 2 also compares the time history of the pressure required to maintain acidized CO_2 flow through the six Indiana Limestone and a Rudna Sandstone sample. and Rudna Sandstone samples.

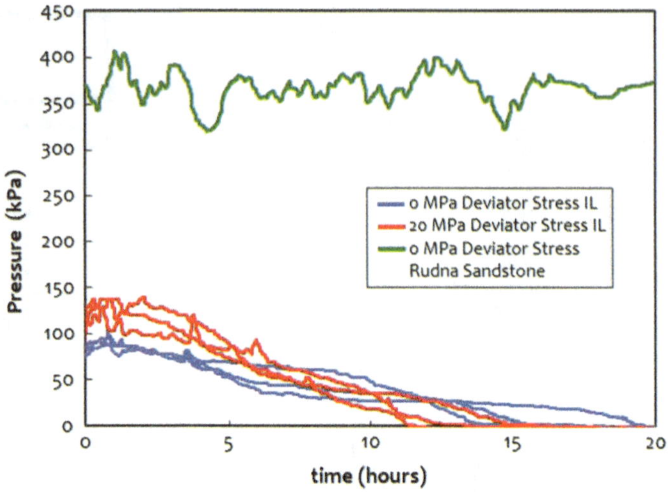

Fig. 2 Time history of upstream pressure variation during steady flow of CO_2-acidized water through Indiana Limestone and Rudna Sandstone samples (*Indiana Limestone used contains nearly* 98.31% *calcium carbonate with traces of Silica, Titanium oxide, Alumina and oxides of Iron, Magnesium, Potassium and Calcium. The porosity from mercury intrusion porosimetry is approx.* 14.4% *and from vacuum saturation with water approx.* 16%. *A chemical analysis of the sandstone indicted that it consisted* of 62.3% *quartz*, 14.4% *dolomite*, 14.9% *microcline and other minerals and with a porosity* (measured *using vacuum saturation*) *of* 2.3%. *The deviator stress is defined as the difference in the maximum principal stress* (σ_1) *and the minimum principal stress* (σ_3) *applied to the test specimen. In the triaxial testing configuration shown in Fig.* 1, *the ambient pressure within the pressurized chamber is* σ_3 *and the axial stress applied to the sample through piston is the deviator stress* ($\sigma_1 - \sigma_3$). *The deviator stress is a measure of the geostatic stress difference that can be present within a geologic formation*)

3 Concluding Remarks

Defect generation or development of wormhole-type features in calcium carbonate-rich rocs can occurs very rapidly. During a 20 h period, all six samples of Indiana Limestone, irrespective of their initial permeability, developed defects creating pathways approximately 3 orders of magnitude more permeable than the virgin Indiana Limestone. The effective permeability of the wormhole can be estimated by considering a an elementary Poiseuille flow model that connects the shortest pathway through the wormhole, with spatially averaged cross sectional area. A simpler approach is to conduct a falling head-type permeability experiment where water is allowed to percolate through the Indiana Limestone sample. The effective permeability values for the wormholes range from 6.93E-13 to 1.71E-12, which are around three orders of magnitude larger than the permeability of the virgin rock. In these experiments, the dissolved solids tend to accumulate at the back pressure regulator incorporated to control flow. In an actual injection setting, it is possible that dissolved solids emanating from a carbonate-rich zone in a storage formation can result in pore clogging of a predominantly sandstone formation, thus reducing its permeability. This research also investigates the influence of stresses applied to the skeleton of the porous medium on the generation of worm holes. For the deviatoric stress states used in the experiments, there is no appreciable influence of the deviator stress state on the mode of generation of the wormhole features. Researchers have attempted to model the process of wormhole formation through carbonate acidization [4]. These studies indicate the influence of the flow velocity on the development of dissolution fronts. Processes such as *compact dissolution*, *conical wormholes*, *dominant wormholes* and *ramified wormholes* can be influenced by the flow velocities. From the results of CT scans it would appear that the observations of this research confirm the existence of the conical variety of wormholes. Here we note that once a wormhole is nucleated, the acidized dissolution promotes or localizes its growth at the expense of creation of companion wormholes.

Acknowledgements The work described in the paper was supported by the following research grants awarded to the first author: A Discovery Grant awarded by the Natural Sciences and Engineering Research Council of Canada and a Research Grant awarded by Carbon Management Canada.

References

1. Bachu, S., Bonijoly, D., Bradshaw, J., Burruss, R., Holloway, S., Christensen, N.P., Mathiassen, O.M.: CO_2 storage capacity estimation: methodology and gaps. Int. J. Greenhouse Gas Control **1**, 430–443 (2007)
2. Bauer, A., Walle, L., Stenebråten, J., Papamichos, E.: Impact of acidizing-induced wormholes in chalk on rock strength. In: 47th US Rock Mechanics/Geomechanics Symposium, pp. 2959–2964. American Rock Mechanics Association, San Francisco, CA (2013)
3. Birkholzer, J., Tsang, C.F.: Introduction to the special issue on site characterization for geological storage of CO_2. Environ. Geol. **54**, 1579–1581 (2008)
4. Cohen, C.E., Ding, D., Quintard, M., Bazin, B.: From pore scale to wellbore scale: impact of geometry on wormhole growth in carbonate acidization. Chem. Eng. Sci. **63**, 3088–3099 (2008)
5. Ford, D., Williams, P.D.: Karst Hydrogeology and Geomorphology. Wiley, Hoboken, NJ (2013)
6. Girard, J.P., Chiquet, P., Thibeau, S., Lescanne, M., Prinet, C.: Geochemical assessment of the injection of CO_2 into Rousse depleted gas reservoir. Part I: initial mineralogical and geochemical conditions in the Mano Reservoir. Energy Procedia **37**, 6395–6401 (2013)
7. Grigg, R.B., McPherson, B.J., Svec, R.K.: Laboratory and model tests at reservoir conditions for CO_2-brine-carbonate rock systems interactions, In: The Second Annual Conference on Carbon Capture and Sequestration, pp. 1–17. Washington, DC (2003)
8. Haq, F., Haderlein, S.B., Cirpka, O.A., Nowak, M., Blum, P., Grathwohl, P.: Flow-through experiments on water-rock interactions in a sandstone caused by CO_2 injection at pressures and temperatures mimicking reservoir conditions. Appl. Geochem. **58**, 136–146 (2015)
9. Izgec, O., Demiral, B., Bertin, H., Akin, S.: CO_2 injection into saline carbonate aquifer formations I: laboratory investigation. Transp. Porous Med. **72**, 1–24 (2008)
10. Kim, J., Selvadurai, A.P.S.: Ground heave due to line injection sources. Geomech. Energ. Environ. **2**, 1–14 (2015)

11. Kramer, D.: Scientists poke holes in carbon dioxide sequestration. Phys. Today **65**, 22–24 (2012)
12. LeGuen, Y., Renard, F., Hellmann, R., Brosse, E., Collombet, M., Tisserand, D., Gratier, J.P.: Enhanced deformation of limestone and sandstone in the presence of high PCO_2 fluids. J. Geophys. Res. **112**, B05421 (2007)
13. Luquot, L., Gouze, P.: Experimental determination of porosity and permeability changes induced by injection of CO_2 into carbonate rocks. Chem. Geol. **265**, 148–159 (2009)
14. Luquot, L., Rodriguez, O., Gouze, P.: Experimental characterization of porosity structure and transport property changes in limestone undergoing different dissolution regimes. Transp. Porous Med. **101**, 507–532 (2014)
15. McPherson, B.J., Sundqvist, E.T.: Carbon Sequestration and Its Role in the Global Carbon Cycle. American Geophysical Union, Washington, DC (2013)
16. Metz, B., Davidson, O., De Coninck, H., Loos, M., Meyer, L.: Special Report on Carbon Dioxide Capture and Storage. Cambridge University Press, Cambridge (2005)
17. Pijaudier-Cabot, G., Pereira, J.M.: Geomechanics in CO_2 Storage Facilities. Wiley-ISTE, Hoboken, NJ (2013)
18. Rutqvist, J., Birkholzer, J.T., Tsang, C.F.: Coupled reservoir-geomechanical analysis of the potential for tensile and shear failure associated with CO_2 injection in multilayered reservoir-caprock systems. Int. J. Rock Mech. Min. Sci. **45**, 132–143 (2008)
19. Selvadurai, A.P.S.: Heave of a surficial rock layer due to pressures generated by injected fluids. Geophys. Res. Lett. **36**, L14302 (2009)
20. Selvadurai, A.P.S.: Fluid leakage through fractures in an impervious caprock embedded between two geologic aquifers. Adv. Water Res. **41**, 76–83 (2012)
21. Selvadurai, A.P.S.: Caprock breach: a potential threat to secure geologic sequestration of CO_2. In: Pijaudier-Cabot, G., Pereira, J.M. (eds.) Geomechanics in CO_2 Storage Facilities, pp. 75–94. Wiley-ISTE, Hoboken, NJ (2013)
22. Selvadurai, A.P.S., Kim, J.: Poromechanical behaviour of a surficial geological barrier during fluid injection into an underlying poroelastic storage formation. Proc. R. Soc., A. Math. Phys. Sci. (2016). doi:10.1098/rspa.2015.0418
23. Seo, J.G., Mamora, D.D.: Experimental and simulation studies of sequestration of supercritical carbon dioxide in depleted gas reservoirs. J. Energ. Res. Tech. **127**, 1–6 (2005)
24. Smith, M.M., Hao, Y., Mason, H.E., Carroll, S.A.: Experiments and modeling of variably permeable carbon reservoir samples in contact with CO_2-acidified brines. Energ. Proc. **63**, 3126–3137 (2014)
25. Walle, L., Papamichos, E.: Acidizing of hollow cylinder chalk specimens and its impact on rock strength and wormhole network structure. In: 49th US Rock Mech./Geomech. Symposium, vol. 4, pp. 2500–2503 American Rock Mechanics Association, San Francisco, CA (2015)

An Inversion Framework for Numerical Modelling of Pore Collapse in Soft Porous Rocks

Jack Lin, Mustafa Sari, Thomas Poulet and Manolis Veveakis

Abstract In this work we present an inversion framework to identify material properties defining the plastic behaviour of pore collapse modeled by thermo-hydro-mechanical simulations. This framework is built on the finite element REDBACK numerical simulator, which is capable of solving a multi-physics problem in a tightly coupled, massively parallel manner. We demonstrate the approach by matching the stress-strain response of an Adamswiller sandstone in a drained triaxial experiments.

1 Introduction

Pore collapse has a significant influence in compaction and subsidence of petroleum reservoirs, which in turn is vital for the lifetime and performance of the energy producing formations because of the severe effect on the formation's porosity [2, 7]. Understanding and simulating the mechanism of pore collapse is therefore essential to model the performance of a reservoir.

In this contribution we use an energy based approach to model pore collapse with a pressure-dependant elasto-viscoplastic model [4]. This model uses two material parameters to determine the plastic behaviour of a rock related to the pore collapse mechanism, and those values can be extremely difficult to calibrate since all the parameters are dependent and affect the response of the system in a highly non-linear manner.

We present a computationally assisted inversion workflow to handle this complex and time-consuming task. We demonstrate the approach by simulating a triaxial test on an Adamswiller sandstone and identifying the input parameters to match the stress-strain curve reported by [6].

J. Lin · M. Sari (✉) · M. Veveakis
School of Petroleum Engineering, UNSW, Sydney, Australia
e-mail: m.sari@unsw.edu.au

T. Poulet
CSIRO, Canberra, Australia

E. Papamichos et al. (eds.), *Bifurcation and Degradation of Geomaterials with Engineering Applications*, Springer Series in Geomechanics and Geoengineering, DOI 10.1007/978-3-319-56397-8_40

2 Theoretical Model Description

The underlying physical model for this paper is based on the principles of over-stress plasticity [3], used in a novel elasto-viscoplastic approach [4]. Following the classical considerations of mechanics, the total strain rate is decomposed in an elastic (reversible) part and a plastic (irreversible) part. The reversible component is assumed to obey a linear thermo-elastic relationship. The irreversible element of the strain rate obeys an associative visco-plastic flow law of the form $\dot{\varepsilon}^i_{ij} = \dot{\lambda}\frac{\partial f}{\partial \sigma'_{ij}}$, where f is the yield function and $\dot{\lambda}$ is a (scalar) plastic multiplier (see Fig. 1).

The plastic multiplier follows the relation $\dot{\lambda} = \sqrt{\dot{\varepsilon}^{i\,2}_d + \dot{\varepsilon}^{i\,2}_v}$. In this expression, $\dot{\varepsilon}^i_d$ and $\dot{\varepsilon}^i_v$ are the deviatoric and volumetric parts of the strain rate tensor (see Fig. 1), respectively, following the incremental relations

$$\dot{\varepsilon}^i_d = \dot{\varepsilon}_0 \left\langle \frac{q - q_Y}{\sigma_{ref}} \right\rangle^m \exp\left(-\frac{Q^d_{mech}}{RT}\right), \tag{1a}$$

$$\dot{\varepsilon}^i_v = \dot{\varepsilon}_0 \left\langle \frac{p' - p_Y}{\sigma_{ref}} \right\rangle^m \exp\left(-\frac{Q^V_{mech}}{RT}\right), \tag{1b}$$

where $\dot{\varepsilon}_0$ is a reference strain rate, q is the equivalent deviatoric (or von Mises) stress, p' is the volumetric mean effective stress, q_Y and p_Y are the respective effective stresses at yield, σ_{ref} is a reference stress, R is the universal gas constant, T is the temperature field and $\langle\cdot\rangle$ denote the Macaulay brackets.

These expressions imply that the material is admitting thermal sensitivity expressed through the activation enthalpies for the deviatoric (Q^d_{mech}) and the volumetric (Q^v_{mech}) components. This activation enthalpy incorporates the activation energies of all the micromechanical mechanisms, like frictional initiation [5] or volumetric pore collapse. It is in principle expressed in the generalised form $Q = E + PV_{act}$ where P is a measure of pressure responsible for driving the internal mech-

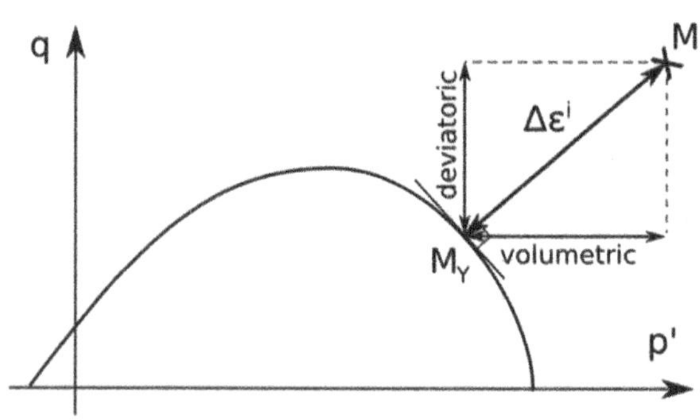

Fig. 1 Schematic for the decomposition of the plastic flow rule in volumetric and Deviatoric components. A point M in the p – q space (mean effective stress shear stress) is shown with its corresponding point MY on the yield envelope

anism, while E and V_{act} are the activation energy and volume of the internal mechanism considered (here pore collapse) and are yet to be determined.

Following the recent work of Poulet and Veveakis [4], and assuming isotropic response of the activation enthalpies ($Q^d_{mech} = Q^v_{mech} = Q_{mech}$), we introduce the excess pore pressure as the driving measure of pressure. This yields the following expression:

$$Q_{mech} = E + \Delta p_f \, V_{act} \tag{2}$$

In the following sections, we match the experimental data provided by [6] on Adamswiller sandstone by adjusting the values of E and V_{act}. To this end, we introduce a swarm particle based inversion workflow based on the finite element numerical simulator REDBACK [4]. For further details on REDBACK (Rock mEchancis with Dissipative feedBACKs), the reader is referred to [4].

3 Pathfinder: An Inversion Workflow for REDBACK

We have developed an optimisation framework for REDBACK simulations, allowing us to find parameter values by matching experimental results. In this section we demonstrate this framework by fitting the values of E and V_{act} of Eq. 2 to match on stress-strain curve obtained from the experimental work of [6], in triaxial testing on Adamswiller sandstone.

This optimization workflow is implemented in a new program called Pathfinder. The software is based on a mathematical optimization module, coupled to another module to generate corresponding REDBACK simulation input files, run the simulations by existing HPC platforms (Leonardi @UNSW) to run large numbers of REDBACK simulations in parallel and pass the results to the controlling engine (see Fig. 2). A cost function is computed from each simulation result, expressing the fit between the numerical and experimental/target results. The controlling engine generates sets of input values to optimize this cost function. The workflow is initialized with arbitrary parameter values for E and V_{act}.

We use a Particle Swarm Optimization (PSO) algorithm as it provides a good trade-off between ease of implementation, performance, and minimization of the number of function evaluations. PSO solves for N unknowns by simulating particles moving through an N-dimensional search space [1]. At each iteration, each particle evaluates its individual cost function and the overall algorithm monitors the global best. The positions of all particles are then adjusted by accounting for their current position, velocity, and the global best.

The core of the algorithm is the custom cost function we need to define, in order to assess the quality of a match between a set of simulation results and some reference strain-stress experimental values. Let us define the simulation stress-strain values as two arrays of the same dimension m: $[\sigma^1_{sim}, ..., \sigma^m_{sim}]$ for the stress values and $[\epsilon^1_{sim}, ..., \epsilon^n_{sim}]$ for the strains. The reference (experimental) values are similarly

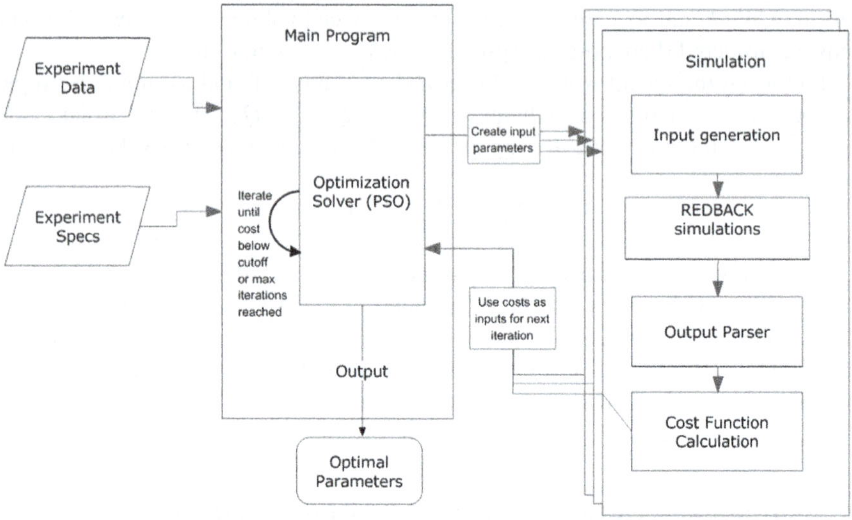

Fig. 2 Overall workflow of Pathfinder

Fig. 3 *Upper* and *lower* cut-offs for cost-function calculation

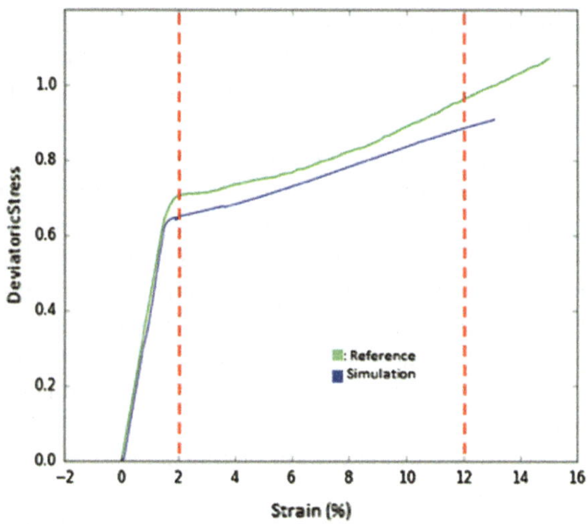

denoted by $[\sigma_{ref}^{1}, \ldots, \sigma_{ref}^{m}]$ and $[\varepsilon_{ref}^{1}, \ldots, \varepsilon_{ref}^{n}]$, but for a number n of values. The strain values for the simulation and reference datasets can potentially be sampled differently (m $\neq$ n) but the range of strain values overlap. To compare the two discrete functions $\sigma_{sim}(\varepsilon_{sim})$ and $\sigma_{ref}(\varepsilon_{ref})$ we use a cubic spline algorithm to resample both datasets to a standard grid resolution of ($\Delta\varepsilon = 0.001$) with N points. Furthermore, the data is truncated to be restricted to the overlapping plastic region of interest, where the starting point (start of plasticity) is inferred manually (see Fig. 3).

The computation of the cost function is performed by combining three different components. Firstly, the Euclidean norm of the difference between the simulated and target responses is calculated as

$$D_{euc} = \| \sigma_{sim} - \sigma_{ref} \|_2 = \sqrt{\sum (\sigma^k_{sim} - \sigma^k_{ref})^2} \quad (3)$$

Secondly, the angular difference D_{ang}, in gradient, between the slopes β_{ref} and β_{sim} of straight-line fits for both curves is calculated using a linear least-squared method. Thirdly, the difference of curvatures D_{crv} between the two datasets is calculated, where the degree of curvature is taken as the coefficient (λ) of a second-order polynomial, $y = \alpha + \beta x + \lambda x^2$. The overall cost is then calculated by combining those three components

$$Cost = (D_{euc} + D^{min}_{euc}) \times (D_{ang} + D^{min}_{ang}) \times (D_{crv} + D^{min}_{crv}) \quad (4)$$

where some numerical coefficients ($D^{min}_{euc} = 0.02$, $D^{min}_{ang} = 0.002$, $D^{min}_{crv} = 0.01$) are used to ensure minimal (positive) values when any of the three sub-values approach zero.

Figure 4 shows a plot of the cost function in the E, V_{act} (normalised as α_2, α_1 respectively in the algorithm) space around the best solution for Adamswiller Sandstone undergoing triaxial compression at fixed 100 MPa pore pressure [6]. Results show a localised zone of better results in that space, with a global minimum of the cost function for the values ($\alpha_1 = 3.0$, $\alpha_2 = 50.5$).

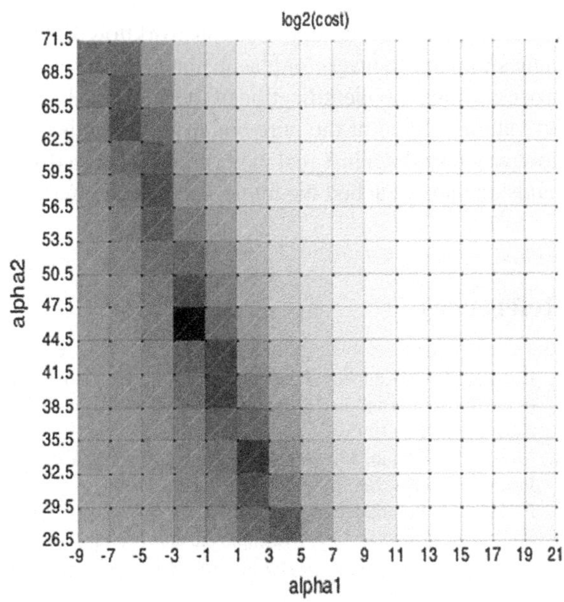

Fig. 4 Grid search of cost function for Adamswiller Sandstone triaxial, 100 MPa [6]. Dark colours represent better fits. Note that $\alpha_1 = V_{act}$ and $\alpha_2 = E$

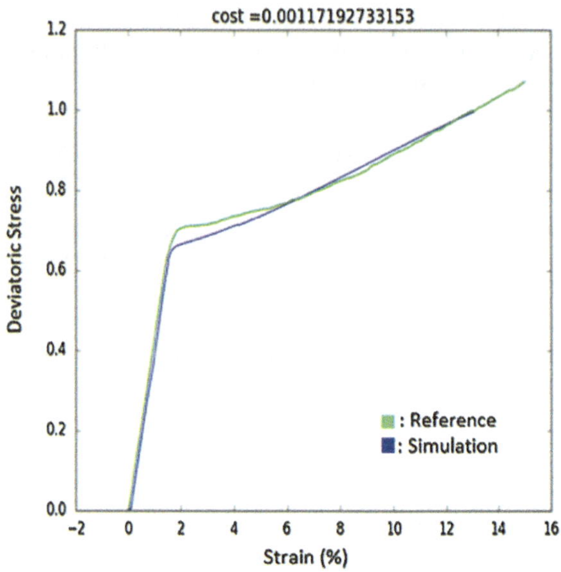

Fig. 5 Stress-strain curves for Adamswiller Sandstone undergoing triaxial compression at fixed 100 MPa pore pressure [6]

Figure 5 shows the best optimised result, comparing the reference stress-strain data against a Pathfinder-optimized simulation.

4 Conclusion

We have presented an optimisation workflow based on the REDBACK finite element simulator to invert for plastic material parameters from experimental data. This process allows the identification of the numerical values for the activation energy and activation volume of the pore collapse mechanism for a given material. The workflow was tested against real experiments reported in the literature for Adamswiller sandstone and matched the numerical values obtained for mudstones by [4].

References

1. Eberhart, R.C., Kennedy, J., et al.: A new optimizer using particle swarm theory. In: Proceedings of the Sixth International Symposium on Micro Machine and Human Science, vol. 1, pp. 39–43. New York, NY (1995)
2. Mowar, S., Zaman, M., Stearns, D.W., Roegiers, J.-C.: Micro-mechanisms of pore collapse in limestone. J. Pet. Sci. Eng. **15**(2), 221–235 (1996)
3. Perzyna, P.: Fundamental problems in viscoplasticity. Adv. Appl. Mech. **9**, 243–377 (1966)
4. Poulet, T., Veveakis, M.: A viscoplastic approach for pore collapse in saturated soft rocks using redback: an open-source parallel simulator for rock mechanics with dissipative feedbacks. Comput. Geotech. **74**, 211–221 (2016)

5. Rice, J.R., Lapusta, N., Ranjith, K.: Rate and state dependent friction and the stability of sliding between elastically deformable solids. J. Mech. Phys. Solids **49**(9), 1865–1898 (2001). The JW Hutchinson and JR Rice 60th Anniversary Issue
6. Wong, T.-F., David, C., Zhu, W.: The transition from brittle faulting to cataclastic flow in porous sandstones: mechanical deformation. J. Geophys. Res.: Solid Earth **102**(B2), 3009–3025 (1997)
7. Zhu, W., Baud, P., Wong, T.-F.: Micromechanics of cataclastic pore collapse in limestone. J. Geophys. Res.: Solid Earth, **115**(B4) (2010)

Thermo-Hydro-Mechanics in Shear Fracturing in Geothermal Reservoirs

Manman Hu, Manolis Veveakis, Thomas Poulet and Klaus Regenauer-Lieb

Abstract This paper presents a novel method to investigate shear stimulation at an injection well in Enhanced Geothermal Systems (EGS). Nowadays, the technique of EGS has been extensively used for extracting thermal energy from the earth. As the intrinsic permeability of the rock is usually too low to allow an economic flow, stimulation for fractures is incorporated. The connectivity of fracture networks around boreholes dominates the system behaviour. In theory, stimulations including both tensile (mode I) and shear (mode II) fracturing are desired, so that sufficient surface area for heat exchange is produced. However, shear stimulation is considered a safer choice than tensile fracturing in terms of possibility of inducing local earthquakes. This study investigates shear fractures only, from a slip-line field point of view. The rock is modelled as elasto-viscoplastic material with damage mechanics coupled. A numerical simulator REDBACK, based on the MOOSE framework, is employed to solve this coupled multi-physics involved problem. With injection pressure imposed on the interior of a borehole, slip lines grows in the form of logarithmic spirals, indicating the potential trace of shear fractures. Imperfections are imposed on the boundary as seeding for the spirals. Cases with and without thermo-mechanical coupling are compared, indicating the essential role of shear-heating feedback in enhancing shear fractures. Bifurcation analysis for various Arrhenius numbers is performed, demonstrating a clear exponential relationship between critical injection pressure and the local temperature of host rock.

M. Hu (✉) · M. Veveakis · K. Regenauer-Lieb
University of New South Wales, Sydney, NSW 2052, Australia
e-mail: manman.hu@unsw.edu.au

T. Poulet
CSIRO Mineral Resources, North Ryde, NSW 2113, Australia

E. Papamichos et al. (eds.), *Bifurcation and Degradation of Geomaterials with Engineering Applications*, Springer Series in Geomechanics and Geoengineering, DOI 10.1007/978-3-319-56397-8_41

1 Introduction

Recently, the technique of Enhanced Geothermal System (EGS) has been employed to unlock thermal energy extraction from low permeability reservoirs. The key is to increase hydraulic connectivity between a pair of injection and production wells (to the depth of 1000–3000 m), in order to allow an economic flow rate. Shear stimulation plays an important role in generating inter-connected fracture networks around boreholes. Examples can be found in literatures, including an undergoing Desert-Peak geothermal field tests in Nevada [1, 2], where formation permeability was enhanced through shear fracturing, Regular Mud Acid (RMA) assisted stimulation in a Soultz-sous-Forêts EGS reservoir in France [5], just to mention a few.

Slip-line field is recognized as an effective mathematical technique for solving plane strain boundary value problem for ideal plasticity. The slip lines are characteristics of the hyperbolic partial differential equations that describe deformation of an ideal rigid-plastic body. For plane strain von Mises plasticity the slip lines are orthogonal lines of sinistral and dextral shear [3]. Here we are interested in the plastic deformation of material around a borehole due to application of an internal pressure in the well such as by injection. Shear localization occurs along slip lines in the form of logarithmic spirals, indicating the potential trace of shear fractures.

In this study, we extend the concept of slip lines into an energy domain. A numerical simulator REDback [6] based on the MOOSE framework is used to solve the coupled equations and to visualise the growth of logarithmic spirals. Once the thermal effect is considered, positive shear heating feedback enhances the weakening/softening of material [4, 7].

2 Problem Formulation

Figure 1a shows a schematic for working process of a pair of geothermal wells. We are interested in how the shear fractures initiate and propagate from the vicinity of injection borehole, as well as the effect of thermo-mechanical coupling. Hence, a deep rock layer around the injection point is chosen for investigation, with the assumption of plane strain condition ($\varepsilon_{zz} = 0$). For computational efficiency, a quarter of the region is chosen as a representative. Then, the boundary constraints for this area can be summarized as follows. Evenly distributed overburden pressure acts on the exterior boundary, injection fluid pressurizing on the interior boundary, and zero circumferential displacement is imposed at the two radial boundaries due to symmetry (as indicated in Fig. 1b).

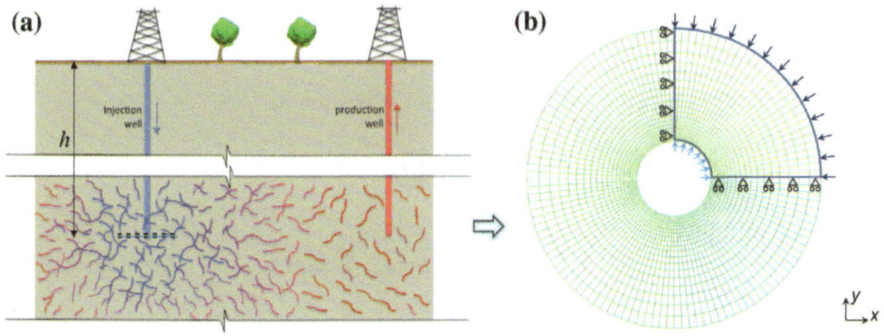

Fig. 1 **a** Schematic for shear stimulation between a pair of wells; **b** Modelling of an injection borehole in a deep rock layer, in plane strain condition

3 Methodology

3.1 Visco-Plasticity Formulation

In the numerical implementation of REDBACK, the irreversible part of the strain rate obeys an associative visco-plastic flow law of the form

$$\dot{\varepsilon}_{ij}^{i} = \dot{\lambda} \frac{\partial f}{\partial \sigma'_{ij}}, \tag{1}$$

where f is the yield function and $\dot{\lambda}$ is a (scalar) plastic multiplier representing the magnitude of irreversible strain rate:

$$\dot{\lambda} = \sqrt{\dot{\varepsilon}_d^{i\,2} + \dot{\varepsilon}_v^{i\,2}}. \tag{2}$$

where $\dot{\varepsilon}_d^{i\,2}$ and $\dot{\varepsilon}_v^{i\,2}$ are the deviatoric and volumetric components of the strain rate tensor respectively, following the incremental power-law relationship:

$$\dot{\varepsilon}_d^{i} = \dot{\varepsilon}_0 \left\langle \frac{q - q_Y}{\sigma_{ref}} \right\rangle^m \exp\left(-\frac{Q_{mech}^d}{RT}\right), \tag{3a}$$

$$\dot{\varepsilon}_v^{i} = \dot{\varepsilon}_0 \left\langle \frac{p' - p_Y}{\sigma_{ref}} \right\rangle^m \exp\left(-\frac{Q_{mech}^V}{RT}\right), \tag{3b}$$

where $\dot{\varepsilon}_0$ is a reference strain rate, q is the equivalent deviatoric (or von Mises) stress, p' is the volumetric mean effective stress, q_Y and p_Y are the respective effective stresses at yield, Notably, this visco-plasticity model can also be derived from the creeping flow law that describes rate-and state frictional rheology by Taylor expansion at the limit of yield stress [8].

3.2 Governing Equations

Neglecting body force, the stress equilibrium is written as

$$\partial_j \sigma'_{ij} - \partial_i \Delta p_f = 0 \tag{4}$$

Energy balance gives

$$\partial_t T - \partial^2_{ii} T - Gr\, \sigma_{ij} \dot{\varepsilon}^{pl}_{ij} = 0 \tag{5}$$

where the plastic strain rate $\dot{\varepsilon}^{pl}_{ij}$ can be obtained from Eq. 1 to Eq. 3. The damage evolution law is described as

$$\dot{D} = \beta_{damage} (\frac{q}{1-D})^n \tag{6}$$

and accordingly, $\sigma_Y = \sigma_{Y0}\,(1-D)$, $E = E_0\,(1-D)$, where D denotes the damage variable, β_{damage} denoting a material constant, q denoting the deviatoric stress. The exponent n can be related to the dislocation power law [4]. Note that damage evolution is null in the absence of initial damage.

Considering the convenience for numerical implementation, all the variables involved in the multi-physics based processes are normalized:

$$t^\star = \frac{c_{th}}{L^2_{ref}} t,\ x^\star = \frac{x}{L_{ref}},\ T^\star = \frac{T - T_{ref}}{\delta T_{ref}},\ \Delta p^\star = \frac{\Delta p_f}{\sigma_{ref}},\ \sigma^\star_{ij} = \frac{\sigma_{ij}}{\sigma_{ref}}, \tag{7}$$

where c_{th} is thermal diffusivity, and L_{ref}, T_{ref} and σ_{ref} are normalization constants for length, temperature and stress, respectively.

4 Simulation Results

In order to observe virtually the growth of logarithmic spirals, imperfections are imposed on the interior boundary as seeding. 16 evenly spaced small defects are imposed at the borehole wall, as an ongoing investigation (outside the scope of this paper) demonstrates that it corresponds to the natural selection of wavelength. Figure 2 shows the comparison between plastic radial strain (as a representative for irriversible deformation) patterns at $t^\star = 0.125$ without and with considering thermal effects. The same difference can also be observed in the energy dissipation pattern (Fig. 3), as the shear-heating feedback accelerates the shear banding/localization.

Figure 4 shows the evolution of the critical ratio $(\frac{\sigma_{in}}{\sigma_{out}})_{cr}$ against various Arrhenius numbers with damage patterns juxtaposed. Clear shear-banding appears with $\frac{\sigma_{in}}{\sigma_{out}}$ beyond the critical value, while smaller value of $\frac{\sigma_{in}}{\sigma_{out}}$ prevents shear localization.

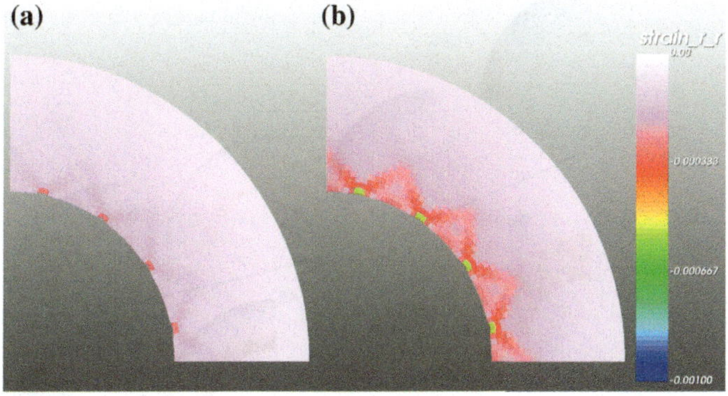

Fig. 2 Shear banding in plastic strain pattern, without and with thermo-mechanical coupling.

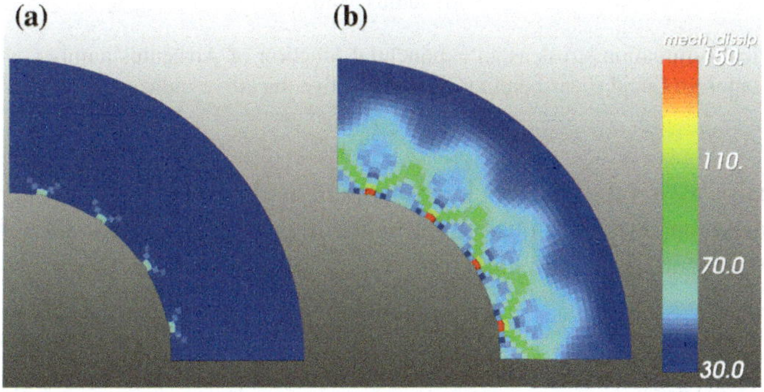

Fig. 3 Shear bands observed in mechanical dissipation pattern, without and with thermo-mechanical coupling.

The concentrated damage implies high porosity in shear bands, indicating effective pathways generated for fluid flow.

By assuming the overburden pressure on the outer boundary constant, this fitting gives an exponential relationship between the critical injection stress and the Arrhenius number ($Ar = Q_{mech}/R\,T_{ref}$):

$$\frac{\sigma_{in}}{\sigma_{out}} \simeq 2.78 * exp(-0.15\ Ar) \tag{8}$$

that is

$$\frac{\sigma_{in}}{\sigma_{out}} \simeq 2.78 * exp(-0.15\ \frac{Q_{mech}}{R\,T_{ref}}) \tag{9}$$

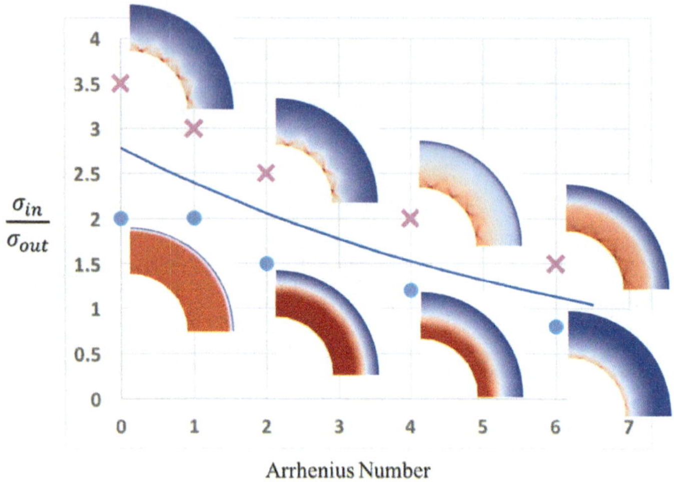

Fig. 4 Critical bifurcation stress as an exponential function of Arrhenius number, with damage patterns juxtaposed. Lucid shear-banding observed above the curve indicating potential flow paths

As usually we set the local rock temperature as the reference temperature, Eq. 9 demonstrates that critical injection pressure required to induce shear fracturing depends exponentially on the local temperature of host rock.

5 Conclusion

A novel method of investigating shear stimulation at an injection well in Enhanced Geothermal Systems (EGS) is presented, from a slip-line field point of view. The main conclusions include: (i) Pure fluid pressure acting on borehole walls can provoke shear failure; (ii) Thermo-mechanical feedback accelerates the propagation of shear bands; (iii) Bifurcation stress is dependent on the Arrhenius number, and thus is affected by host rock temperature.

References

1. Chabora, E., Zemach, E., Spielman, P., Drakos, P., Hickman, S., Lutz, S., Boyle, K., Falconer, A., Robertson-Tait, A., Davatzes, N.C., et al.: Hydraulic stimulation of well 27–15, desert peak geothermal field, Nevada, USA. In: Proceedings of thirty-seventh workshop on geothermal reservoir engineering, Stanford University, Stanford, vol. 30 (2012)
2. Dempsey, D., Kelkar, S., Davatzes, N., Hickman, S., Moos, D.: Numerical modeling of injection, stress and permeability enhancement during shear stimulation at the desert peak enhanced geothermal system. Int. J. Rock Mech. Min. Sci. **78**, 190–206 (2015)

3. Johnson, W., Sowerby, R., Venter, R., Kobayashi, S.: Plane-strain slip-line fields for metal-deformation processes. J. Appl. Mech. **50**, 702 (1983)
4. Karrech, A., Regenauer-Lieb, K., Poulet, T.: Continuum damage mechanics for the lithosphere. J. Geophys. Res.: Solid Earth, **116**(B4) (2011)
5. Portier, S., Vuataz, F.D.: Developing the ability to model acid-rock interactions and mineral dissolution during the rma stimulation test performed at the soultz-sous-forêts egs site, france. Comptes Rendus Geosci. **342**(7), 668–675 (2010)
6. Poulet, T., Paesold, M., Veveakis, M.: Multi-physics modelling of fault mechanics using redback: a parallel open-source simulator for tightly coupled problems. Rock Mech. Rock Eng. 1–17 (2016)
7. Veveakis, E., Alevizos, S., Vardoulakis, I.: Chemical reaction capping of thermal instabilities during shear of frictional faults. J. Mech. Phys. Solids **58**(9), 1175–1194 (2010)
8. Veveakis, E., Regenauer-Lieb, K.: Cnoidal waves in solids. J. Mech. Phys. Solids **78**, 231–248 (2015)

Modelling of the Progressive Failure in a Wellbore Multilateral Junction with Cosserat Model

Panos Papanastasiou

Abstract We employed a higher order continuum based on Mohr-Coulomb Cosserat elastoplasticity in a non-linear finite element analysis capable of modelling the localization of deformation in shear bands that leads to failure near underground openings. The model is used to investigate the stability of wellbore multilateral junctions. The obtained results show a progressive failure mechanism and the computed failure modes are in a good qualitative agreement with laboratory observations. We found that the multilateral junction of a lateral wellbore drilled to the direction parallel to the maximum insitu stress is more stable than in the case of a lateral wellbore drilled to the direction perpendicular to the maximum insitu stress.

Keywords Multilateral junction • Localization • Cosserat • Shear bands • Breakouts

1 Introduction

Drilling horizontal wells through producing reservoir layers can improve significantly the reservoir drainage and hydrocarbon recovery. The horizontal sections are accessed through multiple inclined wells drilled from a relatively small area in different directions, something that allows better exploitation of stationary offshore platforms and land rigs that are under economic and environmental restrictions. Drilling inclined and horizontal well intervals, though, is more difficult and more expensive, due to wellbore instabilities. A particular area of concern is the integrity of the rock near a multilateral (M-L) junction (Fig. 1). The junction is the region where a second wellbore (lateral) takes off from the main wellbore (parent). In the simplest case of M-L, classified as levels 1 and 2, the rock at the junction is not

P. Papanastasiou (✉)
Department of Civil and Environmental Engineering, University of Cyprus, 1678 Nicosia, Cyprus
e-mail: panospap@ucy.ac.cy

E. Papamichos et al. (eds.), *Bifurcation and Degradation of Geomaterials with Engineering Applications*, Springer Series in Geomechanics and Geoengineering, DOI 10.1007/978-3-319-56397-8_42

supported mechanically with cemented casing, so the integrity of the rock around the area of two intersecting holes becomes very important in terms of stability.

In other related work, Aadnoy and Edland [1] considered the stability of two adjacent holes using an analytical solution. This solution is valid for parallel holes embedded in an isotropic stress field. Willson et al. [8], used a 3-D FEM elasto-plastic analysis for assessing the stability of multilateral junctions. Papanastasiou et al. [7] studied the stability of multilateral junctions in a combined experimental and numerical modeling program. The experiments were carried out at Lille University in a true triaxial machine on large size cubical blocks (40 cm) of weak triassic sandstone with two holes intersecting. The experimental results were presented and compared with numerical results obtained with classic finite element analysis developed for engineering assessment of the integrity of rock surrounding a multilateral junction.

Objective of this study is to determine more accurately the failure mechanism at the multilateral junction in relation to the direction that a lateral wellbore branches off. We note that classic elastoplastic stress analyses determines the stress concentration and the yielded zones but is unable to determine the exact failed area. For this reason we use here a higher order continuum with microstructure that can regularize the ill-posed mathematical problem of strain-softening material. This model enables the modelling of progressive localization of deformation in zones of intense shearing that lead to failure of geomaterials and underground structures. The model is based on a Cosserat plasticity model which in addition to the translational degrees of freedom of the classical continuum possesses independent rotation as well. The independent rotation introduces in the model curvatures and coupled stresses. The non-linear finite element analysis is carried out in planes perpendicular to the parent-hole axis (Fig. 1). We do not expect large deviations from a 3-D analysis as long as the build angle of the lateral is small, which is normally the case in the field. Studies on localization of deformation that leads to wellbore breakouts near a single hole were presented earlier by Papanastasiou and Vardoulakis [4], Zervos et al. [9], Papanastasiou and Zervos [6] and Papamichos [3].

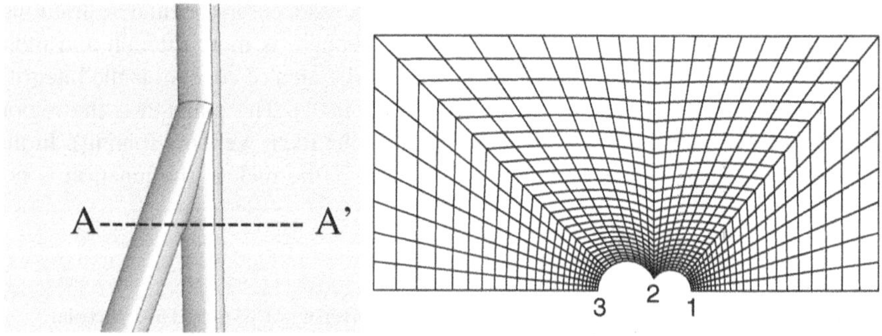

Fig. 1 Wellbore multilateral junction (*left*) and finite element mesh of cross-section A-A' (*right*)

2 Cosserat Flow Theory of Plasticity

The results were obtained using finite element analysis with an advanced Cosserat plasticity model and a robust failure criterion based on localization of deformation in shear bands. Details on the used Cosserat plasticity model and finite element implementation can be found in Papanastasiou and Vardoulakis [4].

As in the classical flow theory of plasticity, plastic strains are generated when the yield condition is satisfied. The Mohr–Coulomb yield criterion can be written as

$$F = \frac{\tau}{p_0 + p} - \mu = 0 \tag{1}$$

where the material parameters are the mobilized friction coefficient μ and the intercept p_0 of the yield surface with the p-axis which is related to the material cohesion, $p_0 = c/\tan\varphi$. In Eq. (1) p and τ are invariant measures of average normal and shear contact tractions over the periphery of a macrocell of the "static" Cosserat model medium given in Muhlhaus and Vardoulakis [2]

$$p = \frac{\sigma_{kk}}{2} \qquad \tau = \sqrt{\left(3s_{ij}s_{ij} - s_{ij}s_{ji}\right)/4 + m_i m_i/R^2} \tag{2}$$

where s_{ij} is the deviatoric stress, $s_{ij} = \sigma_{ij} + p\delta_{ij}$. The internal length R can be identified as an equivalent radius of a typical grain or assembly of grains.

In a friction hardening/softening plasticity model we assumed that the friction coefficient μ is a function of a plastic hardening parameter, $\mu = \mu(\gamma^p)$ and the intercept p_0 of the yield surface with the p-axis is a material constant. Hardening is taking place when μ is monotonously increasing and softening when μ is monotonously decreasing with increasing γ^p. The hardening parameter is defined as the integral over the entire loading history of the increment of the generalized plastic shear strain.

$$d\gamma^p = \sqrt{\left(3d\varepsilon^p_{ij}\varepsilon^p_{ij} + \varepsilon^p_{ij}\varepsilon^p_{ji}\right)/2 + R^2 d\kappa^p_i d\kappa^p_i} \tag{3}$$

where $d\varepsilon^p_{ij}$ is the deviator of the plastic relative deformation. The plastic shear-strain increment, $d\gamma^p$, is defined so that it is energy conjugate to the generalized shear stress intensity τ in Eq. (2). When the couple stresses m_i and curvatures dk^p_i vanish, the definitions of τ and $d\gamma^p$ in Eqs. (2) and (3) coincide with those of the classical plasticity. The actual value of $d\gamma^p$ is determined from Prager's consistency condition, $F = 0$ and $dF = 0$. The plastic strain $d\varepsilon^p_{ij}$ is generated from the flow rule which is assumed to be associative. In the most general case geomaterials obey a non-associated flow rule.

3 Material Parameters

The material parameters for the Mohr–Coulomb elastoplastic constitutive model were derived from triaxial compression tests on Castlegate sandstone and a calibration procedure. The elastic parameters were found to be E = 8,100 MPa and $v = 0.35$. The parameter related to the material cohesion is $p_0 = 9.81$ MPa. The calibration of the friction coefficient, $\mu = \sin\varphi_m$ as a function of the plastic shearing strain γ_p was based on interpolation of pre-peak triaxial data. Post-peak softening behavior was modeled by an additional constant c_0 which controls the rate of softening. The corresponding curve-fit was given by the hyperbolic function

$$\mu = \mu_0 + \frac{(1 - c_0\gamma^p)\gamma^p}{c_1 + c_2\gamma^p} \tag{4}$$

where $\mu_0 = 463$ is the value of the friction coefficient that defines the state of initial yield; $c_1 = 0.0077$, $c_2 = 7.5$ and $c_0 = 12$ is a control parameter of the rate of softening. The dilation coefficient $\beta = \sin\psi_m = \mu$ was taken equal to the friction coefficient μ because the Castlegate sandstone exhibits pronounced dilation.

The internal length required by the Cosserat model is set equal to R = 0.2 mm. This value is related to the microstructure of the sandstone, e.g., the grain size. We emphasize here that the inclusion of the grain size in the constitutive equations provides the Cosserat model with the capability to predict the scale effect.

4 Results and Discussion

Figure 1 (right) shows the mesh used in the analysis. In Fig. 2 we compare the applied load vs deformation at different points around the holes, shown in Fig. 1, for the case of the maximum stress applied in the vertical direction (solid lines) and

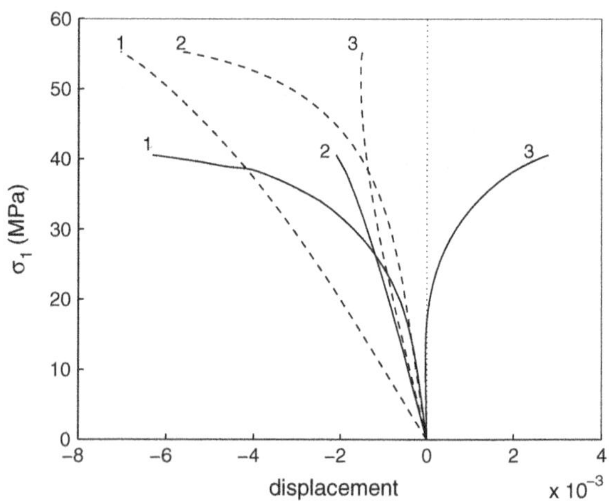

Fig. 2 Load versus displacement curves for points shown in Fig. 1. Solid lines are for maximum stress perpendicular to long axis and dashed lines are for maximum stress parallel to the long axis

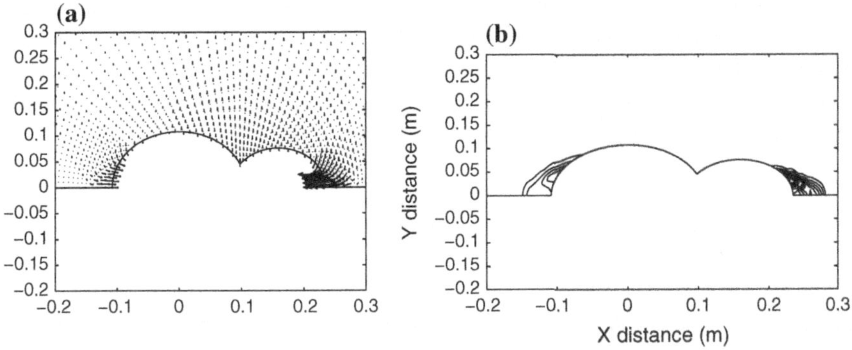

Fig. 3 Failure mechanism for the maximum stress acting in the vertical direction **a** incremental displacement field (*left*) and **b** shear plastic strain (*right*)

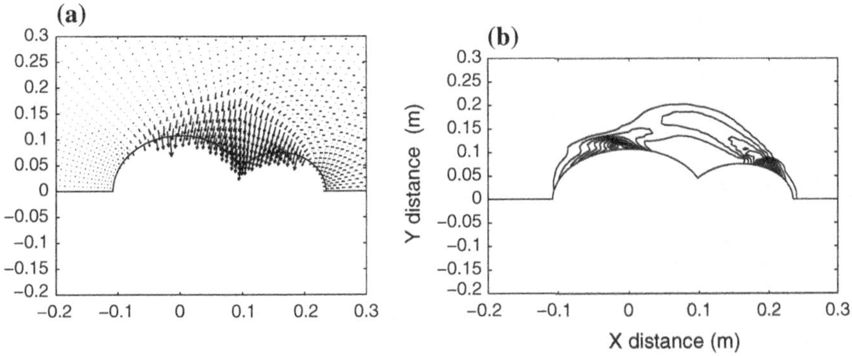

Fig. 4 Failure mechanism for the maximum stress acting in the horizontal direction **a** incremental displacement field (*left*) and **b** shear plastic strain (*right*)

the case of the maximum stress applied in the horizontal direction (dashed lines). It is clear from the peak loads that the M-L junction will be more stable if the lateral hole was drilled in the direction parallel to the maximum horizontal stress. It is worth looking at the failure mechanism at peak loads in both cases. When the maximum load was applied perpendicular to the long axis the failure mechanism consists of breakouts around the holes in the direction parallel to the maximum applied stress, as in the case of a single hole (Fig. 3). In the case of the maximum stress acting parallel to the long axis, localization of deformation initiates parallel to the maximum stress at positions on both parent and lateral holes. The formed shear bands propagate progressively to meet each other resulting in the failure mechanism shown in Fig. 4.

After the formation of the first breakouts in Fig. 3 the hole shape will become more elliptical with the long axis parallel to the minimum insitu stress where as in the case of Fig. 4 the hole shape will become more circular. It has been shown by

Papanastasiou [5] that an elliptical hole with the long axis parallel to the maximum insitu stress is more stable than a circular hole which is more stable than an elliptical hole with the maximum stress perpendicular to the long axis. Therefore, one can conclude that a lateral hole drilled on the side of the parent hole parallel to the maximum insitu stress will be more stable in the entire loading history.

5 Conclusions

In this study we presented results on the failure near a wellbore multilateral junction obtained with a Mohr-Coulomb Cosserat plasticity model that is capable of modelling the progressive shear bands that lead to wellbore breakouts. We found that the direction of a more stable lateral is parallel to the maximum horizontal insitu stress. This finding is consistent with earlier work on elliptical perforations [5]. In all cases alignment of the hole long axis parallel to the maximum compressive stress delays the failure to higher applied load.

Acknowledgements This study was supported by Schlumberger Cambridge Research.

References

1. Aadnoy, B.S., Edland, C.: Borehole stability of multilateral junctions. J. Pet. Sci. Eng. **30**, 245–255 (2001)
2. Muhlhaus, H.-B., Vardoulakis, I.: The thickness of shear bands in granular materials. Geotechnique **37**, 271–283 (1987)
3. Papamichos, E.: Borehole failure analysis in a sandstone under anisotropic stresses. Int. J. Numer. Anal. Method Geomech. **34**(6), 581–603 (2010)
4. Papanastasiou, P., Vardoulakis, I.: Numerical treatment of progressive localization in relation to borehole stability. Int. J. Numer. Anal. Methods Geomech. **16**, 389–424 (1992)
5. Papanastasiou, P.: Localization of deformation and failure around elliptical perforations based on a polar continuum. Comput. Mech. **26**, 352–361 (2000)
6. Papanastasiou, P., Zervos, A.: Wellbore stability analysis: from linear elasticity to post-bifurcation modeling. ASCE Int. J. Geomech. **4**(1), 2–12 (2004)
7. Papanastasiou, P., Sibai, M., Heiland, J., Shao, J.-F., Cook, J., Fourmaintraux, D., Onaisi, A., Jeffryes, B., Charlez, P.: Stability of a multilateral junction: experimental results and numerical modelling. J. SPE Drill. Complet. 4–11, March (2006)
8. Willson, S.M., Dowson, S.L., Plischke, B.: Numerical Simulation of the Stability of Multilateral Well Junctions. ISRM Symposium, Paris (2000)
9. Zervos, A., Papanastasiou, P., Vardoulakis, I.: Modelling of localisation and scale effect in thick-walled cylinders with gradient elastoplasticity. Int. J. Solids Struct. **38**, 5081–5095 (2001)

Simulating Enhanced Production in Fractured Formations Using Dual Porosity Model with a Simplified Finite Element Algorithm

Yarlong Wang, Fotios E. Karaoulanis, Xinzhe Zhao and Xiang Li

Abstract Simulating the production process in natural fractured formations can be performed by the well-known dual-porosity model. Factors related to geomechanics can be crucial for production as both the deformation of the formation and the locally induced stresses can contribute to the pressure changes significantly, particularly in low-permeability and tight formations. A fully coupled, dual-porosity reservoir geomechanics model is developed and presented in this article, focusing on the computational efficiency, without however sacrificing accuracy. In the proposed model, displacements, the saturation and the two system pressures (matrix/fracture) are considered the primary variables and the system is solved in an iterative manner, following a fully coupled approach for the displacements and one of the system pressures, so that the overall problem dimensions to be reduced from four to two. Calculated stresses, pressures and production rates are analyzed and implications to hydraulic fracturing and stimulated reservoir volume (SRV) calculations are discussed and addressed.

1 Introduction

Flow and pressure changes in fractured formations are key issues in hydrogeology, geothermal and petroleum engineering [9]. Dual-porosity and permeability models are used to simulate pressure and stresses changes in the matrix and fractured

Y. Wang · F.E. Karaoulanis (✉)
PGI, Toronto, Canada
e-mail: fotios.karaoulanis@pgi-ca.com; fotios.karaoulanis@gmail.com

Y. Wang
e-mail: yarlong.wang@pgi-ca.com

X. Zhao · X. Li
PGI, Beijing, China
e-mail: bjszhaoxinzhe@pgi-ca.com

X. Li
e-mail: brianxlca@pgi-ca.com

E. Papamichos et al. (eds.), *Bifurcation and Degradation of Geomaterials with Engineering Applications*, Springer Series in Geomechanics and Geoengineering, DOI 10.1007/978-3-319-56397-8_43

system, respectively. One of the most important issues in the fracture-matrix system is the geomechanics effects on the flow [2, 6]. The focus here is on how can we quantify the permeability change under the induced stresses and the volumetric deformations coupled to the mass balance in each system [3]. On one hand, the deformation and stress changes follow different laws in the different systems. On the other hand, induced stresses control the permeability change, which can dominate both flow and stresses change at the same time [1].

In this study, a general dual-porosity and permeability model is developed and a simplified numerical algorithm is proposed and implemented to calculate pressures, stresses and flow rates near a wellbore and a fracture. Stresses and flow inside the two systems (fracture and matrix) are calculated and analyzed.

Our objectives are to study the pressure and the effective stresses changes between these two systems so that we can later analyze the wellbore stability, sanding problems and hydraulic fracture designs.

A procedure for simulating and solving a problem as such is proposed in this article. First of all we assume the permeability is insensitive to the stress, the saturation and the interporosity transfer coefficient remain unchanged. Commonly, the permeability is defined as a function of the confining stress, which is in the principal stress direction. In our formulation we define the permeability as a function of the normal stress to the principal flow direction, which is dictated by the natural fractures and the geological bedding. Furthermore, if plasticity and dilatancy occur, the shear component of the stress tensor should be also used. Secondly saturation is related to the relative permeability, which is sensitive to the capillary pressure and the deformation. Thirdly, the interporosity coefficient is typically defined as a combination of fracture network parameters and quasi-static pressure differences.

2 Fundamental Study on Reservoirs and Numerical Model Development

A general constitutive relationship for an extended Darcys law with a dual-porosity and dual-permeability system is developed in the following. The equilibrium, the mass balance, the momentum and the kinematic equations are defined as (see also [4, 8]):

$$\sigma_{ij,j} = 0 \tag{1}$$

$$\nabla \cdot (\rho_m S_{nm} \phi \mathbf{v}_n) + \frac{\partial (S_{nm} \rho_m \phi_n)}{\partial t} + (-1)^n \rho_m \Gamma = 0 \tag{2}$$

$$S_{nm} \phi_n (\mathbf{v}_n - \mathbf{v}_s) = -\frac{k_{rm} k_n}{\mu} \nabla p_n \tag{3}$$

$$\nabla \cdot (\rho_s (1 - \phi_t) \mathbf{v}_s) + \frac{\partial [\rho_s (1 - \phi_t)]}{\partial t}, \quad \phi_t = \phi_1 + \phi_2 \tag{4}$$

$$\underbrace{\sigma_{ij} = 2G\varepsilon_{ij} + \frac{2G\nu}{1-2\nu}\varepsilon_{kk}\delta_{ij}}_{\sigma'_{ij}} - (a_1p_1 + a_2p_2)\delta_{ij}, \quad \varepsilon_{ij} = \frac{1}{2}\left(\frac{\partial u_i}{\partial x_j} + \frac{\partial u_j}{\partial x_i}\right) \tag{5}$$

$$\alpha_1 = \frac{\alpha^* c_b^*}{c_b}, \; \alpha_2 = \alpha - \alpha_1, \; \alpha^* = 1 - \frac{c_s^*}{c_b^*}, \; \alpha = 1 - \frac{c_s}{c_b} \tag{6}$$

In the above $n = 1$ (matrix) or 2 (fracture) and $m = w$ (wetting) or o (non-wetting, oil). Furthermore star (*) refers to the single porosity problem. The above equations define a 4×4 coupled system for u, p_1 (matrix), p_2 (fracture) and saturation S, by dropping into the equilibrium Eq. (1) and into the mass balance for the fluid phases (2), the rest of the equations. The solution approach can be simplified, by solving the pressure and the displacement equations in the matrix first, assuming the pressures in the fracture system as known and solving explicitly the saturation equation. Then go back to the rest of the system's equations in an iterative way and solve it in a fully coupled form. This way, reliable, single phase solvers can be employed to solve the reduced system.

3 Numerical Simulations, Results, Field Example and Implications

A numerical example is examined next, based on the data found in [5], as summarized in Table 1.

Pore pressures changes in both the fractured and matrix systems are displayed and compared in Fig. 1. In general a flat pressure gradient can be generated in the system where a higher permeability exists as expected. The effective stresses in this higher permeability system can be smaller than that with a lower permeability. This indicates that the pore pressure coupling effect due to the volumetric strain in the fractured system can be weakened by a lower effective stresses, even though an overall effective stress is defined.

Table 1 Simulation data

Property	Value	Property	Value (MPa)
Young's modulus (E)	35 GPa	In situ stresses $\sigma_x = \sigma_y = \sigma_z$	30
Poisson ratio (ν)	0.2	Bottomhole pressure	4
Matrix permeability (k_1)	1 mD	Initial pore pressure	5
Fracture permeability (k_2)	1000 mD	Drawdown pressure	1
Matrix porosity (ϕ_1)	0.1		
Fracture porosity (ϕ_2)	0.01		

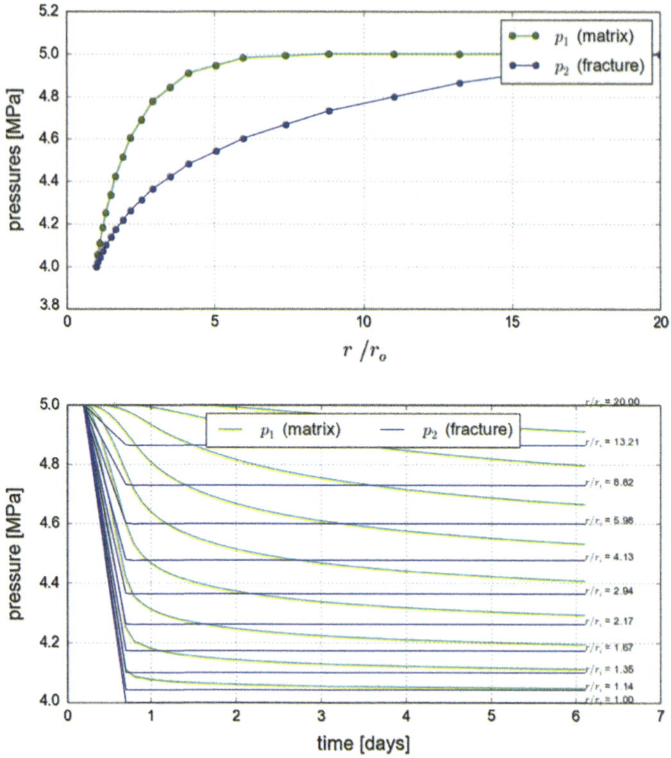

Fig. 1 Pressures for both systems on early stage of production (*top*) and their evolution through time at different distances from the well (*bottom*)

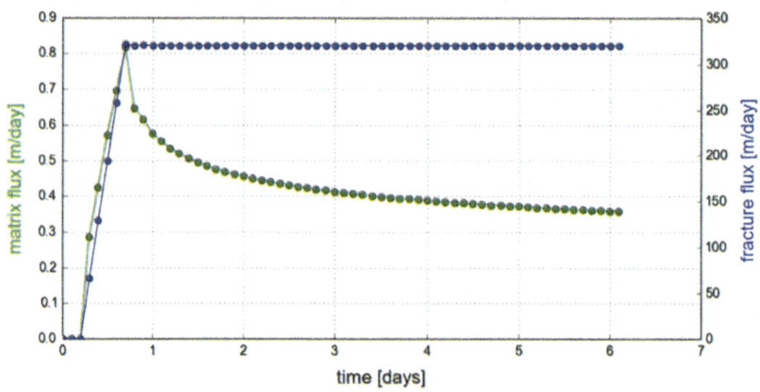

Fig. 2 Fluxes over time for different permeabilities

The fluxes by the two systems are compared and displayed in Fig. 2. Depending on the difference in permeability, the flow rate contribution by the fractured system can be reduced when stress-sensitive effect dominates the permeability change.

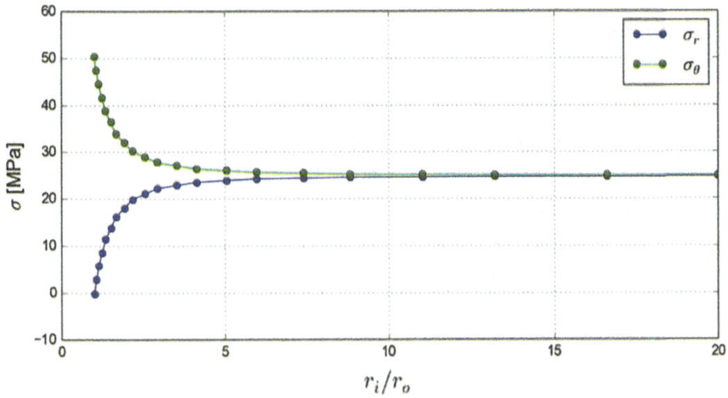

Fig. 3 Extreme effective stresses near the borehole (no support conditions)

The worst case scenario of stress distributions near the well is shown in Fig. 3. Stresses in an elastic, poroelastic and that in a coupled fracture-matrix system can be significantly different from each other, implying a wellbore stability, flow and fracturing processes must be understood, simulated and designed differently.

3.1 *Pressure and Stresses Distribution Near a Borehole and Fracture*

At a given time, pore pressure gradient normally can dissipate rapidly near a borehole and a fracture in a high-permeability formations. Thus, the pore pressure in these formations such as along the fractures can transmit farther away from the borehole and fracture, leading to a low effective stress region. This suggests that a stress perturbation can be more significant in a poroelastic and fractured media which may be favorable for SRV and enhanced permeability change. Such an effect can be further enhanced by a permeability change near the borehole or fracture when a low effective stress region is created.

3.2 *Boosting Production by Increasing Reservoir Pressure*

For stress-sensitive reservoirs, enhancing production by increasing drawdown can defeat the purpose if not applied carefully. One must examine the sensitivity of the formation, either by maintaining the minimum flowing pressure or lifting the far-field pressure in order to keep a significant drawdown. A preliminary simulation by lifting far-field pressure in a dual porosity environment indicates that the sweeping ratio and arrival time for the driving pressure front are the keys for such a strategy to be effective. A typical study on stress sensitivity formations can be found in [7]. Further studies are required and will be conducted elsewhere.

4 Conclusions

A simplified numerical procedure is implemented in this article. A 4×4 system of equations can be reduced into 2×2 and an efficient coupled solver can be used to simulate a fully coupled two phase dual porosity model, in which the geomechanics coupling is rigorously defined. The effective stresses and pore pressure near a wellbore are critical for production and hydraulic fracturing process, during production enhancement in fractured reservoir stimulation.

References

1. Bai, M., Meng, F., Elsworth, D., Roegiers, J.C.: Analysis of stress-dependent permeability in nonorthogonal flow and deformation fields. Rock Mech. Rock Eng. **32**(3), 195–219 (1999)
2. Barenblatt, G.I., Zheltov, I.P., Kochina, I.N.: Basic concepts in the theory of seepage of homogeneous liquids in fissured rocks [strata]. J. Appl. Math. Mech. **24**(5), 1286–1303 (1960)
3. Barton, N., Bandis, S., Bakhtar, K.: Strength, deformation and conductivity coupling of rock joints. Int. J. Rock Mech. Min. Sci. Geomech. Abstr. **22**(3), 121–140 (1985)
4. Chen, H.Y., Teufel, L.W.: Coupling fluid-flow and geomechanics in dual-porosity modeling of naturally fractured reservoirs. In: SPE Annual Technical Conference and Exhibition. Society of Petroleum Engineers (1997)
5. Elsworth, D., Bai, M.: Flow-deformation response of dual-porosity media. J. Geotech. Eng. **118**(1), 107–124 (1992)
6. Huyakorn, P.S., Pinder, G.F.: Computational Methods in Subsurface Flow. Academic Press (1983)
7. Kikani, J., Pedrosa Jr., O.A.: Perturbation analysis of stress-sensitive reservoirs. SPE Form. Eval. **6**(03), 379–386 (1991)
8. Wang, Y., Chen, C.C.: Enhanced oil production owing to sand flow in conventional and heavy-oil reservoirs. SPE Reserv. Eval. Eng. **4**(05), 366–374 (2001)
9. Warren, J.E., Root, P.J.: The behavior of naturally fractured reservoirs. Soc. Pet. Eng. J. **3**(03), 245–255 (1963)

Part VI
Micromechanics and Multi-scale Analysis of Instabilities and Degradation

Hierarchical Multiscale Modeling of Strain Localization in Granular Materials: A Condensed Overview and Perspectives

Jidong Zhao

Abstract This paper presents a brief overview on hierarchical multiscale modeling (HMM) of granular media and its application in simulating and understanding the phenomenon of strain localization. The general principles, solution procedures and advantages of existing HMM approaches are reviewed and compared. Focuses are devoted to the new cross-scale findings and insights offered by recent HMM studies on identification of key micro-structural origins and micro-mechanisms underpinning different deformations bands in granular materials. Limitations, challenges and opportunities pertaining to multiscale modeling of granular media are discussed.

1 Introduction

Multiscale modeling tops the trending words across many disciplines of engineering and science for over two decades, and has become a focal topic of interest in the study of granular materials recently. Granular media represent a wide range of materials that are of tremendous importance for many branches of engineering and industry, including cohesion-less sand, unconsolidated rocks, crushed coal, agricultural grains and chemical powders. Our fundamental understanding on the behavior of granular media, especially their mechanical behavior such as strength, stiffness and localized deformation patterns, has been largely built upon phenomenological characterizations, empirical descriptions and continuum-based modeling. While these traditional approaches have gained certain success in helping us meet the primary needs on design and operation of engineering problems, there are numerous occasions where the multiscale nature of granular media

J. Zhao (✉)
Hong Kong University of Science and Technology, Kowloon, Hong Kong
e-mail: jzhao@ust.hk

J. Zhao
Hong Kong University of Science and Technology Shenzhen
Research Institute, Kowloon, Hong Kong

E. Papamichos et al. (eds.), *Bifurcation and Degradation of Geomaterials with Engineering Applications*, Springer Series in Geomechanics and Geoengineering, DOI 10.1007/978-3-319-56397-8_44

has to be rigorously taken into account in order to attain accurate, comprehensive understanding, *strain localization* being a representative one of them.

There has been compelling recent experimental evidence, especially from test data based on advanced modern facilities such as X-ray micro computerized tomography (X-ray micro CT) and Digital Imaging Correlation (DVC), indicting that the initiation, formation, (sometimes) transition and finalization of strain localization bear intimate correlations with intricate microstructural mechanisms and controlling factors originated from the grain scale of a granular material [1–5, 14]. The corresponding theoretical and computational developments on this subject have apparently lagged behind their experimental counterpart. There has been a strong urge for the development of next-generation modeling and analysis tools, which, if ideally available, may not only retain the robustness and rigor of conventional continuum-based approaches in dealing with practical boundary value problems while voiding being phenomenological, but also incorporate fully the grain-scale granular mechanics showcased by prevailing micromechanics approaches and experimental observations. A special class of multiscale modeling methods, namely, the Hierarchical Multiscale Modeling (HMM) approaches, has been developed recently to take a step forward towards this direction. This paper serves as a condensed review of the latest advance of the HMM methods, with an emphasis placed on their unique roles in helping us gain new cross-scale understanding of strain localization in granular media.

2 Hierarchical Multiscale Modeling (HMM) of Granular Media

2.1 Principle and Methodology

Principle: The concept of HMM typically exploit a hierarchical discrete-continuum coupling scheme to solve a boundary value problem for granular media (see e.g., [8]). In the HMM framework, a continuum approach (e.g., FEM) is employed to discretize the concerned domain and find its solution subjected to the prescribed boundary conditions according to similar ways as conventional continuum approaches. It departs from conventional ways in that a HMM approach does not require a constitutive model to be assumed at the material point of the continuum domain. Rather, the constitutive relations are derived from the solution of a discrete-particle assembly of Representative Volume Element (RVE) attached to each material point of the continuum domain whereby the RVE receives deformations and other state variables as boundary condition from the macro scale and is solved by a discrete-based method (e.g., DEM). In essence, the HMM retains the robustness and flexibility of FEM in handling BVPs (which is a pitfall for DEM), while avoid assuming phenomenological constitutive models as required by conventional continuum approaches. Meanwhile, it fully capitalises the strength of discrete-based methods in reproducing the highly nonlinear, path-dependent mechanical behavior of a granular material. Moreover, the hierarchical structure offers a direct pathway to link the macro observations with their underpinning microstructural mechanisms, a feature highly desirable for multiscale understanding of strain localization in granular media.

Methodology: A dominant majority of existing HMM approaches for granular media have employed FEM to solve the continuum boundary value problem, and DEM for the solution of the boundary value problem of the particle assembly [3, 6, 8, 9, 17, 18, 21–25], Nitka et al. [23] and Liu et al. [20]. Figure 1a demonstrates a typical flowchart for the sequential iterative macro-micro solution procedure of HMM of dry granular media based on a hierarchical FEM-DEM coupling [8]. Since the DEM computation for each RVE is independent, the HMM approach can fully adopt parallel computing techniques for each RVE in the macro domain to enhance the computational efficiency. In existing HMM studies, the DEM commonly uses deformable particles of either spherical or elliptical shapes with linear or nonlinear contact and friction laws governing their contacts. Only exception is the study by Kaneko et al. [16] where the called Granular Element Method (GEM) based on rigid, frictional circular disks was adopted for the RVE solution. Notably, while most of the studies have been formulated under 2D conditions for the sake of simplicity, [9] and Liu et al. [20] have presented 3D HMM formulations and demonstrative examples.

Global-local solution schemes: In deriving the global-local solutions of the HMM approach, the implicit Newton-Raphson (NR) solution scheme with sequential macro-micro iterations has been commonly adopted to solve the non-linear problems for granular media [8, 21]. In particular, Meier et al. [21] used the elastic modulus as the tangent operator for frictionless particles based on Taylor's assumption. Note that the Taylor's assumption may potentially render the material response unrealistically stiffer and less dissipative. Guo and Zhao [8] found that the use of a secant elastic modulus for general granular particles (with friction) could work efficiently and robustly, noting that the perturbation method as advised in Nitka et al. [23] could cause issues of potential non-convergence and relatively low computational efficiency. Shahin et al. [25] suggested that a modified Newton-Raphson approach could be more robust when the quadratic convergence of a NR scheme was lost. Liu et al. [20] have recently proposed an explicit solution procedure for HMM which was claimed to be more efficient than the implicit iterative procedures aforementioned.

2.2 Selection and Benchmark of RVE for HMM

The selection of proper Representative Volume Element (RVE) is critical to the HMM approaches in two-fold. *First*, a proper RVE should faithfully represent the typical material behavior of the simulated granular material observed in the lab. Indeed, a RVE can be considered a virtual specimen equivalent to that used in a

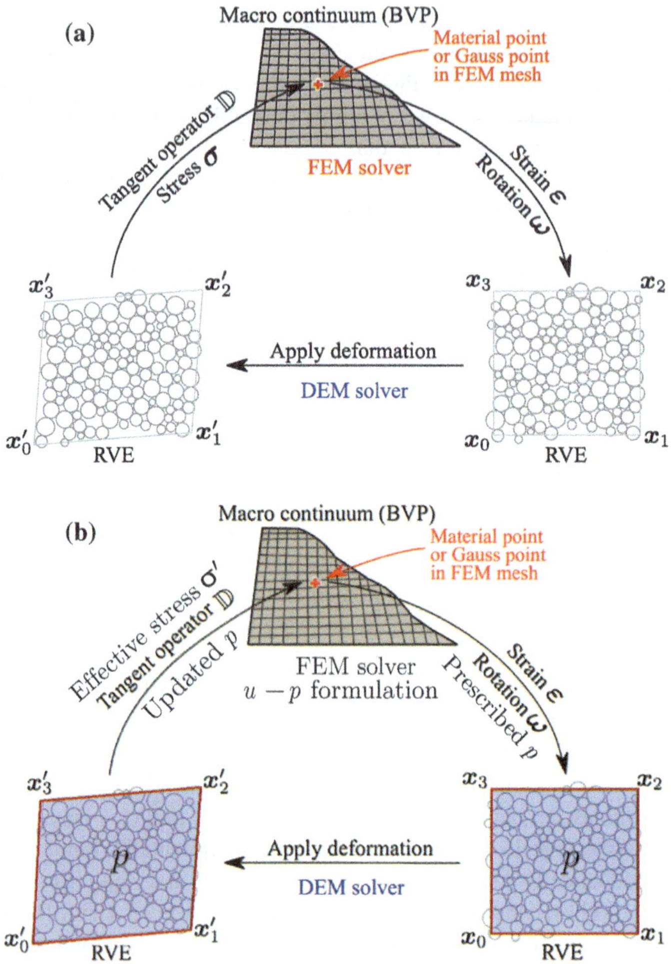

Fig. 1 Illustration of Hierarchical Multiscale Modeling of granular media with *FEM* × *DEM* coupling. **a** *HMM of Dry granular media*: FEM discretizes the macroscale BVP and passes the deformation gradients as boundary conditions for the meso-scale RVE modeled by DEM. The deformed RVE returns tangent operators and stresses for FEM to advance its global solution [8]. **b** *HMM of saturated granular media*: The FEM solves the BVP based on *u-p* formulation. It passes the deformation gradients and prescribes pore water pressure on the RVE. The DEN solution of the RVE returns the effective stress based on Terzaghi's principle of effective stress, the tangent operator and an updated value for the pore water pressure to FEM for iterative global solution [9]

typical laboratory element tests. *Second*, the selection of a RVE should make it computationally affordable as its DEM solution constitutes a typical HMM may involve hundreds of thousand RVE to be computed at each loading step. While constitutive features governing the DEM including contact models and geometry of particles have been extensively discussed in the DEM community, the most critical issue regarding HMM computation boils down to the number of particles to be included in a typical RVE. For 2D case, a size of RVE with a few hundred particles appears to be widely agreed by most study to offer good predictions of material behavior with reasonable computational efficiency. For example, Kaneko et al. [16] examined 16 different RVE size (2D) and aspect ratio in terms of their fabric isotropy and concluded that 200 particles would make an acceptable RVE for their multiscale approach. Meier et al. [21] examined the contact normal density function and average Cauchy stress of RVEs containing particle numbers from 70350 to 700 and concluded that the 700-particle RVE could offer reasonable results for 2D simulations. Guo and Zhao [8] and Nguyen et al. [24] both recommended 400 particles would serve a balanced choice for HMM of 2D problems. For 3D case, Guo and Zhao [11] suggested that a RVE with 1000 particles would be a reasonable unit cell for HMM simulations, while the number suggested Liu et al. [20] was 4000. Note that Shahin et al. [25] have more recently examined the effects of inhomogeneity and imperfection in a RVE on the prediction of strain localization in biaxial compression test simulations. In the future, if faster, more advanced parallel computing facilities can be inexpensively accessible, the number of particles considered in a RVE may be reasonably increased to yield more representative material responses.

Equally important is the benchmarking of the chosen RVE to ensure its prediction is valid and representative of typical granular responses. Guo and Zhao [8] and Liu et al. [20] compared the single-scale RVE response with the Gauss point response of HMM single element tests under either biaxial compression or simple shear to benchmark their methods. Note that Guo and Zhao [10] and Guo et al. [12] further employed their HMM approach and investigated some classic geomechanics problems, including retaining wall, footing and cavity expansion in thick-walled hollow cylinder.

2.3 Hydro-Mechanical Coupling for Saturated Granular Media

The presence of pore fluids in a porous, saturated granular medium may lead to strong hydro-mechanical coupling effects underpinning many aspects of its engineering performance. Micromechanics-based approaches for direct modeling of particle-fluid interactions, such as those based on DEM-CFD (Computational Fluid Dynamics) or DEM-LBM (Lattice Boltzmann Method) coupling, are too expensive to be affordably paired with HMM approaches to solve a practical problem. Moreover, these approaches commonly require complicated considerations and treatments of moving boundaries at the interface of fluid and particles. Guo and Zhao [9, 13] recently proposed adapting the fixed-stress split method with the seminal $\boldsymbol{u}$-p formulation to solve the global governing equations in the HMM, and exploiting the classic Terzaghi's effective stress principle to derive effective stress from the RVE. Wang and Sun [26] presented a similar method. The stress tensor homogenized from a typical DEM assembly according to Love's formula is based on the interparticle contact forces of the soil particles. It is hence a clear measure of the effective stress as defined by Carl Terzaghi. This micro-scale based effective stress added by the macro pore water pressure (as an unknown) makes up the total stress in the global equilibrium equation. Therefore, the entire hydro-mechanical coupling problem can be solved in the HMM framework efficiently and robustly, without resorting to expensive micromechanically based coupling. The entire solution procedure can be readily adapted from the one proposed for dry granular media by Guo and Zhao [8]. As illustrated in Fig. 1b, the pore water pressure is passed down to the RVE as an additional prescribed boundary condition on its calculation, before being updated and passed up to the macro material point again together with the homogenized effective stress and tangent stiffness matrix, and an updated permeability based on new void ratio and fabric structure of the RVE. All the rest of the solution procedure remains the same as in the dry case. Guo and Zhao [9, 13] benchmarked their coupled hydro-mechanical HMM approach with classic Terzaghi's 1D consolidation problem and the 2D consolidation under a strip footing, while Wang and Sun [26] verified their formulation on 1D consolidation problem.

2.4 Non-conventional Continuum Enrichment and Finite Strain

To resolve mesh-dependency issue and to capture size effects, Li et al. [18] presented a micro-macro homogenization method for Cosserat continuum, and Li et al. [17] further developed a mixed FEM of gradient Cosserat continuum with second-order computational homogenization for granular media based on Hu-Washizu variational principle. Liu et al. [20] developed a coupled FEM and DEM nonlocal multiscale method wherein the macro finite element solution was based on a nonlocal strain formulation. Finite strains were considered by Miehe et al. [22] and Liu et al. [20]. A side note on finite strain is made here. When the stress is homogenized from a RVE from interparticle contact forces, both translation and rotation of all particles of the entire RVE are totally considered. Consequently, the attained stress tensor may have already included contributions of co-rotational terms for the assembly (e.g., according to the manner of Jaumann stress rate). Therefore, it appears that the general HMM formulation (e.g., [8]) may well consider finite strain case already.

3 Multiscale Modeling of Strain Localization in Granular Media

The various HMM approaches have been applied to simulating strain localization in granular media. Biaxial compression tests have been a popular example in most of these HMM studies (see, e.g., [3, 8, 9, 13, 16, 22, 23]). Strain localization in 3D problems such as triaxial compression and extension on cubic and cylindrical samples have also been investigated by HMM approaches [11, 26]. The following provides a brief summary of major new cross-scale findings pertaining to strain localization revealed by the various HMM studies.

3.1 *Alternative Localization Indicators*

Cumulative deviatoric and void ratio have long been considered good indicators for localized shear bands in granular media. In addition, Zhao and Guo [28] found that cumulative particle rotation extracted from their HMM simulations can be an equally good localization too (see Fig. 2a, b, d. Meanwhile, both Guo and Zhao [8] and Zhao and Guo [28] demonstrated that the localized pattern for the intensity of fabric anisotropy defined by interparticle contact-normals was not consistent with the accumulated deviatoric strain, and suggested that the contact-normal-based fabric anisotropy was as good as an indicator for strain localization in granular media (c.f. Figure 2e with the rest figures). However, Zhao and Guo [28] employed clumped particles in RVE and showed that the intensity distribution of particle-orientation-based fabric anisotropy was highly consistent with that of deviator strain, and hence can be used as a localization indicator (see, Fig. 2c).

3.2 *2D Versus 3D Loading Conditions*

Guo and Zhao [11] employed 3D HMM approach to predict the occurrence of granular media subjected to 3D loading conditions. The multiscale modeling concluded that localized shear failure could be easier to occur under plane-strain conditions, and triaxial extension conditions could prohibit deformation localization and lead to diffuse failure mode instead. Indeed, as summarized in Fig. 3, their multiscale modeling indicates a cylindrical sample may tend to undergo bulging

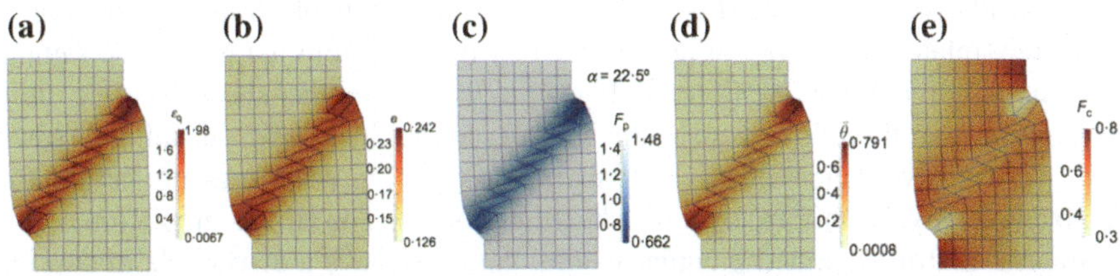

Fig. 2 Localization patterns predicted by HMM simulations of biaxial shear tests (Zhao and Guo [28]), in terms of **a** cumulative deviatoric strain, **b** void ratio, **c** intensity of particle-orientation-based fabric anisotropy, **d** cumulative particle rotation and **e** intensity of contact-normal-based fabric anisotropy

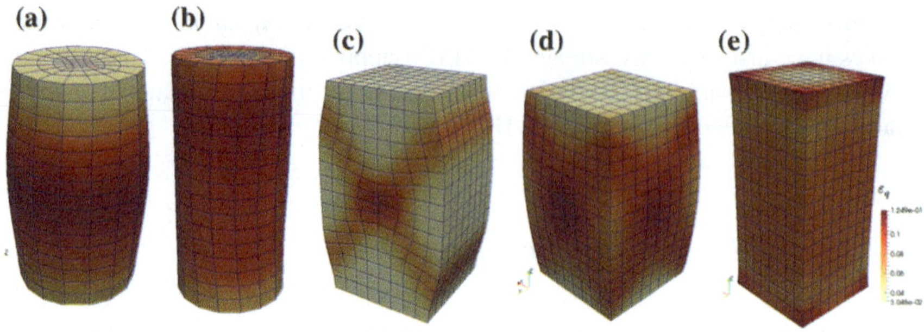

Fig. 3 Shear strain contours predicted by HMM tests simulations of uniform **a** cylindrical sample under triaxial compression condition (CTC), **b** cylindrical sample under triaxial extension condition (CTE), **c** cubic sample under plane-strain biaxial compression condition (PBC), **d** cubic sample under triaxial compression condition (CTC) and **e** cubic sample under triaxial extension condition (CTE) [11]

failure under triaxial compression (Fig. 3a) but fail diffusively under triaxial extension condition (Fig. 3b), whilst a cubic specimen is prone to develop octopus-shaped localizations under triaxial compression condition Fig. 3d and cross planar shear bands under plane-strain biaxial compression Fig. 3c, while it fails in diffuse mode under triaxial extension Fig. 3e. The 3D HMM simulations also showed that, if shear banding occurs, the shear band angle (relative to the minor principal stress direction) decreases from CTC, to PBC and further to CTE loading conditions (or equivalently with the increase of intermediate principal stress ratio). These findings are indeed consistent with experimental observations (e.g., [15]).

3.3 Pore Water Pressure Dissipation and Shear Banding

Guo and Zhao [9] employed a hydro-mechanical coupling HMM formulation to simulate a globally undrained biaxial compression test on saturated dense sand. A striking finding is as follows: (a) An obvious flux flow pattern was found proceeding to a clear strain localization incepts when the porosity of the entire sample remains relatively uniform (Fig. 4a). (b) The formation of strain localization is intimately related to a surge of pore water flux flowing from the rest of the sample into the dilative shear band (e.g., Fig. 4b). (c) When the shear band is fully developed, the flux flow across the entire sample becomes vanishingly small (Fig. 4c, d). Nevertheless, the overall distribution of pore water pressure field in the sample is relatively uniform over the shearing process. The HMM simulation seems to suggest that in saturated granular media, the flow plays a driver role in causing localized dilation and inducing strain localization, not the opposite process that the pore dilation acts a local sucking sink for local flux flow.

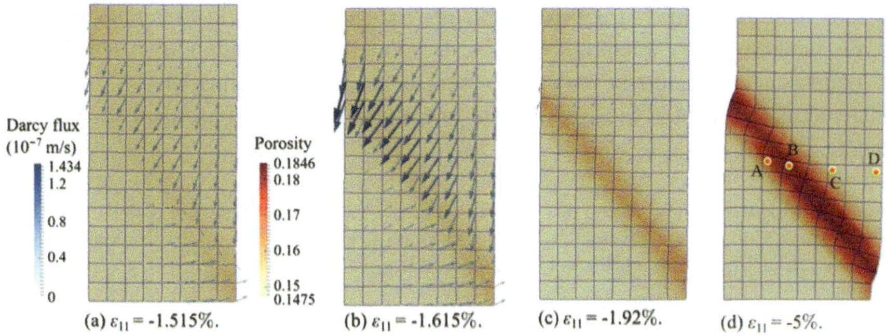

Fig. 4 Augmented illustration of strain localization and Darcy flux patterns of a saturated dense sand sample during globally undrained biaxial shear [9]. Note in the figures the thicker and longer arrows indicate stronger flux of pore fluid flow

3.4 Critical State Reached Within Shear Bands

Guo and Zhao [8] found the material response extracted from the RVE at Gauss point within the shear band formed under drained biaxial shear reached critical state, with constant stress, fabric anisotropy and volumetric strain. The finding confirms early pure DEM simulations by Fu and Dafalias [7] and the anisotropic critical state theory [19, 27]. Indeed, only material points located within the shear band(s) can possibly undergo excessively large shear deformation (e.g., in [7], the shear strain amounts to over 400%) to reach the critical state. While in conventional laboratory tests, such high shear strain levels are difficult to attain under proportional loading conditions.

4 Conclusions and Outlook

A condensed overview was made on hierarchical multiscale modeling of granular media, in particular relation to the simulation of strain localization in these materials. The principle, methodology and solution schemes on HMM were summarized. Recent new advances of HMM in simulating saturated granular media and further gradient or nonlinear enrichments were discussed. Some interesting new findings attained by various HMM studies on strain localization were highlighted. As shown, hierarchical multiscale modeling provides a new, effective toolbox for us to model and understand the multiscale nature of granular media, and opens up new grounds for exciting research. Exploratory future directions on hierarchical multiscale modeling of granular media include: (1) experimental microstructural characterization, verification and validation of the RVE; (2) characterization of realistic particle shapes for RVEs; (3) fully micromechanically-based modeling of fluid-particle interactions for the RVE to receive and pass information to the macro-scale domain; (4) consideration of particle crushing in hierarchical multi-scale modeling; (5) integrated multiscale modeling of continuous-discontinuous domains/stage transitions; (6) Hierarchical multiscale modeling of cohesive granular media; (7) Hierarchical multiscale modeling of dynamic problems in granular media.

Acknowledgements The study was partially supported by Research Grants Council of Hong Kong via a Collaborative Research Fund Project (No. C6012-15G) and by National Science Foundation of China through Project No. 51679207.

References

1. Andò, E., Hall, S.A., Viggiani, G., Desrues, J., Bésuelle, P.: Grain-scale exper- imental investigation of localised deformation in sand: a discrete particle tracking approach. Acta Geotech. **7**(1), 1–13 (2012)
2. Andò, E., Hall, S., Viggiani, G., Desrues, J.: Experimental micro-mechanics of granular media studied by x-ray tomography: recent results and challenges. Géotechnique Lett. **3**, 142–146 (2013)
3. Andrade, J.E., Avila, C.F., Hall, S.A., Lenoir, N., Viggiani, G.: Multiscale modeling and characterization of granular matter: From grain kinematics to continuum mechanics. J. Mech. Phys. Solids **59**, 237–250 (2011)
4. Desrues, J., Viggiani, G.: Strain localization in sand: an overview of the experimental results obtained in Grenoble using stereophotogrammetry. Int. J. Numer. Anal. Methods Geomech. **28**(4), 279–321 (2004)
5. Desrues, J., Andò, E.: Strain localization in granular media. C R Phys. **16**(1), 26–36 (2015)
6. Desrues, J., Nguyen, T.K., Combe, G., Caillerie, D.: FEM × DEM multi-scale analysis of boundary value problems involving strain localization. In: Bifurcation and Degradation of Geomaterials in the New Millennium, pp. 259–265 (2015)
7. Fu, P., Dafalias, Y.F.: Fabric evolution within shear bands of granular materials and its relation to critical state. Int. J. Numer. Anal. Method Geomech. **35**(18), 1918–1948 (2011)
8. Guo, N., Zhao, J.: A coupled FEM/DEM approach for hierarchical multiscale modelling of granular media. Int. J. Numer. Methods Eng. **99**, 789–818 (2014)
9. Guo, N., Zhao, J.D.: Parallel hierarchical multiscale modelling of hydro-mechanical problems for saturated granular soils. Comput. Methods Appl. Mech. Eng. **305**, 37–61 (2016)
10. Guo, N., Zhao, J.D.: 3D multiscale modeling of strain localization in granular media. Computers and Geotechnics. **80**, 360–372 (2016)
11. Guo, N., Zhao, J.D.: Multiscale insights into classical geomechanics problems. Int. J. Numer. Anal. Method Geomech. **40**(3), 367–390 (2016)
12. Guo, N., Zhao, J., Sun, W.C.: Multiscale analysis of shear failure of thick-walled hollow cylinder in dry sand. Géotechnique Lett. **6**, 77–82 (2016)
13. Guo, N., Zhao, J.D.: Hierarchical multiscale modeling of fluid-saturated soils. The 15th Asian Regional Conference on Soil Mechanics and Geotechnical Engineering (15ARC). 9–13 Nov 2015, Fukuoka, Japan (2015)
14. Hall, S.A., Wood, D.M., Ibraim, E., Viggiani, G.: Localized deformation patterning in 2d granular materials revealed by digital image correlation. Granul. Matter **12**(1), 1–14 (2010)
15. Hall, S.A., Bornert, M., Desrues, J., Pannier, Y., Lenoir, N., Viggiani, G.: Discrete and continuum analysis of localised deformation in sand using X-ray μCT and volumetric digital image correlation. Geotechnique **60**(5), 315–322 (2010)

16. Kaneko, K., Terada, K., Kyoya, T., Kishino, Y.: Global-local analysis of granular media in quasi-static equilibrium. Int. J. Solids Struct. **40**, 4043–4069 (2003)
17. Li, X.K., Liang, Y.B., Duan, Q.L., Du, Y.Y.: A mixed finite element procedure of gradient Cosserat continuum for second-order computational homogenization of granular materials. Comput. Mech. **54**(5), 1331–1356 (2014)
18. Li, X.K., Liu, Q.P., Zhang, J.B.: A micro-macro homogenization approach for discrete particle assembly—Cosserat continuum modeling of granular materials. Int. J. Solids Struct. **47**(2), 291–303 (2010)
19. Li, X., Dafalias, Y.: Anisotropic critical state theory: the role of fabric. J. Eng. Mech. **138**(1), 263–275 (2012)
20. Liu, Y., Sun, W.C., Yuan, Z.F., Fish, J.: A nonlocal multiscale discrete-continuum model for predicting mechanical behavior of granular materials. Int. J. Numer. Method Eng. **106**(2), 129–160 (2016)
21. Meier, H.A., Steinmann, O., Kuhl, E.: Towards multiscale computation of confined granular media—contact forces, stresses and tangent operators. Technische Mechanik **28**(1), 32–42 (2008)
22. Miehe, C., Dettmar, J., Zah, D.: Homogenization and two-scale simulations of granular materials for different microstructural constraints. Int. J. Numer. Method Eng. **83**, 1206–1236 (2010)
23. Nitka, M., Combe, G., Dascalu, C., Desrues, J.: Two-scale modeling of granular materials: a DEM-FEM approach. Granul. Matter **13**, 277–281 (2011)
24. Nguyen, K., Combe, G., CAillerie, D., Desrues, J.: FEM x DEM modeling of cohesive granular materials: numerical homogenization and multi-scale simulations. Acta Geophys. **62** (5), 1109–1126 (2014)
25. Shahin, G., Desrues, J., Pont, S.D., Combe, G., Argilaga, A.: A study of the influence of REV variability in double-scale FEM x DEM analysis. Int. J. Numer. Method Eng. **107**, 882–900 (2016)
26. Wang, K., Sun, W.C.: A semi-implicit discrete-continuum coupling method for porous media based on the effective stress principle at finite strain. Comput. Methods Appl. Mech. Eng. **304** (1), 546–583 (2016)
27. Zhao, J., Guo, N.: Unique critical state characteristics in granular media considering fabric anisotropy. Geotechnique **63**(8), 695–704 (2013)
28. Zhao, J.D., Guo, N.: The interplay between anisotropy and strain localisation in granular soils: a multiscale insight. Géotechnique. **65**(8), 642–656 (2015)

Constitutive Modeling of Granular Geo-Materials Under High-Speed Impact

Takashi Matsushima and Kan Sato

Abstract A constitutive model of crushable granular geomaterials subjected to high pressure/high speed compression is proposed in this study. The proposed model is based on a micromechanical elasticity model of mono-disperse granular materials [2], and it is extended into well-graded ones in a simple manner. A geometrical recursive equation is introduced to describe the confined crushing process of granular materials. High-pressure Equation of States of solid [21] is also incorporated into the model. The basic response of the proposed model is compared with some experimental data.

1 Introduction

High speed impact problem on geomaterials has been of great concern in various fields such as earth and planetary sciences [18], materials science [30], energy engineering, military and civil engineering [32]. The term "high speed" is often used when the impact speed exceeds the elastic compressive wave speed of a medium. In such an impact, the density of the medium cannot be regarded as constant any more, and an instant density increase leads to a shock wave transmission. This phenomenon is often described by the Rankine-Hugoniot equation [12, 25], and the related Hugoniot Equation of State (EOS) parameters are investigated experimentally for various solid materials such as metals, plastics and rocks [15]. Based on such results, more comprehensive thermodynamics theory of solids has been established [1, 6].

On the other hand, geomaterials on the surface of solid planets, including earth, moon, mars and others, are usually composed of geological grains, and accordingly their bulk mechanical behavior is affected by their grain-scale behavior. In particular,

T. Matsushima (✉) · K. Sato
Department of Engineering Mechanics and Energy, University of Tsukuba, Tsukuba, Japan
e-mail: tmatsu@kz.tsukuba.ac.jp

E. Papamichos et al. (eds.), *Bifurcation and Degradation of Geomaterials with Engineering Applications*, Springer Series in Geomechanics and Geoengineering, DOI 10.1007/978-3-319-56397-8_45

high pressure impact loading results in considerable grain crushing and subsequent grain rearrangement, which leads not only to plastic compression but also to drastic change in their bulk mechanical properties. Such grain crushing effect has been studied extensively in terms of resulting grain size distribution [31, 33], of single grain crushing mechanics [22, 34, 36], of quasi-static one-dimensional compression [23, 26] and shear [5, 19] and of their constitutive modeling [8, 14, 28]. However, to authors' knowledge, such constitutive models are constructed mostly within the framework of quasi-static deformation.

The present paper attempts to construct micromechanics constitutive model for crushable granular materials subjected to high pressure loading. We begin with a well-established micromechanics model of monodisperse granular system [2] and extend it into well-graded granular system. We introduce a geometrical recursive equation to describe the confined crushing process of brittle granular materials. Moreover, high-pressure EOS of solid [21] is also incorporated into the model such that the present model can be applied into the behavior of granular geomaterials under very high pressure. Finally the basic response of the proposed model is compared with some experimental data.

2 Micromechanics Elasticity of Mono-Disperse Grains System

Micromechanical averaging approach to granular systems have been studied since 1980s [2, 4, 7, 35]. In this section we briefly overview the uniform strain elasticity theory following Chang and Misra [2].

Cauchy stress rate tensor is defined by the spatial average for all the contact forces rate $\dot{f}_i$ in a representative volume (after deformation), V_R as follows:

$$\dot{\sigma}_{ij} = \frac{1}{V_R}\sum_{c}^{2N_c} r_i \dot{f}_j \tag{1}$$

where r_i is a branch vector connecting the centroids of a grain and a contact point, and N_c is the total number of contacts in the domain. Note that the summation is made twice for each contact because two grains are involved in one contact. Contact force rate vector $\dot{f}_i$ can be related to the contact displacement rate vector δ_i as:

$$\dot{f}_i = R_{ik}\dot{f}_k^L = R_{ik}K_{kl}^L\dot{\delta}_l^L = R_{ik}K_{kl}^L R_{jl}\dot{\delta}_j \equiv K_{ij}\dot{\delta}_j \tag{2}$$

$$K = k_n n \cdot n^T + k_s s \cdot s^T + k_s t \cdot t^T \tag{3}$$

where R_{ij} is the transformation matrix from the local coordinates which are defined by the contact normal, n, and tangential, s and t, in terms of Euler angles, β and γ to the global coordinates as follows:

$$R^T = \begin{pmatrix} n^T \\ s^T \\ t^T \end{pmatrix}, n = \begin{pmatrix} \cos\beta\,\sin\gamma \\ \sin\beta\,\sin\gamma \\ \cos\gamma \end{pmatrix}, s = \begin{pmatrix} \cos\beta\,\cos\gamma \\ \sin\beta\,\cos\gamma \\ -\sin\gamma \end{pmatrix}, t = \begin{pmatrix} -\sin\beta \\ \cos\gamma \\ 0 \end{pmatrix} \tag{4}$$

and k_n, k_s, k_t ($k_s = k_t$) are the contact stiffness for normal and tangential directions, n, s, t, respectively. Note that the superscript L shows a quantity described in the local coordinates.

Then, we adopt the uniform strain assumption in which δ_i is described uniquely by the bulk strain rate tensor $\dot{\varepsilon}_{ij}$ as $\delta_i = \dot{\varepsilon}_{ij}(2r_j)$. Accordingly we have the relation between the stress rate and the strain rate as $\dot{\sigma}_{ij} = C_{ijkl}\dot{\varepsilon}_{kl}$ and the overall elastic stiffness matrix C_{ijkl} is described by the microstructural properties. For further analysis, the summation of all contacts is transformed into the following integral over all spatial directions $d\Omega = \sin\gamma\, d\gamma\, d\beta$ $(0 \le \gamma \le \pi,\ 0 \le \beta \le 2\pi)$ using the probability density function of n, $F(n)$ [13]:

$$\begin{aligned} C_{ijlm} &= \frac{2}{V_R}\sum_c (r_i K_{jl} r_m) = \frac{2N_c}{V_R}\int_\Omega r_i K_{jl} r_m F(n) d\Omega \\ &= \frac{2N_c}{V_R}\int_0^{2\pi}\int_0^{\pi} r_i K_{jl} r_m F(\beta,\gamma)\sin\gamma\, d\gamma\, d\beta \end{aligned} \tag{5}$$

$F(n)$ is often represented by the spherical harmonics function. Moreover, in the case of mono-disperse spheres system, which leads to $r_i = r\,n_i$ (r is the grain radius), the integration is analytically feasible and we obtain various anisotropic stiffness tensor depending on $F(n)$. The simplest isotropic case $(F(n) = 1/4\pi)$ yields following expression on the overall Young's modulus and the Poisson ratio of the granular system:

$$E_g = \frac{z_c}{2\pi r(1+e)}(2k_n + 3k_s)\left(\frac{k_n}{4k_n + k_s}\right), \tag{6}$$

$$\nu_g = \frac{k_n - k_s}{4k_n + k_s} \tag{7}$$

Here V_R was described by the void ratio, e, and the coordination number, $z_c = 2N_c/N_g$ (the number of contacts per grain), as $N_c/V_R = z_c/\{2V_{grain}(1+e)\}$ where V_{grain} is the single grain volume, which is $V_{grain} = 4\pi r^3/3$ if the compression of the grain itself is sufficiently small, and N_g is the total number of grains.

Finally, contact stiffness k_n and k_s are modeled by Hertz model [11] and Mindlin and Deresiewics model [20] as:

$$k_n = \left(\frac{\sqrt{3}r\,E}{2(1-\nu^2)}\right)^{2/3} f_n^{1/3}, \tag{8}$$

$$k_s = \frac{2(1-\nu)}{2-\nu} k_n \left(1 - \frac{f_s}{\mu f_n}\right)^{1/3} \tag{9}$$

where E and ν are the Young's modulus and the Poisson ratio of the solid material of the grains, and μ is the intergranular friction coefficient. f_n is related to the mean stress p under isotropic condition as follows:

$$p = \frac{\sigma_{ii}}{3} = \frac{1}{3V_R} \sum_c^{2N_c} r_i f = \frac{rN_c}{6\pi V_R} \int_0^{2\pi} \int_0^{\pi} f_n \sin\gamma \, d\gamma \, d\beta = \frac{2rN_c f_n}{3V_R} \tag{10}$$

f_s is assumed to be zero for simplicity in this study. After some manipulation, we obtain the following explicit expressions to connect E_g and ν_g with E and ν:

$$E_g = \sqrt[3]{\frac{3}{2}\frac{5-4\nu}{5-3\nu}} \left(\frac{E}{2\pi(1-\nu^2)}\right)^{2/3} \left(\frac{z_c}{1+e}\right)^{2/3} p^{1/3}, \tag{11}$$

$$\nu_g = \frac{\nu}{10-6\nu} \tag{12}$$

where $4 \leq z_c \leq 12$ depending on the void ratio, based on the isostatic theory [10, 24].

The previous studies showed that the above uniform strain model provides an "upper bound" solution of the elastic properties of granular materials [3]. Matsushima and Kaneko [16] showed the model captures the dependence of void ratio and confining pressure on shear modulus of sands quantitatively.

When one wants to apply this model to well-graded granular materials, the integration in Eq. (5) is difficult to be done analytically even if the grain size distribution $P(r)$ is analytically specified, because the two PDFs, $P(r)$ and $F(n)$, are coupled. In the next section, we propose a simple method to deal with this problem.

3 Evolution of Elastic Modulus of Crushable Granular Materials Under Compression

Based on the power-law grain size distribution observed in the previous one-dimensional compression experiments of sands [23, 26], we assume a simplified prefractal microstructure after grain crushing, which is constructed in the following manner. First, we begin with a monodispersed grains of radius r_0 packed in the volume V_0 with the void ratio e_0, and then fill the void $(V_v)_0 = V_1$ with the grains of radius $r_1 \ll r_0$ with the same void ratio. This process gives us the following relation:

$$\frac{1}{e_0} = \frac{V_0}{V_1} - 1 = \frac{V_1}{(V_v)_1} - 1 \tag{13}$$

and the bulk void ratio after filling the small grains e_1 satisfies:

$$\left(1 + \frac{1}{e_1}\right) = \left(1 + \frac{1}{e_0}\right)^2 \tag{14}$$

Regarding these equations as a recursive one, we have:

$$\frac{V_k}{V_0} = \left(\frac{e_0}{e_0 + 1}\right)^k, \tag{15}$$

$$\left(1 + \frac{1}{e_k}\right) = \left(1 + \frac{1}{e_0}\right)^{k+1} \tag{16}$$

Now we consider to replace the regions of the larger grains generated up to the $(k-1)$-th processes, whose volume is $V_0 - V_k$, by a number of "virtual" grains of radius r_k, the same size as those generated by the k-the process. Then we assume that the contact properties for the "virtual" grains is determined so that the bulk elasticity is the same as the solid material, E and ν. In other words, we regard the well-grade grains system as two-phase monodisperse grains system consisting of the "real" smallest grains region (denoted by region g) and the "virtual" grains region (denoted by region G). The volumes of those regions are denoted by $V_g = V_k$ and $V_G = (V_0 - V_k)$, respectively.

This treatment makes it possible to apply the micromechanics model for monodisperse grains system in the previous section to the well-graded system. Since the ratio of the number of contacts in the two region may correspond to their volume ratio $(\alpha = V_g/V_0 = (N_c)_k/N_c)$, the summation in Eq. (5) is made separately in the two regions, and accordingly we obtain an overall elastic constitutive model for isotropic compression as follows:

$$p = \alpha p_g + (1-\alpha)p_G, \tag{17}$$

$$\varepsilon_v = \alpha\varepsilon_g + (1-\alpha)\varepsilon_G \tag{18}$$

$$\dot{p}_g = K_g \dot{\varepsilon}_g^{EL}, \tag{19}$$

$$\dot{p}_G = K_G \dot{\varepsilon}_G \tag{20}$$

$$K_g = \frac{E_g}{3(1-2\nu_g)}, \tag{21}$$

$$K_G = \frac{E_G}{3(1-2\nu_G)} = \frac{E}{3(1-2\nu)} \tag{22}$$

$$E_g = \sqrt[3]{\frac{3}{2}\frac{5-4\nu}{5-3\nu}}\left(\frac{E}{2\pi(1-\nu^2)}\right)^{2/3}\left(\frac{z_c}{1+e}\right)^{2/3} p_g^{1/3}, \tag{23}$$

$$\nu_g = \frac{\nu}{10-6\nu} \tag{24}$$

In order to close the equations, we need to introduce an additional equation:

$$\dot{\varepsilon}_G = B\dot{\varepsilon}_g + (1-B)\frac{K_g}{K_G}\dot{\varepsilon}_g^{EL} \tag{25}$$

where a parameter B describes the contribution of two extreme conditions; the uniform strain condition or the uniform stress condition as follows:

$$\dot{\varepsilon}_v = \dot{\varepsilon}_g = \dot{\varepsilon}_G \quad (B=0) \quad \text{uniform strain} \tag{26}$$

$$\dot{p} = \dot{p}_g = \dot{p}_G \quad (B=1) \quad \text{uniform stress} \tag{27}$$

Next we model the plastic regime, which is mainly caused by the grain crushing in brittle granular materials. Based on the proposed microstructure, we assume that the grain crushing occurs only in region g, because the larger grains in region G are subjected to isotropic stress from a large number of smaller grains surrounding them. This effect is referred to as "cushioning effect" [8]. In region g, the grain crushing occurs when the contact force exceeds a threshold. Since we assume the prefractal structure, the stress condition in region g is identical during loading process, and accordingly the crushing criterion is simply described as:

$$p_g \begin{cases} = p_Y & \text{crushing} \\ < p_Y & \text{not crushing} \end{cases} \tag{28}$$

Although experimental observations show the grain size dependence on p_Y [17, 22], we assume p_Y is constant in this study for simplicity. In order to keep the pressure constant in region g during plastic compression, the volume ratio α must decrease such that the compression rate is equal to the plastic void reduction rate $-\dot{V}_{gv}^{PL}$ as $\dot{\varepsilon}_g^E = -\dot{V}_{gv}^{PL}/V_g$, where ε_g^E is the plastic Eulerian strain defined by the current configuration and is expressed by the plastic Lagrangian strain ε_g^{PL} as $\varepsilon_g^E = -\ln(1-\varepsilon_g^{PL})$. Since the void ratio in region g is always e_0, we obtain:

$$V_{gv}^{PL} = \frac{e_0}{1+e_0}V_0\alpha \quad \text{or} \quad \dot{V}_{gv}^{PL} = \frac{e_0}{1+e_0}V_0\dot{\alpha} \tag{29}$$

Considering the definition of α ($\alpha = V_g/V_0$ excluding the effect of elastic compression), the evolution law of α is obtained as $\dot{\alpha}/\alpha = -(1+e_0)\dot{\varepsilon}_g^E/e_0$. Accordingly, the integration of this equation yields:

$$\ln \alpha = -\frac{1+e_0}{e_0}\varepsilon_g^E = \frac{1+e_0}{e_0}\ln(1-\varepsilon_g^{PL}) \tag{30}$$

taking care of the initial condition, $\alpha|_{\varepsilon_g=0} = 1$.

Then, the void ratio after plastic compression is described as:

$$\bar{e} = \frac{V_{gv}^{PL}}{V_0 - V_{gv}^{PL}} = \frac{\frac{e_0}{1+e_0}V_g^{PL}}{V_0 - \frac{e_0}{1+e_0}V_g^{PL}} = \frac{\frac{e_0}{1+e_0}\alpha}{1-\frac{e_0}{1+e_0}\alpha} = \frac{\alpha e_0}{1+e_0-\alpha e_0} \tag{31}$$

Note that $\bar{e}$ in Eq. (31) does not contain the effect of volume change of grain solid. In other words, $\bar{e}$ is the void ratio in an unloaded state. The calculation of the void ratio under loading requires the evaluation of the compression of solid. The elastic volumetric strain of solid in region G, ε_{Gs}, is equal to ε_G because it does not contain voids. The one in region g, ε_{gs}, is difficult to obtain because the stress field in a grain is not uniform, and accordingly the resulting strain is not uniform as well.

$$e = \frac{V_G' + V_g'}{V_G' + V_{gs}'} - 1 = \frac{(1-\varepsilon_G)(1-\alpha) + (1-\varepsilon_g)\alpha}{(1-\varepsilon_G)(1-\alpha) + (1-\varepsilon_{gs})\alpha/(1+e_0)} - 1 \tag{32}$$

where the prime indicates the volume after deformation, and

$$1-\varepsilon_{Gs} = 1-\varepsilon_G = \frac{V_G'}{V_G}, \tag{33}$$

$$1-\varepsilon_{gs} = \frac{V_{gs}'}{V_g/(1+e_0)}, \tag{34}$$

$$1-\varepsilon_g = \frac{V_g'}{V_g} \tag{35}$$

ε_{gs} must be modeled such that the following conditions satisfied:

$$e \geq 0 \Leftrightarrow V_g' \geq V_{gs}' \Leftrightarrow (1-\varepsilon_g)(1+e_0) \geq (1-\varepsilon_{gs}) \tag{36}$$

$$\varepsilon_{gs} = \begin{cases} 0 & (\varepsilon_g = 0) \\ 1 & (\varepsilon_g = 1) \end{cases} \tag{37}$$

Therefore, as the simplest model of ε_{gs}, we adopt a bi-linear model as follows:

$$\varepsilon_{gs} = \begin{cases} 0 & (\varepsilon_g \leq e_0/(1+e_0)) \\ (1+e_0)\varepsilon_g - e_0 & (\varepsilon_g \geq e_0/(1+e_0)) \end{cases} \tag{38}$$

4 Modification Under High Pressure Regime

The EOS of elastic solids have been studied extensively in the past decades. Here we use the well-known empirical relation that the bulk modulus of a solid K increases linearly with hydrostatic pressure p [9], that is:

$$K = K_0 + K_0^{'}(p - p_0) \quad K = -v\left(\frac{\partial p}{\partial v}\right)_T \quad K^{'} = \left(\frac{\partial K}{\partial p}\right)_T \tag{39}$$

where K_0 and $K_0^{'}$ are material constants, v is the specific volume ($v = 1/\rho$) and the subscript T shows the isothermal process. Note that $K_0^{'}$, the derivative of K_0 with respect to p, takes the value ranging $3 < K_0^{'} < 7$ for various solids according to Ruoff's relation, $K_0^{'} = 4s - 1$ [27] and the observation ($1 < s < 2$) in shock Hugoniot data [15]. Also, it should be noted that K in Eq. (39) is defined by the Eulerian frame, which relates the pressure to the strain with respect to the current configuration.

When we choose $p = 0$ as a reference state, the following relation holds during the isothermal process:

$$\frac{dv}{v} = \frac{-dp}{K_0 + K_0^{'}p} \tag{40}$$

Integrating it, we obtain:

$$p = \frac{K_0}{K_0^{'}}\left[\left(\frac{v}{v_0}\right)^{-K_0^{'}} - 1\right] = \frac{K_0}{K_0^{'}}\left[\left(\frac{\rho}{\rho_0}\right)^{K_0^{'}} - 1\right] \tag{41}$$

This is well known as Murnaghan EOS during isothermal process [21]. Note that a more general expression including the effect of the change of temperature [6] is not used in this study for simplicity.

Now we apply this relation into the well-graded granular system in the previous section. It is straight forward to do it in region G, which is regarded as the solid grains subjected to homogeneous boundary stress from sufficiently large number of contacts with smaller grains. In place of Eqs. (20) and (22) the following relation is adopted:

$$p_G = \frac{K_0}{K_0'}\left[(1-\varepsilon_G)^{-K_0'} - 1\right] \tag{42}$$

where $\rho_0/\rho = 1-\varepsilon_G$ is used. On the other hand, the relation in region g is difficult because the stress field in a grain is not uniform. However, it can be easily confirmed that the high pressure effect shown in Eq. (42) becomes predominant when the applied pressure satisfies $p/K_0 > 0.1$, which is much higher than p_Y in most of brittle geological solids. Therefore, we assume that there is no high pressure effect in region g in this study.

5 Validation

Figure 1 shows an example of the response of the proposed model. It includes the uniform strain case ($B = 1$), the uniform stress case ($B = 0$) and intermediate case ($B = 0.1$). The plastic (or unloading) void ratio $\bar{e}$ (Eq. (31)) and the total void ratio e (Eq. (32)) are drawn in each case. The parameters used in the calculation are also shown in the figure. Note that the horizontal axis is the pressure normalized by the bulk modulus of the solid, K. On the other hand, Fig. 2 shows an example of the one-dimensional compression of two types of sands; angular mountain silica sand and round river silica sand [29]. The yield stress p_y of these sands are about 3 MPa and 15 MPa, respectively. Although their K s are unknown, the typical value for silica glass is about 30–40 GPa, and accordingly $p_y/K = 10^{-3}$ to 10^{-4}. Comparing the two figures, it turns out that the proposed model well simulates the experimental results by tuning the parameter B. As shown in Eqs. (25)–(27), B represents the contribution of two extreme conditions; the uniform strain

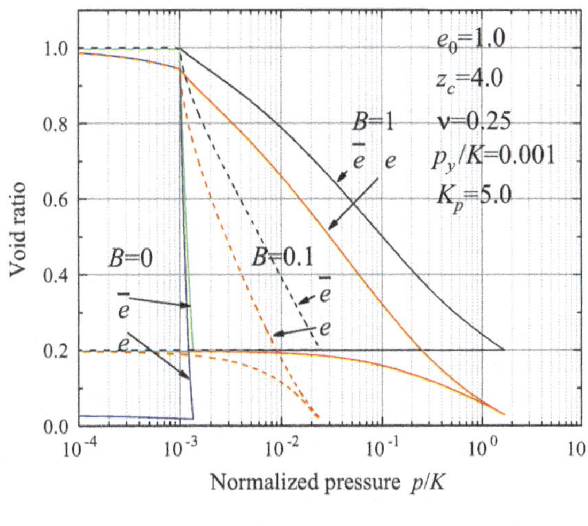

Fig. 1 Model response

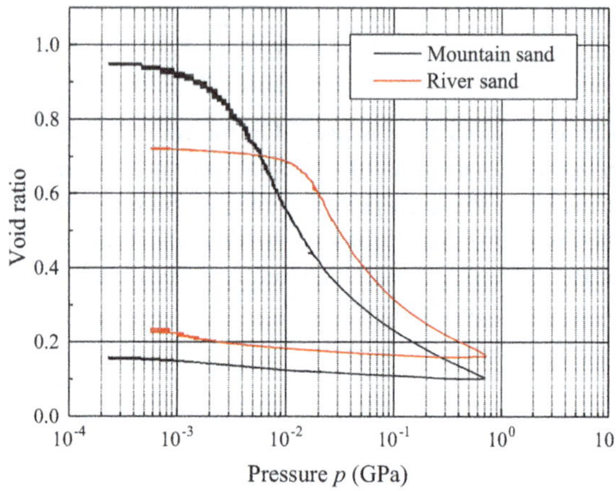

Fig. 2 1-D compression test result [29]

condition or the uniform stress condition, and will be studied in details in the future study.

Please note that the experimental results includes the effect of side wall friction, which may cause the different unloading path from the loading path.

6 Conclusions

The present study attempted to construct a micromechanics-based constitutive model for crushable granular materials subjected to high pressure loading. The proposed model can be applied from quasi-static loading to high-pressure impact loading including the effect of grain crushing and grading change in a reasonable way. Further experimental validation and modification is needed including the grain size effect on single particle crushing strength and the stress contribution parameter B.

References

1. Ahrens, T.J.: Equation of State. In High-Pressure Shock Compression of Solids, pp. 75–113. Springer, New York (1993)
2. Chang, C.S., Misra, A.: Application of uniform strain theory to heterogeneous granular solids. J. Eng. Mech. ASCE **116**(10), 2310–2328 (1990)
3. Chang, C.S., Gao, J.: Kinematic and static hypotheses for constitutive modeling of granulates considering particle rotation. Acta Mech. **115**, 213–229 (1996)
4. Christoffersen, J., Mehrabadi, M.M., Nemat-Nasser, S.: A micro-mechanical description of granular material behaviour. J. Appl. Mech. **48**(2), 339–344 (1981)
5. Coop, M.R., Sorensen, K.K., Freitas, T.B., Georgoutsos, G.: Particle breakage during shearing of a carbonate sand. Géotechnique **54**(3), 157–164 (2004)

6. Davison, L.: Fundamentals of Shock Wave Propagation in Solids. Springer Science & Buisiness Media, New York (2008)
7. Digby, P.J.: The effective moduli of porous granular rock. J. Appl. Mech. **48**, 803–808 (1981)
8. Einav, I.: Breakage mechanics–part I: theory. J. Mech. Phys. Solids **55**, 1274–1297 (2007)
9. Grover, R., Getting, I.C., Kennedy, G.C.: Simple compressibility relation for solids. Phys. Rev. B **7**(2), 567 (1973)
10. Guyon, E., Roux, S., Hansen, A., Bideau, D., Troadec, J.-P., Crapo, H.: Non-local and non-linear problems in the mechanics of disordered systems: application to granular media and rigidity problems. Rep. Progr. Phys. **53**, 373–419 (1990)
11. Hertz, H.: Über die Berührung fester elastischer Körper (On the contact of elastic solids). J. reine und angewandte Mathematik **92**, 156–171 (1882)
12. Hugoniot, H.: Mémoire sur la propagation des mouvements dans les corps et spécialement dans les gaz parfaits (première partie). Journal de l'ecole polytechnique **57**, 3–97 (1887)
13. Kanatani, K.: Distribution of directional data and fabric tensors. Int. J. Eng. Sci. **22**(2), 149–164 (1984)
14. Marks, B., Einav, I.: A mixture of crushing and segregation: the complexity of grainsize in natural granular flows. Geophys. Res. Lett. **42**(2), 274–281 (2015)
15. Marsh, S.P. (ed.): LASL Shock Hugoniot Data, p. 658p. University of California Press, USA (1980)
16. Matsushima, T., Kaneko, K.: Micromechanical constitutive models. In: Elasto-plastic constitutive models of soils, Japanese Geotechnical Society, pp. 120–134 (2009). ISBN-10: 4886449522 (in Japanese)
17. McDowell, G.R., Bolton, M.D.: On the micromechanics of crushable aggregates. Géotechnique **48**(5), 667–679 (1998)
18. Melosh, H.J.: Impact cratering—a geologic process, Oxford monographs on geology and geophysics, no. 11, p. 253. Oxford University Press (1989)
19. Miao, G., Airey, D.: Breakage and ultimate states for a carbonate sand. Géotechnique **63**(14), 1221 (2013)
20. Mindlin, R.D., Deresiewicz, H.: Elastic spheres in contact under varying oblique forces. J. Appl. Mech. ASME **20**(327), 221–227 (1953)
21. Murnaghan, F.D.: Finite deformations of an elastic solid. Am. J. Math. **49**, 235–260 (1937)
22. Nakata, Y., Hyde, A.F.L., Hyodo, M., Murata, H.: A probabilistic approach to sand particle crushing in the triaxial test. Geotechnique **49**(5), 567–583 (1999)
23. Nakata, Y., Hyodo, M., Hyde, A.F.L., Kato, Y., Murata, H.: Microscopic particle crushing of sand subjected to high pressure one-dimensional compression. Soils Found. **41**(1), 69–82 (2001)
24. Ouaguenouni, S., Roux, J.-N.: Compaction of well-coordinated lubricated granular pilings. Europhys. Lett. **32**(5), 449 (1995)
25. Rankine, W.J.M.: On the thermodynamic theory of waves of finite longitudinal disturbances. Philos. Trans. R. Soc. Lond. **160**, 277–288 (1870)
26. Pestana, J.M., Whittle, A.J.: Compression model of cohesionless soils. Geotechnique **45**(4), 611–631 (1995)
27. Ruoff, A.L.: Linear shock-velocity-particle-velocity relationship. J. Appl. Phys. **38**(13), 4976–4980 (1967)
28. Russell, A.R., Wood, D.M., Kikumoto, M.: Crushing of particles in idealised granular assemblies. J. Mech. Phys. Solids **57**(8), 1293–1313 (2009)
29. Sato, K., Seta, M., Kitajima, H., Takahashi, M., Matsushima, T.: Evolution of grain size distribution during high-pressure compression and shear. In: The Twenty-Eight KKHTCNN Symposium on Civil Engineering, Bangkok, Thailand (2015)
30. Sawaoka, A.B. (ed.): Shock Waves in Materials Science. Springer Science & Business Media, New York (2012)
31. Steacy, Sandra J., Sammis, Charles G.: An automaton for fractal patterns of fragmentation. Nature **353**(6341), 250–252 (1991)

32. Taylor, T., Fragaszy, R.J., Ho, C.L.: Projectile penetration in granular soils. J. Geotech. Eng. **117**(4), 658–672 (1991)
33. Turcotte, D.L.: Fractals and fragmentation. J. Geophys. Res.: Solid Earth **91**(B2), 1921–1926 (1986)
34. Tavares, L.M., King, R.P.: Single-particle fracture under impact loading. Int. J. Miner. Process. **54**(1), 1–28 (1998)
35. Walton, K.: The effective moduli of a random packing of spheres. J. Mech. Phys. Solids **35** (3), 213–226 (1987)
36. Zhang, Y.D., Buscarnera, G., Einav, I.: Grain size dependence of yielding in granular soils interpreted using fracture mechanics, breakage mechanics and Weibull statistics. Géotechnique **66**(2), 149–160 (2016)

Modelling Fabric Evolution of Granular Materials Along Proportional Strain Paths

Jinshan Shi and Peijun Guo

Abstract A series of biaxial tests are conducted along proportional strain paths using DEM to investigate the evolution of internal structure in the overall contact network and sub-networks (the strong and weak contact networks). The evolution of fabric in the overall contact network along various strain paths varies with the imposed strain ratio and the resulting stress ratio. Along dilatant strain paths, a unique fabric-stress relation is obtained for the strong network as $\phi_1^s - \phi_2^s = \kappa(t/s)$ with κ being a constant. The weak contact network and its evolution are affected by both the stress ratio and the imposed strain ratio. Deformation instability occurs only along dilatant strain paths and can be related to the degradation of weak network, even though the strong network dominates the strength of the material.

1 Introduction

The internal structure of a granular material can be altered when the material is subjected to shear distortion, due to the rearrangement of particles via relative particle movements including rolling and sliding at particle contacts. When deformation and shear resistance are concerned, not all contacts play the same role. According to Radjai et al. [4], the overall contact network can be separated into a sub-network of "strong contacts" that carry a force larger than the average force and another sub-network of "weak contacts" with a force lower than the average. The strong network is the "loading-bearing" network in which the contacts are non-sliding, whereas the weak network is the dissipative network and contributes to the average pressure only. Intuitively, the strong and weak contact networks play different roles in the deformation and failure process of granular materials.

J. Shi · P. Guo (✉)
Department of Civil Engineering, McMaster University,
Hamilton, ON L8S 4L7, Canada
e-mail: guop@mcmaster.ca

J. Shi
e-mail: shij9@mcmaster.ca

E. Papamichos et al. (eds.), *Bifurcation and Degradation of Geomaterials with Engineering Applications*, Springer Series in Geomechanics and Geoengineering, DOI 10.1007/978-3-319-56397-8_46

Various relations have been proposed to quantify fabric evolution. Results of biaxial tests on stack of photoelastic rods by Satake [6] demonstrate $\sigma_1/\sigma_2 \approx \sqrt{\phi_1/\phi_2}$. Antony et al. [1] propose a relation between the shear stress ratio and the fabric tensor of strong network as $q/p \approx (1/2)\sqrt{\phi_2/\phi_1}$. Under triaxial stress conditions, Ng [3] reveals $\ln(\phi_1/\phi_3) = k\sigma_1/\sigma_3$, while Wan and Guo [8] assume that the ratio of deviatoric fabric components is proportional to the deviatoric stress ratio. Sazzad [7] relates the deviatoric stress ratio to the fabric tensor for strong contact network rather than the whole contact network.

This paper examines the evolution of fabric in granular assemblies during biaxial compression along proportional strain paths through a series of DEM simulations. The results reveal that the deviator of fabric tensor based on the strong network has a unique relation with the deviatoric stress ratio, particularly for specimens along proportional strain paths corresponding to forced dilation. The maximum value of the deviator of fabric tensor varies with the applied strain ratio, or the maximum dilation rate in stress-controlled tests. The examination of the second-order work shows that the instability of deformation is a consequence of internal structure collapse indicated by the decrease of fabric anisotropy of the weak contact network.

2 DEM Simulation of Biaxial Tests Along Imposed Strain Paths

A rectangle packing of polydisperse disks is considered as the DEM specimen, which is constrained by four rigid sidewalls throughout the deformation history. A linear force-displacement contact law is employed where the contact behavior is governed by the normal stiffness k^n, tangential stiffness k^s and the friction coefficient μ. Here we assume $k^n/k^s = 1.0$ and $\mu = 0.5$. The specimens are sheared along proportional strain path with constant strain ratio $\mathfrak{R} = -\dot{\varepsilon}_1/\dot{\varepsilon}_2$. Herein $\dot{\varepsilon}_1$ and $\dot{\varepsilon}_2$ are the strain rate in the vertical (the direction of σ_1) and horizontal (the direction of σ_2) direction, respectively. The moving velocities of the boundary walls are slow enough ($\dot{\varepsilon}_1 = 10^{-5}$/s) to maintain a quasi-static condition. The vertical strain rate ($\dot{\varepsilon}_1$) is kept positive (compression), but $\dot{\varepsilon}_2$ can be either positive or negative depending on the value of $\mathfrak{R}$. The proportional strain paths can also be characterized by parameter $\vartheta = \dot{\varepsilon}_v/\dot{\gamma}$ that is related to $\mathfrak{R}$ via $\vartheta = (\mathfrak{R} - 1)/(\mathfrak{R} + 1)$. All the proportional strain paths can be categorized into forced dilation with $\vartheta < 0$ (or $0 < \mathfrak{R} < 1$) and forced contraction when $\vartheta > 0$ (or $|\mathfrak{R}| > 1$).

Following Radjai et al. [4], we evaluate geometrical anisotropy of granular assemblies by determining the fabric tensor for both the network of all contacts and subsets of contacts (the strong and weak contact networks). The fabric tensors of different contact networks are defined as Satake [5] $\phi_{ij} = (1/N_c)\Sigma n_i^k n_j^k$, $\phi_{ij}^s = (1/N_c^s)\Sigma n_i^k n_j^k$ and $\phi_{ij}^w = (1/N_c^w)\Sigma n_i^k n_j^k$, in which n_i^k is the i-th component of the unit vector along the direction of contact normal at contact k and the summation/averaging is taken over all N_c contacts in the contact network. The superscripts s and w represent quantities for strong and weak network respectively. The degree of anisotropy of different contact networks can be characterized by the deviator fabric $q_\phi = \phi_1 - \phi_2$, $q_\phi^s = \phi_1^s - \phi_2^s$ and $q_\phi^w = \phi_1^w - \phi_2^w$, respectively. The superscripts s and w stand for quantities for the strong and weak network respectively.

3 Fabric Evolution

3.1 *Internal Structure Evolution of Overall Contact Network*

The evolution of $q_\phi = \phi_1 - \phi_2$ in the whole contact network, which is near isotropic initially, with respect to the stress ratio (t/s) and shear strain is presented in Fig. 1. Prior to the state of peak stress ratio, q_ϕ increases with the increase of (t/s) (see Fig. 1a). After the peak, a reduction of q_ϕ is observed as t/s decreases and gradually approaches the critical state. Significant anisotropy develops when the specimen deforms along dilatant strain paths and it is considerably affected by the imposed strain rate. For a given vertical strain ε_1, the higher the lateral expansion, the stronger the induced anisotropy. This implies that the reduced lateral constraint allows more particles to rearrange themselves easily to form columns in the direction of σ_1, resulting in strong fabric anisotropy. At the same time, reduced constraint provides less support in the direction of σ_2 to the primary force chains, which may collapse when the deviator stress reaches a certain level. On the other hand, increased lateral constraint helps to build horizontal contacts, which tends to increase the number of contacts carrying low forces. Consequently, the fabric anisotropy become lower in tests along contractant strain paths. Prior to the peak stress state, the variation of q_ϕ with t/s and the imposed strain ratio can be described as $\phi_1 - \phi_2 = A(t/s)^n$, with $A = 1.60 - 0.94\vartheta$ and $n = 0.50 - 0.76\vartheta$.

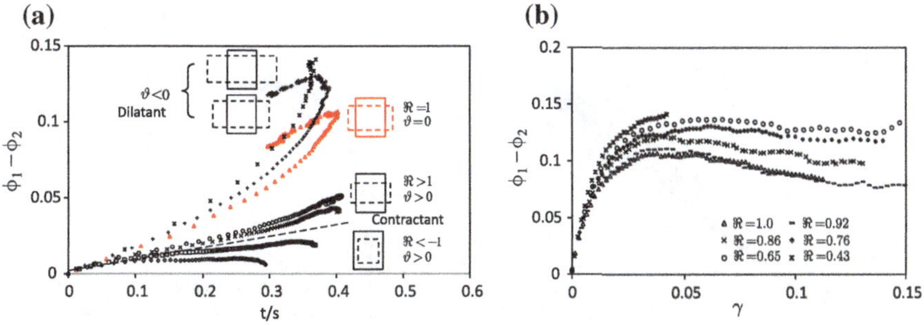

Fig. 1 Fabric evolution in the whole contact network along imposed strain paths

3.2 Evolution of Strong Network Fabric/Anisotropy

Figure 2 presents the evolution of $q^s_\phi = \phi^s_1 - \phi^s_2$ in the strong network with the *t/s* ratio and shear strain along imposed dilatant strain paths of $\mathfrak{R} = 0.43 - 1.0$. For dilatant strain paths, a unique relation $\phi^s_1 - \phi^s_2 = \kappa(t/s)$ with $\kappa \approx 1.0$ is observed prior to the peak deviator stress ratio, which is different from the fabric evolution of the whole contact network shown in Fig. 1a. The maximum induced anisotropy $q^s_{\phi,peak}$ appears simultaneously with $(t/s)_{\max}$. The post-peak data are slightly scattered, but are in a narrow band that almost overlaps the pre-peak data. Moreover, the $q^s_\phi \sim (t/s)$ relation is independent of the porosity of specimens or the confining pressure, and a critical state of internal structure exists at large deformation.

For contractant strain paths with $\vartheta > 0$ and $\mathfrak{R} > 1$, the $q^s_\phi \sim (t/s)$ curves are in a narrow band confined by the lines of $\mathfrak{R} = 1.0$ and $\mathfrak{R} = +\infty$; see Fig. 3a. The results suggest a strong dependency of q^s_ϕ in the strong network on (*t/s*) along contractant strain paths with $\mathfrak{R} > 1$. For imposed contractant strain paths with $\vartheta > 0$ and $\mathfrak{R} < -1$, q^s_ϕ increases monotonically with (*t/s*) and strongly depends on the value of $\mathfrak{R}$ or ϑ. The sustainable high level of anisotropy is owing to the increasing constraint in the direction of σ_2 along contractant strain paths.

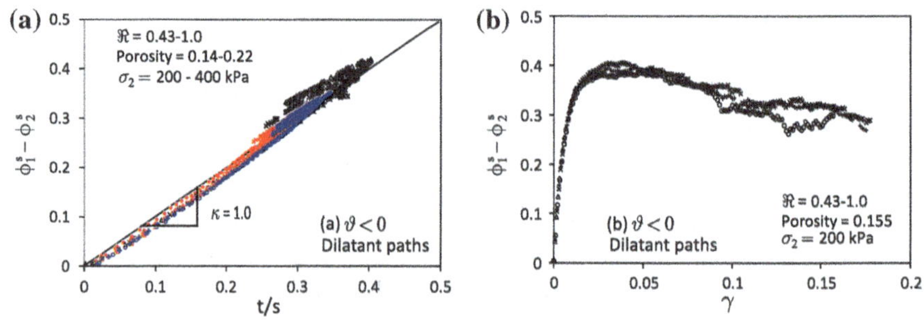

Fig. 2 Fabric evolution in strong contact network along imposed dilatant strain paths

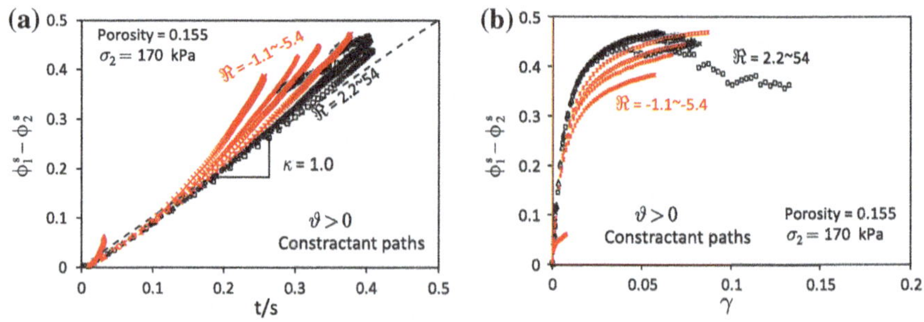

Fig. 3 Fabric evolution in strong contact network along imposed contractant strain paths

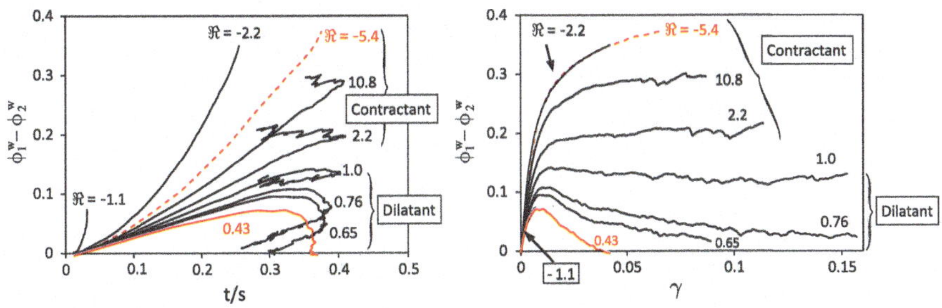

Fig. 4 Fabric evolution in weak contact network along various imposed strain paths

3.3 Evolution of Weak Network Fabric/Anisotropy

The evolution of anisotropy in the weak network with the (*t/s*) ratio and shear strain along different strain paths are presented in Fig. 4. For deformation along dilatant strain paths ($\Re < 1$), minor anisotropy in the weak network develops, with the principal direction perpendicular to the direction of σ_1 and hence the major principal direction of fabric tensor in the strong network. Following an initial buildup of anisotropy with the increase of the *t/s* ratio and shear strain, the value of q_ϕ^w reaches its maximum and then decreases with continuous deformation. The initial buildup of q_ϕ^w along dilatant strain paths can be described by $\phi_1^w - \phi_2^w = \kappa^w (t/s)^m$, with κ^w and m are functions of $\Re$ or ϑ. For high dilation rate or low $\Re$ values (e.g., $\Re = 0.43 \sim 0.65$), the anisotropy diminishes gradually after the peak anisotropy. The level of anisotropy of the weak network is much lower than that developed in the strong contact network along imposed dilatant strain paths.

For imposed contractant strain paths, q_ϕ^w develops intensively as (*t/s*) increases. With the increase of shear strain, the value of q_ϕ^w approaches a limit depending on the imposed strain ratio $\Re$ (see Fig. 4b). As a special case, no significant induced-anisotropy is observed in the weak network when a specimen is subjected to isotropic compression ($\Re = -1$). In general, the fabric evolution in the weak network does not depend on the stress state only, and the weak network could be highly anisotropic along imposed contractant strain paths.

4 Fabric Evolution and Deformation Instability

For deformation along an imposed proportional strain path, the second-order work is expressed as $d^2W = d(\sigma_1 - \sigma_2/\Re)d\varepsilon_1$, which implies that instability starts at $(\sigma_1 - \sigma_2/\Re) = \max$ [2, 9]. Figure 5a presents the correlation between q_ϕ^s and $(\sigma_1 - \sigma_2/\Re)$ along dilatant strain paths with $\Re = 0.43 \sim 1.0$. The thicker dotted segment on each curve represents states of $d^2W \le 0$. One immediate observation is

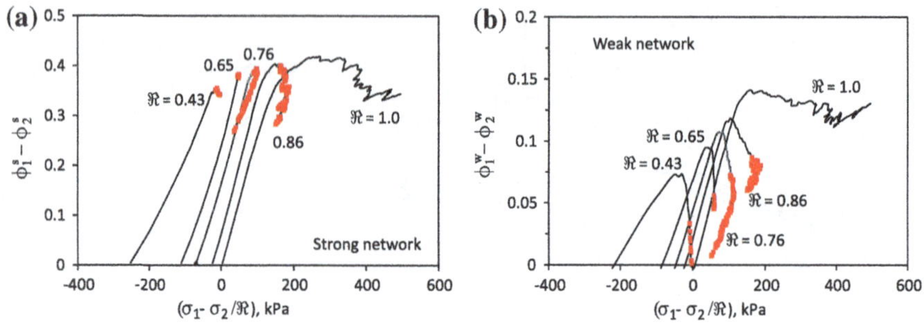

Fig. 5 Fabric evolution with $(\sigma_1 - \sigma_2/\mathfrak{R})$ in strong and weak sub-networks

that the strong sub-network is responsible for deformation instability since the peak fabric deviator of the strong subnetwork is reached when $d^2W = 0$. Degradation of strong contact network with reduced degree of anisotropy takes place following the onset of deformation instability. These observations indicate that the onset of deformation instability is accompanied by a degradation or progressive collapse of strong contact network. However, the degradation of strong contact network may not guarantee unstable deformation, as can be observed for the case of $\mathfrak{R} = 1$ in Fig. 5a.

Figure 5b shows the evolution of q_ϕ^w with $(\sigma_1 - \sigma_2/\mathfrak{R})$. For $\mathfrak{R} = 0.43 \sim 0.86$, $(q_\phi^w)_{\max}$ is not associated with $(\sigma_1 - \sigma_2/\mathfrak{R})_{\max}$ that is achieved after the peak of q_ϕ^w. In all these cases, significant decrease of q_ϕ^w takes place when deformation becomes unstable. For the strain path of isochoric compression at $\mathfrak{R} = 1.0$, deformation is always stable and degradation of q_ϕ^w does not seem to occur. The results for the contractant strain path of $\mathfrak{R} = 2.2$ have the same features as those of $\mathfrak{R} = 1.0$. Consequently, one may conclude that degradation of strong network seems to be a necessary but not sufficient condition for deformation instability, while the degradation of the weak network structure will eventually result in deformation instability.

5 Conclusions

A series of biaxial tests are conducted along different strain paths using DEM to investigate the evolution of internal structure characterized by the overall contact network and sub-networks (the strong- and weak-force chain networks). The evolution of fabric based on the overall contact network prior to the peak stress ratio *t/s* along various strain paths can be described as $\phi_1 - \phi_2 = A(t/s)^n$, with *A* and *n* being functions of imposed strain ratio. Along dilative strain paths, a unique fabric-stress relation is obtained for the strong network as $\phi_1^s - \phi_2^s = \kappa(t/s)$ with κ being a constant. The value of $(\phi_1^s - \phi_2^s)_{\max}$ within the strong sub-network varies with the imposed strain ratio. Deformation instability is observed only along dilatant strain paths and can be related to the degradation of weak network, even though the strong contact network dominates the strength of the material.

References

1. Antony, S.J., Momoh, R.O., Kuhn, M.R.: Micromechanical modelling of oval particulates subjected to bi-axial compression. Comput. Mater. Sci. **29**(4), 494–498 (2004)
2. Guo, P., Su, X.: Shear strength, inter-particle locking, and dilatancy of granular materials. Can. Geotech. J. **44**(5), 579–591 (2007)
3. Ng, T.T.: Fabric evolution of ellipsoidal arrays with different particle shapes. J. Eng. Mech. **127** (10), 994–999 (2001)
4. Radjai, F., Jean, M., Moreau, J.J., Roux, S.: Force distributions in dense two-dimensional granular systems. Phys. Rev. Lett. **77**(2), 274–277 (1996)
5. Satake, M.: Constitution of mechanics of granular materials through graph representation. Continuum mechanical and statistical approaches in the mechanics of granular materials, 47–62 (1978)
6. Satake, M.: Anisotropy in ground and soil materials. Tsuchito Kiso **32**(11), 5–12 (1984)
7. Sazzad, M.M.: Micro-scale behavior of granular materials during cyclic loading. Particuology, **16**, 132–141 (2014)
8. Wan, R., Guo, P.: Stress-dilatancy and fabric dependence of sand behaviour. J. Eng. Mech. **130** (6), 635–645 (2004)
9. Wan, R., Guo, P., Al-Mamun, M.: Behaviour of granular materials in relation to their fabric dependencies. Soils Found. **45**(2), 77–86 (2005)

Localised Deformation of Weakly Cemented Sands: A Case Study

Ilaria Soriano, Elli-Maria Charalampidou, Helen Lewis, Gioacchino Viggiani, Jim Buckman and Gary Couples

Abstract Deformation band occurrence in cemented granular materials is intimately related to grain and cement properties. In this work we focus on a Cretaceous sand deposit (France), in which numerous deformation bands have been locally observed. Material from this deposit is used to fabricate artificially cemented samples at the laboratory scale. We principally use two experimental techniques to visualise the inner structure of the natural and artificial samples and comment on the deformation processes: the Scanning Electron Microscopy (SEM) and the X-Ray Computed Tomography (XRCT). So far, microstructural observations on the material far from the deformation bands have revealed that the medium-sized sand grains are held together by weak menisci bonds made of clays. Inside one deformation band we have discovered, instead, sand grains cemented with clays and syntaxial quartz overgrowths. We argue that grain fragmentation in the deformation band is the main silica source for quartz precipitation. The syntaxial overgrowths formation, however, is still weakly justified by the geological reconstructed pressure and temperature conditions within the sand depositional area. Furthermore, in the clay cement we have identified quartz micro-fragments, derived probably from the mentioned grain breakage, and micro-quartz, which may have generated from the interaction between free silica and clay in the pore space.

1 Introduction

The way grain texture and cement characteristics affect the occurrence of strain localisation in sandstones has not yet been exhaustively explored. Grain size seems to control the density of deformation band formation [8] and grain mineralogy tends

---

I. Soriano (✉) · E.-M. Charalampidou · H. Lewis · J. Buckman · G. Couples
Heriot-Watt University, Edinburgh, UK
e-mail: is24@hw.ac.uk

I. Soriano · G. Viggiani
Université Grenoble Alpes – Laboratoire 3SR, Grenoble, France

E. Papamichos et al. (eds.), *Bifurcation and Degradation of Geomaterials with Engineering Applications*, Springer Series in Geomechanics and Geoengineering, DOI 10.1007/978-3-319-56397-8_47

to define the strength of a rock [6], whereas the dependence of deformation bands on cement properties has been fewly discussed so far.

The role of grain cementation in the laboratory scale has been principally studied in artificially cemented samples. However, these studies often do not take into account the nature of the cement that may also affect the mechanical response of cemented sands [16]. Therefore, it can be argued that artificially cemented granular materials could lead to different behaviours with respect to the naturally cemented materials, if the cementing agent is not similar (or at least a good analogue) to the natural cement. Previous works have documented the influence of different cement types in the mechanical behaviour of artificially cemented sands [10]. Some other studies comment on the intrinsic properties of the cement, e.g. cement bonding and strength with respect to the grain [5]. Mineralogy and degree of cementation [3] together with variations in cement distribution [2] and morphology can mechanically differentiate samples made of the same granular material.

In this work we aim at a deeper understanding of the interplay between grain texture and cement characteristics in deformation processes occurring during strain localisation in a weakly cemented sand deposit. By doing this, we can well approximate the analysis of deformation band occurrence and evolution in sandstones, since the two materials share similar deformation features [11]. The field study is coupled with a laboratory investigation on artificially cemented sand samples, which consist of grains coming from the sand outcrops being cemented similarly to the field case cementing agents.

2 Study Area and Material Properties

The material used in this work comes from a site located near Bédoin, southern France [14, 17]. This is an undersaturated Cretaceous sand deposit of weakly cemented sands, which according to geological estimations should have reached a maximum burial depth of 500 m. The sand grains have an average size of 350 μm (Fig. 1) and are sub-rounded to round. This material is rather poorly sorted because of the presence of clay particles mixed with the sand grains. Along the field outcrops the sand grains are held together by a mixture of clays and water, creating a matrix with a low degree of cementation.

Although the cemented sands are generally weak, numerous sets of conjugate deformation bands developed within the outcrops of the French sand deposit. There is a number of deformation bands with a thickness of 1 to 10 cm (Fig. 2a), with one reaching up to 50 cm in thickness (Fig. 2b) crossing the deposit sub-vertically with a strike direction 265.

The study material has been collected from three different sites: (i) from places far from the deformation bands; (ii) from the thin deformation bands commonly spread within the field (Fig. 2a); and (iii) from the thick deformation band (Fig. 2b).

Fig. 1 Sand grains collected from Bédoin, cemented and rimmed with flocculated clay

Fig. 2 Deformation bands observed in the site on horizontal surfaces: **a** thin deformation bands; **b** a thick deformation band (*red* tool as scale, l = 15 cm)

3 Methodology and Tools

The work is divided in three interconnected parts: the first part consists of a thorough inspection of the natural cementation, by analysing the material collected from the field; the second part involves the reproduction at the laboratory scale of cementation features (e.g. cement type, morphology) observed in the field case; with the third part we aim at connecting the two previous parts through the comparison of the mechanical behaviour and the assessment of the occurred deformation processes between the natural (collected from the field) and artificially cemented (reproduced in the laboratory) sand assemblies.

Table 1 Selected parameters of the sample components and of the assembly that will be object of measurement

Properties of	
Grains	**Cement**
Mineralogy	Mineralogy
Roundness, sorting	Morphology
Size	Content
Orientation	Distribution
Assembly	
Porosity	
Permeability	

Two different experimental methods are used throughout this study: the Scanning Electron Microscopy (SEM), which is a destructive method, and the X-Ray Computed Tomography (XRCT), which is a non-destructive one. Note that the two imaging techniques are distinct and complementary in their outcomes: on the one hand, with SEM we can produce 2D images with a resolution of nm, necessary to analyse the role of the cement texture and of the grain characteristics (Table 1); on the other hand, XRCT cannot reach the same resolution but this technique provides a 4D visualisation of the sample with a resolution of some µm. This allows us not only to measure the cement distribution and porosity within the entire sample but also to quantify grain breakage, through density variation, and grain rotation occurring during localised deformation (triaxial compression tests).

We are currently working on the understanding of the naturally and artificially cemented samples using primarily the SEM so as to characterise qualitatively both materials. The next step is to perform triaxial compression experiments on both natural and artificially cemented sand assemblies coupled with pre-, syn- and post-deformation XRCT (not discussed here). The latter are combined with Digital Image Correlation (DIC) leading to both qualitative and quantitative investigation of the occurred deformation processes.

4 Cementation Procedures

The laboratory work consists, among others, in the cementation of artificially sand samples that can be representative of the reference case study. We use sand with the same mineralogy and grain roundness (Table 1) as the sand in the deposit (collected directly from the site, far from deformation bands); however, the grain sorting and size slightly vary since we clean the grains from the clay rimmed around them (Fig. 1) using initially a water solution at 10% of Sodium Hexametaphosphate, (surface inhibitor). The grains are, then, washed with distilled water and filtered with a sieve of small aperture to flush the small particles away, which results in slightly bigger average grain size sand than the natural one.

Two different cements have been used for the artificial cementation: kaolin clay and calcium carbonate. The former has a similar mineralogy to the natural cement detected in the field (i.e. kaolinite minerals among all the other clay types); the latter, although it has not been detected in our field case, is of equal importance since it can reproduce a bonding morphology similar to our naturally cemented sands (i.e. [1, 15]. Moreover, calcite cementation is often found in nature as cement of granular materials and rocks, sometimes even associated with quartz formation [18]. Procedures for calcite cementation are proposed in the literature [10, 13]. We are currently investigating potential alternative procedures.

Kaolin clay cementation

The kaolin clay cement is directly injected from the top of the sample in the sand matrix through a water suspension containing 20% of clay with respect to the sand weight. The water quantity is a function of the sample dimensions: for cementing a sample with a diameter of 11 mm and a length of 22 mm, we use 1 ml of water. A key point for the cementation procedure is to keep the sample under unsaturated conditions so as to prevent the clay precipitation at the bottom of the sample, which systematically occurs in a saturated environment driven by the gravity force. The applied procedure favours the clay adherence to the sand grains. Our unsaturated samples resemble the sand deposit, where clay menisci occupy the unsaturated pore space filled with air, connecting the grains in a weak structure.

To avoid saturation of the sample, the suspension is split in two and is injected twice in an interval of 1–2 days. Certainly, the double injection creates cement heterogeneity inside the sample (Fig. 3), which may influence the onset and evolution of strain localisation during the triaxial compression experiments; however, the heterogeneously cemented sample is not an important issue as long as such heterogeneity can be identified by the experimental methods (i.e. XRCT). We are currently working on evolving further the present cementation procedure.

Fig. 3 Cylindrical sample cemented with clay: *bottom* and *top* of the sample result being cemented with different degree of cementation

5 Results and Discussion

Initial results concentrate mainly on the better understanding of the sand assemblies outside and within some of the deformation features in the sand outcrop. Herein, we focus on the cement type and morphology.

In places far from the deformation bands, the sands are cemented with a mixture of clays, which should have filled the pore space in an undersaturated environment, where the presence of air probably led to formation of menisci (Fig. 4). Similarly, Cheng et al. [1], while focusing on calcite bio-cementation, have shown that partial saturation favours the deposition of cement at the inter-granular contacts, forming menisci shapes, whereas in fully saturated conditions, calcite crystals precipitate all around the pore space and not only at the grain contacts. The presence of menisci bonds in poorly consolidated sands has been already documented [4] and it is consistent with our hypothesis that this sand deposit should not have reached a burial depth greater than 500 m, i.e. it has not experienced consolidation.

Although clay cementation globally characterises the sand outcrops, the cementation within the thick deformation band (Fig. 2b) is somehow different. A multiple generation of syntaxial quartz overgrowths has grown around the sand grains. Figures 5a presents a number of quartz overgrowth layers, which reflect changes in water chemistry during precipitation stages. Up to nine overgrowths generations can be observed (see different grey levels). The quartz layers are locally broken and detached from the surface they were formed on, which could be the effect of deformation.

Grain fragmentation is also observed inside the thick deformation band, which is directly linked with deformation processes. Moreover, the presence of numerous quartz micro-fragments is identified between the broken sand grains. These micro-fragments are slightly elongated and particularly smooth with a size of 1–2 μm (Fig. 5b). It is quite unclear the reason for such roundness, since grain breakage and subsequent grinding of the angular fragments is not likely to produce such smooth fragment contours.

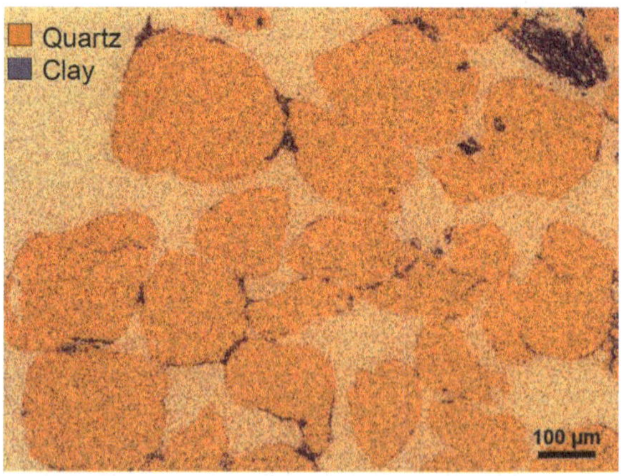

Fig. 4 Energy Dispersive X-ray (EDX) image of quartz sand grains (*orange*) cemented to each other throughout clay menisci (*purple*)

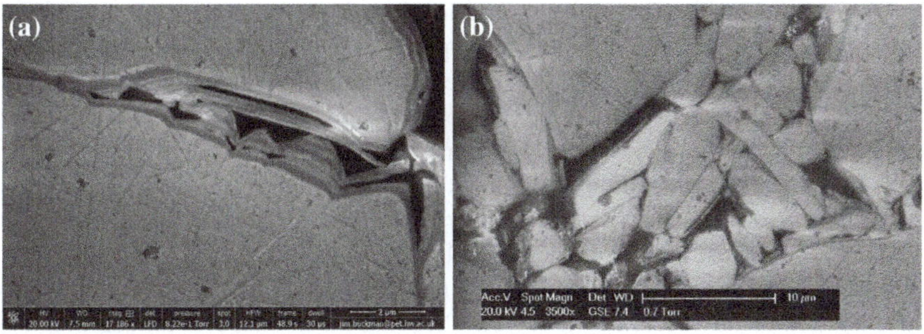

Fig. 5 Charge contrast imaging (CCI) from the thick deformation band of Fig. 2b showing: **a** "pyramidal" syntaxial quartz overgrowths; **b** elongated and smooth quartz micro-fragments surrounded by overgrowths

Inside the thick deformation band, clays are also observed in the pore space holding together all the grain micro-fragments. Clay particles form a flocculated structure that produces a high surface tension. Some clay particles create also a very thin film around the sand grains, where the free silica should have precipitated under the right conditions within the silica mobility window. Usually, the silica in the overgrowths follows the orientation of the quartz crystals. Due to the presence of the clay, though, the silica is unable to recognise the existing orientation, therefore it does not precipitate as syntaxial quartz overgrowths, in agreement with Haszeldine et al. [9]. Consequently, we have identified locally authigenic micro-quartz (Fig. 6) instead of quartz overgrowths together with the clay coats.

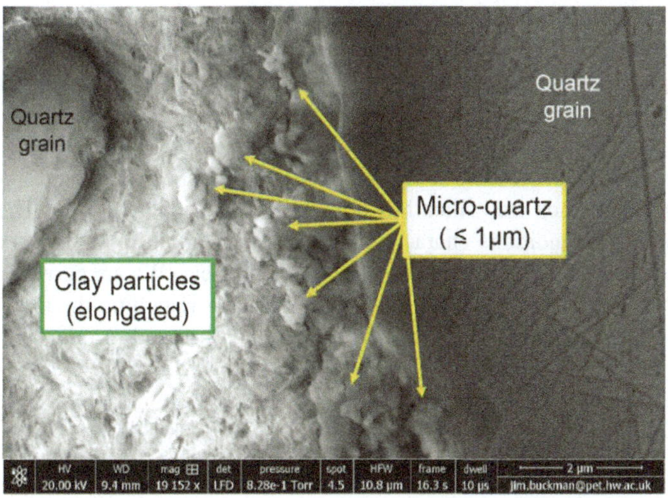

Fig. 6 Charge contrast imaging (CCI) of a detail from the thick deformation band representing the cement between two quartz grains: flocculated clay and granular micro-quartz

Within sandstones [12, 18] the main silica sources for syntaxial overgrowths and micro-quartz are: pressure dissolution of quartz grains; decomposition of feldspar; clay-mineral transformations; carbonate replacement of silicate minerals; and dissolution of amorphous silica. In the case of the French sand deposit, we argue that the silica comes from the intense grain breakage [7], as a result of grain-to-grain pressure during the generation of the thick deformation band. Thus, we look at the quartz overgrowths as post-dated with respect to the strain localisation since so far we have not located them anywhere else but inside the thick deformation band.

Specific conditions of pressure and temperature in the silica mobility window have stimulated the precipitation of free silica into layers that coat all quartz grains and fragments, i.e. syntaxial quartz overgrowths (Fig. 5). We are still investigating the applied pressure and temperature conditions for this specific sand deposit. According to thermic gradient data, a maximum temperature of 30 °C has been reached at shallow burial ($\leq$500 m). Moreover, the proximity to the Pleistocene Glaciated Alps probably favoured a decrease in temperature in the south-eastern France. However, this estimated temperature is not enough for enhancing quartz deposition because higher temperatures (70–80 °C) are usually required. Given the great variability in the conditions that determine the quartz deposition, the origin of the quartz overgrowths in this French deposit is still open.

6 Conclusions

From our current analysis on samples collected from the French sand deposit it has emerged that clay cementation mainly characterises this deposit, with clays creating menisci bridges between the sand grains. A number of deformation bands has been identified in the deposit. Microstructural investigations on one of these bands have demonstrated local intense grain breakage, supported by numerous elongated and smooth quartz micro-fragments that filled tightly the pores between the bigger (non-fractured) sand grains. The free silica, probably produced by the grain fracturing, has precipitated into syntaxial quartz overgrowths, surrounding the sand grains and quartz micro-fragments. In some cases, especially in the presence of flocculated clay, such free silica has transformed into micro-quartz. The pressure and temperature conditions characterising the silica mobility window for the syntaxial quartz overgrowth precipitation are not fully clear. We are currently working on a further understanding of the deformation history regarding the field case by a thorough investigation of other deformation bands observed in this sand deposit.

References

1. Cheng, L., Cord-Ruwisch, R., Shahin, M.A.: Cementation of sand soil by microbially induced calcite precipitation at various degrees of saturation. Can. Geotech. J. **50**, 81–90 (2013). doi:10.1139/cgj-2012-0023
2. David, C., Menéndez, B., Bernabé, Y.: The mechanical behaviour of synthetic sandstone with varying brittle cement content. Int. J. Rock Mech. Min. Sci. **35**, 759–770 (1998)
3. den Brok, S.W.J., David, C., Bernabé, Y.: Preparation of synthetic sandstones with variable cementation for studying the physical properties of granular rocks. C R Acad. Sci. Earth Planet Sci. **325**, 487–492 (1997)
4. Du Bernard, X., Eichhubl, P., Aydin, A.: Dilation bands: a new form of localized failure in granular media. Geophys. Res. Lett. **29**, 21–29 (2002). doi:10.1029/2002GL015966
5. Dvorkin, J., Yin, H.: Contact laws for cemented grains: implications for grain and cement failure. Int. J. Solids Struct. **32**, 2497–2510 (1995). doi:10.1016/0020-7683(94)00279-6
6. Exner, U., Tschegg, C.: Preferential cataclastic grain size reduction of feldspar in deformation bands in poorly consolidated arkosic sands. J. Struct. Geol. **43**, 63–72 (2012). doi:10.1016/j.jsg.2012.08.005
7. Fischer, C., Waldmann, S., Von Eynatten, H.: Spatial variation in quartz cement type and concentration: An example from the Heidelberg formation (Teufelsmauer outcrops), Upper Cretaceous Subhercynian Basin, Germany. Sediment Geol. **291**, 48–61 (2013). doi:10.1016/j.sedgeo.2013.03.009
8. Griffiths, J., Faulkner, D.R., Edwards, A.P., Worden, R.H.: Deformation band development as a function of intrinsic host-rock properties in Triassic Sherwood Sandstone. Geol. Soc. Lond. Spec. Publ. **435**, 1–16 (2016)
9. Haszeldine, R.S., Cavanagh, A.J., England, G.L.: Effects of oil charge on illite dates and stopping quartz cement: calibration of basin models. J. Geochem. Explor. **78–79**, 373–376 (2003). doi:10.1016/S0375-6742(03)00151-1
10. Ismail, M.A., Joer, H.A., Sim, W.H., Randolph, M.F.: Effect of cement type on shear behavior of cemented calcareous Soil. J. Geotech. Geoenv. Eng. **128**, 520–529 (2002). doi:10.1061/(ASCE)1090-0241(2002)128:6(520)
11. Kaproth, B.M., Cashman, S.M., Marone, C.: Deformation band formation and strength evolution in unlithified sand: the role of grain breakage. J. Geophys. Res. Solid Earth **115**, 1–11 (2010). doi:10.1029/2010JB007406
12. McBride, E.F.: Quartz cement in sandstones: a review. Earth Sci. Rev. **26**, 69–112 (1989)
13. Molenaar, N., Venmans, A.A.M.: Calcium carbonate cementation of sand: a method for producing artificially cemented samples for geotechnical testing and a comparison with natural cementation processes. Eng. Geol. **35**, 103–122 (1993). doi:10.1016/0013-7952(93)90073-L
14. Saillet, E., Wibberley, C.A.J.: Permeability and flow impact of faults and deformation bands in high-porosity sand reservoirs: Southeast Basin, France, analog. Am. Assoc. Pet. Geol. Bull. **97**, 437–464 (2013)
15. Tengattini, A., Das, A., Nguyen, G.D., et al.: A thermomechanical constitutive model for cemented granular materials with quantifiable internal variables. Part I—Theory. J. Mech. Phys. Solids **70**, 382–405 (2014)
16. Wang, Y.-H., Leung, S.-C.: A particulate-scale investigation of cemented sand behavior. Can. Geotech. J. **45**, 29–44 (2008). doi:10.1139/T07-070
17. Wibberley, C.A.J., Petit, J.-P., Rives, T.: The mechanics of fault distribution and localization in high-porosity sands, Provence, France. Geol. Soc. Lond. Spec. Publ. **289**, 19–46 (2007). doi:10.1144/SP289.3
18. Worden, R.H., Morad, S.: Quartz cementation in oil field sandstones: a review of the key controversies. Spec. Publ. Int. Assoc. Sedimentol. **29**, 1–20 (2000). doi:10.1002/9781444304237

Compaction Bands in a Porous Sandstone Sample with Pre-induced Shear Bands

Elli-Maria Charalampidou, Sergei Stanchits and Georg Dresen

Abstract Understanding how different modes of deformation bands may interact at the same rock mass is of crucial importance, since such interaction may affect the flow properties within the reservoir rock. In this work we focus on the compaction band nucleation and evolution within a porous sandstone, in which a shear-band has been previously developed. For such a purpose, we performed at the laboratory scale triaxial compression experiments under 20 and 185 MPa confining pressures on a single Bentheim sandstone specimen. Acoustic Emissions (AE) were recorded throughout all experimental stages. AE hypocentre locations and AE source mechanisms were used to describe the spatiotemporal evolution of the developed deformation bands. The shear band evolution was AE controlled. Shear type sources were the prevailing mechanisms up to the peak stress, whereas, the shear band growth was mainly dominated by compressive type sources. The compaction band nucleated at the tip of the pre-existing shear band and evolved towards the circumference of the specimen. A second compaction band nucleated with increasing axial strain at the top part of the specimen and not far from the already developed shear and compaction bands. The dominant mechanisms during the compaction band initiation and formation were compressive type sources.

E.-M. Charalampidou (✉)
Institute of Petroleum Engineering, Heriot Watt University, Edinburgh Campus, Edinburgh, Scotland EH14 4AS, UK
e-mail: e.charalampidou@hw.ac.uk

S. Stanchits
Schlumberger Research, 1935 Fremont Dr., Salt Lake City, UT 84104, USA
e-mail: sstanchits@slb.com

G. Dresen
Helmholtz Centre Potsdam, GFZ German Research Centre for Geosciences, 14473 Telegrafenberg, Potsdam, Germany
e-mail: dre@gfz-potsdam.de

E. Papamichos et al. (eds.), *Bifurcation and Degradation of Geomaterials with Engineering Applications*, Springer Series in Geomechanics and Geoengineering, DOI 10.1007/978-3-319-56397-8_48

1 Introduction

Compaction bands in porous sandstones outcrops have been observed only in few field studies [4, 6, 11]. These deformation features have been mainly described as tabular zones of localised deformation that accommodate pure compaction, with no macroscopic evidence of shear [11]. Compaction bands are formed normal or subnormal to the maximum principal stress direction and are accompanied by localised porosity loss and permeability reduction [6, 15]. The involved micro-processes are mainly characterised by grain crushing and pore collapse [5]. Compaction bands are important deformation features, since they can become baffles to fluid flow and therefore have outstanding implications for hydrocarbon production or CO_2 storage in subsurface geological reservoirs.

On the laboratory scale, a number of experimental studies have been conducted in order to replicate compaction bands on a much smaller scale so as to better understand the conditions of their development within sedimentary rocks, i.e. porous sandstones. In most of these studies intact cylindrical sandstone specimens have been used [1, 7, 8, 17]. Some other studies introduced pre-machined notches along the mid- height of the cylindrical specimens to focus on the role of local stress concentration on the formation of compactions bands [2, 3, 14, 16].

In this work, we investigate at the laboratory scale the impact of a pre-induced shear band on the initiation and propagation of compaction bands. At the field scale, this is a very probable scenario, since a variety of different orientation faults may exist in a subsurface reservoir in which compaction bands may be triggered under the appropriate stress conditions. The goal of this work is to gain further insights in the formation and evolution of compaction bands together with the accompanied micro-processes in cases where pre-existing shear bands are present, which is a crucial step in expanding knowledge of the larger scale systems. We initially present the material used and the experimental set-up. Results are discussed comparing the spatiotemporal evolution and the involved micro-processes of the pre-induced shear and the subsequently developed compaction bands.

2 Material and Experimental Set-up

A cylindrical Bentheim sandstone specimen with a diameter of 50 mm and a length of 125 mm was tested without any lubrication at its top and bottom edges. This sandstone has an average porosity of 23%, a mean grain size of 210 μm and contains 95% quartz, 3% kaolinite and 2% orthoclase [7].

Triaxial compression experiments were carried out in a servo-hydraulic loading frame from Material Testing Systems (MTS). To reduce friction between the specimen and the end-caps, we installed a slim Teflon disk in-between the above mentioned interfaces. The loading of the sandstone was performed in two stages. First loading stage included a triaxial compression test under 20 MPa confining pressure. The axial load was applied to the specimen with an Acoustic Emission control in order to stabilise the failure process [10]. Second loading stage included a triaxial compression test under 185 MPa confining pressure. The axial load was applied to the specimen with a displacement control rate of 20 μm/min. In both experimental stages the applied force was measured with an external load cell and calibrated with an internal sensor.

Ultrasonic transmission signals and Acoustic Emissions (AE) were recorded throughout the experiments using: (a) a set of sixteen P-wave piezoelectric sensors glued directly to the surface of the specimen and sealed in a neoprene jacket with two-component epoxy and (b) two P-wave piezoelectric sensors embedded in two metallic spacers placed at the bottom and top ends of the sandstone specimen. The sensors' network geometry provided a good azimuthal coverage of the AE events. All P-wave sensors had a resonant frequency of 1 MHz. AE signals were amplified by 40 dB, using Physical Acoustic Corporation (PAC) preamplifiers [13]. Ultrasonic transmission measurements took place every 30 s. During the velocity measurements, half of the sensors were disconnected from the preamplifiers and were emitting electrical pulses of 100 V amplitude and 3 μs duration. The remaining sensors were recording these pulses.

AE waveforms and ultrasonic signals were stored continuously in a 16 channel transient recording system (DAXBox, PRÖKEL, Germany) with an amplitude resolution of 16 bit at 10 MHz sampling rate [13, 14]. After the experiments both ultrasonic signals and AE waveforms were automatically discriminated. P-wave onset times were calculated by applying an automatic picking algorithm based on the Akaike information criterion [9]. 4D AE hypocentre locations were calculated by minimising travel-time residuals using the downhill simplex algorithm [12], considering time dependent variations in P-wave velocities and employing an anisotropic heterogeneous ultrasonic velocity model, consisting of five horizontal layers [13]. AE events were classified according to their polarity (i.e., the mean of all signed first motion amplitudes) as tensile ($-1 \leq pol < -0.25$), shear ($-0.25 \leq pol \leq +0.25$) and compressive ($+0.25 < pol \leq +1$) type sources [18].

3 Experimental Results

The mechanical behaviour and failure modes of the Bentheim sandstone, under triaxial compression and at a range of confining pressures has been previously well-characterised by [1, 7, 14, 16, 17]. However, all previous laboratory works were carried out on intact specimens, i.e. specimens that did not have any laboratory induced deformation feature that could be viewed as an induced structural heterogeneity within the sample, although some specimens had a mid-high circumferential notch, which constituted a geometrical heterogeneity. In this work, we consecutively loaded a single Bentheim specimen, which initially developed a shear band (first loading stage) acting as stress concentrator for the subsequent compaction bands (second loading stage) formation.

3.1 Shear Band

Figure 1 shows the loading curve from the Bentheim specimen, which was initially subjected to triaxial compression under 20 MPa. The specimen failed by shear localisation along a single non-planar shear band. Differential stress (q) is plotted as a function of axial strain ($q = (\sigma_a - \sigma_r)$, where σ_a is the axial stress and σ_r is the confining pressure, respectively; the latter being constant during the experiment). During this loading stage, the failure process was stabilised by controlling the axial stress to maintain a constant rate of AE. The post-peak stress curve can be followed quasi-statically [10], extending the shear band growth process to almost 2 h, which normally would take place abruptly in a couple of seconds.

Three 2D projection of the 4D AE hypocentre distributions (one perpendicular and two parallel to the specimen axis) are shown in Fig. 2a–f, corresponding to the equivalent time intervals highlighted by red circles in Fig. 1. Only AE events with amplitudes larger than 1 V are plotted. The colour map represents the polarities of these AE events. The accuracy of AE hypocentre location is about 2 mm.

AE activity has been initially recorded at the top edge boundary, however, soon after the peak stress (Fig. 2a onwards), it concentrates along the shear band that progressively evolves from the top to the mid-section (~65 mm from the bottom edge) of the specimen. The resolved shear band developed at an angle of ~60° to the maximum principal stress direction and it is mainly characterised by AE events (amplitudes larger than 1 V) with polarities ranging from 0 to 1 (i.e. shear and compressive type sources).

3.2 Compaction Band

After the shear band creation (Sect. 3.1), the confining pressure was increased from 20 to 185 MPa. Subsequently, the Bentheim sandstone specimen was loaded in triaxial compression (under 185 MPa confining pressure) at a fixed displacement

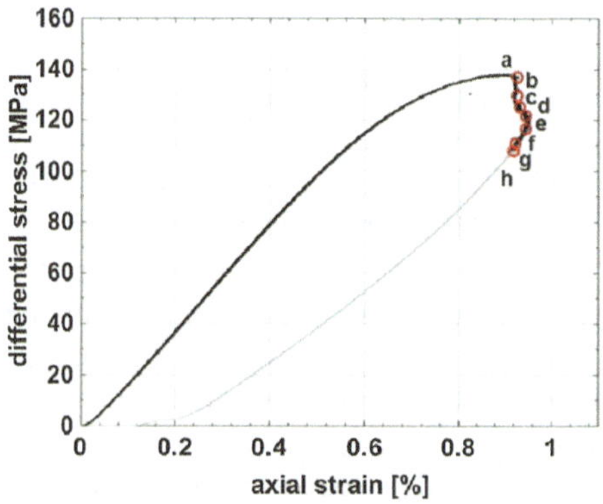

Fig. 1 Differential stress as a function of axial strain. The *dark* and the *light colour lines* highlight the loading and the unloading of the specimen, respectively

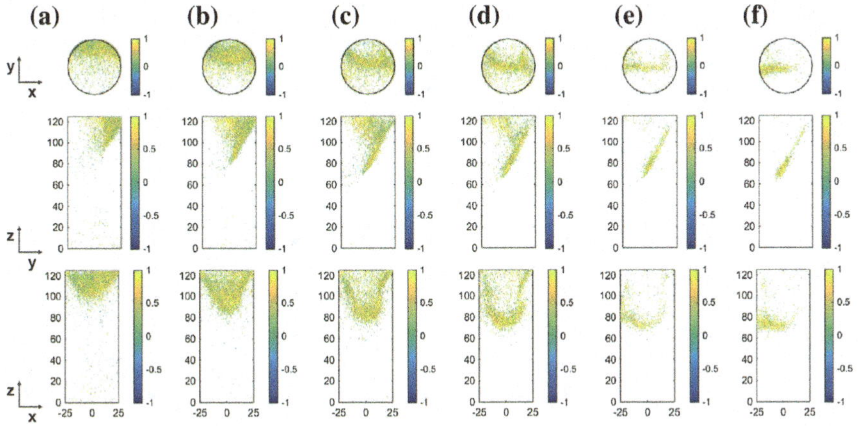

Fig. 2 AE spatiotemporal distribution during triaxial compression under 20 MPa confining pressure: **b–g** maps of AE event locations for the same time intervals viewed perpendicular and parallel to the specimen axis

rate corresponding to a nominal strain rate of 2 × 10^{-3} s^{-1}. Axial loading was then stopped close to 1% of maximum axial strain. Figure 3 shows the differential stress as a function of axial strain. The specimen had an overall strain hardening behaviour with local episodic stain softening. In general, the stress drops were small (with amplitudes ~0.5 MPa) and only during the last two time intervals (Fig. 3g, h) the stress drop reached a maximum amplitude of 2 MPa.

Three 2D projection of the 4D AE hypocentre distributions are shown in Fig. 4a–f, corresponding to the equivalent time intervals highlighted by red circles in Fig. 3. AE events with amplitudes larger than 1 V are plotted in grey colour, whereas the polarities of AE events with amplitudes larger than 3 V are plotted in colour (see colour map). The accuracy of AE hypocentre location is about 2 mm.

Diffuse damage is initially observed within the whole specimen with higher amplitude AE events (see colour map, Fig. 4) being principally located at its top

Fig. 3 Differential stress as a function of axial strain. The *dark* and the *light colour lines* highlight the loading and the unloading of the specimen, respectively

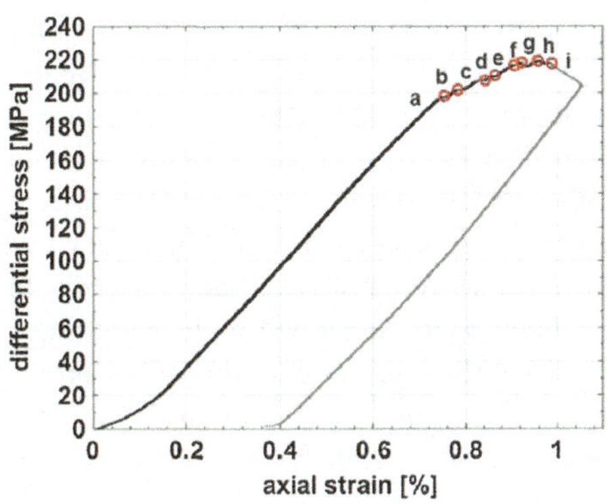

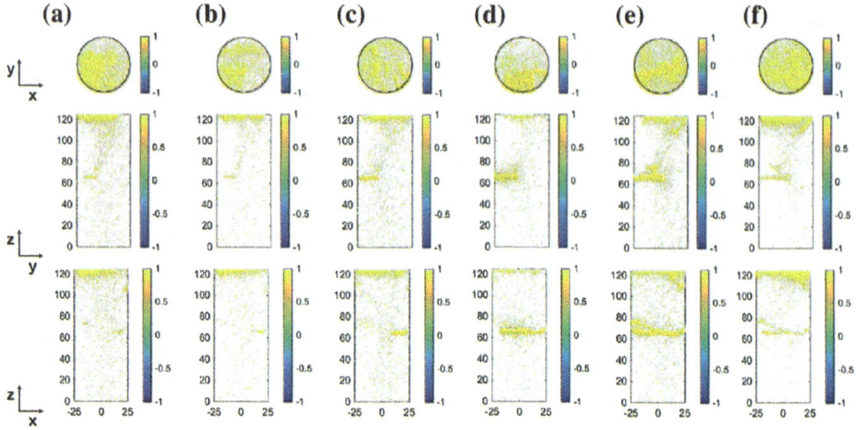

Fig. 4 AE spatiotemporal distribution during triaxial compression under 185 MPa confining pressure: **c–h** maps of AE event locations for the same time intervals viewed perpendicular and parallel to the specimen axis

edge. With increasing stress AE activity concentrates also at the tip of the pre-existing shear band forming progressively a semi-horizontal compaction band that propagates towards the circumference of the specimen. During later time intervals of loading (Fig. 4e, f) a network of deformation bands evolves towards the top part of the specimen. The compaction bands are primarily characterised by AE sources with polarities ranging from 0.7 to 1 (compressive type sources).

4 Discussion

The Bentheim sandstone specimen failed by shear localisation when subjected to a triaxial compression under 20 MPa confining pressure. The recorded peak stress falls within the stress range characterising the brittle failure of the Bentheim sandstone reported by [7, 8]. The AE control facilitated a rather slow shear band growth beyond the peak stress. A compaction band initiated at the tip of the pre-existing shear band, when the Bentheim specimen was loaded under triaxial compression at 185 MPa confining pressure. Although discrete compaction bands had developed in arrays subnormal to the maximum principal stress direction in other intact cylindrical specimens subjected to similar stress conditions [7, 16], the pre-induced deformation band, in our case, triggered the onset of the compaction band in a similar way as the pre-fabricated notches in [14, 16]. A second slightly inclined band developed with increasing cumulative strain (see also [1]).

To further investigate the micro-processes that took place during both loading stages that have been previously described, AE sources were classified as compressive, shear and tensile sources according to [18]. Figure 5 shows the moving

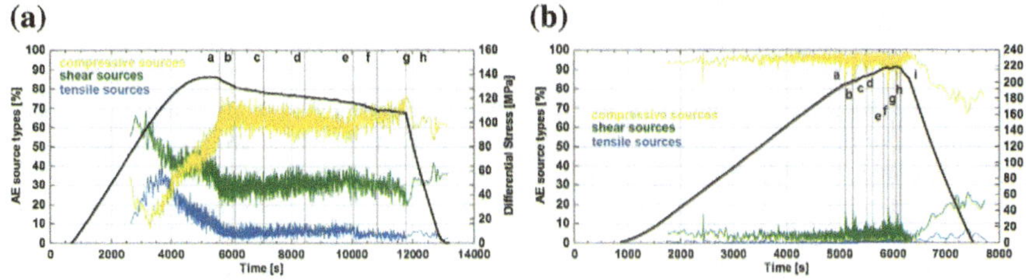

Fig. 5 Temporal evolution of AE source types and differential stress for the case of: **a** shear band formation and evolution; and **b** compaction band onset and propagation

average of the different source type distribution, calculated in a window of 200 events shifted each time by 100 events. The percentage of AE source types was defined by the number of each source type normalised by the total number of events within the window.

For the shear band case (Fig. 5a), the shear sources are the prevailing mechanisms during the initial loading stages, with the tensile sources being higher than the compressive sources. Approaching macroscopic failure, the situation somehow changes with the shear and the tensile sources to decrease and the compressive sources to considerably increase. AE control (slow shear band growth) promotes the compressive mechanisms, which reach a plateau of 65%, while the shear mechanisms reach an average value of 30%.

For the compaction band case (Fig. 5b), the compressive sources are dominant with an average value well above 95%. Both the shear and the tensile sources are rather small. The shear sources slightly increase during the small stress drops at the expense of the compressive sources, however the decrease of the latter is relatively small.

5 Conclusions

Subsurface reservoir rocks can be populated by a number of deformation features among which faults and compaction deformation zones. In this work we investigate the impact of a pre-existing shear band on the formation and evolution of compaction bands that developed in a Bentheim sandstone specimen. Shear band development was AE controlled, which enabled a slower band growth process. The resolved AE micro-mechanisms indicated a shear dominant nucleation and a compactant dominant growth of the shear band. A compaction band developed at the tip of the pre-induced shear band. This deformation band evolved towards the circumference of the specimen. With increasing axial strain a second compaction band developed at the top part of the specimen and in proximity with the pre-existing shear and compaction bands. Compactant sources were the prevailing mechanisms during the onset and evolution of compaction bands.

References

1. Baud, P., Klein, E., Wong, T.-F.: Compaction localisation in porous sandstones: spatial evolution of damage and acoustic emission activity. J. Struct. Geol. **26**, 603–624 (2004)
2. Charalampidou, E.M., Hall, S., Stanchits, S., Lewis, H., Viggiani, G.: Characterisation of shear and compaction bands in a porous sandstone deformed under triaxial compression. Tectonophysics **503**, 8–17 (2011)
3. Charalampidou, E.M., Hall, S., Stanchits, S., Viggiani, G., Lewis, H.: Shear-enhanced compaction band identification at the laboratory scale using acoustic and full-field methods. Int. J. Rock Mech. Mining Sci. **67**, 240–252 (2014)
4. Eichhubl, P., Hooker, J.N., Laubach, S.E.: Pure and shear-enhanced compaction bands in Aztec sandstone. J. Struct. Geol. **32**(12), 1873–1886 (2010)
5. Fossen, H., Schultz, R., Torabi, A.: Conditions and implications for compaction band formation in the Navajo sandstone, Utah. J. Struct. Geol. **33**, 1477–1490 (2011)
6. Holcomb, D., Rudnicki, J.W., Issen, K., Sternolf, K.: Compaction localisation in the Earth and the laboratory: state of research and research directions. Acta Geotech. **2**, 1–15 (2007)
7. Klein, E., Baud, P., Reuschlé, T., Wong, T.-F.: Mechanical behaviour and failure mode of Bentheim sandstone under triaxial compression. Phys. Chem. Earth. (A) **26**(1–2), 21–25 (2001)
8. Klein, E., Reuschlé, T.: A model for the mechanical behaviour of Bentheim sandstone in the brittle regime. Pure. Appl. Geophys. **160**, 833–849 (2003)
9. Leonard, M., Kennett, B.L.N.: Multi-component autoregressive techniques for the analysis of seismograms. Phys. Earth Planet. Inter. **113**, 247–263 (1999)
10. Lockner, D.A., Byerlee, J.D., Kuksenko, V., Ponomarev, A., Sidorin, A.: Quasi-static fault growth and shear fracture energy in granite. Nature **350**, 39–42 (1991)
11. Mollema, P.N., Antonellini, M.A.: Compaction bands: a structural analog for antimode I cracks in Aeolian sandstone. Tectonophysics **267**, 209–228 (1996)
12. Nelder, J., Mead, R.: A simplex method for function minimization. Comput. J. **7**, 308–312 (1965)
13. Stanchits, S., Vinciguerra, S., Dresen, G.: Ultrasonic velocities, acoustic emission characteristics and crack damage of basalt and granite. Pure. Appl. Geophys. **163**(5–6), 975–994 (2006)
14. Stanchits, S., Fortin, J., Guéguen, Y., Dresen, G.: Initiation and propagation of compaction bands in dry and wet Bentheim sandstone. Pure. Appl. Geophys. **166**, 843–868 (2009)
15. Sun, W.C., Andrade, J.E., Rudnicki, J.W., Eichhubl, P.: Connecting microstructural attributes and permeability from 3D tomographic images of in situ shear-enhanced compaction bands using multiscale computations. Geophys. Res. Lett. **38**, 201 (2011). doi:10.1029/2011GL047683
16. Vajdova, V., Wong, T.-F.: Incremental propagation of discrete compaction bands: acoustic emission and microstructural observations on circumferentially notched samples of Bentheim. Geophys. Res. Lett. **30**(14), 1775 (2003). doi:10.1029/2003GL017750
17. Vajdova, V., Baud, P., Wong, T.-F.: Permeability evolution during localised deformation in Bentheim sandstone. J. Geophys. Res. **109**, B10406 (2004)
18. Zang, A., Wagner, F.C., Stanchits, S., Dresen, G., Andresen, R., Haidekker, M.A.: Source analysis of acoustic emissions in Aue granite cores under symmetric and asymmetric compressive loads. Goephys. J. Int. **135**, 1113–1130 (1998)

Modelling Localization on Varying Scales

Eleni Gerolymatou, Carlos Grandas and Theodoros Triantafyllidis

Abstract Localization, both in the form of compaction bands and of shear bands, poses special challenges to finite element simulations, as the results tend to vary with the used discretization. As a rule, so called non simple continua in the sense of Noll are used to resolve the problem. In this way an internal length is introduced, which controls the size of the domain in which strain localizes. Such an approach requires an element length that is significantly smaller than the internal length, making it thus unsuitable for practical applications in the large scale. An alternative was suggested by Pietruszczak and Mróz, as far as shear bands are concerned. In this approach the shear band thickness is assumed to be a material parameter, with an element size that is significantly larger. In the present work this approach is extended to compaction bands and the discretizations required for the two approaches are compared.

1 Introduction

Localization, both in the form of compaction bands and of shear bands, poses special challenges to numerical simulations. As a rule, so called non simple continua in the sense of Noll are used to alleviate the numerical issues arising. Such models introduce in various manners an internal length that controls to a significant extent the width of the band in which the deformation is localized. Examples are non-local, higher gradient or micropolar models. These approaches have been proven successful but are still characterized by a drawback, in so far that they require a discretization with an element size that is at least as large as the characteristic length.

E. Gerolymatou (✉) · C. Grandas · T. Triantafyllidis
Institute of Soil Mechanics and Rock Mechanics, KIT, Engler-Bunte-Ring 14, 76131 Karlsruhe, Germany
e-mail: eleni.gerolymatou@kit.edu

C. Grandas
e-mail: carlos.tavera@kit.edu

T. Triantafyllidis
e-mail: Theodoros.Triantafyllidis@kit.edu

E. Papamichos et al. (eds.), *Bifurcation and Degradation of Geomaterials with Engineering Applications*, Springer Series in Geomechanics and Geoengineering, DOI 10.1007/978-3-319-56397-8_49

An alternative was suggested by Pietruszczak and Mróz [3], as far as shear bands are concerned. In this approach the shear band thickness is assumed to be a material parameter and the element size is selected significantly larger. It is thus possible to consider the mechanical response at the integration point to be that of the composite material, consisting of undamaged material and of a shear band with given thickness and orientation.

The aim of the suggested work is twofold. The first part is concerned with extending the approach suggested by Pietruszczak and Mróz to compaction banding. The second part is concerned with assessing the limits of applicability of the two approaches, as far as the discretization size is concerned.

2 Elastoplastic Description of the Material Behavior

A convex yield surface with horizontal outward vectors at the intersections with the p axis and a flexible form is selected:

$$f = q^2 + M_f^2 h(p)(p - p_t)(p - p_c) \tag{1}$$

where M_f is a constant,

$$h(p) = \exp\left[-\frac{1}{\beta}\left(\frac{p - p_t}{(p_c - p_t)} - a\right)^2\right], \tag{2}$$

p_t is the tensile strength, p_c is the strength at isotropic compression and a and β are material parameters. The yield surface variation is entirely due to the variation of p_t and p_c and follows the approach introduced by Nova and coworkers [1]. It is assumed that

$$p_c = p_s + p_m \tag{3}$$

where p_s reflects the effect of fabric modification and p_m the effect of internal bonding. It is further assumed that

$$\dot{p}_s = \rho_s p_s \left(\dot{\varepsilon}_v^p + \xi_s \dot{\varepsilon}_q^p\right) \tag{4}$$

where ρ_s and ξ_s are material constants. In addition

$$\dot{p}_m = -\rho_m p_m \left(\left|\dot{\varepsilon}_v^p\right| + \xi_m \dot{\varepsilon}_q^p\right) \tag{5}$$

where ρ_m and ξ_m are material parameters controlling the rate of mechanical degradation of bonding. The tensile strength is considered to be analogous to bonding according to

Table 1 Material parameters for gasbeton

E (MPa)	ν	α	β	M_f	m	ρ_s	ξ_m	ρ_m	ξ_m	p_s (MPa)	p_m (MPa)
60	0.15	1.3	2.0	2.0	0.2	10.75	0.0	1.25	−1.0	0.15	3.75

$$p_t = -kp_m \tag{6}$$

where k is a dimensionless parameter.

The parameters, which were calibrated on gasbeton, a highly porous material prone to compaction, are given in Table 1.

3 Nonlocal Model

The averaging function selected for the regularization was

$$h(\mathbf{x}) = \frac{15}{16l_c}\left(1 - \left(||\mathbf{x}||/l_c\right)^2\right)^2 \tag{7}$$

where l_c is the characteristic length. Another common choice is the Gauss function. The main difference between the two lies in the fact that the first has a bounded support, while the second has an unbounded one. The average $\hat{u}$ of the quantity u is defined as

$$\hat{u}(\mathbf{x}) = \frac{\int_S h(\mathbf{x} - \mathbf{s})u(\mathbf{s})ds}{\int_S h(\mathbf{x} - \mathbf{s})d\mathbf{s}} \tag{8}$$

Of course within the frame of a numerical scheme the distribution becomes discrete, so that, assuming $i = 1, \dots, N$ to be the set of integration points within distant l_c of a given point, the above relationship becomes

$$\hat{u}(\mathbf{x}) = \frac{\sum_{i=1}^{N} w_i u_i}{\sum_{i=1}^{N} w_i} \tag{9}$$

where

$$w_i = h(\mathbf{x} - \mathbf{x}_i) \tag{10}$$

In the specific case it was chosen to average the internal variables p_m and p_s.

A further possibility to regularize the problem is introducing viscosity to the material behaviour. For the sake of simplicity the Perzyna model [2] was used here. The strain rates are now decomposed as follows,

$$\underline{\dot{\varepsilon}} = \underline{\dot{\varepsilon}}^e + \underline{\dot{\varepsilon}}^{vp} \tag{11}$$

while the evolution of the viscoplastic strain is given by

$$\underline{\dot{\varepsilon}}^{vp} = \frac{1}{\eta}\left\langle \frac{f}{f_0} \right\rangle^N \frac{\partial g}{\partial \underline{\sigma}} \tag{12}$$

where $f_0 = p_c(t = 0)^2$, N is larger or equal to unity, η is a viscosity parameter with units of time and

$$\langle u \rangle = \begin{cases} u, \text{ if } u > 0 \\ 0, \text{ otherwise.} \end{cases} \tag{13}$$

The above implies that the stress state is not necessarily located on the yield locus, but may lie outside it. For the sake of simplicity, $N = 1$ is selected here.

For the parameter η a value of 40 sec was selected. For the local model this was found to yield the same results with the elastoplastic, i.e. non viscous, model for the oedometric compression of a specimen with an initial height of 20 mm to 50% of axial strain in the space of 30 min. This corresponds to the example used in Sect. 5.

4 Extension of the Pietruszczak and Mróz Formulation

In the above mentioned formulation it is assumed that only the localized zone lies in the plastic domain. A further assumption made is that the stress field is uniform, both in the localized band and outside it. Though for shear banding this does not pose a problem, in order to model the formation of compaction bands, these assumptions need to be raised.

The non localized and localized domains are viewed as consisting of different materials, (1) and (2) respectively, and the deformation band is assumed to be planar with a unit normal along the $\mathbf{e}_3$ direction. Considering the problem in three dimensions, the stress equilibrium requires that

$$\sigma_{13}^{(1)} = \sigma_{13}^{(2)}, \quad \sigma_{23}^{(1)} = \sigma_{23}^{(2)}, \quad \sigma_{33}^{(1)} = \sigma_{33}^{(2)} \tag{14}$$

In order to determine the local stresses and strains on the basis of the corresponding global quantities, three more conditions are required. A perfect contact between the layers is assumed to this purpose:

$$\varepsilon_{11}^{(1)} = \varepsilon_{11}^{(2)}, \quad \varepsilon_{22}^{(1)} = \varepsilon_{22}^{(2)}, \varepsilon_{12}^{(1)} = \varepsilon_{12}^{(2)} \tag{15}$$

The procedure introduced in [4] is used to evaluate the stress state in each of the zones as a function of the global stresses, as well as the mechanical response of the composite material. The only difference introduced here is the fact that the elasto-viscoplastic stiffnesses are used for both materials, rather than using the elastic stiffness for the material outside the zone of localized deformation.

The orientation of the zone of localized deformation is determined as usual using bifurcation analysis, while the width of the band is considered to be a material parameter, l_d, as suggested by Pietruszczak and Mróz [3]. For the sake of brevity in the present work we restrict ourselves to compaction localization, which is the case not covered by the previous analysis. The one further arising difficulty lies in the propagation criterion to be used for compaction bands. It is assumed that the compaction band propagates, when the intact material reaches the softening limit. Requiring the stress in the intact material to remain the same, the vertical elastic strain needs to be maintained as well, so that the vertical strain increment needs to be absorbed by the propagating compaction band. It must hold therefore that

$$\delta\varepsilon H = \delta H_c(\varepsilon_c^{(p)} - \varepsilon_{uc}^{(p)}) \tag{16}$$

where H is the height of the element, H_c the height of the compaction band and $\varepsilon_c^{(p)}$ and $\varepsilon_{uc}^{(p)}$ the plastic strain in the compacted and the uncompacted domain respectively. All strains in the above equation are considered in the direction normal to the compaction zone.

5 Results

An oedometric test on gasbeton is considered with an initial height of 20 mm and lasting 30 min. For both approaches the same elasto-viscoplastic model with the same parameters was used. The results from both approaches are presented and compared below. In Fig. 1 the results for the local and the nonlocal model are compared. The mesh dependence in the case of the local model is clearly visible, while it is absent in the case of the nonlocal model, as would be expected.

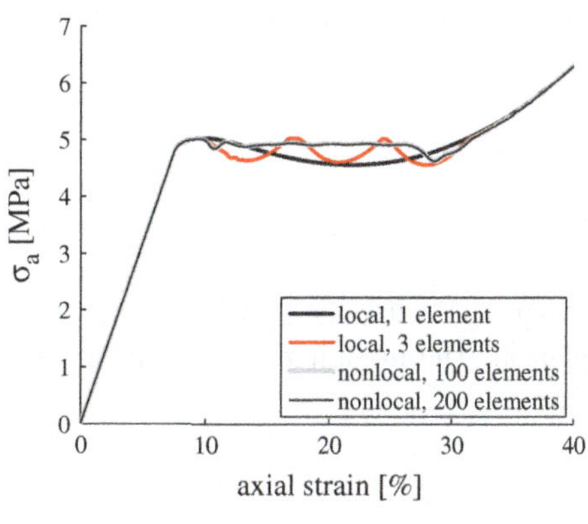

Fig. 1 Stress versus strain for an oedometric test on gasbeton. The local response is shown in *black* for one element and in *red* for three elements. The nonlocal response is shown in *grey* and *black* for 100 and 200 elements respectively

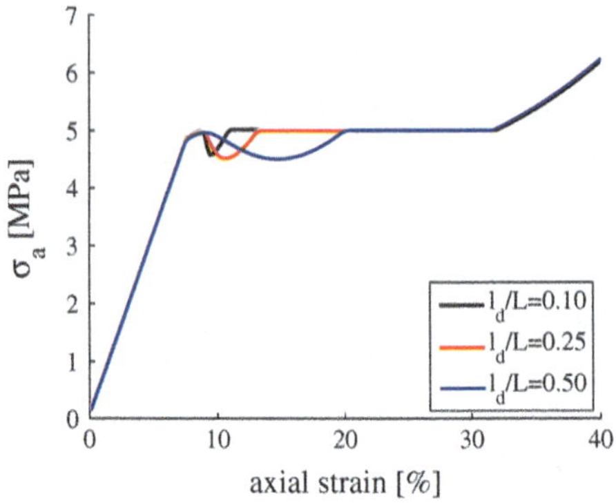

Fig. 2 Stress versus strain for an oedometric test on gasbeton. The response was evaluated using the modification suggested here of the method introduced by Pietruszczak and Mróz [3]

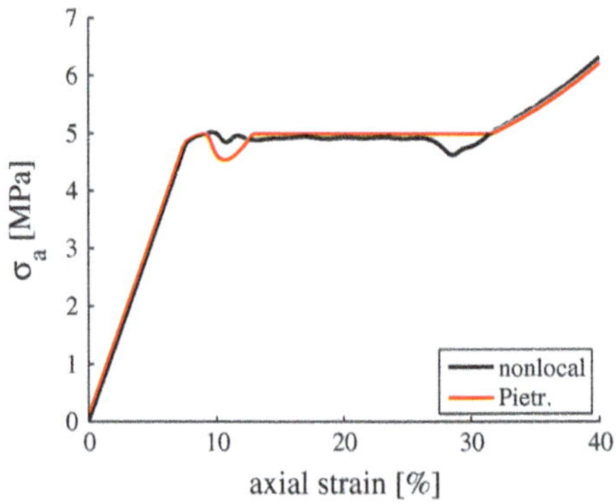

Fig. 3 Comparison of the modification suggested here of the method introduced by Pietruszczak and Mróz [3] and of the nonlocal model

In Fig. 2 the response evaluated using the modification suggested here of the method introduced by Pietruszczak and Mróz [3] is shown for different internal lengths. As may be seen, the model functions well, under the requirement that the internal length be smaller than the size of the element.

Finally in Fig. 3 a comparison between the results of the two methods is presented. It is known that the relation between the internal length of the nonlocal model and the band thickness is on approximation

$$l_d = \frac{\sqrt{2}}{2} l_c \tag{17}$$

For $l_c = L/4$, where L is the height of the specimen, the response of both models is given. As may be seen, the responses are similar. The nonlocal model predicts two global softening domains, one at the beginning and one at the end of the stress plateau. The modified model predicts only one, rather more pronounced, drop in stress.

6 Conclusions

In the present work an extension of the method of Pietruszczak and Mróz [3] to compaction banding was presented in parallel with a nonlocal model. Both approaches were found suitable for the stable and mesh independent prediction of localization. The first approach is suitable for large scale problems, requiring a discretization with an element height that is significantly larger than the band thickness. It seizes however to be valid when the zone becomes as large as the element or larger. In such a case, which may occur with propagating compaction bands, the mesh dependence returns. The nonlocal model on the other hand is not restricted by limitations to the thickness of a propagating compaction band, but requires a fine discretization that is not feasible for practical applications.

References

1. Nova, R., Castellanza, R., Tamagnini, C.: A constitutive model for bonded geomaterials subject to mechanical and/or chemical degradation. Int. J. Numer. Anal. Methods Geomech. **27**(9), 705–732 (2003)
2. Perzyna, P.: Fundamental problems in viscoplasticity. Adv. Appl. Mech. **9**(C), 243–377 (1966) (cited By 848)
3. Pietruszczak, S., Mróz, Z.: Finite element analysis of deformation of strain-softening materials. Int. J. Numer. Methods Eng. **17**(3), 327–334 (1981)
4. Triantafyllidis, T., Gerolymatou, E.: Estimation of the strength of stratified rock mass. Rock Mech. Rock Eng. **47**(2), 535–547 (2014)

Failure of Crushable Grains Using a Three-Dimensional Discrete Element Model

François Nader, Claire Silvani and Irini Djeran-Maigre

Abstract A three-dimensional grain model based on the concept of particles assembly using cohesive bonds is presented. A script is developed to generate polyhedral grains, by assembling tetrahedral particles using cohesive bonds. A series of single grain crushing simulations is performed, and the effect of the intra-granular cohesion on the grain's strength is studied. Oedometric compression tests are simulated. The effect of grain breakage on the macroscopical behaviour is evaluated.

1 Introduction

Rockfill material is more and more used in civil engineering, due to the advantages it exhibits on practical and economical aspects. The stresses inside a rockfill structure can lead to grain breakage, thus leading to changes in the grain size distribution of the material, therefore changing the overall behaviour of rockfill under different loading circumstances.

The discrete element method is the most adapted to simulate the discrete nature of rockfill and grain breakage. Many models are suggested in previous works in the literature. The first category of models considers a breakable grain as a single object. When the stress inside the grain reaches a predefined threshold value, the grain is replaced by a certain number of smaller grains. Åstrom and Hermann [1] and Tsoungui et al. [12] modelled grains as discs using the smooth contact method. When the stress inside the disc reaches the threshold value, it is replaced by a number of overlapping smaller discs, having the same total mass as the broken grain. A repulsive force then allows to get rid of the overlap. Ben-Nun and Einav [2] computed the mean normal contact force acting on a disc, and when the value exceeds the threshold, the broken disc is replaced by discs small enough to avoid overlapping. To reestablish mass equilibrium, the discs are then expanded. McDowell and De Bono [8] model the grain as a sphere, applying the same breakage configuration as Åstrom and Her-

F. Nader (✉) · C. Silvani · I. Djeran-Maigre
LGCIE SMS-ID – INSA Lyon, Université of Lyon, Lyon, France
e-mail: francois.nader@insa-lyon.fr

E. Papamichos et al. (eds.), *Bifurcation and Degradation of Geomaterials with Engineering Applications*, Springer Series in Geomechanics and Geoengineering, DOI 10.1007/978-3-319-56397-8_50

mann [1]. Cantor et al. [3] considered rigid polygons to model grains, and when the induced tensile stress inside the polygon reaches a threshold value, the polygon is split in the direction of the principal stress. Lobo-Guerrero and Vallejo [7] used the largest contact force applied to a disc as a breakage criterion, replacing a broken disc by a number of smaller grains, losing part of the mass during the process.

The second category of models considers a breakable grain as a assembly of a number of smaller particles joined together using cohesive bonds. When the stress applied to the grain leads to internal stresses exceeding cohesive bonds strengths, bonds start to break, leading to the breakage of the grain. Cheng et al. [4] joined small spheres using cohesive bonds, and removed 20% of the spheres to induce defects inside the grain, reducing its strength. Silvani et al. [11] joined together rigid discs using a Mohr-Coulomb law to form circular grains and simulated oedometric tests. Nguyen et al. [10] split a circular grain into smaller polygonal particles using a two dimensional Voronoï tessellation.

The same concept of particle assembly is used to generate three dimensional grains for our study. The goal is to evaluate the effect of breakage and the complexity of the behaviour induced by the grains' angular aspect, phenomena that cannot be clearly seen in spherical nor in two dimensional models.

In order to find a model as close as possible to reality, rigid tetrahedral particles (or subgrains) are joined together to form a three dimensional polyhedral grain.

In this paper, the three dimensional grain generation algorithm is presented. Then a series of single grain crushing simulations is shown to study the effect of intra-granular cohesion on grain strength. Furthermore, a series of oedometric compression tests is conducted on breakable grains to evaluate the influence of breakage on the macroscopic behaviour.

2 Grain Generation

Grain shape is a characteristic that plays a major role in the force distribution and compaction of rockfill media, and thus the overall macroscopic behaviour. In our study, this physical parameter is taken into consideration and polyhedral grains are generated to study the mechanical aspects overlooked by spherical models. To do that, a script that generates three dimensional polyhedra is written, taking as input parameters the maximal grain dimension $2 \times r_{max}$, the grain elongation $e = \frac{r_{max}}{r_{min}}$ where $2 \times r_{min}$ is the minimal grain dimension, the number of half-planes $N_{half-planes}$ and the number of vertices $N_{vertices}$ in each half-plane.

First, the center point of the polyhedron is placed, then two opposite vertices. A vertical half-plane is then considered, in which $N_{vertices}$ vertices are placed at a random distance r from the center, such as $r_{min} < r < r_{max}$ (Fig. 1). A rotation around the vertical axis is done to move to the next vertical half plane and vertices are placed. The same process is repeated until the $N_{half-plane}$ half-planes are filled. Vertices are then joined to form edges and then faces. Figure 2 shows a specimen of grains generated with random input parameters.

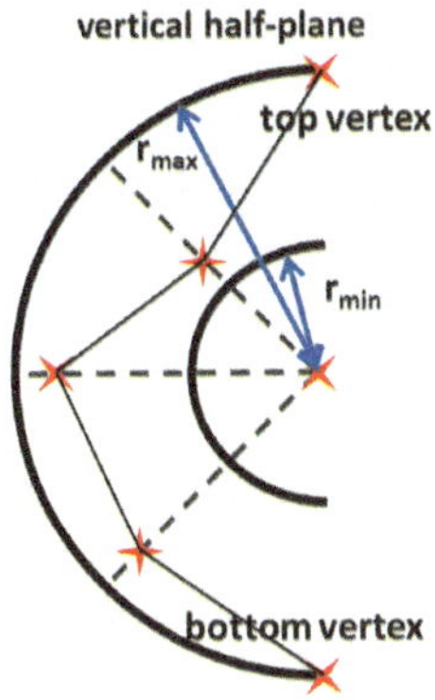

Fig. 1 Grain generation

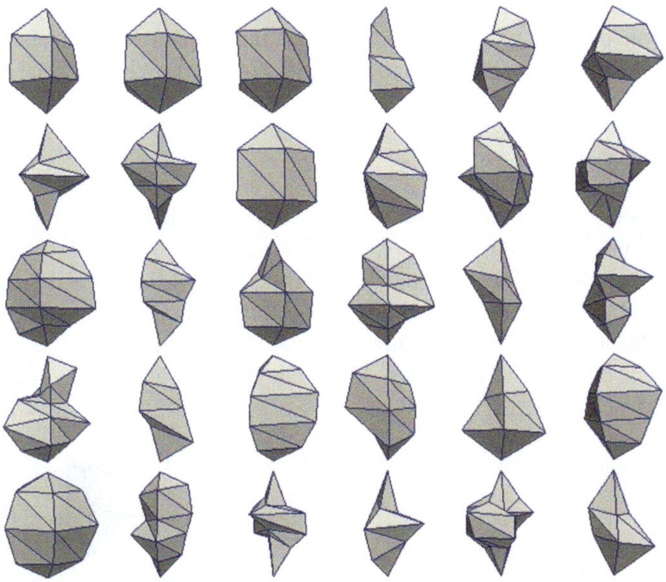

Fig. 2 Sample of 30 randomply generated polyhedra

Once the polyhedron is generated, the *Gmsh* meshing software is used to divide the grain into tetrahedral particles. These particles are then considered each as a rigid object, and joined together using a Mohr-Coulomb law to form a polyhedral grain. Changing the meshing parameters in *Gmsh* allows the generation of different subgranular configurations.

3 Single Grain Crushing

A first evaluation of the model is an evaluation of the grains' strength, using single grain crushing simulations.

The chosen numerical method is the Discrete Element Method (DEM), and more precisely the Non-Smooth Contact Dynamics (NSCD) method [6, 9] implemented in the software $LMGC^{90}$.

The goal of the NSCD method is to obtain the overall behaviour of a collection of objects by considering the dynamics of each element, taking into account the interactions between bodies. The problem is studied at two levels: movement equations of the particles expressed using global variables (kinematic variables) and the interactions described using local variables (contact variables).

Two rigid plates are considered between which the grain is placed. The bottom plate is fixed, and a downward increasing load is applied on the upper plate until the grain is crushed (Fig. 3). Once the grain's breaking force is determined, the characteristic stress σ_b is computed using the formula proposed by Jaeger [5]:

$$\sigma_b = \frac{F}{d^2} \tag{1}$$

where F is the grain's breaking force, and d the grain size considered here as the grain's maximal dimension.

A single generated grain, of a maximum dimension of 0.04 m is divided into 12 subgrains and subjected to multiple crushing simulations while varying the intragranular cohesion between 1 kPa and 10^4 kPa. During the simulation, the applied load increases until it reaches a peak. This peak value is considered equal to grain's

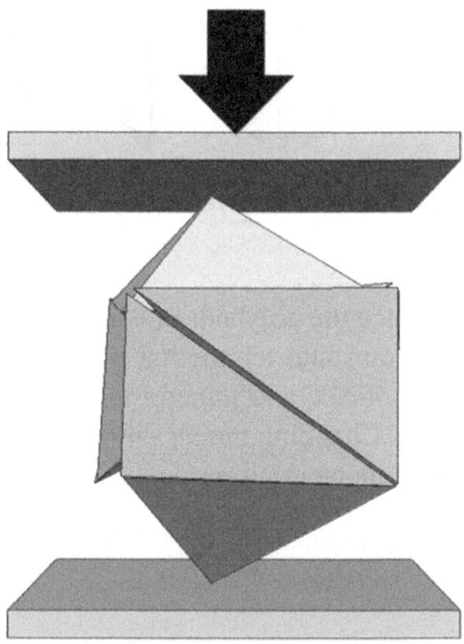

Fig. 3 Single grain crushing

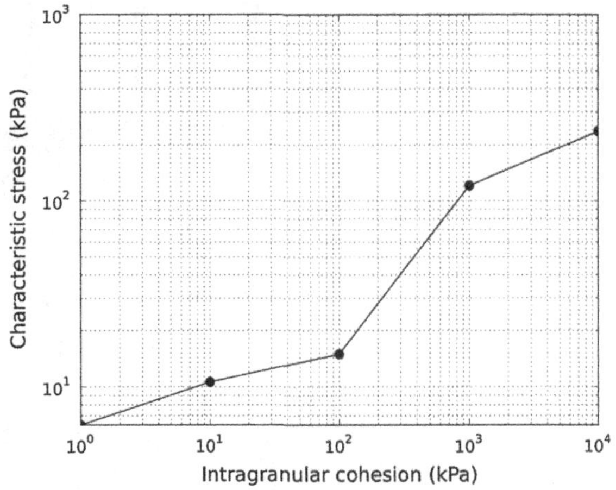

Fig. 4 Characteristic stress as a function of intra-granular cohesion

breaking force. Figure 4 shows the evolution of the grain's characteristic stress (Eq. 1) for the different values of the cohesion. The characteristic stress of the grain increases from 6.25 kPa to 2.38×10^2 kPa. This clear increase in the grain's characteristic stress at breakage validates the expected result. When the intra-granular cohesion increases, the stress needed to break the internal bonds joining the subgrains increases, therefore the applied load needed to break the grain is higher. This means that we can use experimental data to calibrate the model, in order to simulate and study the behaviour of a particular material.

4 Oedometric Simulation on a Multigrain Sample

An oedometric compression simulation is conducted on a multigrain sample. 855 grains are deposited by gravity in a rectangular box with a horizontal section of 30 cm $\times$ 30 cm, and a height of 26 cm. The initial grain size distribution is uniform, with a maximal grain dimension of 0.04 m, and all grains have regular polyhedral geometries (elongation = 1). Each grain is divided into 8–12 tetrahedral subgrains, and a cohesion of 10^3 kPa is applied to join the subgrains together (Fig. 5).

The reaction on the upper plate and its displacement are monitored throughout the simulation, and the oedometric curve showing the evolution of the void ratio as a function of the applied stress is plotted (Fig. 6). A fast decrease in the void ratio is observed at the beginning of the compression, up to a stress value of 10 kPa, where the slope of the curve changes, which means a decrease in the compressibility of the sample.

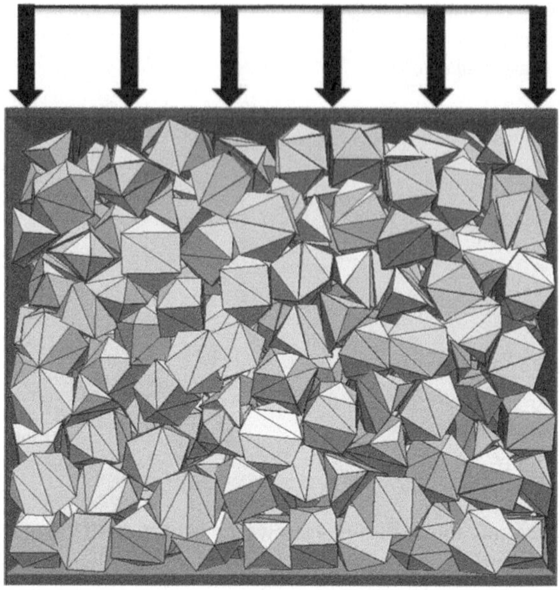

Fig. 5 Sample of breakable grains under oedometric compression

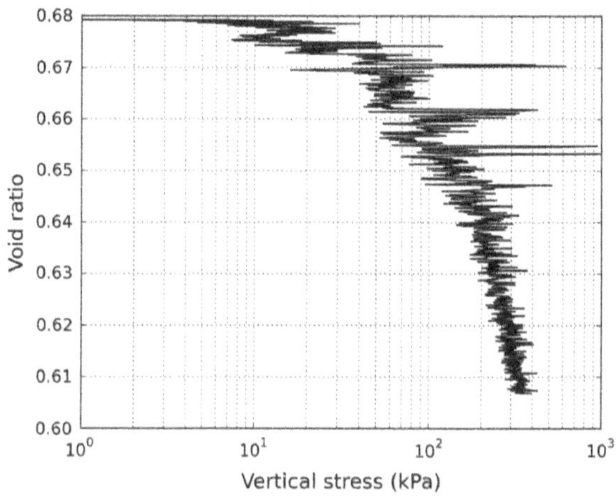

Fig. 6 Oedometric curve for an intra-granular cohesion of 10^3 kPa

5 Conclusion

The main conclusions of the study can be summarized as follows:

A specific algorithm is created to generate different types of polyhedral grains.

Regular polyhedra are selected for the single grain crushing and the oedometric compression simulations, even though the algorithm allows the generation of non-convex polyhedra with complex shapes.

The effect of the intra-granular cohesion is demonstrated in a series of single grain crushing simulations: higher values of the cohesion lead to more resistant grains. The numerical model is capable of reproducing the behaviour of granular materials under compression.

References

1. Åström, J.A., Herrmann, H.J.: Fragmentation of grains in a two-dimensional packing. Eur. Phys. J. B **5**(3), 551–554 (1998)
2. Ben-Nun, O., Einav, I.: The role of self-organization during confined comminution of granular materials. Philos. Trans. R Soc. Lond. Ser A **368**, 231–247 (2010)
3. Cantor, D., Estrada, N., Azema, E.: New approach to grain fragmentation for discrete element methods. Geomech. Micro Macro **1**, 257–262 (2014)
4. Cheng, Y.P., Nakata, Y., Bolton, M.D.: Discrete element simulation of crushable soil. Géotechnique **53**(7), 633–641 (2003)
5. Jaeger, J.C.: Failure of rocks under tensile conditions. Int. J. Rock. Mech. Min. Sci. **4**, 219–227 (1967)
6. Jean, M.: The non-smooth contact dynamics method. Comput. Methods Appl. Mech. Eng. **177**(3), 235–257 (1999)
7. Lobo-Guerrero, S., Vallejo, L.E.: Crushing a weak granular material: experimental numerical analyses. Géotechnique **55**(3), 245–249 (2005)
8. McDowell, G.R., De Bono, J.P.: On the micro mechanics of one-dimensional normal compression. Géotechnique **63**(11), 895–908 (2013)
9. Moreau, J.J.: Unilateral contact and dry friction in finite freedom dynamics. In: Nonsmooth Mechanics and Applications, pp. 1–82. Springer, Vienna (1988)
10. Nguyen, D.H., Sornay, P., Azema, E., Radjai, F.: Evolution of particle size distributions in crushable granular materials. Geomech. Micro Macro **1**, 275–280 (2015)
11. Silvani, C., Desoyer, T., Bonelli, S.: Discrete modelling of time-dependent rockfill behaviour. Int. J. Numer. Anal. Methods Geomech. **33**(5), 665–685 (2009)
12. Tsoungui, O., Vallet, D., Charmet, J.C.: Numerical model of crushing of grains inside two-dimensional granular materials. Powder Technol **105**(1), 190–198 (1999)

A Micro-Mechanical Analysis of Induced Anisotropic Damage in Initially Anisotropic Materials

Mei Qi, Albert Giraud, Jean-Baptist Colliat and Jian-Fu Shao

Abstract This paper is devoted to a micro-mechanical study of induced anisotropic micro-cracks in initially anisotropic materials. The effective elastic properties of cracked materials are determined by a rigorous up-scaling homogenization procedure based on an efficient numerical method to calculate Hill polarization tensor. A linear damage criterion is defined in the framework of irreversible thermodynamics to describe the growth of damage. Multiple families of micro-cracks in different orientations are taken into account by using the Gauss-type numerical integration method. Numerical assessments are proposed by using a *PCW* estimation and a set of 33 families of micro-cracks for uniaxial tension and compression tests. Induced crack density distributions are investigated for both isotropic and anisotropic materials. Effects of the initial anisotropy on crack propagation process are clearly demonstrated.

1 Introduction

Due to nucleation and propagation of cracks, damage is an essential dissipation mechanism in a class of brittle materials. The induced damage affects not only mechanical properties of materials such as the degradation of elastic stiffness, induced anisotropy, unilateral effects, but also transport properties such as the

M. Qi · J.-B. Colliat · J.-F. Shao (✉)
Laboratory of Mechanics of Lille, University of Lille, 59655 Villeneuve D'ascq, France
e-mail: jian-fu.shao@polytech-lille.fr

M. Qi
e-mail: mei.qi@etudiant.univ-lille1.fr

A. Giraud
GeoRessources Laboratory, UMR7359CNRS, University of Lorraine, 54501 Vandoeuvre-les-nancy, France
e-mail: albert.giraud@univ-lorraine.fr

E. Papamichos et al. (eds.), *Bifurcation and Degradation of Geomaterials with Engineering Applications*, Springer Series in Geomechanics and Geoengineering, DOI 10.1007/978-3-319-56397-8_51

enhancement of permeability. Mechanical approaches based on rigorous homogenization techniques provide an efficient way to establish relationships between micro-crack propagation and macroscopic properties.

In this paper, based on a homogenization method, a numerical micro-mechanical damage model is developed for initially transversely isotropic materials with induced anisotropic micro-cracks based on the extension of [6, 7]. The representative elementary volume (REV) of cracked materials is represented by different families of oblate cracks which are embedded in an initially transversely isotropic elastic solid matrix. The key step of the homogenization procedure is to determine the strain concentration law which is based on the Hill's polarization tensor or equivalently Eshelby's tensor [2]. Using the numerical integration of the exact Green's function provided by [4] and the coordinate frame rotation method proposed by [3], an efficient numerical method is proposed and it is able to determine the Hill tensor for an arbitrarily oriented family of cracks embedded in a transversely isotropic matrix. Effective elastic properties of cracked materials are first determined. By the definition of a linear damage criterion, crack propagation processes are investigated for two different loading path. Effects of initial material anisotropy are clearly demonstrated.

2 Effective Behavior of Anisotropic Cracked Materials

Consider a REV of cracked material occupying a geometric domain Ω limited by its external boundary surface $\partial\Omega$. The REV is composed of a transversely isotropic linear elastic solid and a number of oblate ellipsoidal cracks. All micro-cracks with the same normal vector are put into the same family, which is defined by its normal vector $\underline{n}$.

Assume the crack growth as the unique dissipation process, the free energy of cracked material can be expressed by $W = \frac{1}{2}\boldsymbol{E} : \mathbb{C}^{hom} : \boldsymbol{E}$, with $\boldsymbol{E}$ being the macroscopic strain. The effective elastic stiffness tensor of the homogenized equivalent medium (HEM) can be obtained by a linear homogenization method [6]:

$$\mathbb{C}^{hom} = \mathbb{C}_m + \sum_{r=1}^{N} \varphi_r(\mathbb{C}_{c,r} - \mathbb{C}_m) : \mathbb{A}_{c,r} \tag{1}$$

$\mathbb{C}_m$ denotes the elastic stiffness tensor of solid matrix, and $\mathbb{C}_{c,r}$ denotes the elastic stiffness of the rth family of micro-cracks. The volume fraction per unit volume of micro-cracks is given $\varphi_r = \frac{4}{3}\pi\varepsilon d$. ε is the aspect ratio of micro-cracks and d is the crack density parameter which is used as a damage variable. $\mathbb{A}_{c,r}$ is a concentration tensor which accounts for crack interactions and spatial distribution effects. The Ponte-Castaneda and Willis (*PCW*) scheme is used in this work to determined the concentration tensor:

$$\mathbb{A}^{pcw}_{c,r} = [\mathbb{I} + \mathbb{P}^r_\varepsilon : (\mathbb{C}_{c,r} - \mathbb{C}_m)]^{-1} : \left[\mathbb{I} + \sum_{j=1}^{N} \varphi_j [\mathbb{I} + (\mathbb{P}^j_\varepsilon - \mathbb{P}_d) : (\mathbb{C}_{c,j} - \mathbb{C}_m)] : [\mathbb{I} + \mathbb{P}^j_\varepsilon : (\mathbb{C}_{c,j} - \mathbb{C}_m)]^{-1}\right]^{-1} \tag{2}$$

Note that two independent tensorial functions have been introduced: the first one $\mathbb{P}_d$ allows to accounting for the effects of spatial distribution, the second one $\mathbb{P}^r_\varepsilon$ denotes the so-called Hill tensor involved in the classical Eshelby's solution by $\mathbb{S}^r_\varepsilon = \mathbb{P}^r_\varepsilon : \mathbb{C}_m$ and characterizes the interaction between cracks.

The key step of homogenization of cracked materials is the calculation of the Eshelby or Hill polarization tensor. In the present work, the objective is to study anisotropic damage evolutions in initially anisotropic materials. In this case, no analytical solutions are available. Therefore, we shall adopt the efficient numerical method proposed by [3] to calculate the Hill tensor which is based on the exact Green's function proposed by [4]. This numerical method has been presented in [3] and applied in [5].

3 Damage Criterion

The damage criterion is formulated at the local scale in the framework of irreversible thermodynamics to describe the evolution of damage:

$$f^r(F^{d_r}, d_r) = F^{d_r} - \mathscr{R}(d_r) \leqslant 0 \tag{3}$$

F^{d_r} is the thermodynamic force which can be determined by $-\frac{\partial W}{\partial d_r}$. It is affected by interactions between different cracks families. The function $\mathscr{R}(d_r)$ represents the current material resistance against damage growth at microscopic scale and it is a function of the damage variables d_r, $r = 1, \ldots, N$. For the present study, a linear form $\mathscr{R}(d_r) = c_0 + c_1 d_r$ is proposed. The parameter c_0 denotes the initial damage threshold and c_1 is controlling the kinetics of damage evolution.

4 Numerical Applications

In this study, some numerical assessments are performed by using the *PCW* estimation and a set of 33 families of micro-cracks by neglecting frictional sliding along cracks surfaces. The Gauss-type integration method (with a selected number of orientations defined on the surface of a semi-sphere) is used [1]. The numerical results obtained respectively for initially isotropic and anisotropic materials are presented and compared. A representative set of parameters are used: for the isotropic material ($E_m = 38000$ MPa, $\nu_m = 0.19$), and for the transversely isotropic material ($E_1 = 38000$ MPa, $E_3 = 19000$ MPa, $G_{13} = 8000$ Mpa, $\nu_{12} = 0.19$, $\nu_{31} = 0.19$), as well as for the damage evolution law ($c_0 = 0.74 \times 10^{-3}\,Jm^{-2}$, $c_1 = 0.084\,Jm^{-2}$), the initial damage density $d_0^r = 0$, the aspect ratio of micro-cracks $\varepsilon = 0.01$. A spherical spatial distribution of micro-cracks is considered for the *PCW* scheme. The Cartesian coordinate system (e_1, e_2, e_3) is used and the axial stress is applied along the axis $\underline{e}_3$ in all numerical calculations.

4.1 Uniaxial Tension Test

An uniaxial tension test is firstly studied. Both isotropic and anisotropic matrix weakened by 33 families of penny-shaped micro-cracks are considered. Note that in this loading condition, all induced cracks are considered to have an open state ($\mathbb{C}_{c,r} = 0$). The boundary conditions of macroscopic stresses are given by $\boldsymbol{\Sigma} = \Sigma \underline{e}_3 \otimes \underline{e}_3$. The loading is driven by the prescribed tensile strain ($E_{33} > 0$).

In Figs. 1a, b, the numerical results of axial stress-strain curves and damage evolutions are presented. It is found that the macroscopic behavior of cracked mater-

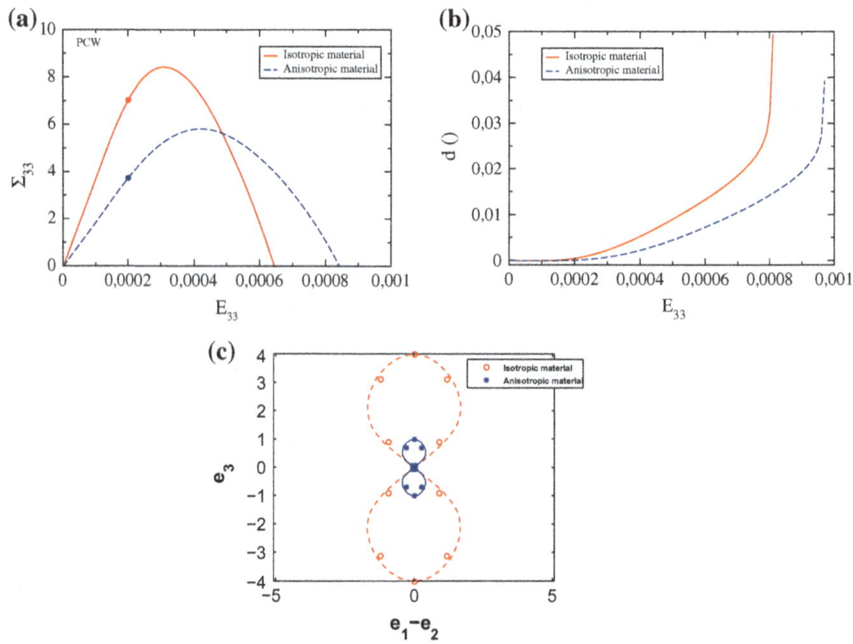

Fig. 1 Comparison of mechanical behaviors of isotropic and transversely isotropic material in uniaxial tension test: **a** stress-strain curves, **b** damage evolution, **c** rosette surface of damage density distribution in the plane $(\underline{e}_1, \underline{e}_3)$ at the two points of the figure (**a**)

ial is significantly influenced by the initial material anisotropy. It has an interaction between the initial anisotropy and induced crack propagation. In Fig. 1c, one can see the comparison of damage density distribution between isotropic and anisotropic materials in the plane $(\underline{e}_1, \underline{e}_3)$ at the same strain values in Fig. 1a. The rosette surface is obtained in normalizing the crack density of all families of micro-cracks with respect to the biggest one in the anisotropic material, which is that along the normal (0, 0, 1). This Figure shows that the micro-crack with the normal (0, 0, 1) plays a dominant role in the damage evolution, not only in the isotropic material but also in the anisotropic material. With the influence of the initial anisotropy, the crack density of the anisotropic material is 4 times smaller than that of the isotropic material at the same macroscopic deformation.

4.2 Uniaxial Compression Test

An uniaxial compression test is then considered. All micro-cracks are assumed to be closed ($\mathbb{C}_{c,r} = \eta \underline{n} \otimes \underline{n} \otimes \underline{n} \otimes \underline{n}$). In Figs. 2a, b, the numerical results of axial stress-strain curves and damage evolution are presented. Again, the macroscopic mechan-

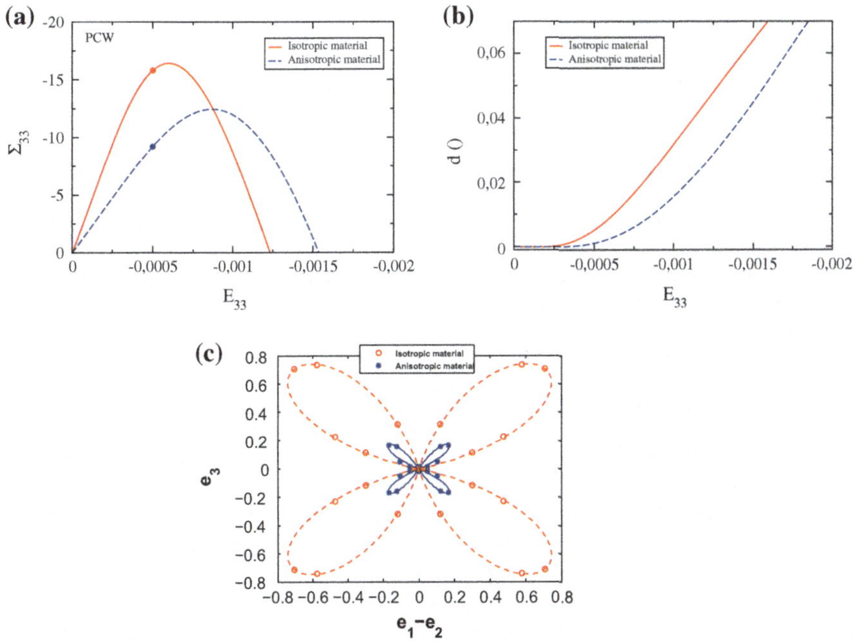

Fig. 2 Comparison of mechanical behaviors of isotropic and transversely isotropic material in uniaxial compression test: **a** stress-strain curves, **b** damage evolution, **c** rosette surface of damage density distribution in the plane $(\underline{e}_1, \underline{e}_2)$ at the two points of the figure (**a**)

ical behavior of cracked material is affected by the initial anisotropy of the solid matrix and then by an interaction between the initial anisotropy and induced crack propagation. With the same manner as Fig. 1c, Fig. 2c shows that the value of crack density of the anisotropic material is 3 times smaller than that of isotropic material at the same deformation because of the influence of the initial anisotropy. Further, the damage propagation in (45° 0°) seems to be the most important direction for either isotropic or anisotropic material.

5 Conclusion

In this work, a numerical micro-mechanical damage model has been proposed for a class of brittle materials. Comparing with existing models, the proposed model is able to describe macroscopic behaviors of initially anisotropic materials with induced anisotropic micro-cracks. Unilateral effects between open and closed cracks have also been considered. It is found that the induced damage process is clearly affected by its interaction with the initial material anisotropy. In our ongoing work, coupling between damage growth and frictional sliding along crack families is investigated.

References

1. Bazant, Z.P., Oh, B.H.: Efficient numerical integration on the surface of a sphere. ZAMM **66**, 37–49 (1986)
2. Eshelby, J.D.: The determination of the elastic field of an ellipsoidal inclusion and related problems. Proc. R. Soc. Lond. **241**(1226), 376–396 (1957)
3. Giraud, A., Huynh, Q., Hoxha, D., Kondo, D.: Effective poroelastic properties of transversely isotropic rock like composites with arbitrarily oriented ellipsoidal inclusions. Mech. Mater. **39**, 1006–1024 (2007)
4. Pan, Y., Chou, T.: Point force solution for an infinite transversely isotropic solid. J. Appl. Mech. **43**, 608–612 (1976)
5. Qi, M., Shao, J.F., Giraud, A., Zhu, Q.Z., Colliat, J.B.: Damage and plastic friction in initially anisotropic quasi brittle materials. Int. J. Plast. **82**, 260–282 (2016)
6. Zhu, Q.Z., Kondo, D., Shao, J.F., Pensee, V.: Micromechanical modeling of anisotropic damage in brittle rocks and application. Int. J. Rock Mech. Mining Sci. **45**, 467–477 (2008)
7. Zhu, Q.Z., Shao, J.F.: A refined micromechanical damage friction model with strength prediction for rock-like materials under compression. Int. J. Solids Stuct. **60**, 75–83 (2015)

Strain Localization in High Performance Fiber Reinforced Cementitious Composites

Marta Miletić and Dunja Perić

Abstract High Performance Fiber Reinforced Cementitious Composite (HPFRCC) is a cementitious composite, which consists of a specifically tailored cementitious matrix reinforced with short discrete fibers that have a proper geometry and enhanced bond properties in order to improve tensile properties of the overall composite. Despite all of its beneficial properties HPFRCC has not yet found its way into the engineering practice largely due to the lack of adequate numerical models. To this end, the main objective of this research was to develop and implement a combined analytical-numerical algorithm that can capture a stress-strain response and inception of strain localization in elastic-plastic HPFRCC. Multi-directional fibers are embedded into a matrix and modeled as a linear elastic material, while the resulting composite is described by a two-invariant non-associated non-linear Drucker-Prager hardening plasticity model. A diagnostic strain localization analysis was conducted for several uniaxial tension and uniaxial compression tests on a non-reinforced cementitious composite as well as on the HPFRCC. It was found that the presence of fibers delayed the inception of strain localization in all tests on the HPFRCC. Furthermore, it appears that the onset of strain localization in uniaxial tension on HPFRCC detects the inception of distributed cracking, while the onset of strain localization in uniaxial compression detects the onset of more localized cracking.

M. Miletić (✉)
Department of Civil Engineering, Auburn University, 238 Harbert Engineering Center, Auburn, AL 36849-5337, USA
e-mail: mmiletic@ksu.edu

D. Perić
Department of Civil Engineering, Kansas State University, 2111 Fiedler Hall, Manhattan, KS 66506-5000, USA

E. Papamichos et al. (eds.), *Bifurcation and Degradation of Geomaterials with Engineering Applications*, Springer Series in Geomechanics and Geoengineering, DOI 10.1007/978-3-319-56397-8_52

1 Introduction

HPFRCC was in this study modelled by embedding multidirectional fibers into a matrix. The resulting composite was described by a two-invariant exponential Drucker-Prager (EDP) plasticity model with isotropic hardening. Furthermore, linear and nonlinear hardening was used in order to examine effects of hardening type on the inception of strain localization. A simple volume-based homogenization procedure was used to derive the corresponding macroscopic tangent stiffness moduli tensor of the fiber reinforced composite. Actual uniaxial tension (UT) and uniaxial compression (UC) tests on non-reinforced mortar and on the HPFRCC were modelled.

2 Fiber Contribution

In this study, a representative volume element (RVE) was chosen to have a cubical shape and it is depicted in Fig. 1. The side length of the cubical RVE is two times greater than the length of a fiber L_f. Fibers are isotopically distributed throughout the RVE. Fibers were assumed to be cylindrical with a diameter d_f, length L_f, and an aspect (length to diameter) ratio η_f. Thus, a macroscopic tangent stiffness moduli tensor of the composite was developed by consistently homogenizing the contribution of fibers in a RVE Miletić and Perić [1].

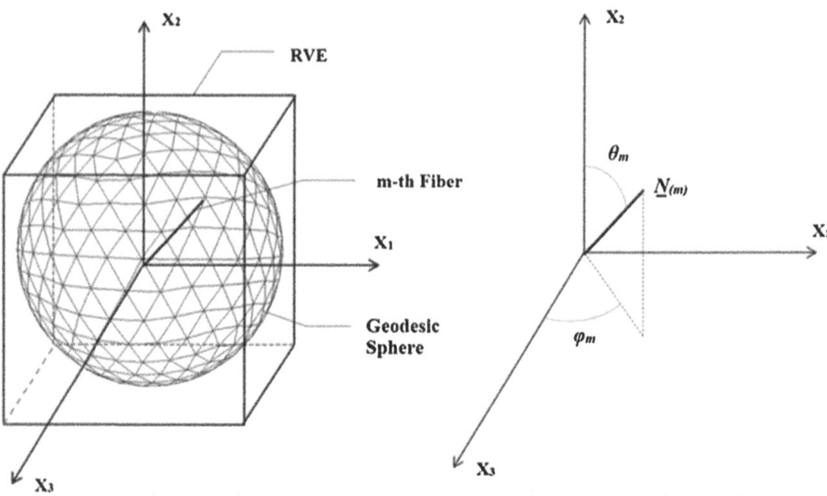

Fig. 1 Fiber distribution and orientation in the cubical RVE

$$D^e_{ijkl,f} = \frac{E_f}{n_f} \sum_{m=1}^{n_f} N_{i(m)} N_{j(m)} N_{k(m)} N_{l(m)} \tag{1}$$

where E_f is the elastic modulus of a fiber, n_f is the number of fibers contained in RVE and m is the summation index. A unit normal vector in the direction of fiber m is denoted by $N_{i(m)}$.

3 Tangent Stiffness Moduli Tensor

In this study, it was assumed that the cementitious composite is an elastic-plastic material undergoing an infinitesimal strain and obeying a general non-associative flow rule. The corresponding elastic stiffness moduli tensor of the homogenized equivalent isotropic material is denoted by D^e_{ijkl}. It is obtained as a weighted sum of the elastic stiffness moduli tensors of the matrix and of fibers and it is given by

$$D^e_{ijkl} = (1 - \chi_f) D^e_{ijkl,m} + \chi_f D^e_{ijkl,f} = \mu(\delta_{ik}\delta_{jl} + \delta_{jk}\delta_{il}) + \lambda\delta_{ij}\delta_{kl} \tag{2}$$

where χ_f is a volumetric fiber content and elastic stiffness moduli tensor of the matrix is given by

$$D^e_{ijkl,m} = \mu_m(\delta_{ik}\delta_{jl} + \delta_{jk}\delta_{il}) + \lambda_m\delta_{ij}\delta_{kl} \tag{3}$$

and δ_{ij} is Kronecker delta, μ and λ are Lamé's constants of the composite, while μ_m and λ_m are Lamé's constants of the matrix.

4 Conditions for Onset of Strain Localization

An onset of strain localization may be considered as a loss of stability of the constitutive relation governing the homogeneous deformation. Therefore, once the inception of strain localization occurs a local constitutive level response is not valid anymore.

A classical Rice-Rudnicki condition for onset of strain localization Rudnicki and Rice [2] was used to obtain analytical solution for a critical hardening modulus H_{cr} and a unit vector n_i perpendicular to the direction of deformation band. The tangent stiffness moduli tensor for plane stress loading was obtained from Eq. (2), thus enabling use of the analytical solution given by Runesson et al. [3], which also gives the corresponding eigenvector z_i.

Table 1 Material properties

Property	Mortar	HPFRCC-1	HPFRCC-1.5	HPFRCC-2
X_f (%)	0	1	1.5	2
E (GPa)	26.59	26.66	26.70	26.72
ν	0.2	0.2007	0.2011	0.2014
a ($x10^{-4}$)	4	7.3	7.4	6.5
b	2.8	2.65	2.64	2.67
p_{t0} (MPa)	0.27	0.38	0.4	0.41
ψ (°)	22	15	13	11.5

5 Application to Drucker-Prager Model

The high performance fiber reinforced composite was described by two-invariant exponential Drucker-Prager (EDP) plasticity model with isotropic hardening. The corresponding yield and plastic potential functions can be found in ABAQUS [4]. The model parameters are also defined in ABAQUS [4]. The constitutive model was calibrated against the actual experimental data Sirijaroonchai [5] and the resulting parameters are shown in Table 1. Parameter E represents the modulus of elasticity of the HPFRCC, ν is the Poisson's ratio of the HPFRCC, and ψ corresponds to a dilatation angle measured at a high confining pressure. The rest of the material parameters are described in ABAQUS [4].

Onset of strain localization (OSL) was detected in case that the critical hardening modulus is equal to the actual hardening modulus.

Two types of isotropic hardening were investigated in this study: linear and nonlinear. In the case of non-reinforced cementitious composites, the assumption was made that the deviatoric stress at the onset of yielding coincided with the peak deviatoric stress in UT, thus resulting in no hardening.

6 Results

Figure 2 depicts the comparisons between the numerically predicted and experimentally observed responses Sirijaroonchai [5] in plane stress UT tests for the EDP model. It can be seen that addition of fibers to the mortar had a minor effect on the increase of the yield stress in UT while its major contribution was in increase in the peak stress and tensile strain capacity. For example, the yield stress of mortar was approximately 0.77 MPa. However, for the χ_f of only 1% the yield stress increased to approximately 1.1 MPa. Moreover, the peak stress of mortar was at least four times lower than the peak stresses of HPRFCCs. The predicted axial strain level at

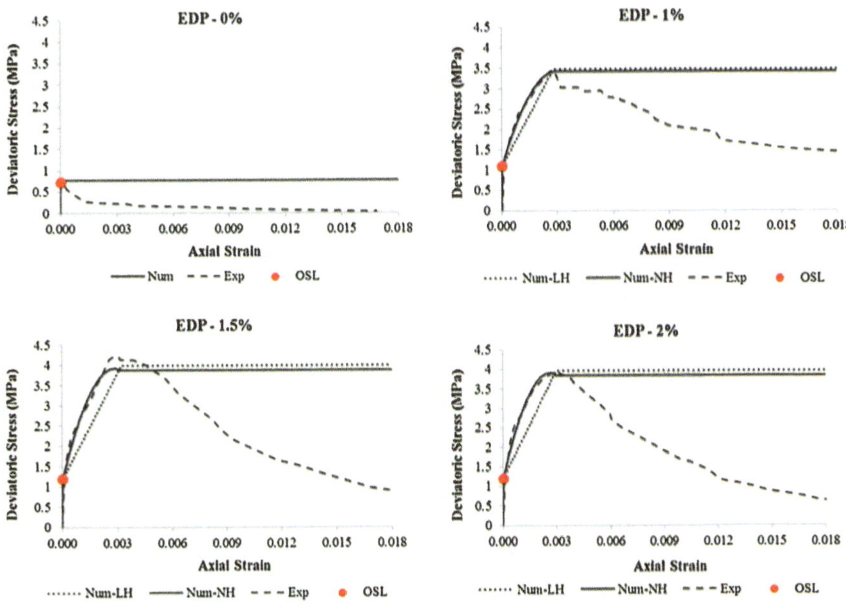

Fig. 2 Experimentally observed and numerically predicted responses for EDP model with linear and nonlinear hardening (UT tests)

the OSL in the plain mortar coincided with the axial strain level at which a severe post peak drop in deviatoric stress began. On the contrary, the predicted OSL in fiber reinforced materials was followed by a significant hardening leading to significantly increased peak stresses as compared to the plain mortar. Thus, it appears that the principal mechanism of the delayed OSL in the UT tests was fiber-induced increase in the yield stress. Furthermore, the numerical predictions seem to indicate that OSL in plain mortar coincides with the formation of a localized major crack. On the contrary, the OSL in HPFRCC indicated the inception of distributed cracking that manifests itself in the development of a global level hardening mechanism. Thus, the major crack development appears to have been delayed through the mechanism of fiber induced distributed cracking.

Figure 3 shows numerically predicted stress-strain responses for plane-stress UC tests according to EDP model. Experimentally observed responses were not included in Fig. 3, because the experimental data Sirijaroonchai [5] are available only for axisymmetric UC tests. In the case of the plain cementitious composite, the predicted OSL coincided closely with the peak stress. For HPFRCC, stress-strain responses exhibited more pronounced strain hardening prior to the predicted OSL, which also coincided closely with the peak stress. Moreover, results showed a 25% increase in the peak stress in HPFRCC compared to the plain mortar.

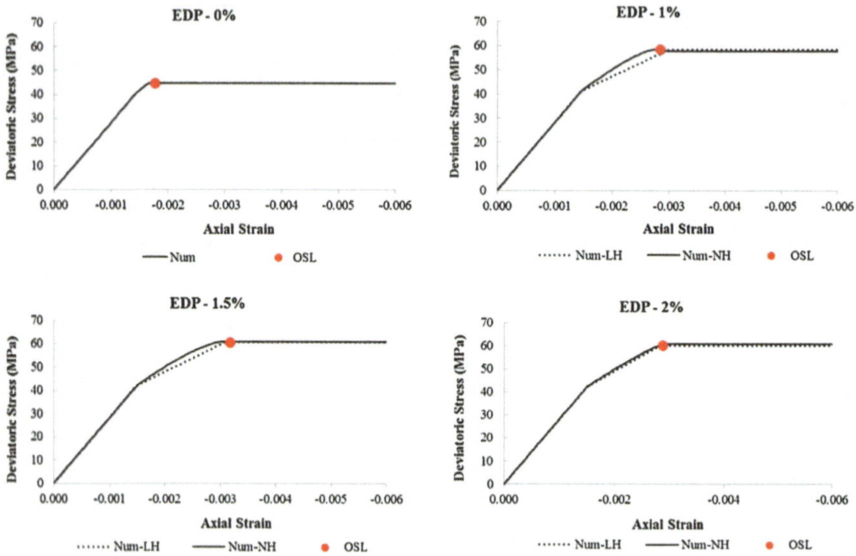

Fig. 3 Numerically predicted responses for EDP model with linear and nonlinear hardening (UC tests)

7 Conclusions

It was found that the presence of fibers delayed the OSL in all tests, which appears to be in agreement with the experimental data. The delay was more pronounced in UC tests than in UT tests.

References

1. Miletić, M., Perić, D.: Onset of Strain Localization in Fiber Reinforced Composites Subjected to Plane Stress oading. In Chau, K.T., Zhao, J. (eds.) Bifurcation and Degradation of Geomaterials in the New Millennium, Proceedings of the 10th International Workshop on Bifurcation and Degradation in Geomaterials, pp. 279–283 (2015)
2. Rudnicki, J.W., Rice, J.R.: Conditions for the localization of deformation in pressure-sensitive dilatant materials. J. Mech. Phys. Solids **23**, 371–394 (1975)
3. Runesson, K., Ottosen, N.S., Perić, D.: Discontinuous bifurcations of elastic-plastic solutions at plane stress and plane strain. Int J Plast. **7**, 99–121 (1991)
4. ABAQUS: Version 6.13-1. Dassault Systemes Simulia Corporation. Providence, Rhode Island (2013)
5. Sirijaroonchai, K.: A macro-scale plasticity model for high performance fiber reinforced cement composites Ph.D. thesis, The University of Michigan, United States of America (2009)

On the Three-Dimensional Extension of the Micromechanically-Based H-Model

Hao Xiong, François Nicot and Zhenyu Yin

Abstract Constitutive modeling of granular materials is still a central issue. Granular materials react with complicated mechanical responses when subjected to external loading paths. This leads to sophisticated constitutive formulations requiring large number of parameters. In order to avoid this, a micromechanically based model embedding both micro-scale and meso-scale is developed. The 3D-H model takes an intermediate scale (meso-scale) into account, which makes it possible to describe a variety of constitutive features in a natural way. The conventional triaxial compression tests are selected to examine the model performance.

1 Introduction

Granular materials are significant constituents involved in many industrial processes and geophysical phenomena. However, no fundamental statistical theory is currently available to describe their properties. The behavior of one single grain is easily understood, but the properties of a granular collection are much more complex. Several focal topics are wildly discussed such as strain localization, instability occurrence and the existence of a bifurcation domain within the plastic limit surface, in which a variety of failure modes can be encountered [1]. These macroscopic features not only

H. Xiong (✉) · F. Nicot
Université Grenoble Alpes, IRSTEA, UR ETGR, 2 rue de la Papeterie-BP 76, F-38402 St-martin-d'Hères, France
e-mail: xionghao19529@gmail.com

F. Nicot
e-mail: francois.nicot@irstea.fr

Z. Yin
LUNAM University, Ecole Centrale de Nantes, UMR CNRS GeM, Nantes, France
e-mail: zhenyu.yin@gmail.com

Z. Yin
Department of Geotechnical Engineering, Key Laboratory of Geotechnical and Underground Engineering of Ministry of Education, College of Civil Engineering, Tongji University, Shanghai 200092, China

E. Papamichos et al. (eds.), *Bifurcation and Degradation of Geomaterials with Engineering Applications*, Springer Series in Geomechanics and Geoengineering, DOI 10.1007/978-3-319-56397-8_53

need to be observed, but also need to be understood. Thus, increasing researches on the micro-scale are conducted to explain the physical background, including the failure occurrence [4], the force chain buckling [5] and the meso-structure evolution [6]. Recently, a series of micro-mechanical models have been proposed by considering the average behavior of all contacts along each contact direction [3], or considering a meso-structure oriented along all the directions of the physical space [2]. In this paper, a three-dimensional extension of the H model is developed based on a multi-scale approach. The model responses on confining stress dependent stress-strain relationships are then examined, based on which the critical state can be described.

2 The Three-dimensional-H Model

In 3D conditions, a global coordinate system $(\vec{x}_1, \vec{x}_2, \vec{x}_3)$ is required, where $\vec{x}_1, \vec{x}_2, \vec{x}_3$ axes stand for the principal stress directions on the macro-scale. The 3D-H model is derived from the incremental macroscopic strain tensor $\delta\bar{\bar{\varepsilon}}$ to the incremental macroscopic stress tensor $\delta\bar{\bar{\sigma}}$ by transforming the micro-scale and meso-scale information. No rotation of the principal axis of both stress and strain tensors is supposed to take place. The kinematic localization is expressed $\delta\vec{L} = \bar{\bar{P}}\delta\bar{\bar{\varepsilon}}\bar{\bar{P}}^{-1}\vec{L}$, where $\vec{L} = [l_1, l_2, l_3]^T$ and $\bar{\bar{P}}$ is rotation matrix.

The meso-structure is composed of ten spherical grains with the identical radius r_g (Fig. 1b). This meso-structure is selected because it includes a grain cluster, large enough to contain four force chains, and enabling grain rearrangement. It can be described analytically, and solved by hand, which is a great advantage. The grain

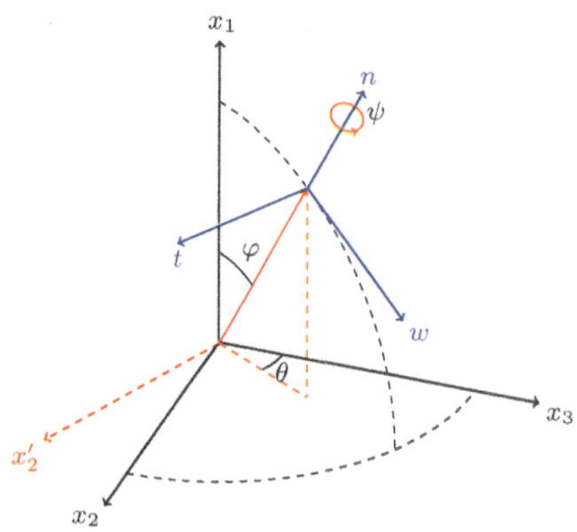

(a) Global and local coordinate systems of 3D-H model.

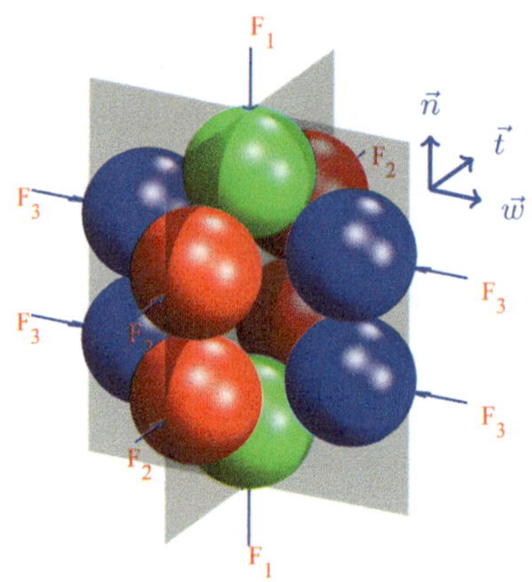

(b) Meso-Structure of 3D-H model.

Fig. 1 Coordinate systems and meso-structure of 3D-H model centers respectively belong to two mutually perpendicular planes. Meanwhile, the centers of grains make up two mutually perpendicular hexagon patterns. Thus, the meso-structure can be decomposed into two independent hexagon patterns: Hexagon A (shown in Fig. 2) and Hexagon B (similar to Hexagon A). Inheriting the features from the H model, this meso-structure is able to undergo complicated kinematic mechanisms, including local dilatant and contractant behaviors. To simplify the derivation, it is assumed that each meso-structure is subjected to an external symmetric load. Consequently, only two contacts between grains 1 to 2 and grains 2 to 3, respectively denoted by contact 1 and contact 2, are considered in Hexagon A. T_1 is the tangential component and N_1 is the normal component of the contact force of contact 1. The contact force applied by grain 3 to grain 2 only involves a normal component N_2, the tangential component is nil due to the symmetry. The relative incremental displacement between grains 2 and 3 is composed of a normal component δu_n^1 and a tangential component δu_t^1. Likewise, the relative incremental displacement of contact 2 is composed of a single normal component. Employing the notations given in Fig. 2a, it follows that

$$\begin{aligned} \delta u_n^1 &= \delta d_1 \\ \delta u_t^1 &= d_1 \delta \alpha_1 \\ \delta u_n^2 &= \delta d_2 \end{aligned} \tag{1}$$

The geometrical compatibility yields:

$$\begin{aligned} l_1 &= d_2 + 2d_1 \cos \alpha_1 \\ l_2 &= 2d_1 \sin \alpha_1 \end{aligned} \tag{2}$$

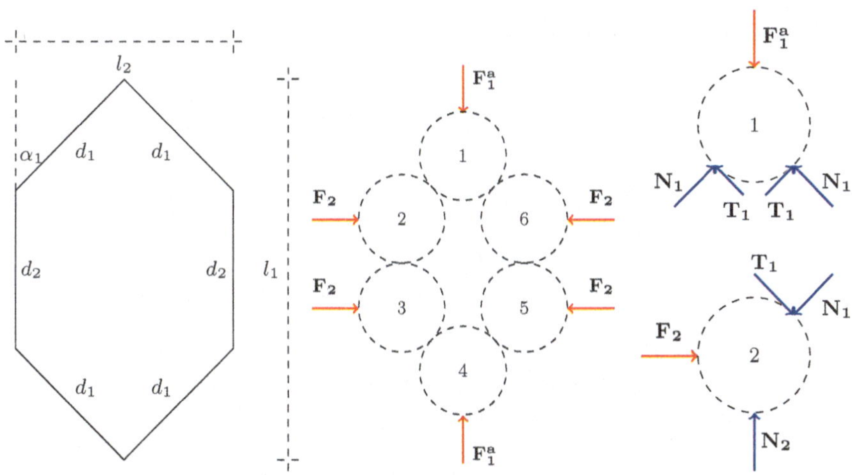

(a) Branch vector (b) External forces (c) Force balance

Fig. 2 Mechanical description of hexagon pattern A

Force balance of grain 1 along direction $\vec{n}$ and of grain 2 along direction $\vec{w}$ reads:

$$\begin{aligned} F_1^a &= N_1 \cos\alpha_1 + T_1 \sin\alpha_1 \\ F_2 &= N_1 \sin\alpha_1 - T_1 \cos\alpha_1 \end{aligned} \tag{3}$$

A elastic-perfect plastic inter-particle contact law is selected:

$$\begin{aligned} \delta F_n^c &= k_n \delta u_n^c \\ \delta \vec{F}_t^c &= \min\left\{ \left\| \vec{F}_t^c + k_t \delta \vec{u}_t^c \right\|, \tan\varphi_g \left(F_n^c + \delta F_n^c\right) \right\} \times \frac{\vec{F}_t^c + k_t \delta \vec{u}_t^c}{\left\| \vec{F}_t^c + k_t \delta \vec{u}_t^c \right\|} - \vec{F}_t^c \end{aligned} \tag{4}$$

Thus, differentiating Eq. 3 and combining with Eq. 4 gives:

$$\begin{aligned} \delta F_1^a &= -k_n \cos\alpha_1 \delta u_n^1 + k_t \sin\alpha_1 \delta u_t^1 - F_2 \delta\alpha_1 \\ \delta F_2 &= -k_t \cos\alpha_1 \delta u_t^1 - k_n \sin\alpha_1 \delta u_n^1 + F_1^a \delta\alpha_1 \end{aligned} \tag{5}$$

Differentiating Eq. 2 and combining with Eq. 5 leads to the following algebraic system expressing the incremental changes in δd_1, δd_2 and $\delta\alpha_1$ with respect to the incremental changes in δl_1 and δl_2

$$\begin{bmatrix} 2\cos\alpha_1 & 1 & -2d_1 \sin\alpha_1 \\ 2\sin\alpha_1 & 0 & 2d_1 \cos\alpha_1 \\ \cos\alpha_1 + \frac{A_1}{k_n}\sin\alpha_1 & -1 & \frac{F_2 - B_1 \sin\alpha_1}{k_n} \end{bmatrix} \begin{bmatrix} \delta d_1 \\ \delta d_2 \\ \delta\alpha_1 \end{bmatrix} = \begin{bmatrix} \delta l_1 \\ \delta l_2 \\ 0 \end{bmatrix} \tag{6}$$

where: in elastic regime $A_i = 0$, $B_i = k_t d_i$; in plastic regime $A_i = k_n \xi_i \tan\varphi_g$, $B_i = 0$.

Thus, combining Eq. 1, 5 and 6, the incremental constitutive relation for Hexagon A can be expressed as follows:

$$\frac{1}{|D|^a} \begin{bmatrix} K_{11}^a & K_{12}^a \\ K_{21}^a & K_{22}^a \end{bmatrix} \begin{bmatrix} \delta l_1 \\ \delta l_2 \end{bmatrix} = \begin{bmatrix} \delta F_1^a \\ \delta F_2 \end{bmatrix} \tag{7}$$

Similarly, the incremental constitutive relation for Hexagon B reads:

$$\frac{1}{|D|^b} \begin{bmatrix} K_{11}^b & K_{12}^b \\ K_{21}^b & K_{22}^b \end{bmatrix} \begin{bmatrix} \delta l_1 \\ \delta l_3 \end{bmatrix} = \begin{bmatrix} \delta F_1^b \\ \delta F_3 \end{bmatrix} \tag{8}$$

Superimposing Hexagon A and Hexagon B, the total incremental force along direction $\vec{n}$ is $\delta F_1 = \delta F_1^a + \delta F_1^b$. The incremental constitutive relation of the 3D meso-structure can be obtained as follows:

$$\bar{\bar{K}} \delta\vec{l} = \delta\vec{F} \tag{9}$$

$$\text{with } \bar{\bar{K}} = \begin{bmatrix} \frac{1}{|D|^a}K^a_{11} + \frac{1}{|D|^b}K^b_{11} & \frac{1}{|D|^a}K^a_{12} & \frac{1}{|D|^b}K^b_{12} \\ \frac{1}{|D|^a}K^a_{21} & \frac{1}{|D|^a}K^a_{22} & 0 \\ \frac{1}{|D|^b}K^b_{21} & 0 & \frac{1}{|D|^b}K^b_{22} \end{bmatrix}$$

3 Performances of the Three-dimensional-H Model

Drained triaxial tests at different confining pressures are selected to show how the proposed model simulates the behavior observed for sands. For demonstration purpose, the simulations at 100, 200 and 400 kPa confining pressures are compared in Fig. 3 from $\varepsilon_a = 0$ to $\varepsilon_a = 25\%$. The parameters $k_n = 1.78 \times 10^6$ N/m, $k_t = 1.06 \times 10^6$ N/m, $\varphi_g = 35°$ and $e_0 = 0.65$ are adopted. It can be stressed that the H model does not require more than these 4 parameters.

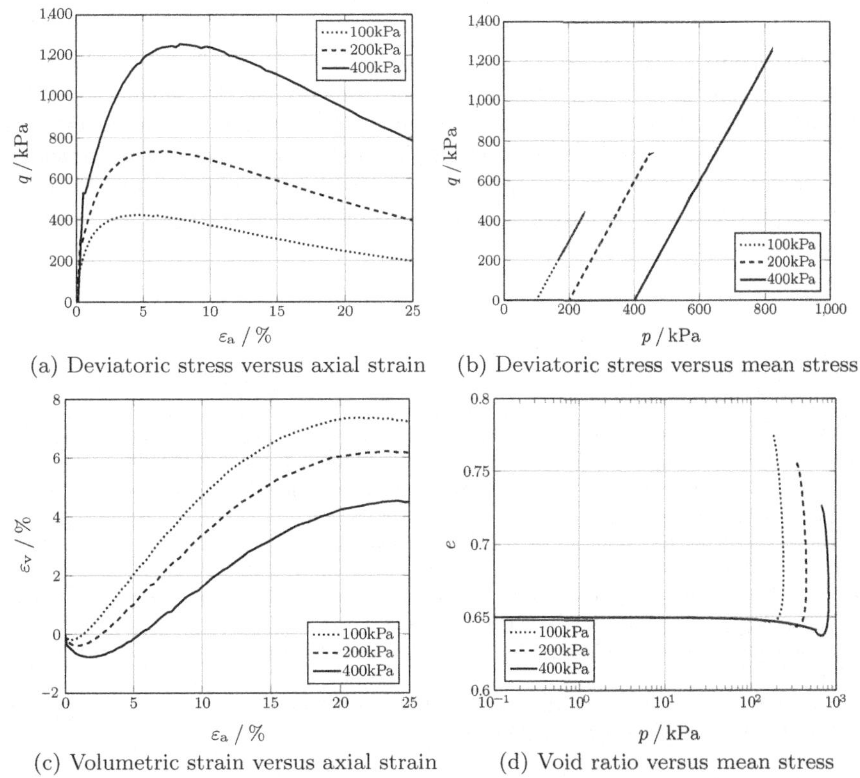

Fig. 3 Effect of different initial confining pressures on model responses along an axisymmetric drained triaxial loading path

Fig. 3a shows that the deviatoric stress q reaches a peak at a small strain about $\varepsilon_a = 5\%$ and undergoes a subsequent softening regime until critical state. As shown in Fig. 3b, the deviatoric stress q triple increases as the mean stress p during the triaxial loading path. The volumetric strain responses at different confining pressures are displayed in Fig. 3c, where a dilatancy can be observed before the critical state, except a sensible contractancy captured at the beginning. Figure 3d plots the void ratio e versus the mean stress p in logarithms form. The curves are increasing until the critical state line.

4 Conclusions

Based on the multi-scale approach, the H-model is extended to 3D conditions by replacing the 2D hexagonal pattern with a 3D meso-structure. For this purpose, a decomposition and superposition approach is introduced to analyze the local behavior of the meso-structure. The 3D-H model only introduces four parameters, wherein k_n, k_t and φ_g stem from the elastic-perfect plastic inter-particle contact law. The other one (the opening angle α_0) is a key microscopic geometrical parameter which can be estimated from the initial void ratio. It is noteworthy that the granular assembly rearrangement can be reflected by the evolution of the opening angle α_0. Drained triaxial tests were considered to examine the model performance. The hardening and softening regimes are well reproduced in the stress-strain response. Moreover, the critical state can be approximately described from void ratio evolution without involving any critical state related parameters.

References

1. Darve, F., Servant, G., Laouafa, F., Khoa, H.D.V.: Failure in geomaterials: continuous and discrete analyses. Comput. Methods Appl. Mech. Eng. **193**(27), 3057–3085 (2004)
2. Nicot, F., Darve, F.: The H-microdirectional model: accounting for a mesoscopic scale. Mech. Mater. **43**(12), 918–929 (2011)
3. Nicot, F., Darve, F., Group, R.: A multi-scale approach to granular materials. Mech. Mater. **37**(9), 980–1006 (2005)
4. Nicot, F., Daouadji, A., Laouafa, F., Darve, F.: Second-order work, kinetic energy and diffuse failure in granular materials. Granular Matter **13**(1), 19–28 (2011)
5. Tordesillas, A., Walker, D.M., Lin. Q.: Force cycles and force chains. Phys. Rev. E **81**(1), 011-302 (2010)
6. Zhu, H.X., Nicot, F., Darve, F.: Meso-structure evolution in a 2D granular material during biaxial loading. Granular Matter **18**(1), 1–12 (2016)

Borehole Instabilities in Granular Rocks Revisited: A Multiscale Perspective

Huanran Wu, Ning Guo and Jidong Zhao

Abstract We employ a recently developed hierarchical multiscale approach based on Finite-Discrete Element Method (FEM/DEM) coupling to revisit the borehole stability problem for granular rocks. The progressive failure of the borehole is simulated by gradually reducing the support pressure. A Representative Volume Element (RVE) with dual-porosity structure is adopted to characterize the material response of high-porosity quartz-rich sandstone without considering particle crushing. Compaction bands, or so-called 'fracture-like' breakouts are reproduced which initiate almost from the crossing points of σ_0 springline and the borehole wall and penetrate into the rock matrix along radial direction. Further development of the borehole is characterized by curved shear bands forming 'V-shaped' patterns.

1 Introduction

The stability of borehole is of important relevance to many applications in geotechnical, mining and petroleum engineering. A wide variety of methods have been developed to study this topic, including theoretical analysis based on bifurcation theory [8], numerical simulations using gradient-enhanced models [11] and Discrete Element Method (DEM) [9] and experimental means (e.g., [5]). In particular, miniature drilling experiments have been conducted to examine the failure mechanism of borehole in rock. Frequently, 'V-shaped' or 'dog-eared' breakouts have been observed as a result of progressive extensile or shear failure [5]. Interestingly, fracture-like borehole breakout has more recently been found in the

H. Wu · N. Guo · J. Zhao (✉)
Hong Kong University of Science and Technology, Clearwater Bay, Kowloon, Hong Kong
e-mail: jzhao@ust.hk

H. Wu
e-mail: hwuan@connect.ust.hk

E. Papamichos et al. (eds.), *Bifurcation and Degradation of Geomaterials with Engineering Applications*, Springer Series in Geomechanics and Geoengineering, DOI 10.1007/978-3-319-56397-8_54

drilling experiments of highly porous sandstones, such as high-porosity Berea sandstones (porosity: 25%) and St. Peter sandstone (porosity: 16–22%) [5]. This failure pattern for borehole shares great similarities with the 'compaction band' observed in these sandstones, including their host-rock properties and their shear-free, contractive failure mechanism.

In this study, we re-examine the instability patterns and failure mechanisms of boreholes in granular rocks by employing a hierarchical multiscale modelling approach based on rigorous coupling of Finite Element Method (FEM) and DEM [3, 4]; see also [1, 2, 7]. The multiscale approach adopts FEM to solve a boundary value problem (BVP), and attaches a Representative Volume Element (RVE) to each Gauss point of the FEM mesh to solve by DEM for the local material constitutive responses. It retains the robustness of FEM in handling BVPs, while helping by-pass the necessity of assuming phenomenological constitutive models in conventional continuum approaches. Importantly, this hierarchical approach offers a viable way to directly link the macro observations with their underpinning microstructural mechanisms, a key feature to be employed to revisit the borehole stability problem in highly porous sandstones.

2 Multiscale Analysis of Borehole Stability

We simplify the borehole in a highly porous sandstone into to a plane-strain ring with the borehole radius of $r_I = 15$ mm and the outer radius $r_O = 150$ mm (Fig. 1a). The whole domain is discretized by eight-noded serendipity finite element elements (80 layers in the hoop direction and 20 layers in the radial direction). Finer mesh is adopted for near borehole zone to highlight the deformation patterns around it. A similar domain and discretization has been used by Papanastasiou and Vardoulakis [8]. We simulate the progressive failure of the borehole by gradually decreasing the inner support pressure following a scheme illustrated in Fig. 1b. The support pressure is normalized by minimal principal stress (σ_{I0}), and the time is normalized by the total time used. In the simulation, the maximum principal stress (σ_{I1}) is reduced at a constant rate, while σ_{I0} is kept unchanged until σ_{I1} is reduced to be equal to σ_{I0}. After that, both σ_{I1} and σ_{I0} are further reduced to zero at the same rate. Detail formulation and solution procedures of the coupled FEM/DEM multiscale approach can be found in Guo and Zhao [3].

A recent study by the authors shows compaction bands can be reproduced in biaxial compression tests on highly porous sandstone with no particle crushing [10]. Considering the great similarities between the fracture-like breakout and 'compaction band', we employ a similar dual-porosity structure to generate the RVE for the multiscale analysis in this study. The typical RVE at an initial state of $\sigma_0 = 30\,\text{MPa}, \sigma_1 = 45\,\text{MPa}$ is depicted in Fig. 2. Each RVE contains 749 sphere particles, with an initial porosity of 0.326. A bonded contact model is employed

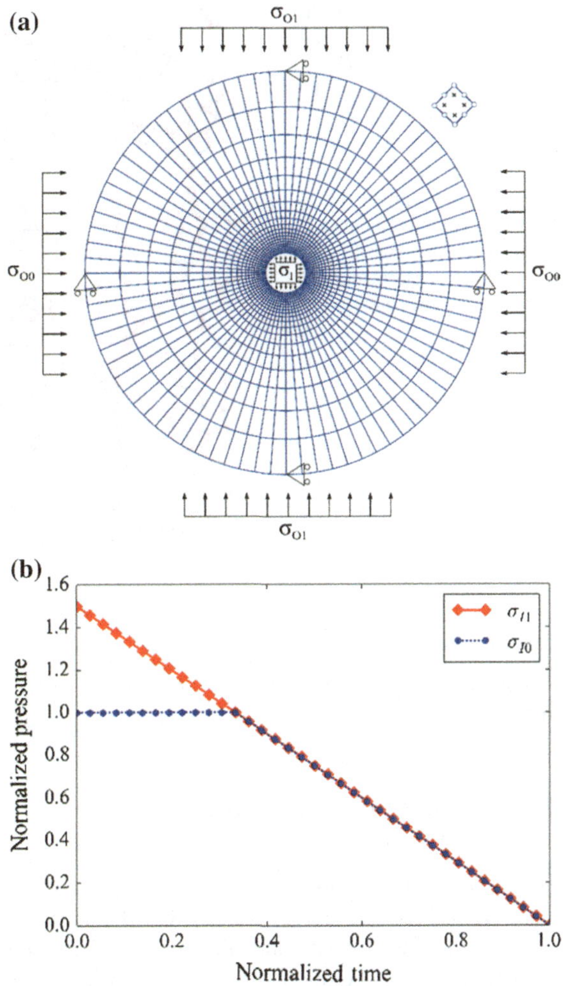

Fig. 1 **a** FE discretization by eight-noded serendipity elements of four Gauss points (reduced integration) and the prescribed boundary conditions for 2D borehole stability analysis and **b** Illustration of the simulating scheme by reducing the inner support pressure (σ_{I1} is reduced at a constant rate with σ_{I0} being kept constant until the two are equal, after which both are reduced to zero at the same constant rate)

with linear force-displacement contact law and a Coulomb-type friction criterion. The normal and tangential contact stiffness are expressed as $k_n = E_c r^*$ (where $r^* = 2r_1 r_2/(r_1 + r_2)$, r_1 and r_2 are the radii of the two contacted particles) and $\mathbf{k_t} = \boldsymbol{\nu}_\mathbf{c}\mathbf{k_n}$. Normal and tangential bonds are assigned to each contact as $a_n = C_n \min(r_1, r_2)^2$ and $a_t = C_t \min(r_1, r_2)^2$. If the tensile force exceeds the threshold $F_n^{max} = a_n$ or tangential force exceeds the bond strength $F_t^{max} = a_t + F_n \tan\phi$, the bond will be eliminated, leaving the contact a pure frictional one. The typical micro parameters adopted for multiscale modeling of Berea sandstone of 25% porosity (BS25) are summarized in Table 1.

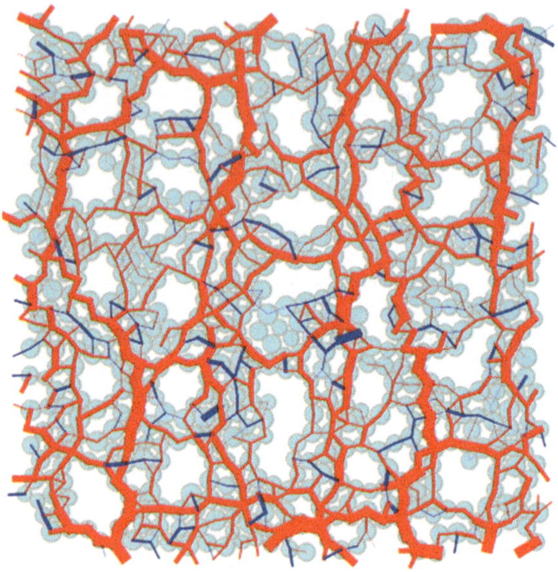

Fig. 2 Initial structure of the typical RVE prepared for the subsequent borehole study. The figure is an augmented illustration of bonded particles and macro pores with the interparticle contact force network (online: *red lines*—compressive contact forces; *blue lines* tensile forces) after consolidation at the confining pressures

Table 1 Micro parameters for the RVE

Radii (mm)	Density (kg/m^3)	E_c (GPa)	ν_c	ϕ (°)	C_n, C_t (GPa)
0.2–0.3	2,650	950	1.0	35	6.8

3 Results and Discussion

To highlight the failure mechanism, a typical case with an initial stress state at $\sigma_{I0} = \sigma_{O0} = 30$ MPa and $\sigma_{I1} = \sigma_{O1} = 45$ MPa is demonstrated. The RVE shown in Fig. 2 is adopted to simulate the material behavior of Berea sandstone of 25% porosity (BS25). The inner support pressure is reduced following the scheme illustrated in Fig. 1b at a rate of 1.25 MPa/S (S is the abbreviation for Step, not second). The whole simulation is terminated when $T = 36$ s. The initiation and development of the failure is presented in Fig. 3. The deviatoric strain is defined as $\epsilon_q = \sqrt{2\boldsymbol{e}:\boldsymbol{e}}$, where $\boldsymbol{e}$ is the deviatoric strain tensor. The particle rotation is defined as $\theta = \sum_{N_p} \theta_p / N_p$, where N_p is the number of particles within the packing; θ_p is the accumulated rotation of individual particles. Anti-clockwise rotation is taken as positive.

3.1 Initiation of Fracture-Like Breakout

Localized failure of the borehole incepts from $T = 15$ s when $\sigma_{I0} = \sigma_{I1} = 26.25$ MPa. The contours of N, n, ϵ_q and θ are illustrated in Fig. 3 (I). The failure starts almost

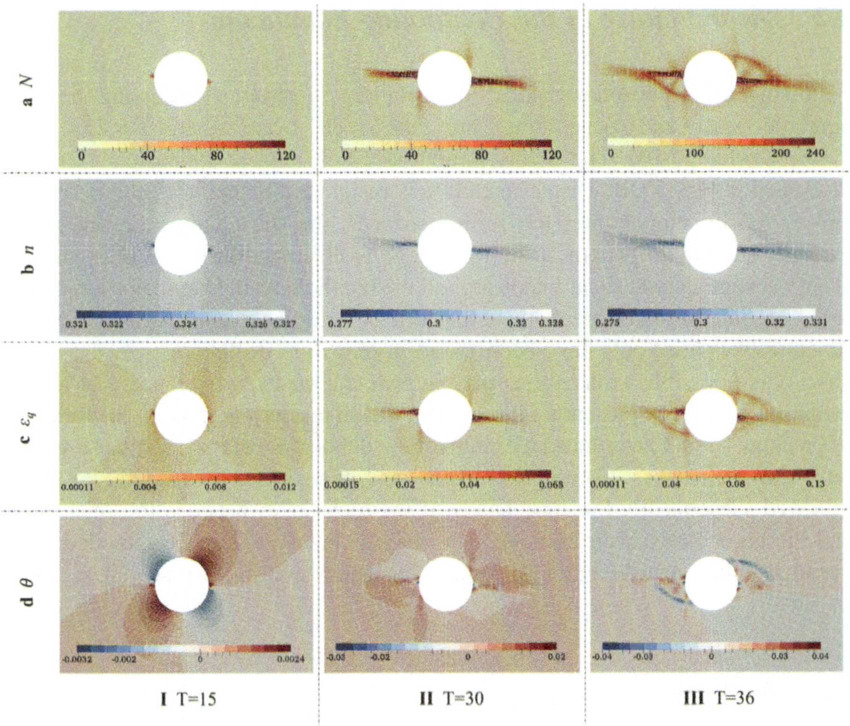

Fig. 3 Contours of N (debonding number), n (porosity), ϵ_q (deviatoric strain), θ (particle rotation) at different time step demonstrating the development of localization

from the crossing point of σ_0 springline and the borehole wall. The porosity of the failure zone is reduced to a minimum of 0.317 from the original 0.326 (the minimum of the color bar is adjusted in the figure to make the failure zone more obvious). The contour of θ also shows a typical compaction failure pattern that the rock mass outside rotates towards the band. A clearer pattern could be observed in Fig. 3 (II) (d) along the x-axis showing a mature compaction band. With further decrease of the support pressure, the fracture-like failure zone penetrates into the rock mass along the radial direction. As inter-particle debonding happens, debonded particles are compacted into a much denser state, resulting in significant porosity reduction without even considering particle crushing. Figure 3 (II) (b) shows the reduction of porosity from the maximum of 0.328 in the matrix (slightly larger than the initial value due to unloading) to a minimum of 0.277 within the mature compaction band.

3.2 Shear Failure in the Continuing Simulation

Following the compaction bands, shear failure is observed initiating from the borehole wall and penetrating into the rock mass since $T = 25\,\text{s}$ when $\sigma_{I0} = \sigma_{I1} = 13.75\,\text{MPa}$. The failure pattern at $T = 30\,\text{s}$ when $\sigma_{I0} = \sigma_{I1} = 7.5\,\text{MPa}$ is illustrated in Fig. 3 (II). The shear failure is easily identifiable in contours of N, ϵ_q, and θ. As the support pressure further reduces, the shear failure becomes more evident. The failure pattern at the final stage is illustrated in Fig. 3 (III), showing clearly several curved shear bands around the borehole wall. One major issue in the study of 'fracture-like' breakout is the lack of field observation. All the reported 'fracture-like' breakouts have been found in miniature drilling tests in the laboratory. While possible explanations may include the inadequate accuracy of logging technologies [6], the present multiscale simulation results suggest that the formation of breakouts could be induced by interparticle debonding and pore collapse without removing rock flakes. Compaction bands may appear as a premier failure pattern followed subsequently by shear bands. If the rock flakes between the shear bands and the borehole wall are removed, 'V-shaped' breakouts can be recorded as the eventual failure pattern although compaction bands may have appeared at first.

4 Conclusions

Compaction bands, or so-called 'fracture-like' breakouts have been reproduced in borehole stability analysis in highly-porosity sandstone based on a coupled FEM/DEM multiscale modeling approach. Under progressive reduction of the support pressure, compaction bands have been found initiating almost from the crossing points of σ_0 springline and the borehole wall and penetrating into the rock matrix along radial direction. Further reduction of the support pressure triggers curved shear bands starting from the borehole wall and penetrating into the matrix to form a 'V-shaped' pattern. If the rock flakes between the shear bands and the borehole wall are removed, 'V-shaped breakouts' can eventually be formed, which offers a possible explanation on the lack of field report of 'fracture-like' breakouts.

Acknowledgements This study was partially supported by University Grants Council of Hong Kong through a Theme-based Research Project T22-603/15N.

References

1. Andrade, J.E., Avila, C.F., Hall, S.A., Lenoir, N., Viggiani, G.: Multiscale modeling and characterization of granular matter: from grain kinematics to continuum mechanics. J. Mech. Phys. Solids **59**, 237–250 (2011)

2. Desrues, J., Nguyen, T.K., Combe, G., Caillerie, D.: FEM × DEM multi-scale analysis of boundary value problems involving strain localization. In: Bifurcation and Degradation of Geomaterials in the New Millennium, pp. 259–265 (2015)
3. Guo, N., Zhao, J.: A coupled FEM/DEM approach for hierarchical multiscale modelling of granular media. Int. J. Numer. Methods Eng. **99**, 789–818 (2014)
4. Guo, N., Zhao, J., Sun, W.C.: Multiscale analysis of shear failure of thick-walled hollow cylinder in dry sand. Géotech. Lett. **6**, 77–82 (2016)
5. Haimson, B.: Micromechanisms of borehole instability leading to breakouts in rocks. Int. J. Rock Mech. Min. Sci. **44**, 157–173 (2007)
6. Haimson, B., Klaetsch, A.: Compaction bands and the formation of slot-shaped breakouts in St. Peter sandstone. Geol. Soc. Lond. Spec. Publ. **284**, 89–105 (2007)
7. Nitka, M., Combe, G., Dascalu, C., Desrues, J.: Two-scale modeling of granular materials: a DEM-FEM approach. Granul. Matter **13**, 277–281 (2011)
8. Papanastasiou, P.C., Vardoulakis, I.G.: Numerical treatment of progressive localization in relation to borehole stability. Int. J. Numer. Anal. Methods Geomech. **16**, 389–424 (1992)
9. Rahmati, H., Nouri, A., Chan, D., Vaziri, H.: Simulation of drilling-induced compaction bands using discrete element method. Int. J. Numer. Anal. Methods Geomech. **38**, 37–50 (2014)
10. Wu, H., Zhao, J., Guo, N.: Multiscale modeling of compaction band in highly porous sandstone. In: The First International Conference on Geo-Energy and Geo-Environment. Hong Kong (2015)
11. Zervos, A., Papanastasiou, P., Vardoulakis, I.: Modelling of localisation and scale effect in thick-walled cylinders with gradient elastoplasticity. Int. J. Solids Struct. **38**, 5081–5095 (2001)

Measuring Force-Chains in Opaque Granular Matter Under Shear

Eloïse Marteau and José Andrade

Microstructural information, such as inter-particle forces and particle kinematics, plays a key role in understanding the continuum behavior of complex granular structure. Although micromechanical techniques have provided tremendous insights, they still lack quantitative accuracy and, associated with this, capacity to predict macroscopic behavior. We report here a set of experiments performed on a novel mechanical device in which we have successfully extracted particle-scale kinematics and inter-particle forces in a two-dimensional idealized granular assembly. This mechanical device allows a specimen composed of a two-dimensional analogue granular assembly to be subjected to quasi-static shear conditions over large deformation. Digital Image Correlation (DIC) is employed to measure particle kinematics. The inter-particle forces are inferred using the Granular Element Method (GEM), provided that average particle strains are measured and that the location of the contact points in the array is known. DIC combined with GEM allow us to observe and assess the force distribution in the complex granular assembly. These results represent an important step in our understanding of the micromechanical response of a complex granular assembly to applied macroscopic strains and stresses.

1 Introduction

The ability to measure inter-particle forces in natural granular materials is essential to understanding how forces are transmitted in complex granular assembly. Inter-particle forces in granular assemblies were first observed optically in packings

E. Marteau · J. Andrade (✉)
Division of Engineering & Applied Science,
California Institute of Technology, Pasadena, CA 91125, USA
e-mail: jandrade@caltech.edu

E. Papamichos et al. (eds.), *Bifurcation and Degradation of Geomaterials with Engineering Applications*, Springer Series in Geomechanics and Geoengineering, DOI 10.1007/978-3-319-56397-8_55

of idealized photoelastic particles, which have the property to develop birefringence by the application of stress [2, 3, 5]. Although photoelasticity is a useful experimental technique to study granular materials, it has many limitations: it requires the use of birefringent grains with simple geometries and is difficult to implement in three dimensions. To overcome many of these limitations, in this work, we employ the Granular Element Method (GEM) [1, 4] to infer contact forces in granular assembly with grains of arbitrary shape, texture and opacity. The GEM methodology consists of two stages. First, Digital Image Correlation (DIC) [6–8] is used to extract information about the average grain stresses while geometric arrangement is determined by mean of image processing techniques. Secondly, this information is taken as input for the constrained optimization problem, whose solution is the desired contact forces. More details on the method can be found in Hurley, 2014. In this paper, we present a novel mechanical device capable of reproducing quasi-static shear conditions. We report results for a set of experiments where we have directly observed the so-called 'force chains' using DIC and successfully applied GEM to measure inter-particle forces in opaque, 2D, frictional granular materials.

2 Experimental Setup

We have designed and built a novel experimental device for investigating the continuum behavior of granular material by incorporating microscopic experimental quantities. This mechanical apparatus, pictured in Fig. 1, is capable of reproducing (quasi-static) shear conditions over large deformation on a two-dimensional analogue granular assembly. The shear cell consists of a horizontal deformable parallelogram of size 50 cm by 50 cm in the undeformed configuration. The grains stand on a glass plate and are individually tracked by a camera that sits above the apparatus. A linear actuator provides the shear mechanism while a constant normal stress σ_n is applied. We measure the shear angle γ and the strain in the y-direction ε_{yy} using image processing whereas the strain in the x-direction ε_{xx} remains equal to zero. We performed experiments on a two-dimensional analogue opaque granular assembly composed of 400 cylinders of two different diameters (20 and 30 mm) and height of 25 mm. The grains are made of polyurethane that has a Young's modulus of 50 MPa and a Poisson's ratio of approximately 0.5.

3 Experimental Results

We present here results of experiments performed using the shear apparatus described in the previous section where a constant normal stress σ_n = 14.2 kN is applied. The ε_{xy} component of the strain field extracted using DIC and the

Fig. 1 Experimental setup

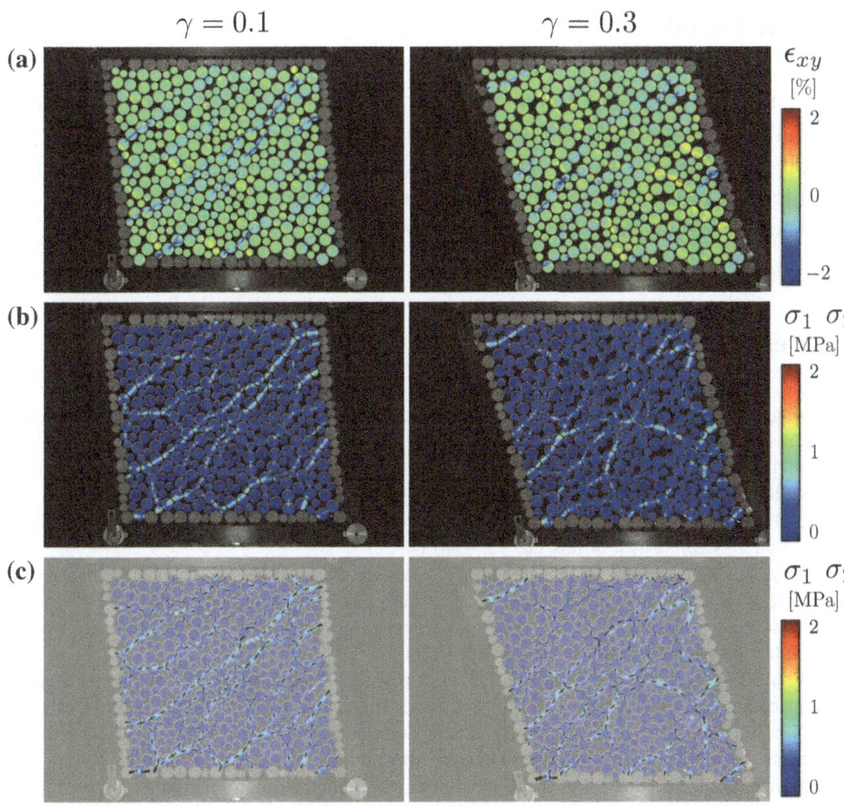

Fig. 2 Contour distribution of DIC measurements and GEM results for shear test (σ_N = −28.5 kPa) at γ = 0.1 and 0.3. **a** Strain field ε_{xy}. **b** Difference of principal stresses $\sigma_1 - \sigma_2$. **c** Contact forces inferred with GEM superimposed on difference of principal stresses $\sigma_1 - \sigma_2$

difference of principal stresses $\sigma_1 - \sigma_2$ is shown in, respectively, Fig. 2a, b for two different shear angle γ. The difference of principal stresses $\sigma_1 - \sigma_2$ is calculated from the strain field measured with DIC using generalized Hooke's law. We emphasize that the contour distribution $\sigma_1 - \sigma_2$ is directly related to isochromatics observed with photoelasticity [5]. It can be seen from Fig. 2a, b that DIC measurements provide a qualitative information of the force distribution in an opaque granular assembly. GEM was then applied to these data sets to obtain the force distribution as a function of macroscopic load. The resulting inter-particle forces is superimposed to the field $\sigma_1 - \sigma_2$ in Fig. 2c. Figure 2c shows that the force magnitude matches really well measured stress contour distribution and that force chains are clearly seen to develop. GEM enable a quantitative assessment of force network. Furthermore, these results showcase that the entire topology of granular forces can be obtained without resorting to photoelasticity.

4 Conclusion

DIC combined with GEM has allowed the observation and inference of the force distribution in a complex granular assembly under shear deformation. These preliminary experimental results demonstrate a potential application of the method for constructing physic-based constitutive models and for understanding the micromechanical origin of macromechanical phenonema.

References

1. Andrade, J.E., Avila, C.F.: Granular element method (GEM): linking inter-particle forces with macroscopic loading. Granul. Matter **14**, 51–61 (2012)
2. Daniels, K.E., Hayman, N.W.: Force chains in seismogenic faults visualized with photoelastic granular shear experiments. J. Geophys. Res. **113**, 11411 (2008)
3. Drescher, A.: Photoelastic verification of a mechanical model for the flow of a granular material. J. Mech. Phys. Solids **20**, 337–340 (1972)
4. Hurley, R., Marteau, E., Ravichandran, G., Andrade, J.E.: Extracting inter-particle forces in opaque granular materials: beyond photoelasticity. Geotechnique **29**, 47–65 (2014)
5. Majmudar, T.S., Behringer, R.P.: Contact force measurements and stress-induced anisotropy in granular materials. Nature **435**, 1079–1082 (2005)
6. Pan, B., Qian, K., Xie, H., Asundi, A.: Two-dimensional digital image correlation for in-plane displacement and strain measurement: a review. Meas. Sci. Technol. **20**, 062001 (2009)
7. Sutton, M.A., Orteu, J.J., Schreier, H.: Image Correlation for Shape, Motion and Deformation Measurements: Basic Concepts, Theory and Applications. Springer, New York (2009)
8. Viggiani, G., Hall, S.A.: Full-field measurements, a new tool for laboratory experimental geomechanics (2008)

Investigations of Vortex-Structures in Granular Bodies Based on DEM and Helmholtz-Hodge Flow Field Decomposition

Jan Kozicki and Jacek Tejchman

Abstract The paper presents some two-dimensional simulation results of vortex-structures in cohesionless initially dense sand during a quasi-static passive wall translation. The sand behaviour was simulated using the discrete element method (DEM). Sand grains were modelled by spheres with contact moments to approximately capture the irregular grain shape. In order to detect vortex-structures, the Helmholtz-Hodge decomposition of a flow displacement field from DEM calculations was used. This approach enabled us to distinguish both incompressibility and vorticity in the granular displacement field.

1 Introduction

Granular vortex structures (swirling motion of several grains around its central point) were frequently observed in experiments on granular materials [1, 8] and in calculations using the discrete element method (DEM) [5, 7, 11]. The dominant mechanism responsible for the vortex formation was the buckling of force chains [7, 11]. The collapse of main force chains lead to the formation of larger voids and their build-up to the formation of smaller voids [7]. The vortices have been mainly observed in shear zones [7] which are the fundamental phenomenon in granular bodies [9].

The objective of this paper is to report the results of comprehensive 2D studies by DEM on vortex-structures in sand behind a rigid wall during its quasi-static passive translation by using the Helmholtz-Hodge decomposition of the displacement vector field [2, 10]. This detection method has the following advantages over other techniques: (1) it operates directly on displacement vectors, (2) it is designed

J. Kozicki · J. Tejchman (✉)
Faculty for Civil and Environmental Engineering, Gdańsk University of Technology, Narutowicza, 11/12, 80-952 Gdańsk-Wrzeszcz, Poland
e-mail: tejchmk@pg.gda.pl

J. Kozicki
e-mail: jkozicki@pg.gda.pl

E. Papamichos et al. (eds.), *Bifurcation and Degradation of Geomaterials with Engineering Applications*, Springer Series in Geomechanics and Geoengineering, DOI 10.1007/978-3-319-56397-8_56

for both 2D and 3D flow and (3) it does not need any additional parameters for calculations. The analyses were carried out with spheres with contact moments [4, 5, 7] to approximately capture the irregular grain shape. In order to accelerate the computation time, some simplifications were assumed in analyses: larger spheres, linear sphere distribution and linear normal contact model.

2 DEM Results of Passive Earth Pressure Model Tests

In order to simulate the behaviour of real sand, the 3D spherical discrete model YADE, developed at University of Grenoble [3] was used by taking advantage of the so-called soft-particle approach (i.e. the model allows for particle deformation which is modelled as an overlap of particles).

The DEM calculation results were described in detail in [7]. The quasi-static simulations were performed for a 2D initially dense sand body of 0.40 m length and 0.20 m height in order to compare with experiments with Karlsruhe sand (d_{50} = 0.5 mm) [9]. The vertical retaining wall and the bottom of the granular specimen were assumed to be stiff and very rough, Since the experiments were idealized as a 2D boundary value problem, in order to significantly accelerate simulations, the computations were performed with the specimen depth equal to the grain size (i.e. one layer of spheres was simulated along the depth only). The spheres with d_{50} = 1.0 mm, characterized by a linear grain size distribution, were assumed (grain size range 0.5–1.5 mm, 62,600 spheres). The initial void ratio of sand was $e_o = 0.62$ (volumetric weight γ = 25.5 kN/m^3).

Figure 1A shows the evolution of the resultant normalized horizontal earth pressure force (earth pressure coefficient) $K_p = 2E_h/(\gamma h^2 d_{50})$ versus the normalized horizontal wall displacement u/h (h = 0.2 m, E_h—the horizontal force acting on the wall) for d_{50} = 1 mm. The specimen exhibited the initial strain hardening up to the peak (u/h = 0.038), followed by some softening before a common asymptote was reached. The earth pressure coefficient was $K_p^{max} = 30$ for d_{50} = 1 mm.

The distribution of single sphere rotations ω during wall translation is presented in Fig. 1B (red denotes the sphere rotation $\omega > +30°$ and blue $\omega < -30°$, dark grey is related to the sphere rotation in the range $5° \leq \omega \leq 30°$ and light grey to the range $-30° \leq \omega \leq -5°$, positive sign means clockwise rotation). Accepting such a colour convention makes shear zones clearly observable (only particles within shear zones significantly rotate). There exists a clear grain separation regarding clockwise (red) and anti-clockwise (blue) rotation. Except of two main shear zones (curved and radial shear zone), some secondary shear zone might also form [7]. The curved shear zone started to develop along the specimen bottom for the normalized wall translation u/h = 0.02 (Fig. 1B(b)). It was fully developed for u/h = 0.06. Its thickness was t_s = 20 mm (20d_{50}). The radial shear zone began to develop for u/h = 0.04 (Fig. 1B d) and for u/h = 0.06 connected the curved shear zone. Its thickness was t_s = 10 mm. There is a qualitative agreement between DEM simulation results and real experimental outcomes [6, 9].

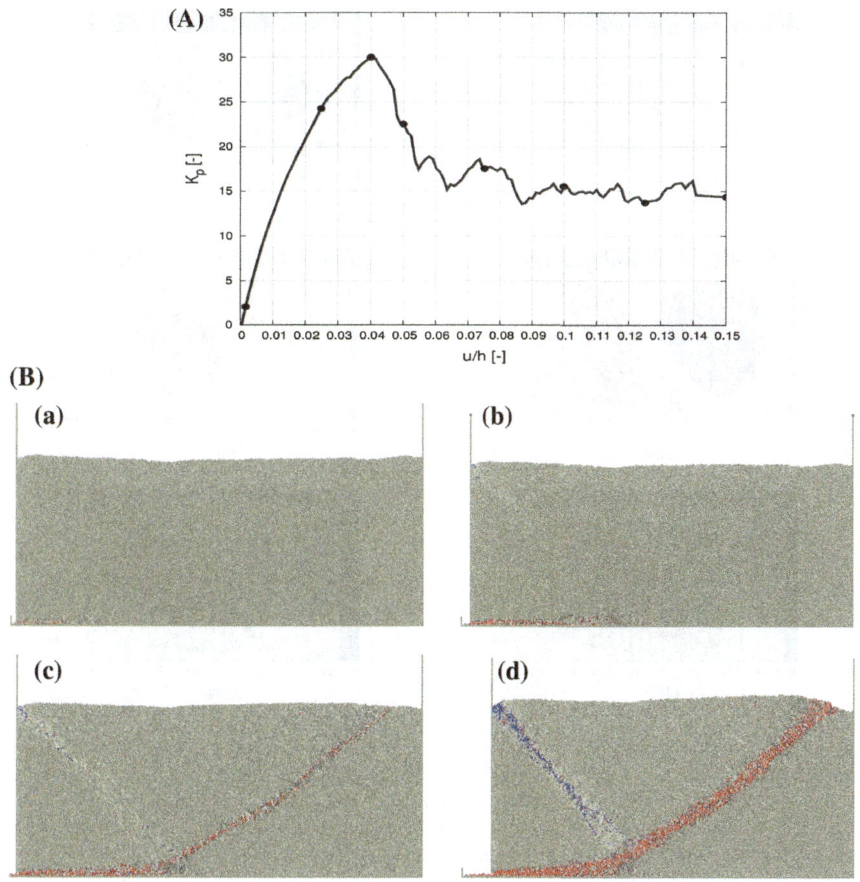

Fig. 1 DEM results (passive case, translating wall, initially medium dense sand, height $h = 0.2$ m): **A** evolution of resultant normalized horizontal earth pressure force $2E_h/(\gamma h^2 d_{50})$ versus normalized horizontal wall displacement u/h and **B** deformed granular body 0.2×0.4 m^2 with distribution of single sphere rotations: **a** $u/h = 0.02$, **b** $u/h = 0.04$, **c** $u/h = 0.06$ and **d** $u/h = 0.15$ (*red* colour denotes clockwise rotation $\omega > +30°$, *blue* colour denotes anticlockwise rotation $\omega < -30°$) [7]

3 Numerical Results Using Helmholtz-Hodge Decomposition

The Helmholtz-Hodge decomposition (HHD) of 3D vector fields is one of the fundamental theorems in fluid dynamics [2, 10]. It describes a vector displacement field $\vec{\xi}$ in terms of its curl-free and divergence-free components based on potential functions:

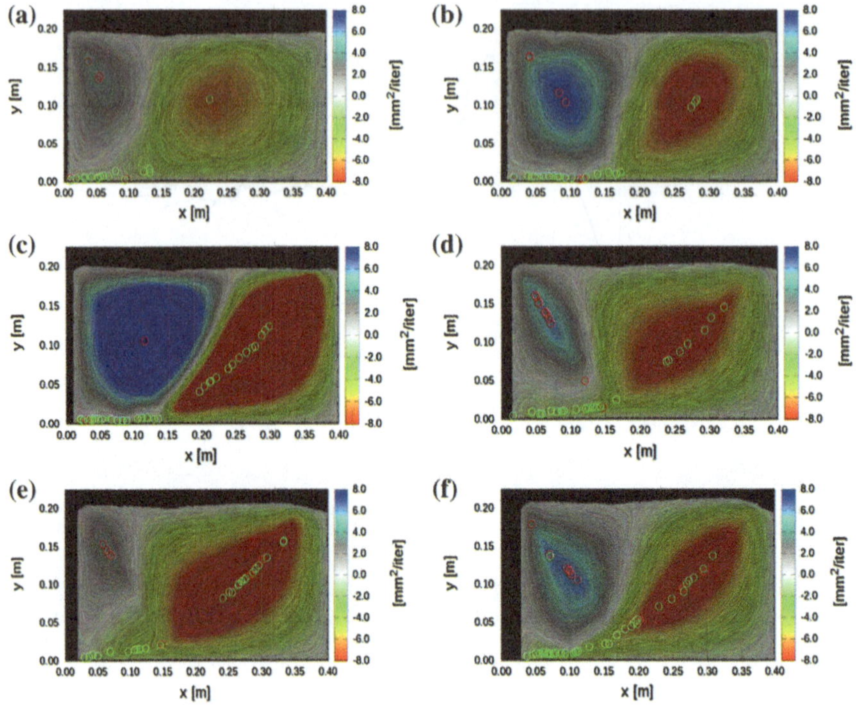

Fig. 2 Evolution of vector field curl $\vec{\nabla}\times\vec{v}$ (divergence-free component related to vorticity) in granular specimen area $x \times y$ during normalized wall translation u/h: **a** $u/h = 0.025$, **b** $u/h = 0.04$, **c** $u/h = 0.05$, **d** $u/h = 0.075$, **e** $u/h = 0.1$ and **f** $u/h = 0.15$ (scale denotes velocity perpendicular to specimen v_z in [mm^2/iteration]), *green circles* describe local minima (*right*-handed vortices) and *red circles* local minima (*left*-handed vortices)

$$\vec{\xi} = \vec{\nabla}u + \vec{\nabla}\times\vec{v} + \vec{h}, \tag{1}$$

where $\nabla = \left(\frac{\partial}{\partial x}, \frac{\partial}{\partial y}, \frac{\partial}{\partial z}\right)^T$ is the gradient, $\nabla\cdot = \left(\frac{\partial}{\partial x} + \frac{\partial}{\partial y} + \frac{\partial}{\partial z}\right)$ denotes the divergence operator, $\nabla\times$ is the curl operator, u denotes the scalar potential field, $\vec{v}$ is the vector potential field and $\vec{h}$ denotes the harmonic vector field. The gradient of the scalar potential function $\vec{\nabla}u$ is called the curl-free component and is related to expansion/contraction (because is irrotational) while the curl of the vector potential function $\vec{\nabla}\times\vec{v}$ is called the divergence-free component and is related to vorticity (because is incompressible). The harmonic component is related to the pure translation. A discrete vector field decomposition was done according the method given in [10]. The system of linear equations was solved using the following general boundary conditions: $\vec{\nabla}\times\vec{v}$ was tangential to the domain boundary $\vec{v}|_{\partial T} = 0$ and ∇u was orthogonal to the boundary domain $u|_{\partial T} = 0$.

Figure 2 presents the evolution of the vector field curl $\vec{\nabla} \times \vec{v}$ (divergence-free component related to vorticity) during normalized wall translation u/h. The green circles describe the local minima (right-handed vortices) and the red circles the local minima (left handed vortices). Vortex-structures appeared from the wall translation beginning. They were immediately concentrated in shear zones only. Right-handed and left-handed vortices alternately occurred. The right-handed vortices dominated in the curved shear zone and left-handed vortices dominated in the radial shear zone. Their distance was different.

Figure 3 shows the vector displacement field $\vec{\xi}$ during the normalized horizontal wall translation u/h. Sphere displacement directions are marked by white arrows. Based on the displacement vector length and direction changes, a curved shear zone between the wall bottom and free upper boundary already started in the first calculation step. Later it moved to the right to reach its ultimate position due to wall friction along the bottom that was close to the maximum resultant normalized horizontal earth pressure force of Fig. 1A ($u/h = 0.04$). Behind the curved shear zone the material was totally rigid. A radial shear zone was created for $u/h = 0.065$.

The evolution of the scalar field gradient $\vec{\nabla} u$ (curl-free component related to compressibility) during normalized wall translation u/h is described in Fig. 4. The green circles describe the sources (local dilatancy minima) and red circles the sinks

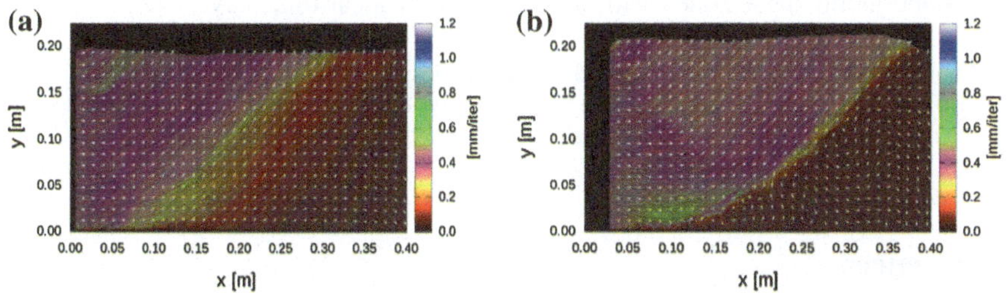

Fig. 3 Vector field $\vec{\xi}$ in granular specimen area $x \times y$ during normalized wall translation u/h: **a** $u/h = 0.025$ and **d** $u/h = 0.15$ (scale denotes vector length in [mm/iteration])

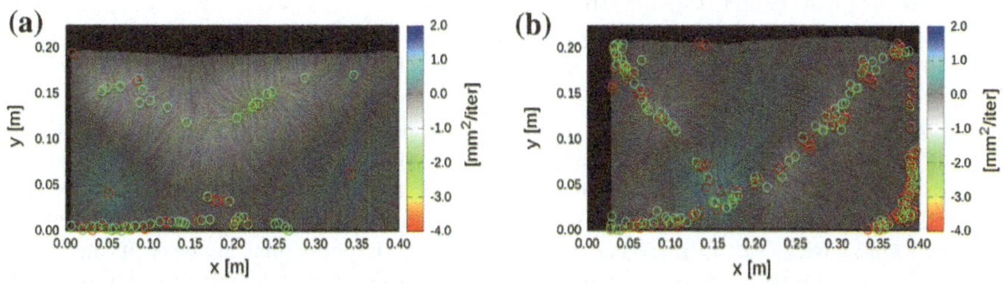

Fig. 4 Scalar field gradient $\vec{\nabla} u$ (curl-free component related to compressibility) in granular specimen area $x \times y$ during normalized wall translation u/h: **a** $u/h = 0.025$ and **d** $u/h = 0.15$ (scale denotes u in [mm^2/iteration] (sign (−)—dilatancy, sign (+)—contractancy, *green circles*—sources (local dilatancy minima), *red circles* (local contractancy maxima))

(local contractancy maxima). Initially global contractancy and later global dilatancy occurred in the granular specimen. The global dilatancy was maximum after the peak (Fig. 4b) and later it diminished. In the residual state ($u/h \geq 0.075$, Fig. 1A), local regions of dilatancy and contractancy alternately happened along both shear zones (with the prevalence of dilatancy).

4 Conclusions

The following main conclusions may be listed from DEM simulations of vortex-structures in sand during a passive earth pressure problem:

The vortex-structures were the precursor of shear localization since they clearly appeared in a curved and radial shear zone from the beginning of the deformation process, i.e. significantly earlier than e.g. based on single grain rotations. They vortex-structures solely appeared in shear zones during the entire deformation process They had a tendency to move along shear zones. Their number varied and was maximum on average at the residual state. The right-handed vortices were dominant in the curved shear zone and left-handed ones were dominant in the radial shear zone.

In the residual state, local regions of dilatacy and contractancy alternately happened along shear zones with a dominance of local dilatancy.

Acknowledgements The authors would like to acknowledge the support by the grant 2011/03/B/ST8/05865 "Experimental and theoretical investigations of micro-structural phenomena inside of shear localization in granular materials" financed by the Polish National Science Centre.

References

1. Abedi, S., Rechenmacher, A.L., Orlando, A.D.: Vortex formation and dissolution in sheared sands. Granul. Matter **14**, 695–705 (2012)
2. Bhatia, H., Pascucci, V.: The Helmholtz-Hodge decomposition—a survey. IEEE Trans. Visual Comput. Graph. **19**(8), 1386–14404 (2013)
3. Kozicki, J., Donze, F.V.: A new open-source software developed for numerical simulations using discrete modelling methods. Comput. Methods Appl. Mech. Eng. **197**, 4429–4443 (2008)
4. Kozicki, J., Tejchman, J., Műhlhaus, H.B.: Discrete simulations of a triaxial compression test for sand by DEM. Int. J. Num. Anal. Methods Geomech. **38**, 1923–1952 (2014)
5. Kozicki, J., Tejchman, J.: DEM investigations of two-dimensional granular vortex- and anti-vortex- structures during plane strain compression. Granul. Matter **2**(20), 1–28 (2016)
6. Niedostatkiewicz, M., Leśniewska, D., Tejchman, J.: Experimental analysis of shear zone patterns in sand for earth pressure problems using particle image velocimetry. Strain **47**(s2), 218–231 (2011)
7. Nitka, M., Tejchman, J., Kozicki, J., Leśniewska, D.: DEM analysis of micro-structural events within granular shear zones under passive earth pressure conditions. Granul. Matter **3**, 325–343 (2015)

8. Richefeu, V., Combe, G., Viggiani, G.: An experimental assessment of displacement fluctuations in a 2D granular material subjected to shear. Geotech. Lett. **2**, 113–118 (2012)
9. Tejchman, J.: FE modeling of shear localization in granular bodies with micro-polar hypoplasticity. In: Wu, B. (eds.) Springer Series in Geomechanics and Geoengineering. Springer, Berlin, Heidelberg (2008)
10. Tong, Y., Lombeyda, S., Hirani, A.N., Desbrun, M.: Discrete multiscale vector field decomposition. ACM Trans. Graph. (TOG) **22**(3), 445–452 (2003)
11. Tordesillas, A., Pucilowski, S., Walker, D.M., Peters, J.F., Walizer, L.E.: Micromechanics of vortices in granular media: connection to shear bands and implications for continuum modelling of failure in geomaterials. Int. J. Numer. Anal. Methods Geomech. **38**(12), 1247–1275 (2014)

Restoring Mesh Independency in FEM-DEM Multi-scale Modelling of Strain Localization Using Second Gradient Regularization

J. Desrues, A. Argilaga, S. Dal Pont, G. Combe, D. Caillerie and T. kein Nguyen

Abstract Continuum media from classical mechanics cannot appropriately reproduce the evolution of materials exhibiting strong heterogeneities in the strain field, e.g. strain localization. Models without a microscale representation cannot properly reproduce the microscale mechanisms that trigger the strain localization, in addition, first gradient relations don't present any length parameter in the formulation. This results in a model without a characteristic length that cannot exhibit any objective band width. In this paper, techniques to introduce an internal length will be enumerated. Microstuctured materials will be retained and in particular Second Gradient model will be exposed and used along with a FEMxDEM approach. Numerical results showing the abilities of the enriched model will conclude the text.

1 Introduction

The idea of FEM×DEM is to solve a continuum boundary value problem (BVP) at the macroscale while obtaining the constitutive material behaviour from a DEM

J. Desrues (✉)
CNRS - Université Grenoble Alpes, Grenoble, France
e-mail: jacques.desrues@3sr-grenoble.fr

A. Argilaga · S. Dal Pont · G. Combe · D. Caillerie · T. kein Nguyen
Université Grenoble Alpes, Grenoble, France

A. Argilaga
e-mail: albert.argilagaclaramunt@3sr-grenoble.fr

S. Dal Pont
e-mail: stefano.dalpont@3sr-grenoble.fr

G. Combe
e-mail: gael.combe@3sr-grenoble.fr

D. Caillerie
e-mail: denis.caillerie@3sr-grenoble.fr

T. kein Nguyen
e-mail: trungkien.nguyen@3sr-grenoble.fr

E. Papamichos et al. (eds.), *Bifurcation and Degradation of Geomaterials with Engineering Applications*, Springer Series in Geomechanics and Geoengineering, DOI 10.1007/978-3-319-56397-8_57

microscale in a fully coupled hierarchical multiscale method. Some early works [11, 16, 20], have put in evidence the potential of the method to provide a refined description of complex constitutive behaviors. Indeed, FEM×DEM methods allow to couple the advantages of Discrete Elements and the efficiency of Finite Elements. Later works have enhanced and extended this approach to the study of anisotropy [7, 18], granular cohesion [19], material heterogeneity [25], real scale engineering applications [8, 18], more realistic constitutive behaviors using 3D DEM [12, 26], macroscale hydro-mechanical coupling [10, 26]. More recently, [12] have embedded non-local regularization at the macro-scale and [9] has developed a full micro-macro 3D approach.

It is known that for a strain softening material the initiation of strain localization can lead to an ill-posed Partial Differential Equation (PDE) problem [23]. The consequence of this ill posedness are numerical instabilities and strain mesh dependency [22]. Regularization techniques have been developed to overcome these problem; both nonlocal [5] and local [6, 17] approaches exist. An example of nonlocal regularization in a FEM×DEM model is presented in [12]. Local formulations use a local relationship between stress and strains in the same manner as classical constitutive relations are defined. Second gradient model, as a particular case of the Germain theory [6] has been developed [1, 4, 14, 21]. It has been extensively applied in geomechanics and engineering applications with satisfactory results [3, 13, 24].

Previous developments of FEM×DEM could not take advantage of Second Gradient regularization due to the poor solution stability and limited available mesh refinement. Recent improvements concerning stability and computational efficiency allow to build a FEM×DEM model including local Second Gradient. This results in an objective model capable of simulating real scale problems with any mesh refinement.

2 Introduction of an Internal Length

Models using a first-order constitutive relation of classical mechanical continuum cannot properly predict the behaviour of a medium with high strain gradients. Those approaches suffer from non-objectivity due to mesh dependency in localization problems.

Possible causes: the first-order constitutive relation does not give any information about the internal length of the model; due to that, the localization band thickness will tend to shrink to a size proportional to the mesh size. In this way, if the mesh is refined making the size of elements tend to zero in order to get an exact solution, the strain will concentrate in a null size band posing obvious problems.

It's needed to set a relationship between the micro-scale heterogeneity and the macro-scale characteristic length in order to establish a proper micro-scale size effect on the macro-scale. An enriched model with microstructure is proposed: local second gradient model [1, 4, 14, 21].

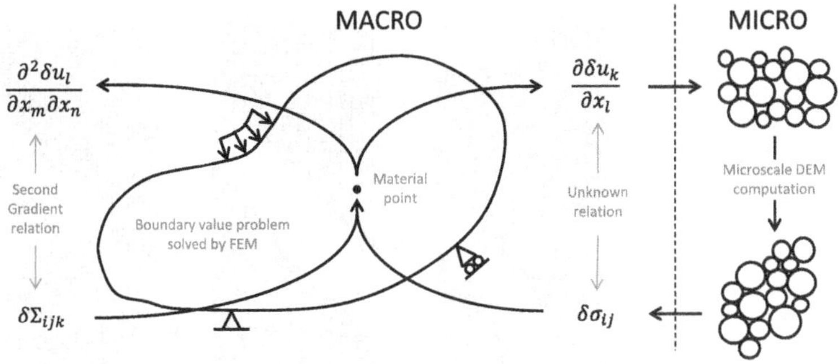

Fig. 1 Computational homogenization scheme with Second Gradient

3 Local Second Gradient Model

Microstructured material descriptions consider a continuum field in the micro-scale enriched by higher order terms [6]. In this way, the kinematics of the media is enriched by its micro-scale introducing a local dependence on an internal length parameter. This characteristic length [1] regularizes the solution making strain localization mesh independent. A local formulation complies with the principle of local action so it states a stress-strain relationship in the same manner as a classical constitutive laws do, this makes the implementation of local regularization in classical models a straight forward procedure (Fig. 1).

We present the weak form of the balance equations written for the strain gradient theory viewed as a particular case of the microstructured continuum theory [2]:

$$\int_{\Omega^t} \left(\sigma^t_{ij} \frac{\partial u_i^\star}{\partial x_j^t} + \Sigma^t_{ijk} \frac{\partial^2 u_i^\star}{\partial x_j^t \partial} x_k^t \right) d\Omega^t - \bar{P}_e^\star = 0 \tag{1}$$

where, superscripts t and $\star$ denote, quantities at a given time t and virtual quantities, σ^t_{ij} is the Cauchy stress tensor, Σ^t_{ijk} is the corresponding double stress tensor, $u_i^\star$ is a kinematically admissible virtual displacement feld, x_i^t are the current coordinates of the points of the studied body and $P_e^\star$ is the external virtual work generated by the corresponding external forces [15].

4 Numerical Simulations

A compression biaxial test with second gradient enrichment shows how the regularization turns the shear band patterns independent from the mesh size (Fig. 2). This regularization not only enriches the physics of the model but also accelerates the simulation because of an improvement of the iterative efficiency, i.e. a regularized

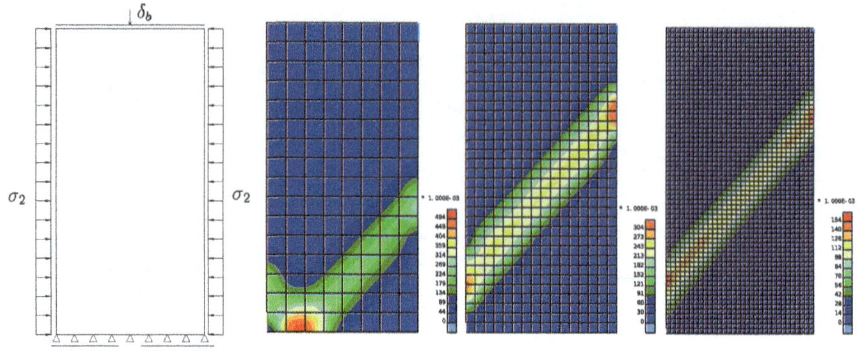

Fig. 2 Biaxial compression test boundary condition. Results with different mesh size: 128, 512 and 2048 elements. Band width independent of mesh size

problem has less possible solutions meaning that the Newton scheme will converge faster.

5 Conclusion

A second gradient regularization has been presented. First gradient mechanical models do not present any internal length, this poses some issues as the problem undergoes softening and strain localization, i.e. mesh dependency. Second gradient is used to provide the model with an internal length and to regularize the problem.

Second gradient is a microstructured local model, this means that the relation can be applied in a material point in the same fashion as a classical constitutive law is. The regularization is of special interest in the FEM×DEM model; because of the noisy behaviour of the DEM constitutive law the problem is very likely to lose its ellipticity leading to an ill posed problem. The second gradient allows to soften the loss of ellipticity not only regularizing the problem but also allowing for a faster convergence.

Results are presented, Compression biaxial test simulations with a second gradient enrichment show that the regularization sets an internal length which makes the model mesh independent as far as the mesh size is fine enough.

References

1. Chambon, R., Caillerie, D., El Hassan, N.: One-dimensional localisation studied with a second grade model. Eur. J. Mech.-A/Solids **17**(4), 637–656 (1998)
2. Chambon, R., Caillerie, D., Matsuchima, T.: Plastic continuum with microstructure, local second gradient theories for geomaterials: localization studies. Int. J. Solids Struct. **38**(46), 8503–8527 (2001)

3. Collin, F., Chambon, R., Charlier, R.: A finite element method for poro mechanical modelling of geotechnical problems using local second gradient models. Int. J. Numer. Methods Eng. **65**(11), 1749–1772 (2006)
4. De Borst, R., Mühlhaus, H.-B.: Gradient-dependent plasticity: formulation and algorithmic aspects. Int. J. Numer. Methods Eng. **35**(521–539), 1992 (1992)
5. Eringen, A.C.: Nonlocal polar elastic continua. Int. J. Eng. Sci. **10**(1), 1–16 (1972)
6. Germain, P.: La méthode des puissances virtuelles en mécanique des milieux continus. J. Mécanique **12**, 236–274 (1973)
7. Guo, N., Zhao, J.: A hierarchical model for cross-scale simulation of granular media. AIP Conf. Proc. **1542**, 1222–1225 (2013)
8. Guo, N., Zhao, J.: Multiscale insights into classical geomechanics problems. Int. J. Numer. Anal. Methods Geomech. (2015)
9. Guo, N., Zhao, J.: 3D multiscale modeling of strain localization in granular media. Comput. Geotech. (2016)
10. Guo, N., Zhao, J.: Parallel hierarchical multiscale modelling of hydro-mechanical problems for saturated granular soils. Comput. Methods Appl. Mech. Eng. **305**, 37–61 (2016)
11. Kaneko, K., Terada, K., Kyoya, T., Kishino, Y.: Global-local analysis of granular media in quasi-static equilibrium. Int. J. Solids Struct. **40**(15), 4043–4069 (2003)
12. Liu, Y., Sun, W., Yuan, Z., Fish, J.: A nonlocal multiscale discrete-continuum model for predicting mechanical behavior of granular materials. Int. J. Numer. Methods Eng. (2015)
13. Marinelli, F.: Comportement couplé des géomatériaux: deus approches de módelisation numérique. Ph.D. thesis (2013)
14. Matsushima, T., Chambon, R., Caillerie, D.: Second gradient models as a particular case of microstructured models: a large strain finite elements analysis. Comptes Rendus de l'Académie des Sciences-Series IIB-Mechanics-Physics-Astronomy **328**(2), 179–186 (2000)
15. Matsushima, T., Chambon, R., Caillerie, D.: Large strain finite element analysis of a local second gradient model: application to localization. Int. J. Numer. Methods Eng. **54**(4), 499–521 (2002)
16. Miehe, C., Dettmar, J., Zäh, D.: Homogenization and two-scale simulations of granular materials for different microstructural constraints. Int. J. Numer. Methods Eng. **83**(8–9), 1206–1236 (2010)
17. Mindlin, R.D.: Second gradient of strain and surface-tension in linear elasticity. Int. J. Solids Struct. **1**(4), 417–438 (1965)
18. Nguyen, T.: Modélisation numérique à double échelle des matériaux granulaires cohésifs: Approche par éléments finis-éléments discrets. Ph.D. thesis (2013)
19. Nguyen, T., Combe, G., Caillerie, D., Desrues, J.: FEM × DEM modelling of cohesive granular materials: numerical homogenisation and multi-scale simulations. Acta Geophysica **62**(5), 1109–1126 (2014)
20. Nitka, M., Combe, G., Dascalu, C., Desrues, J.: Two-scale modeling of granular materials: a DEM-FEM approach. Granul. Matter **13**(3), 277–281 (2011)
21. Pamin, J.K.: Gradient-dependent Plasticity in Numerical Simulation of Localization Phenomena. Delft University of Technology, TU Delft (1994)
22. Pietruszczak, S., Mroz, Z.: Finite element analysis of deformation of strain-softening materials. Int. J. Numer. Methods Eng. **17**(3), 327–334 (1981)
23. Rice, J.R.: The Localization of Plastic Deformation. Brown University, Division of Engineering (1976)
24. Salehnia, F., Collin, F., Li, X.L., Dizier, A., Sillen, X., Charlier, R.: Coupled modeling of excavation damaged zone in boom clay: Strain localization in rock and distribution of contact pressure on the gallerys lining. Comput. Geotech. **69**, 396–410 (2015)
25. Shahin, G., Desrues, J., Dal Pont, S., Combe, G., Argilaga, A.: A study of the influence of rev variability in double scale FEM × DEM analysis. Int. J. Numer. Methods Eng. (2016)
26. Wang, K., Sun, W.: A semi-implicit discrete-continuum coupling method for porous media based on the effective stress principle at finite strain. Comput. Methods Appl. Mech. Eng. **304**, 546–583 (2016)

Strain Localization as a Function of Topological Changes in Mesoscopic Granular Structures

Nejib Hadda, Richard Wan, François Nicot and Félix Darve

Abstract In this paper, the microphysical aspects leading to the occurrence of a localized failure were investigated by examining the initial emergence of ubiquitous potential slip planes and their link to the shear band development. A Discrete Element Model (DEM) representing a dense granular assembly was considered and subjected to a biaxial compression such that it undergoes localized failure into a persistent shear band. The paper includes both energetic and meso-structural analyses based on so-called minimal closed loops.

1 Introduction

One of the most compelling phenomena occurring when a dense granular assembly is subjected to a drained compression is the emergence of slip planes [1] before shear banding. At the approach of the limit stress state, a system of potential slip planes develop into a persistent shear band to the detriment of weaker ones that vanish due to strain localization into one shear band. Accordingly, the micro- and meso-structural aspects of shear band development are directly related to those operating during the initiation of a potential slip plane. In this paper, the development of slip planes within a granular assembly was analyzed from an energetic viewpoint by considering the distribution of the plastic work resulting from particles sliding within the granular assembly. Additionally, meso-structural aspects were examined by exploring the topological evolutions of so-called minimal closed loops [2–4] (n-cycles). By definition, minimal closed loops refer to the minimal simple polygons that are formed

N. Hadda (✉) · R. Wan
Department of Civil Engineering, University of Calgary, Calgary, Canada
e-mail: nhadda@ucalgary.ca

F. Nicot
Geomechanics Group, ETNA, IRSTEA, Grenoble, France

F. Darve
Institut Polytechnique de Grenoble, Université Grenoble Alpes, CNRS, 3SR, 38000 Grenoble, France

E. Papamichos et al. (eds.), *Bifurcation and Degradation of Geomaterials with Engineering Applications*, Springer Series in Geomechanics and Geoengineering, DOI 10.1007/978-3-319-56397-8_58

by joining the centers of particles through the existing contacts. Furthermore, the macroscopic failure was elucidated in terms of the shear strength of each category of loops with respect to their size. This is accomplished by computing the normalized deviatoric stress for each category of loop and analyzing its evolution during loading history. These analyses were made based on numerical simulations performed on a discrete element model [5] using the open source code Yade [6].

2 Discrete Element Model

In this study, a two-dimensional discrete element model representing a dense dry sand assembly is considered. It consists of 100,000 cylindrical particles of unit length packed within four rigid frictionless walls. In accordance with Cundall contact law [5], pairs of interacting particles are connected by linear springs in the normal and tangential directions of the contact. Hence, the normal and tangential contact forces (f_n and f_t respectively) are calculated incrementally as the product of the springs stiffnesses (k_n and k_t respectively) and the relative displacement increments (Δu_n and Δu_t respectively) of the two contacting particles. The sliding between particles is governed by Coulomb friction law through the friction coefficient μ introduced at the contact level. The granular assembly was subjected to a strain-controlled biaxial compression at 300 kPa of confining pressure. The compression was performed in quasi-static regime with zero viscous damping [7]. The physical and mechanical properties of the numerical model are listed in Table 1. The evolutions of the stress deviator $q = \sigma_1 - \sigma_2$ and the volumetric strain $\varepsilon_v = -(\varepsilon_1 + \varepsilon_2)$ in terms of the axial strain ε_1 resulting from the biaxial compression are plotted in Fig. 1a. Such responses demonstrate a behavior of a dense dry sand characterized by a dilatant volumetric strain and a well-marked peak. Moreover, according to Fig. 1b showing snapshots of the deviatoric strain field (ε_d), the granular assembly underwent a localized mode of failure with the development of an intense shear band after failure took place. The deviatoric strain field was computed based on particle displacements at the stress-strain states (2) to (6) (see Fig. 1a) with respect to the isotropic state.

Table 1 Physical and mechanical properties of the two-dimensional numerical model h, w, e and z denote respectively the height, width of the specimen, the void ratio and the coordination number at isotropic state. D, ρ and D_m denote respectively the particle diameter, the mass density and the mean diameter of two particles in contact

h/w	e	z	D (mm)	ρ (kg m^{-3})	k_n/D_m (GPa)	k_t/k_n	μ
1.00	0.168	4.3	2–18	3,000	3.6	0.42	0.7

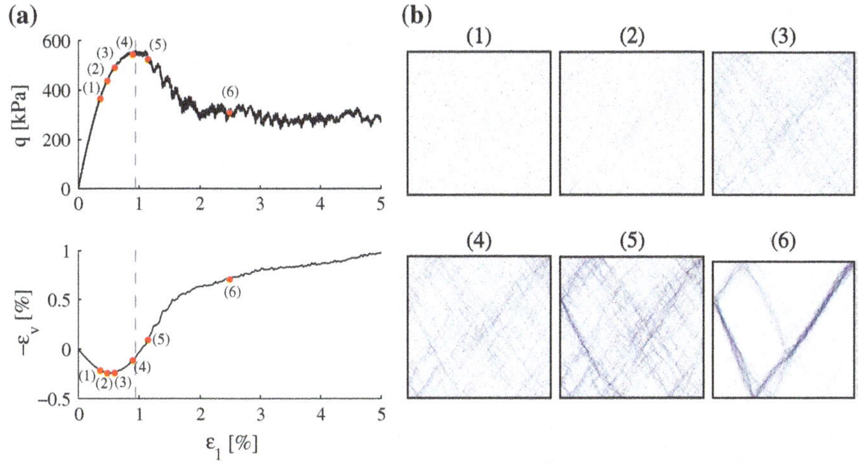

Fig. 1 **a** Evolution of the deviatoric stress q and volumetric strain ε_v in terms of the axial strain ε_1, **b** deviatoric strain fields at states (*2*) to (*6*)

3 Spatial Distribution of Plastic Work

The deviatoric strain fields show that strain localization emerges before the limit stress state is reached (peak of q). Narrow zones of localized strain (slip planes) with various intensities can be observed at state (2) to (4). At their onset, they appear local and tiny (see for instance state (2)), subsequently they grow wider, duplicate and spread across the whole specimen along two opposite main directions as q increases. These two main directions are well-known and defined by the internal (macroscopic) friction angle of the granular assembly.

The slip planes can be also detected by considering the distribution of plastic work (W_{pl}) resulting from particles sliding within the specimen. Figure 2a shows the distribution of the mean plastic work per elementary cell at state (3). The cells are equal in size and they enclose 6×6 particles on average. The highest values of the plastic work are found to be located within the slip planes, which means that sliding between contacts are more concentrated within the slip planes compared to the rest of the specimen. By joining with lines the centers of cells for which $W_{pl} > \overline{W}_{pl}$, where $\overline{W}_{pl}$ is the mean value of the plastic work, one can characterize the potential slip planes as shown in Fig. 2b where also the deviatoric strain field is superimposed. A good agreement is found between the slip planes characterized by both deviatoric strain fields and zones of highest densities of plastic work. Being linked to the zones of highest plastic dissipation, the development of potential slip planes can be considered as part of the shear band formation process where it was shown (see [8]) to involve the majority of inter-particle sliding occurrences.

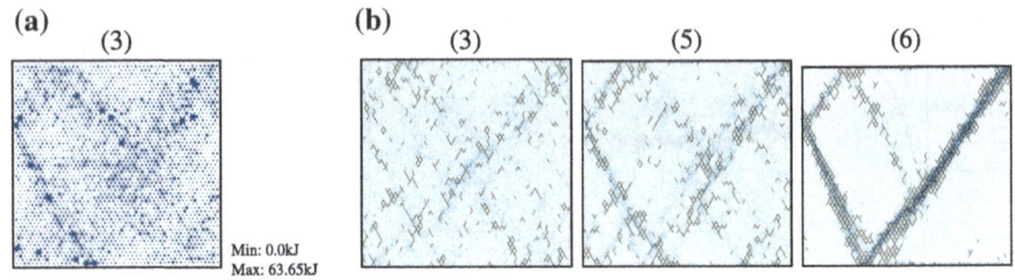

Fig. 2 **a** Density of the plastic work (W_{pl}) within the specimen, the marker size is proportional to the intensity of W_{pl}, **b** line-built slip planes linking the centers of cells for which $W_{pl} > \overline{W}_{pl}$ (superimposed with the deviatoric strain fields)

In the next section, the development of slip planes will be analyzed from a meso-structural viewpoint by examining the stress evolution and the spatial distribution of the n-cycles.

4 Evolution and Spatial Distribution of N-Cycles

The investigations made in this section are based on the evolutions of the minimal polygonal loops *(n-cycles)* with respect to their sizes. The minimal loops are defined by the minimal set of particle pairs in contact having their branch vectors forming a closed path. As such, they are representative samples of the granular assembly mesoscopic structure and their evolution mainly accounts, among others, for particles rearrangement and the variation of the elementary void volume they enclose.

Figure 3a shows the evolution of the number of n-cycles (denoted henceforth by C_n, where n is the size of the cycle) in terms of ε_1 during the biaxial compression. Before the limit stress state is reached, a continuous increase of very large (C_9 and above) and large (C_6 to C_8) loops is observed while, irrespectively, a continuous decrease is observed in the number of small and medium size loops (C_3 to C_5). It is worth noting that the increase in the number of a given set of large loops is delayed with respect to the previous set. Such dependency reveals that the growth process of n-cycles is gradual and continuous.

Figure 3b shows the spatial distribution of large loops within the specimen at states (4) and (6). In accordance with findings in [3], it is found that the large loops are concentrated within localized strain zones prior to failure and within the shear band later. Moreover, the steady increase of large loops is responsible for to the intensification of slip planes and volumetric behavior change (see Fig. 1).

In contrast to their modest presence compared to the medium and small cycles, the role played by large cycles toward the development of the shear band is crucial if the normalized stress deviator q/p, where $p = (\sigma_1 + \sigma_2)/2$, were to be examined at n-cycles level.

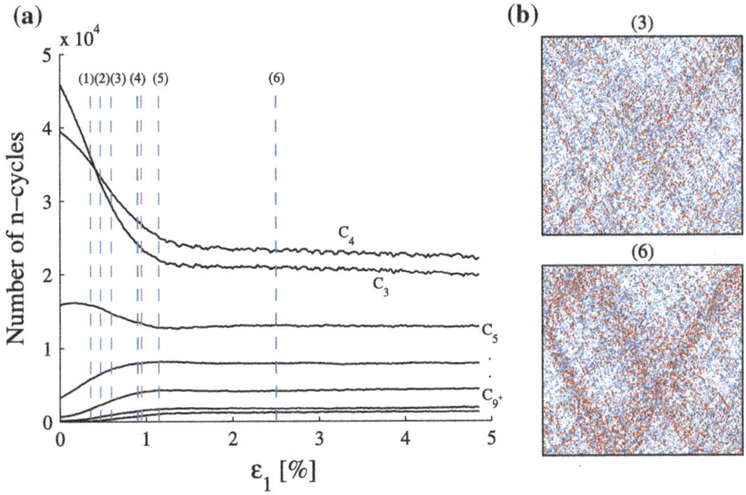

Fig. 3 **a** Evolution of the number of n-cycles during loading, **b** spatial distribution of large n-cycles, C_6, C_7 (*blue*) and C_{8+} (*red*), within the specimen at states (*3*) and (*6*)

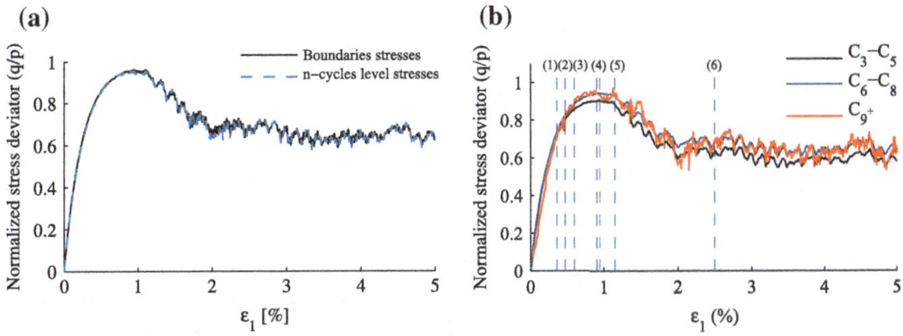

Fig. 4 **a** Comparison between the normalized stress deviator *q*/*p* based on principal stresses computed at boundaries and those computed at n-cycles level, **b** Evolution of *q*/*p* within small, medium and large size loops during loading

In order to compute the stress tensor components at n-cycles level, the stress tensor is first calculated at the particle level [9], i.e., $\sigma_{ij}^{p} = -\frac{1}{v}\sum_{c \in p} f_i^c x_j^c$, where v is the volume associated to the particle, $\bar{f}_c$ is the contact force and $\bar{x}^c$ is the position of the contact point. The volume of the particle v includes both a solid part (v_s) and a void part (v_v). This latter is computed based on the total void volume within the specimen and weighted for each particle with respect to its solid volume v_s. Thereafter, the stress tensor is computed as the average stress tensor of particles involved within the n-cycle i.e., $\sigma_{ij}^{C_n} = \frac{1}{n}\sum_{p \in C_n} \sigma_{ij}^{p}$, where n denotes the number of particles involved within the n-cycle. According to Fig. 4a, a good agreement was found between the stress tensor computed at boundaries and the one computed at n-cycles level.

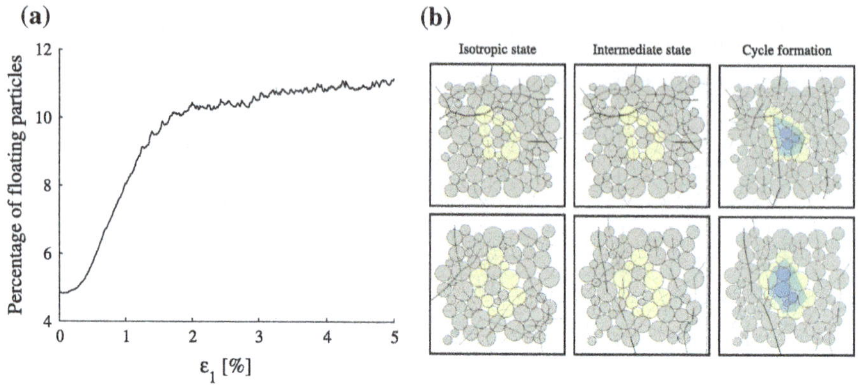

Fig. 5 **a** Evolution of the number of floating particles during the biaxial compression, **b** development of typical large n-cycles during and at the approach of failure, the thickness of the branch vectors is proportional to the intensity of the normal force

Figure 4b shows that the normalized stress deviator is higher for large loops compared to smaller ones. Such result ensues due to the nature of the loops growth process. At the beginning of the compression, the granular assembly discards the contacts whose normals coincide with the lateral direction to build larger cycles elongated in the direction of the axial loading so as to resist the axial loading. As these loops grow, they lose confinement due to more losses of horizontal contacts during loading; thus a higher stress deviator is obtained as the current n-cycle set moves to the next higher set. At the approach of the peak, the largest loops are the first to lose shear strength resulting from low confinement caused by the depletion of smaller loops although that they rise with a higher one compared to the rest of loops. They are also the ones to host floating particles excluded by the granular assembly in order to undergo large deformations (see Fig. 5a). Figure 5b shows a brief development process of two typical mesoscopic structures constituting the slip planes and/or the shear band. A C_9 and a C_{10}, both enclosing, upon their full development, a set of floating particles. These comprised particles preventing the sudden collapse of large cycles during such non-effective failure.

5 Conclusion

In this study, a discrete element model representing a dense dry sand was considered in order to achieve a localized failure resulting into the appearance a well-marked shear band. In view of the findings presented in this paper, potential slip planes, emerging before failure, were shown to encounter similar energetic and meso-structural aspects during their development process likewise those observed during shear banding. The highest concentrations of plastic dissipation were proven to be located only within the dominant slip planes which ultimately lead to the development of the shear band. The n-cycles based investigations not only showed that the void ratio is higher within such slip planes compared to the rest of the specimen, but also provided a comprehensive description of the mesoscopic shear strength evolution toward the full development of the shear band by virtue of the stress tensor computed at the n-cycles level.

Acknowledgements This work is jointly funded by the Natural Science and Engineering Council of Canada and the Foundation Computer Modeling Group (FCMG).

References

1. Kuhn, M.R.: Structured deformation in granular materials. Mech. Mater. **31**(6), 407–429 (1999)
2. Walker, D.M., Tordesillas, A.: Topological evolution in dense granular materials: a complex networks perspective. Int. J. Solids Struct. **47**(5), 624–639 (2010)
3. Zhu, H., Nicot, F., Darve, F.: Meso-structure organization in two-dimensional granular materials along biaxial loading path. Int. J. Solids Struct. **96**, 25–37 (2016)
4. Tordesillas, A., Lin, Q., Zhang, J., Behringer, R.P., Shi, J.: Structural stability and jamming of self-organized cluster conformations in dense granular materials. J. Mech. Phys. Solids **59**(2), 265–296 (2011)
5. Cundall, P.A., Strack, O.D.: A discrete numerical model for granular assemblies. Geotechnique **29**(1), 47–65 (1979)
6. Šmilauer, V. et al.: DEM formulation. Yade Documentation, 2nd edn. The Yade Project. http://yade-dem.org/doc/ (2015)
7. Cundall, P.A.: A computer model for simulating progressive large scale movements in blocky rocky systems. In: Proceeding of the Symposium of the International Society of Rock Mechanics, pp. 129–136. Nancy, France (1971)
8. Hadda, N., Sibille, L., Nicot, F., Wan, R., Darve, F.: Failure in granular media from an energy viewpoint. Granul. Matter **18**(3), 1–17 (2016)
9. Staron, L., Radjai, F., Vilotte, J.P.: Multi-scale analysis of the stress state in a granular slope in transition to failure. Eur. Phys. J. E **18**(3), 311–320 (2005)

Hill's Lemma of Homogeneity for Granular Materials and Discrete Media

D. Caillerie

Abstract R. Hill, in Hill (J Mech Phys Solids 15:79–95, 1967) [1, p. 80], presents a lemma (without any proof) which is often called Hill's condition of homogeneity in the literature. This lemma states that for homogeneous stress or strain boundary conditions on a domain V, the average of the work developed by the divergence-free stress field in the strain field is equal to the work developed by the mean stress in the mean strain. The purpose of this paper is to present a generalization of the Hill's lemma to discrete media such as lattices or nets of beams or granular media and that for different types of boundary conditions.

Keywords Hill's lemma · Homogeneity · Discrete media · Granular materials

1 Introduction

Numerous papers of the literature aim at determining equivalent continuous modelings of discrete structures such as lattices of bars or networks of beams or granular materials. One of the key points in those micro-macro processes is the definition of equivalent stress and strain tensors in a discrete structure. Another key point is the nature of the boundary conditions to be applied on a Representative Volume Element in order to get a representative response of a homogeneous equivalent continuous medium. This point arises also in continuum homogenization and many authors refer to the Hill's lemma as a condition of homogeneity to get a representative response of an homogeneous medium. In this paper, we prove an equivalent Hill's lemma for discrete media for equivalent homogeneous stress or displacement gradient on the

D. Caillerie (✉)
University of Grenoble - Alpes, L3SR, 38000 Grenoble, France
e-mail: Denis.caillerie@3sr-grenoble.fr

D. Caillerie
CNRS, L3SR, 38000 Grenoble, France

D. Caillerie
Domaine Universitaire BP 53 X, 38041 Grenoble Cedex, France

E. Papamichos et al. (eds.), *Bifurcation and Degradation of Geomaterials with Engineering Applications*, Springer Series in Geomechanics and Geoengineering, DOI 10.1007/978-3-319-56397-8_59

boundary of the medium. First the Hill's lemma for continuous media is recalled in Sect. 2 with the proof which is used to establish the equivalent lemma for discrete media. In Sect. 3 is given a description of a generic discrete medium, then the definitions of equivalent mean stress and mean displacement gradient in discrete media are recalled. The Hill's lemma for discrete media is stated and proved in Sect. 3.4.

Notations
$\vec{a}.\vec{b}$ is the scalar product of the vectors $\vec{a}$ and $\vec{b}$. $A@\vec{b}$ is the image of the vector $\vec{b}$ by the linear application (second order tensor) A. A^T, resp. tr A, are the transposed tensor, resp. the trace, of the tensor A. $A \circ B$, resp. $A : B$, are the composition product, resp. the scalar product, of the second order tensors A and B; $(A \circ B) @\vec{a} = A@\left(B@\vec{a}\right)$, $A : B = \operatorname{tr}\left(A \circ B^T\right)$. $\vec{a} \otimes \vec{b}$ is the tensor product of the vectors $\vec{a}$ and $\vec{b}$ defined by $\left(\vec{a} \otimes \vec{b}\right) @\vec{c} = \left(\vec{b}.\vec{c}\right)\vec{a}$ and $\operatorname{tr}\left(\vec{a} \otimes \vec{b}\right) = \vec{a}.\vec{b}$. $\mathbb{I}$ is the identity tensor. Let V be a domain of either a 2D or 3D space, $\vec{x}$ is the space variable in V and ∂V denotes the boundary of V. $|V|$ being the volume (or surface in 2D) of V, $\langle f \rangle = \frac{1}{|V|} \int_V f \, \mathrm{d}v$ is the mean value of the field f over V.

2 Hill Lemma for Continuous Media

Lemma 1 (Hill's lemma) *Let $\vec{u}$ and σ be a differentiable displacement field and a divergence free stress tensor field defined in a domain V. If, on the boundary ∂V either σ or $\vec{u}$, satisfies the conditions:*

$$\sigma @\vec{n} = \Sigma @\vec{n} \tag{1}$$

$$\vec{u} = E @\vec{n} \tag{2}$$

Σ and E are constant 2nd order tensors and $\vec{n}$ is the outer normal to ∂V, then

$$\left\langle \sigma : \nabla \vec{u} \right\rangle = \langle \sigma \rangle : \left\langle \nabla \vec{u} \right\rangle \tag{3}$$

Because it is used to establish the lemma for discrete media, the proof of the Hill's lemma is given in what follows. It makes use of the following results:

$$\forall \sigma, \operatorname{div} \sigma = 0\,, \; \langle \sigma \rangle = \frac{1}{|V|} \int_{\partial V} \sigma \circ \left(\vec{n} \otimes \vec{x}\right) \mathrm{d}s \tag{4a}$$

$$\forall \vec{w}\,, \; \left\langle \nabla \vec{w} \right\rangle = \frac{1}{|V|} \int_{\partial V} \vec{w} \otimes \vec{n} \, \mathrm{d}s \tag{4b}$$

These results, as well as the Hill's lemma, are based on the virtual power formulation of the equation $\operatorname{div} \sigma = 0$ which reads:

$$\forall \vec{w}, \int_V \sigma : \nabla \vec{w} \, dv = \int_{\partial V} (\sigma @ \vec{n}) . \vec{w} \, ds = \int_{\partial V} \sigma : (\vec{w} \otimes \vec{n}) \, ds \tag{5}$$

(4a), resp. (4b), are obtained by taking $\vec{u} = E @ \vec{x}$, resp. $\sigma = \Sigma$, in (5), E and Σ being any constant 2nd order tensors (so $\operatorname{div} \Sigma = 0$). Moreover, taking $\vec{u}$ constant, resp. $\vec{u} = \vec{x}$, in V in (4b) yields respectively:

$$\int_{\partial V} \vec{n} ds = 0 \tag{6a}$$

$$\frac{1}{|V|} \int_{\partial V} \vec{n} \otimes \vec{x} \, ds = \frac{1}{|V|} \int_{\partial V} \vec{x} \otimes \vec{n} \, ds = \mathbb{I} \tag{6b}$$

Proof of the Hill's lemma Taking $\sigma @ \vec{n} = \Sigma @ \vec{n}$ on ∂V in (4a) where Σ is a constant 2nd order tensor and using (6b), we get $\langle \sigma \rangle = \frac{1}{|V|} \Sigma \circ \int_{\partial V} \vec{n} \otimes \vec{x} \, ds = \Sigma$ now, from (5) with $\vec{w} = \vec{u}$ we have:

$$\int_V \sigma : \nabla \vec{u} \, dv = \int_{\partial V} (\Sigma @ \vec{n}) . \vec{u} \, ds = \int_{\partial V} \Sigma : (\vec{u} \otimes \vec{n}) \, ds = \Sigma : \int_{\partial V} (\vec{u} \otimes \vec{n}) \, ds$$

which, using (4b), yields (3).

Taking $\vec{u} = E @ \vec{x}$ on ∂V with E being a constant second order tensor and using (6b), we get $\langle \nabla \vec{u} \rangle = \frac{1}{|V|} E : \int_{\partial V} \vec{x} \otimes \vec{n} \, ds = E$ now, from (5) with $\vec{w} = \vec{u}$ we have:

$$\int_V \sigma : \nabla \vec{u} \, dv = \int_{\partial V} (\sigma @ \vec{n}) . (E @ \vec{x}) \, ds = \left(\int_{\partial V} \sigma \circ (\vec{n} \otimes \vec{x}) \, ds \right) : E$$

which, dividing by $|V|$ and using (4a), yields (3). □

3 Description of Discrete Media—Mean Stress and Displacement Gradient

3.1 Equilibrium and Virtual Powers

The discrete media considered in this study are rather various, they can be nets of bars or of beams, or granular materials (see Fig. 1 for an example of a lattice of bars). The discrete media are assumed to be such that their equilibrium comes down to that of a set of points, numbered by n, m, ..., which interact by forces (and possibly by torques in a granular medium or in a net of beams). Those points are named nodes as in a lattice. Let B denote the set of the peripheral nodes (in red on Fig. 1).

Let $\vec{f}^{m/n}$ be the force applied by the node m on the node n therefore $\vec{f}^{n/m} = -\vec{f}^{m/n}$. The structure is loaded only on its peripheral nodes by forces denoted by $\vec{f}^{e/n}$. The

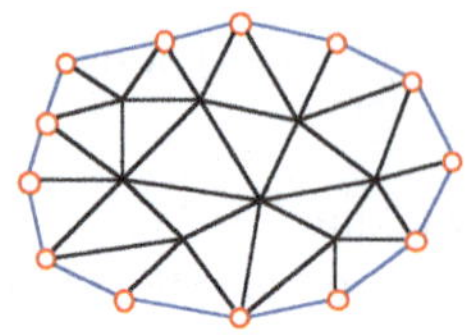

Fig. 1 A lattice of bars—in *red* the set B of the peripheral nodes, in *blue* the (2D) facets forming the boundary ∂V of the discrete medium

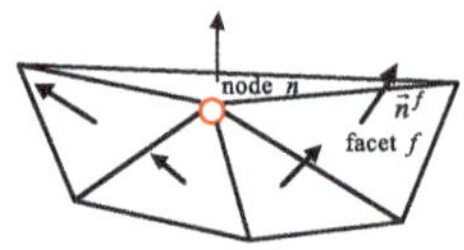

Fig. 2 The set of triangular facets of the boundary of which n is a vertex

exterior forces $\vec{f}^{e/n}$ are supposed to be balanced by the interior forces $\vec{f}^{m/n}$, the virtual power formulation of the corresponding equilibrium reads:

$$\forall \vec{w}^n\,,\ -\frac{1}{2}\sum_{(n,m)} \vec{f}^{m/n}.\left(\vec{w}^m - \vec{w}^n\right) + \sum_{n \in B} \vec{f}^{e/n}.\vec{w}^n = 0 \tag{7}$$

where the first sum runs over the whole set of pairs (n, m) of interacting nodes; (m, n) is different from (n, m) so the coefficient 1/2. Let $\mathcal{P}^i = -\frac{1}{2}\sum_{(n,m)} \vec{f}^{m/n}.\left(\vec{w}^m - \vec{w}^n\right)$ be the inner virtual power.

Remark 2 From the previous equation, it is clear that the resultant force $\sum_{n \in B} \vec{f}^{e/n}$ of the exterior forces $\vec{f}^{e/n}$ has to be equal to 0.

We then define the boundary ∂V of the discrete medium by considering the triangular facets of the—for instance Delaunay's—tessellation of the peripheral nodes of the medium (see Figs. 1 and 2). The domain V, of volume $|V|$, occupied by the discrete medium is the domain bounded by ∂V. The triangular boundary facets that form the boundary ∂V are numbered by f, their exterior normals are denoted by $\vec{n}^f$ and there areas by A^f. The normal vectors $\vec{n}^f$ define a piecewise constant exterior normal $\vec{n}$ to the boundary ∂V. Let $S\,(f)$ denote the set of the peripheral nodes ($n \in B$) which are vertices of the triangular facet f and $S^{-1}\,(n)$ denote the set of the facets f of which n is a vertex.

3.2 Equivalent Stress in Discrete Media

There are different ways to define the mean stress in a discrete medium, one way is to draw inspiration form the mean stress in a continuous medium (4a) and set:

$$\langle \sigma \rangle = \frac{1}{|V|} \sum_{n \in B} \vec{f}^{e/n} \otimes \vec{x}^{n} \tag{8}$$

That expression can be obtained by taking $\vec{w}^n = E@\vec{x}^n$ in (7) (E being constant) and by identifying the inner virtual power $\mathcal{P}^i$ with that of the continuum, which leads to equivalently define the mean Cauchy stress by $\langle \sigma \rangle = \frac{1}{2} \sum_{(n,m)} \vec{f}^{m/n} \otimes \left(\vec{x}^m - \vec{x}^n \right)$.

3.3 Mean Displacement Gradient in Discrete Media

Analogously to what is done in Sect. 3.2, the average displacement gradient can be define by drawing inspiration from the relation (4b). That relation needs the displacement to be defined at least on the discrete medium boundary ∂V as an extension of the displacements of the peripheral nodes, for that those displacements are interpolated on the facets. Let $\vec{w}^l$, $\vec{w}^m$ and $\vec{w}^n$ be displacements of the vertices l, m, n of the facet f, the interpolation on f reads $\vec{w}\left(\vec{x}\right) = N^l\left(\vec{x}\right)\vec{w}^l + N^m\left(\vec{x}\right)\vec{w}^m + N^n\left(\vec{x}\right)\vec{w}^n$, the N's functions are degree one polynomials such that $N^j\left(\vec{x}^i\right) = \delta_{ij}$, $i, j = l, m, n$ with $\vec{x}^l$, $\vec{x}^m$ and $\vec{x}^n$ being the positions of the vertices of the facet. A calculation of the integrals $\int_f N^j\left(\vec{x}\right) \mathrm{d}s$, $i = l, m, n$ shows that:

$$\int_f \vec{w}\, \mathrm{d}s = \frac{A^f}{3} \sum_{n \in S(f)} \vec{w}^n \tag{9}$$

According to the relation (4b), we defined the average displacement gradient in the discrete media by:

$$\left\langle \nabla \vec{w} \right\rangle = \frac{1}{|V|} \int_{\partial V} \vec{w} \otimes \vec{n}\, \mathrm{d}s = \frac{1}{|V|} \sum_{f \subset \partial V} \int_f \vec{w} \otimes \vec{n}^f\, \mathrm{d}s = \frac{1}{|V|} \sum_{f \subset \partial V} \left(\int_f \vec{w}\, \mathrm{d}s \right) \otimes \vec{n}^f \tag{10}$$

that is to say, from (9), $\left\langle \nabla \vec{w} \right\rangle = \frac{1}{3|V|} \sum_{f \subset \partial V} A^f \left(\sum_{n \in S(f)} \vec{w}^n \right) \otimes \vec{n}^f$ which yields $\left\langle \nabla \vec{w} \right\rangle = \frac{1}{|V|} \sum_{n \in B} \vec{w}^n \otimes \frac{1}{3} \sum_{f \in S^{-1}(n)} A^f \vec{n}^f$ and which can be written:

$$\left\langle \nabla \vec{w} \right\rangle = \frac{1}{|V|} \sum_{n \in B} \vec{w}^n \otimes \vec{A}^n \tag{11}$$

where the mean surface vector $\vec{A}^n$ at node n is defined by:

$$\vec{A}^n = \frac{1}{3} \sum_{f \in S^{-1}(n)} A^f \vec{n}^f \tag{12}$$

Remark 3 We have $\sum_{n \in B} \vec{A}^n = 0$ indeed $\sum_{n \in B} \vec{A}^n = \frac{1}{3} \sum_{n \in B} \sum_{f \in S^{-1}(n)} A^f \vec{n}^f$, that is to say $\sum_{n \in B} \vec{A}^n = \frac{1}{3} \sum_{f \subset \partial V} \sum_{n \in S(f)} A^f \vec{n}^f$ which, as $A^f \vec{n}^f$ does not depend on $n \in S(f)$, reads $\sum_{n \in B} \vec{A}^n = \sum_{f \subset \partial V} A^f \vec{n}^f$. Now, taking (6a) into account, it comes $\sum_{n \in B} \vec{A}^n = \sum_{f \subset \partial V} \int_f \vec{n} \, \mathrm{d}s = \int_{\partial v} \vec{n} \, \mathrm{d}s = 0$.

3.4 Hill's Lemma for Discrete Media

Lemma 4 *Let $\vec{u}^n$ be the displacements of the nodes of the discrete medium and $\vec{f}^{m/n}$ be forces between these nodes balancing the exterior forces $\vec{f}^{e/n}$ and let $|V|$ be the volume of V (see Sect. 3.1). If, on the nodes of the boundary B of the medium the applied exterior forces $\vec{f}^{e/n}$, resp. the displacements $\vec{u}^n$ satisfy the conditions:*

$$\forall n \in B, \vec{f}^{e/n} = \Sigma @ \vec{A}^n \ \textit{resp.} \tag{13a}$$

$$\forall n \in B, \vec{u}^n = E @ \vec{x}^n \tag{13b}$$

Σ and E being two constant second order tensors and $\vec{A}^n$ being by (12), then

$$\frac{1}{|V|} \frac{1}{2} \sum_{(n,m)} \vec{f}^{m/n} . \left(\vec{u}^m - \vec{u}^n \right) = \langle \sigma \rangle : \left\langle \nabla \vec{u} \right\rangle \tag{14}$$

which means that the mean of the internal work is equal to the work of the means of the stress and of the gradient of $\vec{u}$.

Proof Let's begin with the condition (13a) on the stress. Σ being a stress tensor, we consider external forces $\vec{f}^{e/n}$ being given on peripheral nodes by $\vec{f}^{e/n} = \Sigma @ \vec{A}^n$. According to the Remark 3 (Sect. 3.3), the resultant of all these forces is equal to zero ensuring that the exterior forces $\vec{f}^{e/n}$ can be balanced by interior forces $\vec{f}^{m/n}$ (see the Remark 2). From the definition of the equivalent mean stress (8), we have $|V| \langle \sigma \rangle = \sum_{n \in B} \vec{f}^{e/n} \otimes \vec{x}^n = \Sigma \circ \sum_{n \in B} \vec{A}^n \otimes \vec{x}^n$ now, using the same calculation as that of Sect. 3.3 for the definition of $\left\langle \vec{w} \right\rangle$, we get $|V| \langle \sigma \rangle = \Sigma \circ \int_{\partial V} \vec{n} \otimes \vec{x} \, \mathrm{d}s$ which, according to the relation (6b) proves that:

$$|V| \langle \sigma \rangle = \Sigma \tag{15}$$

The formulation (7) for $\vec{w}^n = \vec{u}^n$ reads $\frac{1}{2}\sum_{(n,m)}\vec{f}^{m/n}.\left(\vec{u}^m - \vec{u}^n\right) = \Sigma : \sum_{n\in B}\vec{u}^n \otimes \vec{A}^n$ which, according to the definition (11) for $\vec{w}^n = \vec{u}^n$ and the Eq. (15) entails $\frac{1}{|V|}\frac{1}{2}\sum_{(n,m)}\vec{f}^{m/n}.\left(\vec{u}^m - \vec{u}^n\right) = \langle\sigma\rangle : \left\langle\nabla\vec{u}\right\rangle$.

Now, let's consider displacements on B such that $\forall n \in B$, $\vec{u}^n = E@\vec{x}^n$. From (11) for $\vec{w}^n = \vec{u}^n$ we then have $\left\langle\nabla\vec{u}\right\rangle = \frac{1}{|V|}E\circ\left(\sum_{n\in B}\vec{x}^n \otimes \vec{A}^n\right)$ which, using Eq. (11) for $\vec{w}^n = \vec{x}^n$, reads:

$$\left\langle\nabla\vec{u}\right\rangle = E \tag{16}$$

The formulation (7) for $\vec{w}^n = E@\vec{x}^n$ reads $\frac{1}{2}\sum_{(n,m)}\vec{f}^{m/n}.\left(\vec{u}^m - \vec{u}^n\right) = \left(\sum_{n\in B}\vec{f}^{e/n} \otimes \vec{x}^n\right) : E$ which according to the definition (8) and the Eq. (16) reads $\frac{1}{|V|}\frac{1}{2}\sum_{(n,m)}\vec{f}^{m/n}.\left(\vec{u}^m - \vec{u}^n\right) = \langle\sigma\rangle : \left\langle\nabla\vec{u}\right\rangle$. □

Acknowledgements This paper is dedicated to Ioannis Vardoulakis.

Reference

1. Hill, R.: The essential structure of constitutive laws for metal composites and polycrystals. J. Mech. Phys. Solids **15**, 79–95 (1967)

Framework for Multiscale Flow Simulation of Deformable Rocks

Martin Lesueur, Maria Camila Casadiego, Thomas Poulet and Manolis Veveakis

Abstract A Finite Element implementation is presented to solve for Stokes flow on a deformable rock matrix reconstituted from a stack of computerized tomography images. Tightly coupling this flow solution with a mechanical deformation model exhibits the hydro-mechanical evolution of permeability in a fully saturated rock under compression. The scope of the presented micro-scale computation of permeability is demonstrated through a multi-scale simulation of pore pressure progression within a petroleum reservoir under production.

1 Introduction

When coupling rock deformation with fluid flow through porous media considerations, the solid matrix permeability is a key parameter for hydro-mechanical instabilities [12]. It is therefore of primordial importance to have a framework for computing evolving permeability in a deforming matrix for any type of fluids. Such calculations of permeability can be efficiently performed directly from micro-tomographic images [10]. This technique provides a more accurate estimation of permeability than semi-analytical models like Carman [4], removing the uncertainties introduced by the empirical nature of the scaling factors used to fit the experimental data [1].

Permeability is a homogenised quantity that is only valid at the macro-scale, but is obtained through the upscaling of the Navier-Stokes equations at the micro-scale. Various numerical techniques have been implemented successfully to estimate permeability from digital porous rocks. These techniques use numerical schemes like the Finite Element (FEM) [3], Finite Difference [10], Finite Volume [7], or Lattice Boltzmann (LBM) [9] methods. Methods like LBM are extremely efficient numerically but not necessarily extensible to include more physical processes. FEM, on

M. Lesueur (✉) · M. Camila Casadiego · T. Poulet · M. Veveakis
University of New South Wales, Kensington, NSW 2052, Australia
e-mail: m.lesueur@student.unsw.edu.au

M. Lesueur · T. Poulet · M. Veveakis
CSIRO Mineral Resources, North Ryde, NSW 2113, Australia

E. Papamichos et al. (eds.), *Bifurcation and Degradation of Geomaterials with Engineering Applications*, Springer Series in Geomechanics and Geoengineering, DOI 10.1007/978-3-319-56397-8_60

the other hand, provides a way to deal with complex pore geometries and boundary conditions, this allows a rigorous solution of the Navier-Stokes equations [14] and enables it to handle more complex equations to include additional physical processes.

In this contribution we present a fluid flow module implemented in the RED-BACK simulator [11] that provides a framework to run multiphysics simulations across multiple scales.

2 Flow Simulator

We consider a porous rock at the micro-scale composed of a solid matrix and pore space, which contains a single-phase incompressible fluid. In this section we present the underlying system of governing equations and their numerical implementation within the fluid simulator to compute the permeability of digital rocks.

2.1 *Theoretical Model*

The flow in the pore space is governed by the Navier-Stokes equations [2]. In order to compute permeability, we restrict this study to the case of steady-state laminar flow. Based on these assumptions, the system reduces to the traditional Stokes equations and can be brought to a dimensionless form by introducing the following normalised quantities:

$$p_f^* = \frac{p_f L_{ref}}{\mu_f v_{f,ref}}, \quad \boldsymbol{v}_f{}^* = \frac{\boldsymbol{v}_f}{v_{f,ref}}, \quad Re = \frac{\rho_f v_{f,ref} L_{ref}}{\mu_f} \tag{1}$$

where ρ_f denotes the fluid density, μ_f the fluid viscosity, p_f the fluid pressure, $\boldsymbol{v}_f$ the fluid velocity, and Re the Reynolds number. L_{ref}, $v_{f,ref}$ are reference quantities for the normalisation. The dimensionless system becomes

$$-\frac{1}{Re}\nabla \cdot \boldsymbol{\sigma}_f{}^* + \nabla p_f^* = 0, \tag{2}$$

$$\nabla \cdot \boldsymbol{v}_f{}^* = 0, \tag{3}$$

$$\text{with } \boldsymbol{\sigma}_f{}^* = [\nabla \boldsymbol{v}_f{}^* + (\nabla \boldsymbol{v}_f{}^*)^T] - \frac{2}{3}(\nabla \cdot \boldsymbol{v}_f{}^*)I \tag{4}$$

where $\boldsymbol{\sigma}_f$ denotes the fluid stress tensor for a newtonian fluid (Eq. 4). Equation 2 expresses the momentum balance in the case of laminar, steady-state flow and Eq. 3, the mass balance simplified with the assumption of incompressibility.

2.2 Numerical Model

The dimensionless system of Eqs. 2–4 is implemented with a finite element scheme using the open-source REDBACK simulator [11]. REDBACK is based on the Multiphysics Object-Oriented Simulation Environment (MOOSE) [6], which allows a parallel resolution of equations in a tightly coupled manner. In this study we take advantage of several of MOOSE's functionalities. Firstly, its modularity enables the coupling of different physical simulators in an efficient manner. Secondly, it supports a wide range of mesh elements and mesh adaptivity; this is particularly convenient when working on reconstituted meshes from a stack of computed tomography (CT) scan images. Thirdly, the multi-app functionality allows loose coupling to be performed easily by automatically handling the passing and interpolation of inputs and outputs between the various simulations. The accuracy of the flow simulation was benchmarked successfully against the analytical solution of Poiseuille flow in a cylindrical tube.

3 Permeability Computation on Deformable Digitised Rocks

3.1 Permeability Calculation

We use the Redback implementation presented above to compute the permeability of a rock sample at the micro-scale. Steady-state flow is computed by applying no-slip boundary conditions at the rock-fluid interface and at the sample's boundaries. We impose a pressure-driven flow by setting a higher pressure at a chosen inlet and a lower pressure at the corresponding outlet. Most samples considered are cubic and the inlet and outlets are selected on opposite faces. A low-pressure difference, of $\Delta p^* = 1$, is convenient to remain in the laminar regime.

Darcy first introduced the concept of permeability in 1856 [5] as an empirical quantity, k, which related the fluid flow with the applied pressure gradient, to describe the concept of hydraulic conductivity. Darcy's law was eventually proven to be a homogenisation of the Navier-Stokes equations [13], under a strict set of conditions. Permeability can be derived from the definition of the superficial fluid velocity v_{fs} expressed as

$$v_{fs} = \frac{k \Delta p_f}{\mu_f L_{ref}} \tag{5}$$

Note that v_{fs}, the apparent fluid velocity of a porous volume, does not represent the norm of $\boldsymbol{v}_f$ but its projection on the overall flow direction.

$$v_{fs} = \phi \int_{\Omega} \boldsymbol{v_f}.\boldsymbol{n} \tag{6}$$

Consequently, we compute permeability through the following expression

$$k = L_{ref}^2 \frac{\phi \int_{\Omega} \boldsymbol{v_f}^*.\boldsymbol{n}}{\Delta p_f^*} \tag{7}$$

This computed permeability provides a single average value, over the whole volume, for any given direction along the vector $\boldsymbol{n}$.

3.2 Handling Deformable Digitised Rocks

In this section we present a qualitative example to illustrate the capacity of this framework to capture the evolution of permeability in a fully saturated rock sample under compression. We couple the flow simulation module presented above with other Redback modules to simulate flow within a deformable rock matrix reconstituted from a stack of CT images. We model a triaxial experiment on a cubic sample of carbonate under compression at a constant velocity. We consider an elasto-visco-plastic behaviour of the rock with a von Mises yield criterion. The compression is considered to be slow enough for the deformation to remain quasi-static, in this way the flow can be solved in steady-state for each mechanical time step. The flow simulation and mechanical deformation are solved simultaneously in a tightly coupled manner.

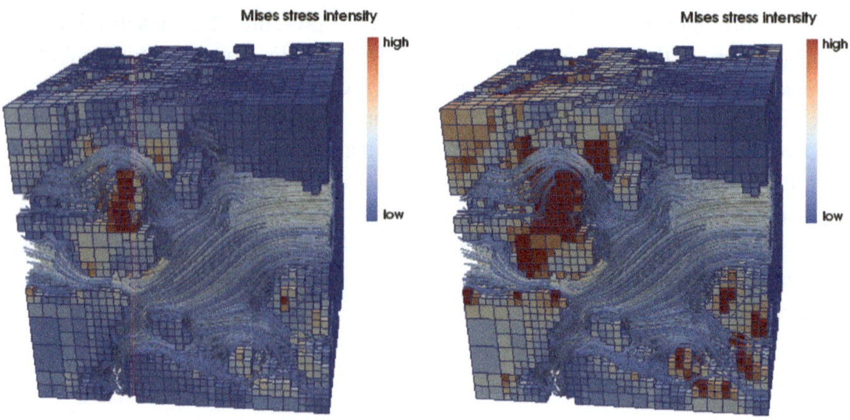

Fig. 1 Two time steps of a Hydro-Mechanically coupled simulation on a carbonate under compression, showing high stresses building up over time at the chain force. The fluid flow is displayed with streamlines. The sample is meshed with 30000 elements

Figure 1 displays two frames of the simulation at different times to show the evolution of the Mises stress during the experiment. It is seen how the sample reaches plasticity in localised areas, mostly at the chain force. The pore throats are slightly reduced in those areas, affecting the overall recomputed permeability of the sample. Given that the initial porosity of this sample is high (25.9%), those changes remain imperceptible over the short duration of the experiment; this was run for illustration purposes.

4 Multi-scale Simulation

The computation of permeability at the micro-scale can be inserted in a mutli-scale framework using MOOSE's mutli-app functionality. We demonstrate this feature and the framework developed by simulating the pore pressure evolution within a petroleum reservoir under production. The reservoir consists of two horizontal layers with distinct geological fabrics, which are represented by using a characteristic sample at the micro-scale for each of them. We investigate the pressure build-up scenario that results from a production well across both layers. For this purpose, we consider a 2D vertical cross section flanked by an injector well on the left hand side. The injection is modelled by applying a normalised excess pore pressure of $p_f^* = 1$, compared to the initial normalised pore pressure of $p_f^* = 0$, over the whole model.

The material chosen for the upper layer is the sandpack sample LV60A [8], with a computed porosity and permeability of 35.7% and 8.5 D respectively. For the lower layer, a carbonate sample with computed porosity of 27.6% and 123 mD permeability. The upscaling is performed in a sequential manner through MOOSE's multi-app, where the permeability of each sample is recomputed at each step. Figure 2 shows a snapshot of the reservoir pore pressure after enough time for the two layers to exhibit different behaviours. The pore pressure is naturally diffusing faster in the upper layer.

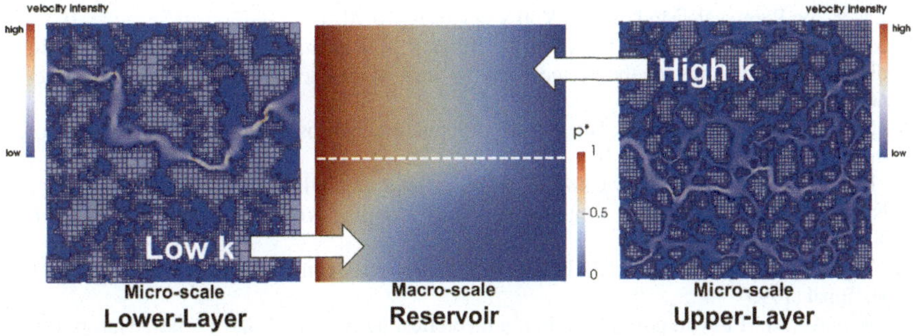

Fig. 2 (*Center*) Pore pressure map of the vertical reservoir cross-section with two *horizontal* layers after some injection time. (*Left*) Velocity field of the carbonate sample characteristic of the lower layer. (*Right*) Velocity field of the sandpack sample characteristic of the *upper* layer

5 Conclusion

The framework presented opens the door for a new type of studies to understand the permeability evolution in deformable porous rocks. It provides the means for other physical processes to be coupled (tightly or not) to the flow simulator and facilitates investigations across different length scales. These features are particular important to study complex mechanisms and hydromechanical instabilities like fault reactivation, where the microstructural geometry and the macroscopic behavior are critically linked. During such slip events, permeability has been shown to evolve drastically [11] and increase by several orders of magnitude. Consequently, when a fault reactivates, it can cause pressure equilibration in deep petroleum reservoirs, across faults that were known to be initially impermeable.

References

1. Andrä, H., Combaret, N., Dvorkin, J., Glatt, E., Han, J., Kabel, M., Keehm, Y., Krzikalla, F., Lee, M., Madonna, C., Marsh, M., Mukerji, T., Saenger, E.H., Sain, R., Saxena, N., Ricker, S., Wiegmann, A., Zhan, X.: Digital rock physics benchmarks—part ii: computing effective properties. Comput. Geosci. **50**, 33–43 (2013). Benchmark problems, datasets and methodologies for the computational geosciences
2. Batchelor, G.K.: An Introduction to Fluid Dynamics. Cambridge University Press (2000)
3. Borujeni, A.T., Lane, N., Thompson, K., Tyagi, M.: Effects of image resolution and numerical resolution on computed permeability of consolidated packing using lb and fem pore-scale simulations. Comput. Fluids **88**, 753–763 (2013)
4. Carman, P.C.: Fluid flow through granular beds. Trans. Inst. Chem. Eng. **15**, 150–166 (1937)
5. Darcy, H.P.G.: Détermination des lois d'écoulement de l'eau à travers le sable (1856)
6. Gaston, D., Newman, C., Hansen, G., Lebrun-Grandie, D.: Moose: a parallel computational framework for coupled systems of nonlinear equations. Nucl. Eng. Des. **239**(10), 1768–1778 (2009)
7. Guibert, R., Nazarova, M., Horgue, P., Hamon, G., Creux, P., Debenest, G.: Computational permeability determination from pore-scale imaging: sample size, mesh and method sensitivities. Transp. Porous Media **107**(3), 641–656 (2015)
8. I. C. C. on Pore-scale Modelling. LV60A sandpack. 10 2014
9. Manwart, C., Aaltosalmi, U., Koponen, A., Hilfer, R., Timonen, J.: Lattice-boltzmann and finite-difference simulations for the permeability for three-dimensional porous media. Phys. Rev. E **66**(1), 016702 (2002)
10. Mostaghimi, P., Blunt, M.J., Bijeljic, B.: Computations of absolute permeability on micro-CT images. Math. Geosc. **45**(1), 103–125 (2013)
11. Poulet, T., Paesold, M., Veveakis, M.: Multi-physics modelling of fault mechanics using redback: a parallel open-source simulator for tightly coupled problems. Rock Mech. Rock Eng. 1–17 (2016)
12. Vardoulakis, I., Sulem, J. (eds.): Bifurcation Analysis in Geomechanics. Blankie Acc. and Professional (1995)
13. Whitaker, S.: Flow in porous media i: a theoretical derivation of darcy's law. Transp. Porous Media **1**(1), 3–25 (1986)
14. Zabaras, N., Samanta, D.: A stabilized volume-averaging finite element method for flow in porous media and binary alloy solidification processes. Int. J. Numer. Methods Eng. **60**(6), 1103–1138 (2004)

Localized Compaction in Tuffeau de Maastricht: Experiments and Modeling

A. Papazoglou, G. Shahin, F. Marinelli, C. Dano, G. Buscarnera and G. Viggiani

Abstract This paper presents an experimental and constitutive study of compaction banding in Tuffeau de Maastricht, a bioclastic sedimentary limestone exhibiting up to 52% of porosity. An elasto-plastic constitutive model is used to simulate the mechanical behavior of the limestone within the compaction localization regime. It is shown that the simulated macroscopic behavior is in good agreement with the experimental data. In addition, image processing tools have been used to perform full-field measurements elucidating the mechanics of initiation and propagation of localized compaction zones. These findings emphasize the complex nature of localized compaction in porous rocks and represent a preliminary step towards the integrated use of multi-scale testing and mechanical modeling for their characterization across scales.

1 Introduction

Compaction bands have been observed in different kinds of materials (e.g., rocks, soils, metallic foams, cellular materials). They are usually promoted by a relatively high porosity, as well as by high mean stress levels. However, the inelastic mechanisms controlling localized compaction at the micro-scale can differ from one material to another. For instance, buckling of thin walls has been observed in metallic foams or honeycomb structures. In the case of geomaterials, due to their natural variability, the microstructural origin of compaction localization is still an open question, often complicated by the coexistence of multiple inelastic processes (e.g., pore

A. Papazoglou (✉) · C. Dano · G. Viggiani
Laboratoire 3SR, Université Grenoble Alpes, Grenoble, France
e-mail: athanasios.papazoglou@3sr-grenoble.fr

G. Shahin · F. Marinelli · G. Buscarnera
Department of Civil and Environmental Engineering, Northwestern University, Evanston, USA
e-mail: ghassanshahin@u.northwestern.edu

E. Papamichos et al. (eds.), *Bifurcation and Degradation of Geomaterials with Engineering Applications*, Springer Series in Geomechanics and Geoengineering, DOI 10.1007/978-3-319-56397-8_61

collapse, grain crushing, degradation of cementation bonds). Full-field non destructive measurements have thus the potential to identify these micro-mechanisms and assess their relative role.

2 Material Tested and Experimental Set-Up

The material studied in this work is Tuffeau de Maastricht, a limestone that was formed by sediments transported and deposited in a marine environment during the Late Cretaceous [2, 3]. The material microstructure consists of carbonate bioclasts and shell fragments. Although the average grain size is about 100–200 μm, bigger fossils and inclusions can also be observed. Such constituents are bonded together by a weak cementation, which mainly occurs at contact points around echinoderm bioclasts. The pore structure of Tuffeau de Maastricht is characterized by high porosity (up to 52%), which is mainly due to the intragranular porosity within the shell fragments (see Fig. 1) [5, 6]. The weak cementation, combined with the high porosity, results in a soft rock with low stiffness and strength (unconfined compressive strength is less than 5 MPa) [11]. Figure 1 illustrates the complex structure of the selected rock at the micro scale, where both intergranular voids and bioclasts can be recognized (black color corresponds to pores).

The testing campaign consisted of an isotropic compression test and a number of triaxial compression tests on dry specimens. For this purpose, cylindrical specimens of 2:1 height-diameter ratio and 10 mm diameter were prepared and tested in a miniaturized triaxial compression apparatus allowing for in situ X-ray scans with a voxel size of 13 μm. Note that since the specimens were tested dry, the volumetric strain was directly measured from the 3D images obtained throughout the tests. The isotropic compression test was conducted by increasing step-by-step the confining pressure. The response obtained in the plane mean stress (p)—volumetric strain (ε_v) can be divided in two segments. The first segment is quasi-linear until a limit stress, at which a sudden reduction in volume is observed (see Fig. 3a).

Fig. 1 CT scan of Tuffeau de Maastricht at 1 μm resolution, revealing intragranular porosity and bioclasts variability

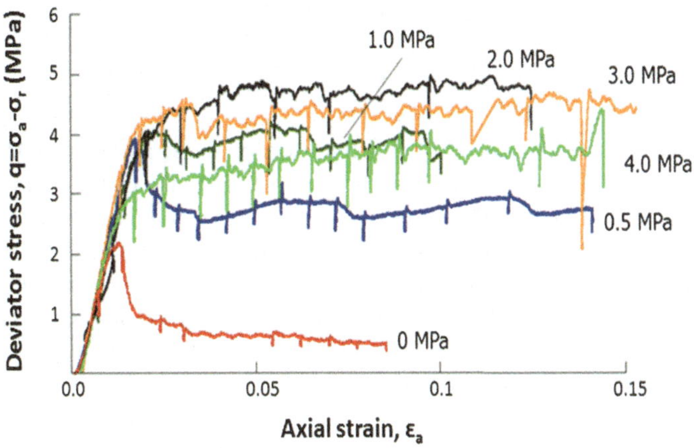

Fig. 2 Curves of deviator stress ($q = \sigma_a - \sigma_r$) versus axial strain obtained from triaxial compression experiments at different confining pressures

The triaxial tests were performed at different confining pressures ranging from 0 (no confinement) to 4.0 MPa (see Fig. 2). Loading was interrupted at various points and X-ray scans were performed. The radiographs obtained were then used to reconstruct 3D tomographic volumes. In addition to the compression tests reported in Fig. 2, a further experiment at 4.0 MPa confinement was conducted, to a much higher axial strain (almost 50%). Figure 5a shows the deviator stress versus axial strain, as well as the volumetric response of the material. The material exhibits an initial linear phase portion and then a plateau of deviator stress. After 14% axial strain, the deviator stress starts to increase again.

3 Constitutive Modeling

A constitutive model based on earlier contributions by Nova et al. [7, 9, 12] is here selected to reproduce the plastic collapse occurring in high-porosity rocks under isotropic and/or shear loading. Yield surface f and plastic potential g are expressed as follows [9]:

$$\left.\begin{matrix} f \\ g \end{matrix}\right\} = \frac{p}{p_c^\star} - \frac{\left(1 + \frac{\eta^\star}{M_h K_2}\right)^{\frac{K_2}{(1-\mu_h)(K_1-K_2)}}}{\left(1 + \frac{\eta^\star}{M_h K_1}\right)^{\frac{K_1}{(1-\mu_h)(K_1-K_2)}}} \qquad \eta^\star = \frac{q}{p + p_t} \tag{1}$$

$$K_{1/2} = \frac{\mu_h (1 - \alpha_h)}{2 (1 - \mu_h)} \left(1 \pm \sqrt{1 - \frac{4\alpha_h (1 - \mu_h)}{\mu_h (1 - \alpha_h)^2}} \right), \tag{2}$$

These functions are characterized by distinct sets of shape parameters M_h, μ_h, and α_h (with the subscript h referring to either f or g). The size of the yield surface is governed by the parameter $p_c^\star = p_s + p_m + p_t$, where p_s reflects the state of material compaction, while p_m simulates the effect of inter-particle bonding. Finally, p_t, which defines the tensile strength of the material, is assumed to be proportional to p_m (i.e., $p_t = \kappa\, p_m$). The mechanical response within the yield surface has been simulated with isotropic linear elasticity, while the hardening/softening characteristics are governed by the following equations:

$$\dot{p}_s = \frac{p_s}{B_p} \dot{\varepsilon}_v^p \qquad \dot{p}_m = -\rho_m p_m (|\dot{\varepsilon}_v^p| + \xi_m \dot{\varepsilon}_q^p) \tag{3}$$

where B_p, ρ_m and ξ_m are constitutive parameters. From Eq. 3 it can be readily noted that the interplay between the hardening of p_s in the compaction regime and the reduction of p_m may result in either contraction or expansion of the yield surface.

An isotropic compression test has been used to calibrate B_p and ρ_m (Fig. 3a). Given the lack of data about the isotropic hardening of the Tuffeau de Maastricht at high pressures, the value of B_p has been defined on the basis of evidences available for a calcareous rock of similar porosity (i.e., the Gravina calcarenite tested by Lagioia and Nova [8]). This choice has allowed us constraining the plastic compressibility of the fully destructured material, as well as the magnitude of ρ_m and the initial values of p_m and p_s. The shape parameters of the yield function (i.e., α_f, μ_f, M_f and κ in Fig. 3a have been calibrated by fitting the yielding points extracted from Fig. 2), and the resulting function is depicted in Fig. 3b. Finally, the shape parameters of the plastic potential (μ_g, α_g, and M_g) have been defined by means of the dilatancy

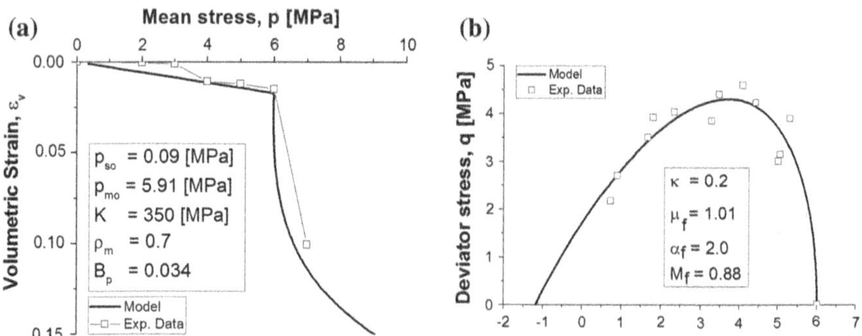

Fig. 3 Calibration of **a** the hardening parameters, and **b** initial yield surface. The values of the calibrated parameters are given in the relevant plot

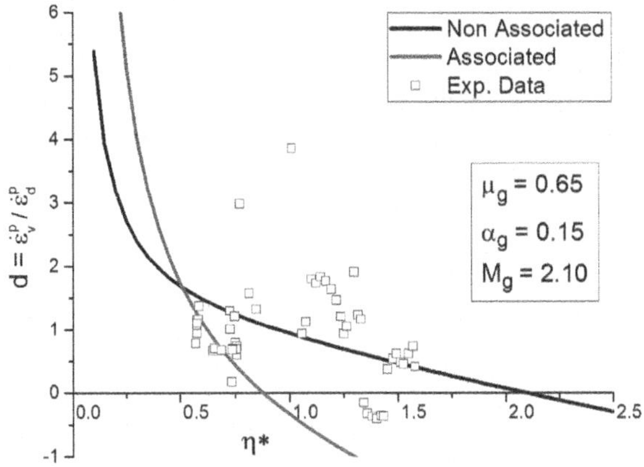

Fig. 4 Calibration of the dilatancy function $d(\eta^\star)$ by using the experimental data plotted in Fig. 5a (the dilatancy function corresponding to associative plasticity is also plotted for comparison purpose)

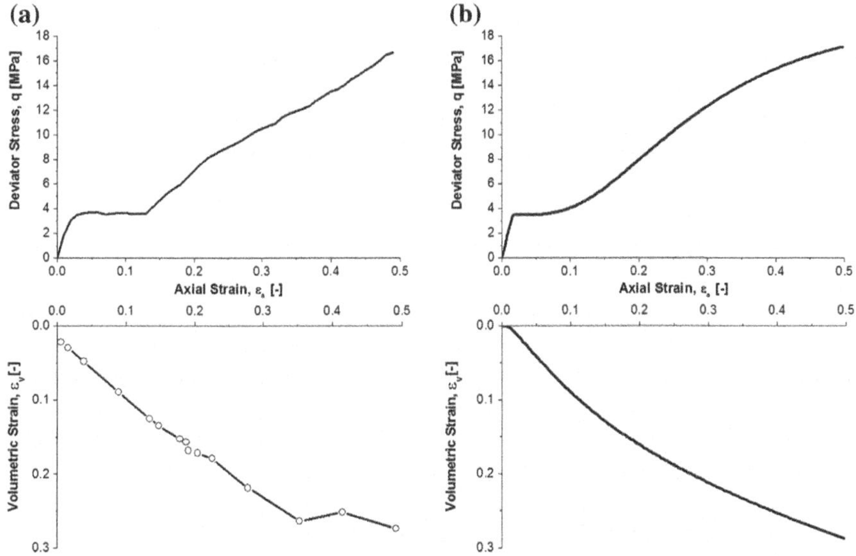

Fig. 5 Comparison between experimental data (**a**) and numerical results (**b**) for a triaxial test performed at 4.0 MPa of confinement pressure; simulation performed with $G = 80$ MPa and $\xi_m = 0.0$

function $d = d(\eta^\star)$, i.e., the ratio between the increments of volumetric and deviatoric plastic strains. Data used for this purpose is extracted from the triaxial test in Fig. 5a. The results are illustrated in Fig. 4 where, for comparison purposes, the $d(\eta^\star)$ relationship which would result from the use of an associated plastic flow rule is also reported. This comparison emphasizes the pronounced non-associativity of the Tuffeau de Maastricht, which usually underscores a non-negligible potential to develop strain inhomogeneity.

The performance of the calibrated model has been tested by simulating a triaxial test at 4.0 MPa confining pressure. The computed material point response in Fig. 5b can be compared to the corresponding experimental result in Fig. 5a. It can be readily noted that by adequately constraining the proportion between hardening and softening terms it is possible to simulate realistically the transition from a perfectly plastic flow in the vicinity of the yielding point to a considerable re-hardening at larger strains. Despite such satisfactory agreement between data and computations, it is worth remarking that this comparison is based on a material-point assessment of the rock response. As a result, it should be regarded as a first-order approximation of the constitutive properties susceptible to further refinements based on the detailed full-field observations discussed in the next section.

4 Measured Fields of Porosity and Incremental Strain

Although the numerical model captures the global mechanical response of the material, the characterization of local inelastic processes requires a more detailed inspection. For this purpose, image analysis was performed on the triaxial test at a 4.0 MPa confining stress, to track the initiation and evolution of localized compaction during the above mentioned compression experiments.

In this work, the porosity was measured locally by working on image gray-scale level through sub-volumes defined within the reconstructed images. These measurements, obtained using a code developed by Andò [1], represent the volume of the voids within the sub-volume. The appropriate (representative) sub-volume size has been decided by analyzing the measurements of porosity and its fluctuations as a function of the sub-volume size. Since the measurements become less sensitive to the cube size and converge to the global porosity of the specimen as the size of the sub-volume increases, a fixed value of 55 voxels has been chosen as representative elementary volume (REV).

3D Digital Image Correlation (DIC) was also performed to measure the evolution of strain field in the specimen while it deforms under loading, using the code "TomoWarp2" (see [13] for details).

Figure 6 shows vertical slices of the 3D strain maps obtained with DIC. The scalar quantity mapped in the figure is the maximum shear strain. Note that this is incremental. The localization appears at the bottom and top of the specimen before the peak at 2–3% of axial strain and extends towards the middle. The porosity map (Fig. 7) indicates the initial high porosity of the material and its evolution during the loading. This densified zone starts from the bottom and top and increases in thickness until it covers the whole specimen. At the end of the deviator stress plateau at 14% axial strain, the entire specimen has a porosity of 38%. It should be noticed that the strain

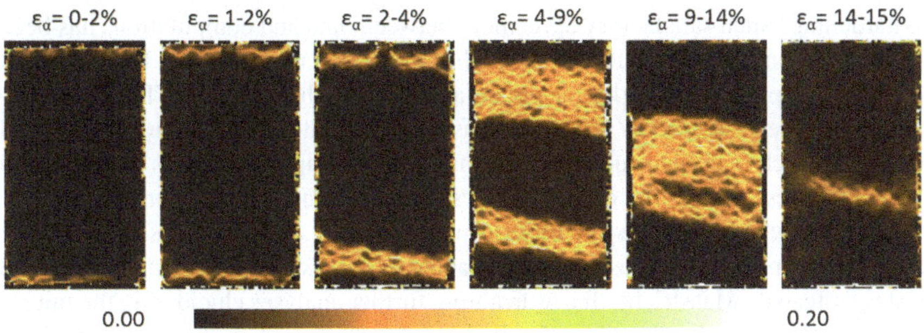

Fig. 6 Vertical slices in different increments of stain, results of the 3D-DIC in terms of maximum shear strain

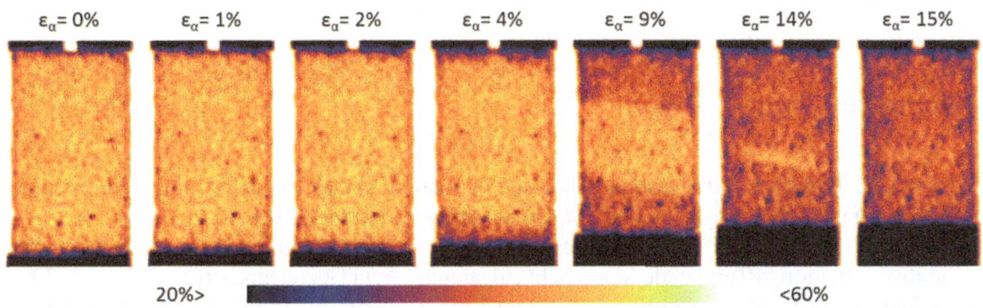

Fig. 7 Vertical slices through the measured 3D field of porosity in different configuration of the triaxial compression test at 4.0 MPa confining pressure. Note that the porosity decreases up to 26% from the initial value (52–38%)

maps clearly point out that the propagation of localized compaction zones coincides with the initial plateau of deviator stress, while the achievement of a homogeneously compacted state marks the onset of re-hardening.

5 Discussion and Conclusions

Localized compaction in Tuffeau de Maastricht has been studied through experiments and constitutive analyses. Triaxial compression tests conducted at different confining pressures have generated a post-yielding response characterized by nearly constant deviator stress and considerable compaction. It has been shown that the continuation of the compression test beyond the post-yielding plateau causes a transition from a perfectly plastic stage to a noticeable re-hardening. These evidences have been interpreted through an elasto-plastic model relying on feedbacks between plastic compaction (hardening terms) and loss of structure (softening terms). The proportion between such feedbacks has been constrained through evidences of plastic collapse under isotropic conditions, thus generating a set of parameters reproducing successfully the transition from perfectly plastic flow to compaction-driven re-hardening. Such satisfactory agreement between data and computations has been achieved by enforcing plastic non-normality in the cap region, i.e., a feature which usually underpins a non-negligible potential for strain inhomogeneity. The latter possibility has been assessed by full-field measurements collected by X-ray imaging and 3D-DIC, which identified successive formations of compaction bands moving from the boundaries towards the center of the sample until achieving a uniformly compacted state which precedes the onset of re-hardening. Such findings illustrate the complex nature of the processes that characterize the response of porous rocks such as the Tuffeau de Maastricht, thus warranting further analyses elucidating the microscopic origin of localized compaction in this material, as well as the enhancement of their constitutive description through microstructural attributes reflecting the alteration of their pore networks.

References

1. Andò, E.: Experimental investigation of microstructural changes in deforming granular media using X-ray tomography. Ph.D. thesis, Université de Grenoble (2013)
2. Baxevanis, T., Papamichos, E., Flornes, O., Larsen, I.: Compaction bands and induced permeability reduction in Tuffeau de Maastricht calcarenite. Acta Geotechnica **1**(2), 123–135 (2006)
3. Bekendam, R.F.: Pillar StabIlity and Large-Sacle Collapse of Abandoned Room and Pillar Limestone Mines inSouth-limburg, The Netherlands. Delft University of Technology, TU Delft (1998)
4. Das, A., Buscarnera, G.: Simulation of localized compaction in high-porosity calcarenite subjected to boundary constraints. Int. J. Rock Mech. Mining Sci. **71**, 91–104 (2014)
5. Dreesen, R., Dusar, M.: Historical building stones in the province of Limburg (NE Belgium): role of petrography in provenance and durability assessment. Mater. Charact. **53**(2), 273–287 (2004)
6. Dubelaar, C., Dusar, M., Dreesen, R., Felder, W., Nijland, T.: Maastricht limestone: a regionally significant building stone in Belgium and the Netherlands. extremely weak yet time-resistant. In: Proceedings of the International Heritage Weathering and Conservation Conference, pp. 9–14. Taylor & Francis Group, London (2006)
7. Gens, A., Nova, R.: Conceptual bases for a constitutive model for bonded soils and weak rocks. Geotech. Eng. Hard Soils-Soft Rocks **1**(1), 485–494 (1993)
8. Lagioia, R., Nova, R.: An experimental and theoretical study of the behaviour of a calcarenite in triaxial compression. Géotechnique **45**(4), 633–648 (1995)
9. Lagioia, R., Puzrin, A., Potts, D.: A new versatile expression for yield and plastic potential surfaces. Comput. Geotech. **19**(3), 171–191 (1996)
10. Marinelli, F., Buscarnera, G.: Parameter calibration for high-porosity sandstones deformed in the compaction banding regime. Int. J. Rock Mech. Mining Sci. **78**, 240–252 (2015)
11. Ngan-Tillard, D., Verwaal, W., Mulder, A., Engin, H., Ulusay, R.: Application of the needle penetration test to a calcarenite, Maastricht, the Netherlands. Eng. Geol. **123**(3), 214–224 (2011)
12. Nova, R., Castellanza, R., Tamagnini, C.: A constitutive model for bonded geomaterials subject to mechanical and/or chemical degradation. Int. J. Numer. Anal. Methods Geomech. **27**(9), 705–732 (2003)
13. Tudisco, E., Hall, S., Charalampidou, E.M., Kardjilov, N., Hilger, A., Sone, H.: Full-field measurements of strain localisation in sandstone by neutron tomography and 3D-volumetric digital image correlation. Phys. Proc. **69**, 509–515 (2015)

Strain Localisation in Sand in Cycles of Triaxial Compression and Extension: Continuum and Grain-Scale Analysis

E. Andò, E. Salvatore, J. Desrues, P. Charrier, J.-B. Toni, G. Modoni and G. Viggiani

Abstract In this work we present selected results from a recent experimental programme where small sand specimens are subjected to cycles of triaxial compression and triaxial extension: the material is "yielded" in extension, after which the loading is reversed and the material is "yielded" in compression—a number of cycles are performed. The ways in which extension and compression-like localisation patterns (i.e., dilatant shear banding, and necking respectively) appear, get activated and disactivated on reversal of loading are measured, and discussed—in terms of both (continuum) strain fields and individual grain rotations.

1 Introduction

X-ray tomography has been used with great success, since the end of the 80s to reveal 3D structures and processes in geomaterials. At the beginning of this adventure, spatial resolution in medical x-ray scanners was not sufficient to resolve individual sand grains, however important findings regarding the *local* significance of the value of critical state void ratio were made.

Since some years, both synchrotrons and laboratory x-ray scanners offer high resolutions and the possibility to perform experiments during x-ray scanning. The usual trade-off between spatial resolution and field-of-view still applies, so to take advantage of the gain in spatial resolution, experiments have had to be miniaturised—in our case specimens of sand for triaxial compression have been reduced to cylinders 11 mm diameter and about 22 mm height, containing tens of thousands of grains.

E. Andò (✉) · J. Desrues · P. Charrier · J.-B. Toni · G. Viggiani
CNRS, 3SR, 38000 Grenoble, France
e-mail: edward.ando@3sr-grenoble.fr

E. Andò · J. Desrues · P. Charrier · J.-B. Toni · G. Viggiani
Université Grenoble Alpes, 3SR, 38000 Grenoble, France

E. Salvatore · G. Modoni
University of Cassino and Southern Lazio, Cassino, Italy

E. Papamichos et al. (eds.), *Bifurcation and Degradation of Geomaterials with Engineering Applications*, Springer Series in Geomechanics and Geoengineering, DOI 10.1007/978-3-319-56397-8_62

Experiments performed on such small specimens have previously allowed grain-scale observations of grain-scale phenomena—such as strain localisation, presented in IWBDG 2011, and some initial observations of grain breakage presented in IWBDG 2014.

In this work we present selected results from a recent experimental programme where the small sand specimens mentioned above are subjected to cycles of triaxial compression and triaxial extension: the material is "yielded" in extension, after which the loading is reversed and the material is "yielded" in compression—a number of cycles are performed. The way in which extension and compression-like localisation patterns (i.e., dilatant shear banding, and necking respectively) appear, get activated and disactivated on reversal of loading are measured, and discussed—in terms of both (continuum) strain fields and individual grain rotations.

2 Tested Material, Experimental Setup and Testing Programme

The experiment is performed in the x-ray scanner of Laboratoire 3SR in Grenoble [1]. Deviatoric loading is applied under strain control, at a constant rate of axial strain equal to 5.7% per hour. A bayonet system enables the sample not only to be compressed, but also to be stretched by the axial loading system—throughout the test the cell pressure, i.e., the radial stress (equal to σ_2 and σ_3 in compression and σ_1 and σ_2 in extension) is held constant. The axial force is recorded with a load transducer, while the axial shortening of the sample is monitored with an LVDT.

The tested material is Hostun sand, an angular quartz sand with mean diameter of 328 µm and a coefficient of uniformity of 1.7. The specimen is prepared by dry pluviation into a 300 µm thick latex membrane stretched in a mould with a drop height of 1 m. The sample is then isotropically loaded up to 430 kPa by increasing confining pressure. While acquiring x-ray tomography data, (which takes around 2 h per state and which requires no change in the specimen), the axial loading system is halted and the piston held in place. The test analysed in this paper includes cycles of compression and extension that are schematically depicted in the inset in the top left of Fig. 1. The macroscopic axial stress-strain response of the specimen is shown on the same figure. The points at which the test is stopped in order to acquire a 3D image of the specimen are labelled 01–20 on the curve. The test will be discussed branch by branch (i.e., according to the direction of loading) showing for each branch results of local measurement of kinematics rendered both as the second invariant of the locally-measured strain tensor (in the case of a "mesoscopic continuum DIC" analysis—see [6]) and as individual grain rotations (in the case of grain-based DIC).

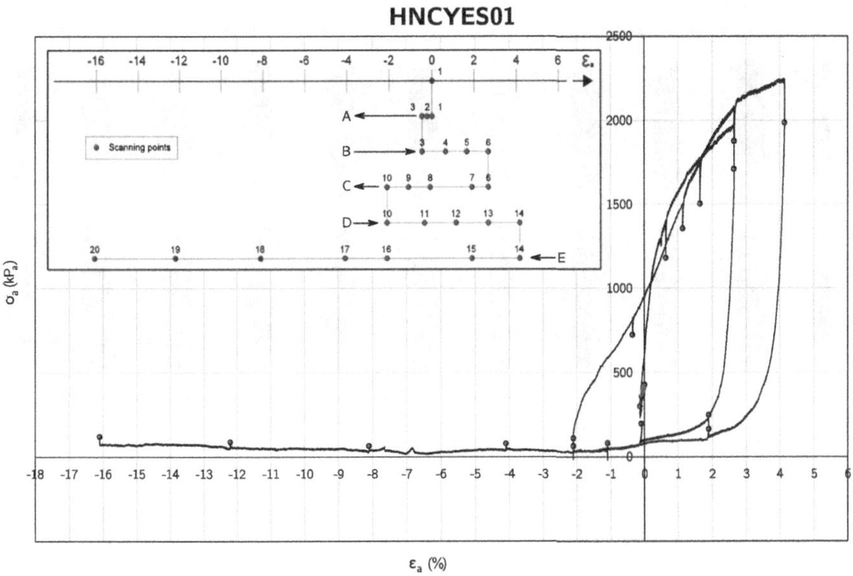

Fig. 1 Macroscopic stress-strain results obtained for the cyclic triaxial compression/extension test on Hostun sand studied herein. *Note* confining pressure is constant (and equal to 430 kPa), and axial stress is always compressive

3 Image Analysis

Figure 2 shows a vertical slice taken from each x-ray tomography reconstruction of each scanned state. These images *are* the microstructural measurement of the specimen in a number of different states at 20 μm/px. For the incremental continuum maximum shear strain field that will be presented below, a local DVC code (TomoWarp2, [7]) is used on a regular grid of points defined in the first of the pair of images compared. Points are 20 px distant from each other in all cardinal directions. Each point is represented by a texture or correlation window centred on the point and measuring 27 × 27 × 27 px. Each correlation window is match in the second image of the pair by maximising the Normalised Correlation Coefficient in a finite search area, with a first approximation to the nearest pixel displacement, and then with sub-pixel accuracy by locally interpolating the field of NCC. The method used of these meso-measurements is therefore rigid and measures displacements only. For a particle-based technique, particles must be identified in the images. Starting from the greyscale images (which represent the field of x-ray attenuation in the scanned area), they are first binarised such that the solid and void phases are unequivocally distinguished. The solid phase is then separated into individual sand grains by using a 3D watershed algorithm. Identified particles are then individually numbered ("labelled"). For the analysis presented herein only the first image is segmented into individual particles, which means that the philosophy is closer to the total approach

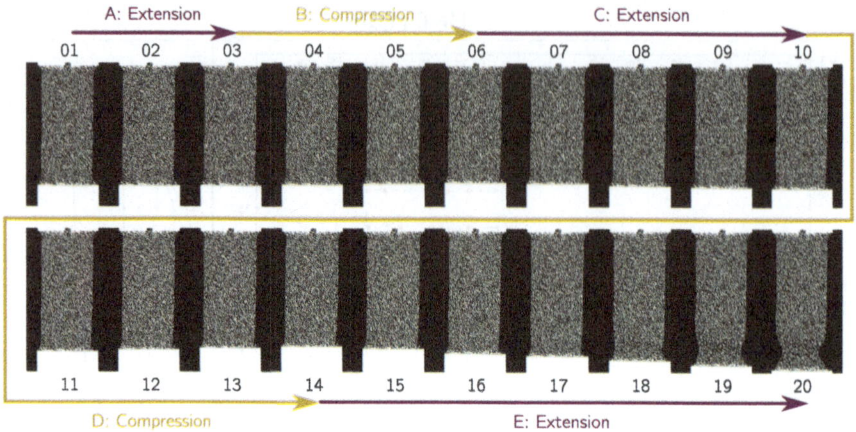

Fig. 2 Vertical slices from the reconstructed x-ray tomography 3D volumes throughout the test, showing the loading direction between states

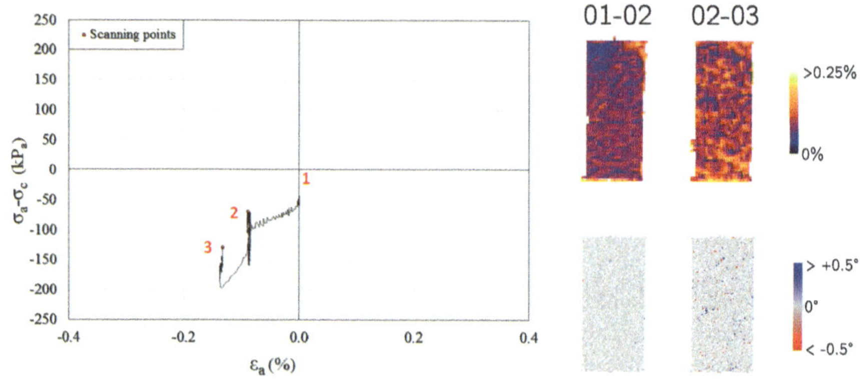

Fig. 3 Branch A—first extension: macroscopic stress-strain response shown alongside incremental fields of maximum shear strain and particle rotations in the axis normal to the slice. Please note that for both types of measurements, given the small size of the externally-applied load increments, the colourmaps have been exceptionally expanded with respect to the other branches discussed

of [4] than our previous work with ID-Track [2, 3] which requires every step to be separated. TomoWarp2 is used to track particles, with points now defined as grain-centres and therefore not on an unstructured grid. Correlation windows are adapted to the size of the grain, and using the labelled image of each particle as a mask, only the greylevel corresponding to the grain is correlated. Particles are tracked always starting from the first image (but using as prior displacements the previous step). In this case the correlation is still rigid, but has translational and rotational degrees of freedom.

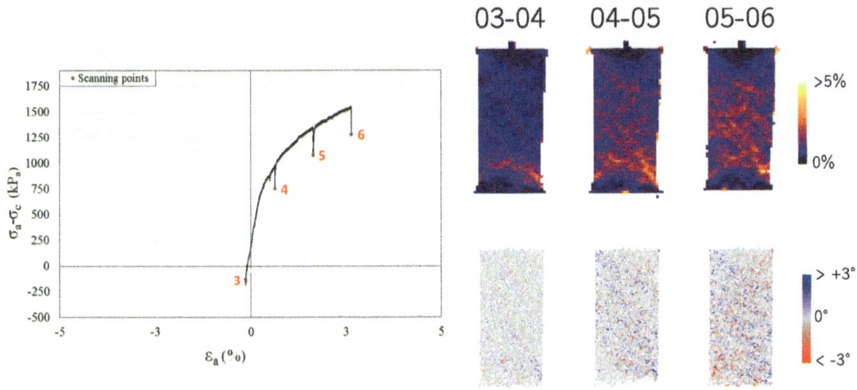

Fig. 4 Branch B—first compression: macroscopic stress-strain response shown alongside incremental fields of maximum shear strain and particle rotations in the axis normal to the slice

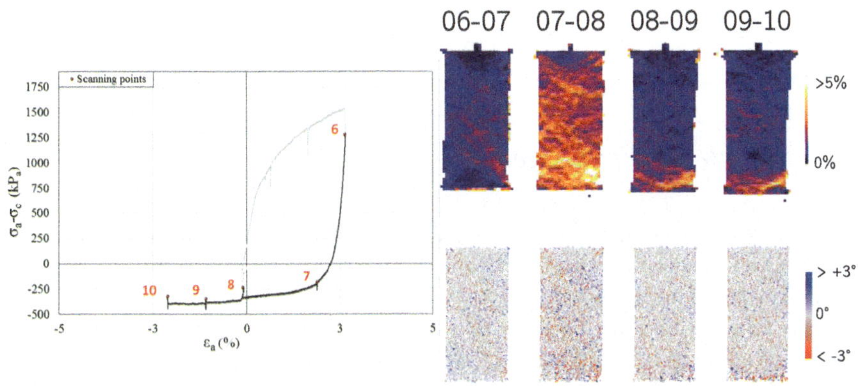

Fig. 5 Branch C: macroscopic stress-strain response shown alongside incremental fields of maximum shear strain and particle rotations in the axis normal to the slice

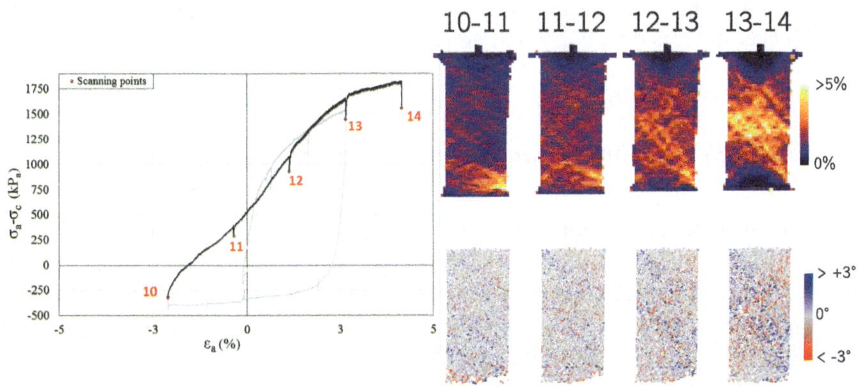

Fig. 6 Branch D: macroscopic stress-strain response shown alongside incremental fields of maximum shear strain and particle rotations in the axis normal to the slice

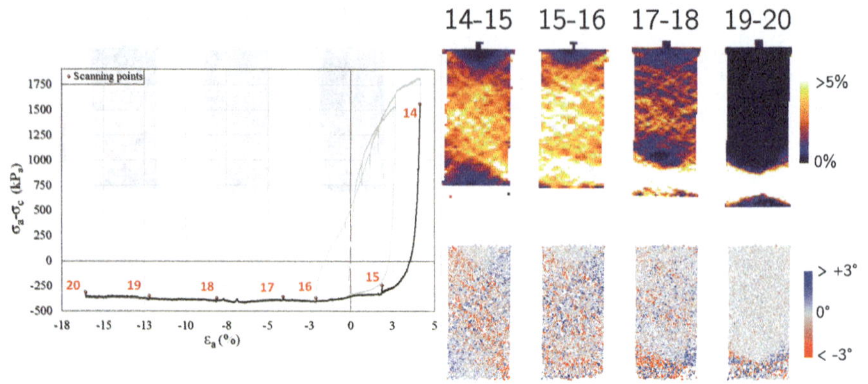

Fig. 7 Branch E: macroscopic stress-strain response shown alongside incremental fields of maximum shear strain and particle rotations in the axis normal to the slice. Please note that in the interest of space not all increments are shown here

4 Results

4.1 Branch A—First Extension

The first two scanned increments 01–02 and 02–03, which go from an isotropic state of stress (01) towards extension (02—$\epsilon_a = 0.08\%$ and 03—$\epsilon_a = 0.13\%$) are very small axial strain increments, meaning that with respect to the sensitivity of the local measurements we're able to make, few mechanisms can be identified, although (with the benefit of hindsight) in the meso-DIC some concentration of shear strain can be seen in the bottom right of the sample in the increment 02–03, in the middle slice chosen (Fig. 3).

On the same slice, grains are coloured with respect to their rotation around the axis normal to the slice (positive clockwise). In both increments, rotations are very small (generally less than 0.2°), and no particular structure can be observed.

4.2 Branch B—First Load Reversal: Towards Compression

In this branch the axial loading direction is reversed (Fig. 4). The first state in this branch (which is the last state of the previous) is in a state of triaxial extension, however the next scanned state is in triaxial compression. It is important to note that the displacement applied in compression is significantly larger what was applied in the previous branch—meaning that the colour maps for the local measurements are now on a different (and forthwith constant) scale.

In these increments a clear picture emerges from both maximum shear strain maps as well as the individual grain rotation: in the first increment activity is concentrated in the bottom right of the slice shown. Between increments 04–05 and 05–06 a more complex pattern of localised strain/rotations starts to develop. As the specimen is pushed further into compression, the kinematic effect of the friction cones is clearly visible—especially on the top of the specimen.

4.3 Branch C—Second Load Reversal: Towards Extension Again

At the beginning and end of this branch, increments 06–07 and 09–10 present localisation patterns with very different orientations (Fig. 5). Increment 06–07 shows a patterns similar to that observed in 05–06 in compression, whereas orientations in 09–10 are tilted at a much lower angle and localised strain is concentrated essentially at the bottom of the specimen, with some activity emerging also in the middle.

The "compression-like" localisation patterns visible in 06–07 are in fact the local reversing of the kinematics seen in increment 05–06—is !!! JD asks: "if" instead of "is" ??? the rotations are studied in detail, it is visible (for example in the !!! blatantly ??? localised friction cone) that grain's rotations change sign in this increment.

The transition that occurs during this branch is made with a large axial loading step—07–08—which despite being all in a state of triaxial extension, clearly manifests memory of the localisation patterns in compression, whereas when the specimen is stretched beyond its initial length, the localised kinematics no longer reflect what was seen in compression but rather "extension-like" arrangements in increment 08–09 and 09–10.

4.4 Branch D—Third Load Reversal: Towards Compression Again

Consistently with what was observed in the load reversal commented above, despite the reversal of loading, the local measurements in the first increment show significant traces of the previous branch. As the sample is compressed more than its initial length, compression-like localisation bands take over and strain starts to localise significantly in the middle of the specimen. Once again, the localisation of maximum shear strain corresponds clearly to the rotations of individual particles (Fig. 6).

4.5 *Branch E—Fourth and Last Load Reversal: Towards Extension and Failure*

Consistent with the previous reversal-into-extension, in the first increment compression-like features are visible, with reversed rotations indicating local kinematical reversal. The sample is stretched very far in extension, until clearly localised necking at the bottom of the specimen dominates all other previously-activated mechanisms of localised strain Fig. 7.

In the increment 17–18 a competing "neck" is visible in the middle of the specimen but is disactivated with increasing strain.

5 Conclusions

In the paper we have presented a micro-mechanical analysis of a single, complex experiment in which the direction of loading is reversed a number of times. Our local image-based kinematical measurements indicate that two distinct patterns of strain localisation are associated with the two loading directions, differing primarily in their orientation of the band of concentrated strain with respect to the axis of the specimen (steeper for compression than for extension—this is supported by previous work with monotonic loading in both directions). The main finding from this preliminary analysis of the cyclic test is that upon load reversal, localisation mechanisms belonging to the previous loading direction surprising persist and "rewind" until the **total strain** changes sign.

Further analysis will include the evolution of fabric quantities throughout these cycles, taking advantage of techniques currently in development by our group within the "SOMEF" ERC project [8]. A clear interest for this kind of rich experimental data is as a test case for advanced micro-mechanical simulations, after they have been calibrated in monotonic loading for example. A perfect example of this is the recently-developed LS-DEM model from Caltech [5] which seems to capture all the macro-scale mechanical subtleties of the behaviour of granular media through a careful treatment of the grain scale (which in some cases has come straight from our 3D images).

Acknowledgements Laboratoire 3SR is part of the LabEx Tec 21 (Investissements d'Avenir—grant agreement n° ANR-11-LABX-0030). The research leading to these results has received funding from the European Research Council under the European Union's Seventh Framework Program FP7-ERC-IDEAS Advanced Grant Agreement no 290963 (SOMEF).

References

1. Andó, E.: Experimental investigation of micro-structural changes in deforming granular media using x-ray tomography. Ph.D. Thesis, Université de Grenoble (2013)
2. Andó, E., Hall, S.A., Viggiani, G., Desrues, J., Bésuelle, P.: Grain-scale experimental investigation of localised deformation in sand: a discrete particle tracking approach. Acta Geotech. **7**(1), 1–13 (2012)
3. Andó, E., Hall, S.A., Viggiani, G., Desrues, J., Bésuelle, P.: Experimental micromechanics: grain-scale observation of sand deformation. Géotech. Lett. **2**(3), 107–112 (2012)
4. Hall, S., Bornert, M., Desrues, J., Pannier, Y., Lenoir, N., Viggiani, G., Bésuelle, P.: Discrete and continuum experimental study of localised deformation in Hostun sand under triaxial compression using X-ray μCT and 3D digital image correlation. Géotechnique **60**(5), 315–322 (2010)
5. Kawamoto, R., Andó, E., Viggiani, G., Andrade, J.E.: Level set discrete element method for three-dimensional computations with triaxial case study. J. Mech. Phys. Solids **91**, 1–13 (2016)
6. Salvatore, E., Andó, E., Modoni, G., Viggiani, G.: Micromechanical study of cyclically loaded sands with x-ray microtomography and digital image correlation. Proc. Eng. **158**, 92–97 (2016)
7. Tudisco, E., Andó, E., Cailletaud, R., Hall, S.A.: TomoWarp2: a local digital volume correlation code. In: SoftwareX (2016)
8. Wiebicke, M., Andó, E., Salvatore, E., Herle, I., Viggiani, G.: Experimental measurement of granular fabric and its evolution under shearing. In: Powders and Grains (2017)

Analysis of Shear Bands in Sand Under Reduced Gravity Conditions

Jason P. Marshall, Ryan C. Hurley, Dan Arthur, Ivan Vlahinic, Carmine Senatore, Karl Iagnemma, Brian Trease and José E. Andrade

Abstract The strength of granular material, specifically sand is of pivotal importance for understanding physical phenomena on other celestial bodies. However, relatively few experiments have been conducted to determine the dependence of strength properties on gravity. In this work, we experimentally investigated three measures of strength (peak, confined flow, and unconfined flow friction angle) in Earth, Martian, Lunar, and near-zero gravity. The angles were captured in a passive Earth pressure experiment conducted on a reduced gravity flight. The results showed no dependence of the peak friction angle on gravity, a weak dependence of the confined flow friction angle on gravity, and no dependence of the unconfined flow friction angle on gravity. These results highlight the importance of understanding strength and deformation mechanisms of granular material at different levels of gravity.

1 Introduction and Background Information

Granular materials are ubiquitous in nature encompassing everything from foods to soils. These particulate materials play a pivotal role in human society on Earth and an increasingly prominent role on other celestial bodies that humanity is exploring. Understanding granular materials is important for future off-world scientific endeavors; these materials impact all functional aspects of missions including penetrator experiments, drilling events, and robotic mobility.

J.P. Marshall (✉) · D. Arthur · I. Vlahinic · J.E. Andrade
California Institute of Technology, Pasadena, CA, USA
e-mail: jmarshal@caltech.edu

R.C. Hurley
Lawrence Livermore National Laboratory, Livermore, CA 94550, USA

C. Senatore · K. Iagnemma
Massachusetts Institute of Technology, Cambridge, MA, USA

B. Trease
University of Toledo, Toledo, OH, USA

E. Papamichos et al. (eds.), *Bifurcation and Degradation of Geomaterials with Engineering Applications*, Springer Series in Geomechanics and Geoengineering, DOI 10.1007/978-3-319-56397-8_63

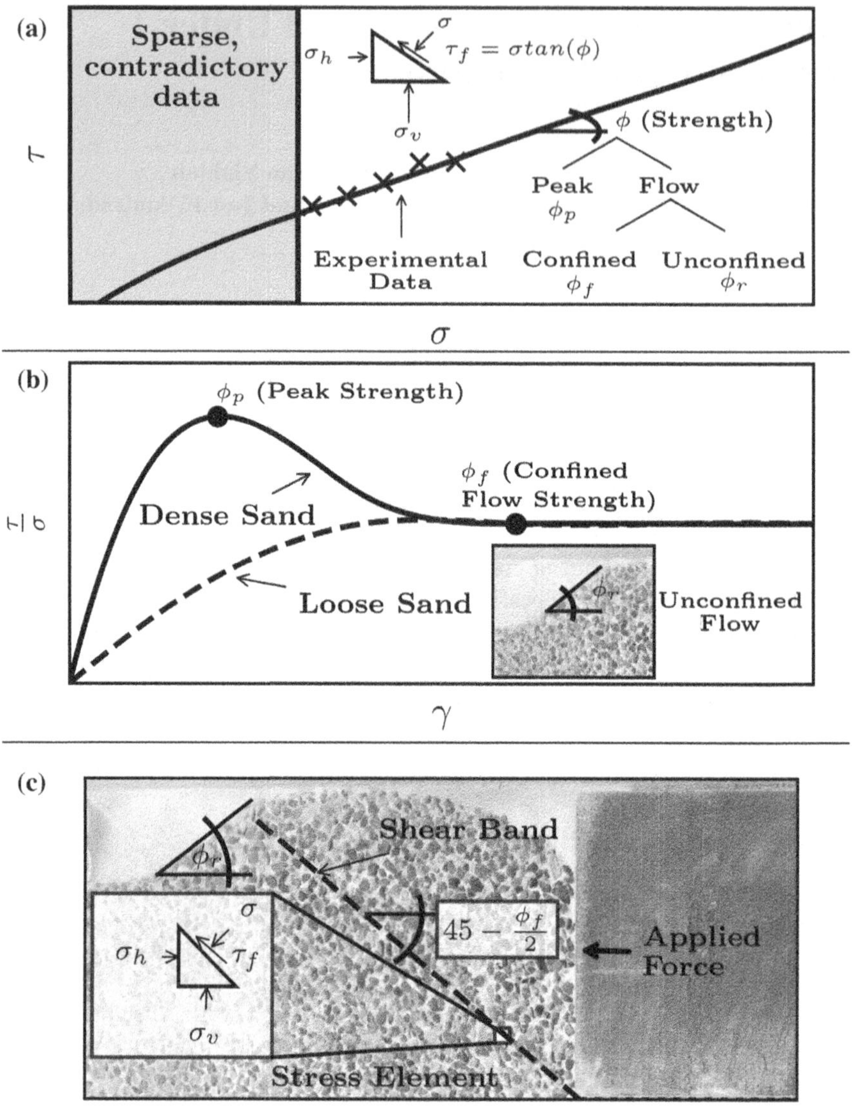

Fig. 1 **a** The Mohr-Coulomb failure envelope is shown. The angle ϕ is used to describe the strength. **b** Strength in granular material can be defined by a peak strength, ϕ_p; confined flow strength, ϕ_f; and an unconfined flow strength, ϕ_r. **c** Failure in granular material along a shear band with an angle of $45° - \frac{\phi}{2}$

Research on granular materials has been conducted for centuries, but almost exclusively under Earth gravity. While many of the conclusions from existing research will remain true in non-Earth gravity environments, it is possible that many will not. We provide experimental evidence of internal friction angles, hereafter called friction angles [11], corresponding to peak strength, confined flow strength, and unconfined flow strength in Earth, Martian, Lunar, and near-zero gravity. These results add data to an under-researched regime of granular materials, namely strengths at low-confinement levels and their dependence on gravity.

In this work, we assumed a Mohr-Coulomb failure envelope [5] given by $\tau = \sigma \tan\phi + c$, where τ is the shear stress, σ is the normal stress, and c is the cohesion. We deal exclusively with a drained granular material and thus assume it is cohesion-less. The Mohr-Coulomb failure envelope defines failure in terms of confining stresses and a friction angle, which is best understood graphically as seen in Fig. 1. Many experimental measurements of friction angles have been made at high confining pressures [2, 3, 6–9], but few studies have been conducted with low confining pressures [1, 4] and even fewer with varying gravity levels [1, 12]. In addition to the friction angle, the shape of the failure pattern in passive Earth pressure experiments is important, for which we assume a log-spiral pattern [11].

2 Experimental Setup

The above properties were investigated at different gravity levels with a passive Earth pressure experiment [11]. The experiment was conducted on a reduced gravity flight with granular material deformed via a push block in a box with a glass face on the front. As the block was pushed in, the sand failed and ruptured in a clearly defined shear band. Data was recorded on 3 Martian, 2 Lunar, and 20 near-zero gravity parabolas in addition to 5 tests at Earth gravity. The sand was fluidized by vibrating and injecting air into the sample between each test, resulting in a relative density of nearly 100%. The free surface was mostly flat with a slight amount of buildup of material on the left side, however, the surface was the same for all tests conducted. A 1920×1080 pixel video of the deformation on the front face was recorded at 60 frames-per-second. Individual images were then extracted and analyzed using digital image correlation (DIC) [10]. The resulting strain field was used to calculate the log-spiral failure surface and the friction angles for each test.

3 Results

3.1 Peak Friction Angle

The peak friction angle was captured by assuming a 2% max shear strain in the sand as the point of failure. While we chose 2%, other reasonably small values will result in nearly the same failure surface. Color plots of all regions of the sand above this max shear strain value resulted in a clear log-spiral failure surface as shown in Fig. 2a. This failure pattern was repeatedly seen at Earth, Martian, and Lunar gravity levels

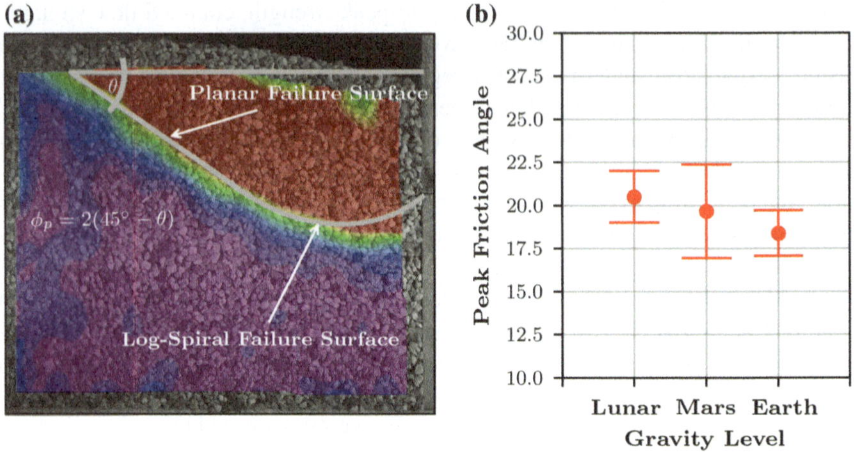

Fig. 2 **a** Log-spiral failure surface for the peak friction angle at lunar gravity. **b** Peak friction angles at different gravity levels with standard *error bars*

and occurred within a fraction of a second of initial loading and with a translation of the push block less then 0.02 in. We calculated the friction angle corresponding to this failure surface and found that a value of around 20° was consistently seen across these gravity levels. Figure 2b shows the friction angle values at the different gravity levels with standard error bars. We found that there was essentially no dependence of the peak friction angle on gravity.

3.2 Confined Flow Friction Angle

Almost immediately after the log-spiral failure surface formed, flow began in a localized shear band. The shear band shape was a planar surface at a lower friction angle between 0° and 10°, as shown in Fig. 3. The angle was clearly lower than the peak values at all gravity levels. Additionally, there was a weak dependence of the confined flow friction angle on gravity level, as shown in Fig. 3b. At near-zero and lunar gravity, the confined flow friction angle essentially dropped to 0°, while angles at the higher gravity levels for Mars and Earth were around 5°–10°. We note that there were a few negative friction angle values in the data counter to theoretical lower bounds at 0°. We attributed these values to random variations of granular contacts within the sample and errors in establishing the angle of the failure surface that led to slight deviations from the theoretically bounded values.

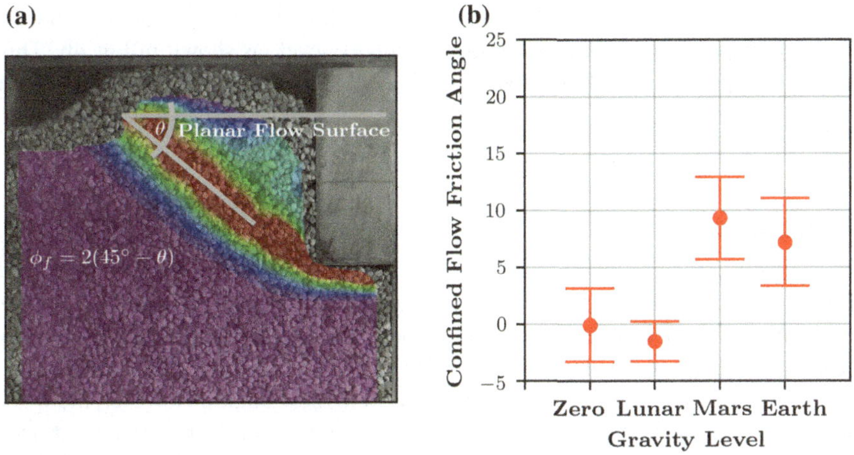

Fig. 3 **a** Planar flow surface for the confined flow friction angle at Martian gravity. **b** Confined flow friction angles at different gravity levels with standard *error bars*

3.3 *Unconfined Flow Friction Angle*

As the push block progressed into the mass of sand, a pile formed. Eventually, the slope of the pile increased beyond the unconfined flow friction angle and failure occurred. Individual grains rolled or slid down the pile causing a reduction of the pile slope to a stable configuration. We captured this angle at all non-zero gravity levels on the left side of the pile, as shown in Fig. 4a. The calculated unconfined

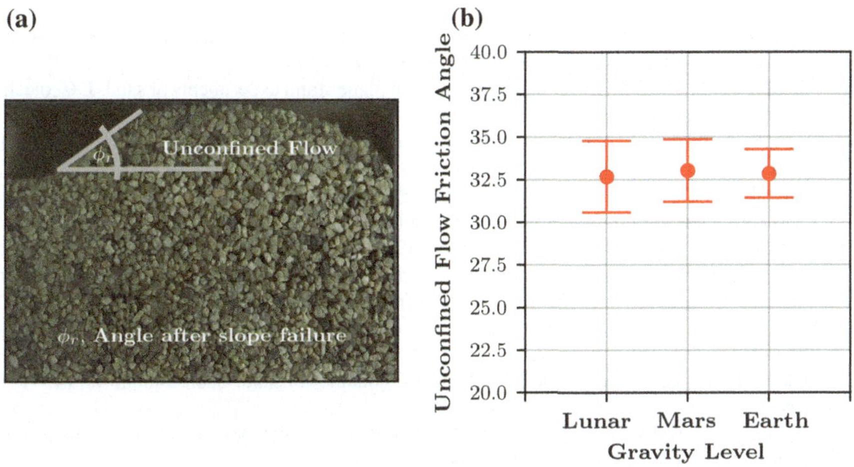

Fig. 4 **a** Example of unconfined flow friction angle at Martian gravity. **b** Unconfined flow friction angles at different gravity levels with standard *error bars*

flow friction angle at Earth, Martian, and Lunar gravities was around 33° and we found no dependence of this value on the gravity level, as shown in Fig. 4b. This value corresponded closely with values calculated at Earth gravity with a standard test.

4 Conclusions

These results have a few important consequences with specific application to space exploration. The lack of dependence of the peak and unconfined flow friction angles on gravity suggests that soil experiments conducted at Earth gravity with representative granular materials will be valid with respect to these angles. However, the weak dependence of the confined flow friction angle on gravity suggests that any extrapolation needs be done carefully as different levels of gravity can affect the soil properties. In general careful consideration must be taken when designing components and experiments for non-Earth celestial bodies that include granular material interactions, as the material strength and deformation mechanisms may be significantly different. These results also suggest that further experiments should be pursued to refine the trends found.

Acknowledgements This research was supported in part by the Keck Institute for Space Studies (KISS) at the California Institute of Technology under the program "x-Terramechanics—Integrated Simulation of Planetary Surface Missions". This support is gratefully acknowledged.

References

1. Alshibli, K.A., Batiste, S.N., Sture, S.: Strain localization in sand: plane strain versus triaxial compression. J. Geotech. Geoenviron. Eng. **129**(6), 483–494 (2003)
2. Alshibli, K.A., Sture, S.: Shear band formation in plane strain experiments of sand. J. Geotech. Geoenviron. Eng. **126**(6), 495–503 (2000)
3. Been, K., Jefferies, M.G., Hachey, J.: The critical state of sands. Geotechnique **41**(3), 365–381 (1991)
4. Chakraborty, T., Salgado, R.: Dilatancy and shear strength of sand at low confining pressures. J. Geotech. Geoenviron. Eng. **136**(3), 527–532 (2010)
5. Holtz, R.D., Kovacs, W.D.: An Introduction to Geotechnical Engineering. Prentice Hall (1981)
6. Negussey, D., Wijewickreme, W.K.D., Vaid, Y.P.: Constant-volume friction angle of granular materials. Can. Geotech. J. **25**(1), 50–55 (1988)
7. Sadrekarimi, A., Olson, S.M.: Critical state friction angle of sands. Géotechnique **61**(9), 771–783 (2011)
8. Schanz, T., Vermeer, P.A.: Angles of friction and dilatancy of sand. Geotechnique **46**, 145–151 (1996)
9. Simoni, A., Houlsby, G.T.: The direct shear strength and dilatancy of sand-gravel mixtures. Geotech. Geol. Eng. **24**(3), 523–549 (2006)

10. Sutton, M.A., Wolters, W.J., Peters, W.H., Ranson, W.F., McNeill, S.R.: Determination of displacements using an improved digital correlation method. Image Vis. Comput. **1**(3), 133–139 (1983)
11. Terzaghi, K., Peck, R.B., Mesri, G.: Soil Mechanics in Engineering Practice. Wiley (1996)
12. Zou, M., Fan, S., Shi, R., Yang, Y., Li, J.: Effect of gravity on the mechanical properties of lunar regolith tested using a low gravity simulation device. J. Terramech. **60**, 11–22 (2015)

Instability Analysis of Granular Media via a Purely Micromechanical Constitutive Model

Mehdi Pouragha, Richard Wan and Xu Gong

Abstract In this paper, we develop a rate independent, incrementally linear constitutive model for 2D granular materials within a robust multiscale framework where micromechanical interactions and microstructural rearrangements between particles are being described statistically. The model when implemented into a finite element code allows us to explore the different microscopic states both inside and outside of the strain localization zone that develops during shearing in a biaxial test. A major finding is the discovery of similarities in microstructural characteristics inside a shear band during localized failure and critical state in diffuse failure.

1 Introduction

Micromechanical studies on granular materials during the past three decades have resulted into up-scaling expressions for stress and strain in terms of contact and force network properties and their statistics [1]. However, a coherent constitutive framework to close the loop of equations so as to relate stress to strain through such micromechanical expressions is still missing. In fact, currently available micromechanical constitutive models often involve ad-hoc evolution laws for macrovariables or continuum-inspired potential functions [2].

Based on a mechanism-specific decomposition of micro-variables evolution [3], the current study proposes a general multiscale framework from which a constitutive model emerges from purely micromechanical considerations. A rate-independent linear incremental model for the simple case of 2D granular materials under proportional loading condition is herein formulated. Comparison with DEM results proves that the model is able to predict not only the stress-strain response, but also the evolution of the principal underlying microvariables. The resulting constitutive model is next incorporated into finite elements (FEM) simulations in order to investigate

M. Pouragha (✉) · R. Wan · X. Gong
Civil Engineering Department, University of Calgary, Calgary, Canada
e-mail: mpouragh@ucalgary.ca

E. Papamichos et al. (eds.), *Bifurcation and Degradation of Geomaterials with Engineering Applications*, Springer Series in Geomechanics and Geoengineering, DOI 10.1007/978-3-319-56397-8_64

strain localization and instability characteristics of granular materials. The results indeed show that the microstructural signature within the shear band matches that of critical state in diffuse failure.

2 Emergence of a Statistical Constitutive Relation

We start by assuming that stress and strain rates in granular assembly can be expressed in terms of the change in statistical descriptors, $\dot{\boldsymbol{S}}$ as follows:

$$\dot{\boldsymbol{\sigma}} = \boldsymbol{K}\dot{\boldsymbol{S}}, \quad \dot{\boldsymbol{\varepsilon}} = \boldsymbol{L}\dot{\boldsymbol{S}} \tag{1}$$

where $\{\boldsymbol{\sigma}, \boldsymbol{\varepsilon} \in \mathbb{R}^m\}$, while $\{\boldsymbol{S} \in \mathbb{R}^p\}$ is a generalized vector containing statistical descriptors, such as anisotropy parameters, coordination number and the likes, with $p \geq m$. As such, the two matrices $\boldsymbol{L}$ and $\boldsymbol{K}$ represent linear $\mathbb{R}^p \to \mathbb{R}^m$ transformations that are not invertible. Furthermore, it is assumed that two different mechanisms, namely dissipative, d, and non-dissipative, nd, contribute to the strain and the change in S throughout the loading. This decomposition slightly differs from elastic/plastic split from total strain in that the dissipative mechanism here refers to the evolution of the material's structure along "Stable Evolution State" (SES) [4] which can also contribute to stress changes. Moreover, the contributions of the two mechanisms are assumed to be decoupled:

$$\dot{\boldsymbol{\varepsilon}} = \dot{\boldsymbol{\varepsilon}}_{nd} + \dot{\boldsymbol{\varepsilon}}_d, \quad \dot{\boldsymbol{S}} = \dot{\boldsymbol{S}}_{nd} + \dot{\boldsymbol{S}}_d, \quad \text{and} \quad \dot{\boldsymbol{\varepsilon}}_{nd} = \boldsymbol{L}_{nd}\dot{\boldsymbol{S}}_{nd}, \quad \dot{\boldsymbol{\varepsilon}}_d = \boldsymbol{L}_d\dot{\boldsymbol{S}}_d \tag{2}$$

In order for the transformation from stress and strain space to the statistical domain to be bijective, $n = p - m$ number of additional relations are needed among the variables. Such "consistency" relations are assumed to describe the interrelation among the evolution of statistical variables, and are assumed to remain unchanged during the loading. Consistency equations are often mechanism-specific as different mechanism are often subjected to different constraints such that:

$$G^i_{nd}(\boldsymbol{S}_{nd}) = 0 \to \frac{\partial G^i_{nd}}{\partial \boldsymbol{S}_{nd}}\dot{\boldsymbol{S}}_{nd} = 0 \quad \text{and} \quad G^i_d(\boldsymbol{S}_d) = 0 \to \frac{\partial G^i_d}{\partial \boldsymbol{S}_d}\dot{\boldsymbol{S}}_d = 0 \quad i = 1 \ldots n \tag{3}$$

Equations (1) and (3) can now be combined through partitioning of the individual elements:

$$\left\{\begin{array}{c}\dot{\boldsymbol{\varepsilon}}_{nd}\\ --\\ \boldsymbol{0}_{n,1}\end{array}\right\} = \left[\begin{array}{c}\boldsymbol{L}_{nd}\\ --\\ \frac{\partial G^i_{nd}}{\partial \boldsymbol{S}_{nd}}\end{array}\right]\dot{\boldsymbol{S}}_{nd}, \quad \left\{\begin{array}{c}\dot{\boldsymbol{\varepsilon}}_{d}\\ --\\ \boldsymbol{0}_{n,1}\end{array}\right\} = \left[\begin{array}{c}\boldsymbol{L}_{d}\\ --\\ \frac{\partial G^i_{d}}{\partial \boldsymbol{S}_{d}}\end{array}\right]\dot{\boldsymbol{S}}_{d} \tag{4}$$

which now express $\mathbb{R}^p \to \mathbb{R}^p$ transformations. The symbol $\boldsymbol{0}_{n,1}$ in Eq. (4) represents the zero vector in $\mathbb{R}^n$. Similar expressions can be derived for stress increments.

Assuming that the inverse of each of the transformation matrices in Eq. (4) exists, stress and strain increments for the two mechanisms can be related to each other, i.e.

$$\dot{\bar{\boldsymbol{\sigma}}}_{nd} = \bar{\boldsymbol{D}}_{nd}\,\dot{\bar{\boldsymbol{\varepsilon}}}_{nd}, \quad \dot{\bar{\boldsymbol{\sigma}}}_{d} = \bar{\boldsymbol{D}}_{d}\,\dot{\bar{\boldsymbol{\varepsilon}}}_{d} \tag{5}$$

with

$$\dot{\bar{\boldsymbol{\sigma}}}_{nd} = \begin{Bmatrix} \dot{\boldsymbol{\sigma}}_{nd} \\ -- \\ \boldsymbol{0}_{n,1} \end{Bmatrix}, \quad \dot{\bar{\boldsymbol{\sigma}}}_{d} = \begin{Bmatrix} \dot{\boldsymbol{\sigma}}_{d} \\ -- \\ \boldsymbol{0}_{n,1} \end{Bmatrix}, \quad \dot{\bar{\boldsymbol{\varepsilon}}}_{nd} = \begin{Bmatrix} \dot{\boldsymbol{\varepsilon}}_{nd} \\ -- \\ \boldsymbol{0}_{n,1} \end{Bmatrix}, \quad \dot{\bar{\boldsymbol{\varepsilon}}}_{d} = \begin{Bmatrix} \dot{\boldsymbol{\varepsilon}}_{d} \\ -- \\ \boldsymbol{0}_{n,1} \end{Bmatrix}$$
$$\bar{\boldsymbol{D}}_{nd} = \begin{bmatrix} \boldsymbol{K} \\ -- \\ \dfrac{\partial G^{i}_{nd}}{\partial \boldsymbol{S}_{nd}} \end{bmatrix} \begin{bmatrix} \boldsymbol{L}_{nd} \\ -- \\ \dfrac{\partial G^{i}_{nd}}{\partial \boldsymbol{S}_{nd}} \end{bmatrix}^{-1}, \quad \bar{\boldsymbol{D}}_{d} = \begin{bmatrix} \boldsymbol{K} \\ -- \\ \dfrac{\partial G^{i}_{d}}{\partial \boldsymbol{S}_{d}} \end{bmatrix} \begin{bmatrix} \boldsymbol{L}_{d} \\ -- \\ \dfrac{\partial G^{i}_{d}}{\partial \boldsymbol{S}_{d}} \end{bmatrix}^{-1} \tag{6}$$

We further assume that the contributions of the different mechanisms to strain are controlled by a parameter α which is a function of the distance to SES [4], i.e.

$$\dot{\boldsymbol{\varepsilon}}_{nd} = \alpha\dot{\boldsymbol{\varepsilon}}, \quad \text{and} \quad \dot{\boldsymbol{\varepsilon}}_{d} = (1-\alpha)\dot{\boldsymbol{\varepsilon}} \tag{7}$$

As such, the two expressions in Eq. (5) can be combined to give:

$$\dot{\bar{\boldsymbol{\sigma}}} = \bar{\boldsymbol{D}}_{ep}\,\dot{\bar{\boldsymbol{\varepsilon}}}, \quad \bar{\boldsymbol{D}}_{ep} = \alpha\bar{\boldsymbol{D}}_{nd} + (1-\alpha)\bar{\boldsymbol{D}}_{d} \tag{8}$$

This represents a $\mathbb{R}^{p} \rightarrow \mathbb{R}^{p}$ relation which can be reduced to $\mathbb{R}^{m} \rightarrow \mathbb{R}^{m}$ recalling that last $p-m$ components in both $\dot{\bar{\boldsymbol{\sigma}}}$ and $\dot{\bar{\boldsymbol{\varepsilon}}}$ are zero. The change in statistical variables can also be found as:

$$\dot{\boldsymbol{S}}_{nd} = \begin{bmatrix} \boldsymbol{L}_{nd} \\ -- \\ \dfrac{\partial G^{i}_{nd}}{\partial \boldsymbol{S}_{nd}} \end{bmatrix}^{-1} \alpha\dot{\boldsymbol{\varepsilon}}, \quad \dot{\boldsymbol{S}}_{d} = \begin{bmatrix} \boldsymbol{L}_{d} \\ -- \\ \dfrac{\partial G^{i}_{d}}{\partial \boldsymbol{S}_{d}} \end{bmatrix}^{-1} (1-\alpha)\dot{\boldsymbol{\varepsilon}} \tag{9}$$

For the simple case of proportional loading in 2D, the statistical description is chosen to include coordination number, fabric anisotropy, average normal contact force, normal force anisotropy, and tangential force anisotropy, $\boldsymbol{S} = \{z, a_c, f_n, a_n, a_t\}^T$. The conversion matrix for stress can be found from the expressions in [1] to be:

$$\boldsymbol{K} = \frac{r}{4v}\begin{bmatrix} 2f_n & 0 & 2z & 0 & 0 \\ f_n(a_c + a_n + a_t) & zf_n & z(a_c + a_n + a_t) & zf_n & zf_n \end{bmatrix} \tag{10}$$

The matrix $\boldsymbol{L}_{nd}$ can be computed from any elastic constitutive matrix that expresses elastic strain components in terms of microvariables. Herein, we use the model developed by the authors in [5]. The expression for $\boldsymbol{L}_{d}$ in terms of coordination number and fabric anisotropy can also be found in [6, 7].

Based on the framework described above, three consistency relations for each mechanism are needed to close the loop of equations. For dissipative mechanisms, the first relation comes from the fact that macrovariables evolution should be contained within SES surface [4]. The second and third relations come from the observation that, beyond the peak stress, the curves of $a_n(1 + a_c)/(1 - ac)$ versus q/p and $a_n^2(1 - a_c)/(1 + ac)$ versus a_t assume the same form beyond the peak strength.

Similar relations are observed for non-dissipative mechanisms: The initial z versus a_c variations during the contact loss and gain regime have been modeled analytically in [3]. Furthermore, the initial variations of a_n versus q/p and $a_n(1 - a_c)/(1 + ac)$ versus a_t are observed to remain unchanged for different loading paths.

The final ingredient of the constitutive model is the proportionality of the two mechanisms, α, which is assumed to be a linear function decreasing from 1 to zero in relation to the distance to SES [4]. By substituting these variables into the constitutive framework, a full constitutive model is obtained whose accuracy has been verified, though not included in this paper, by comparing benchmark tests with DEM simulations.

3 Finite Elements Implementation and Discussion

The developed micromechanical constitutive model has been implemented into FEM to simulate the response of an initially dense sample under biaxial loading condition, with special attention given to the instability characteristics beyond stress peak. By providing access to salient microvariables such as coordination number and fabric anisotropy, the constitutive model enables us, probably for the first time, to virtually look into the evolution of microstructure for different failure modes such as diffuse and localized. Figure 1 shows the stress-strain and volumetric response of the sample where a clear bifurcation emerges shortly after the peak stress. From a micromechanical point of view, the bifurcation takes place exactly when the SES surface is

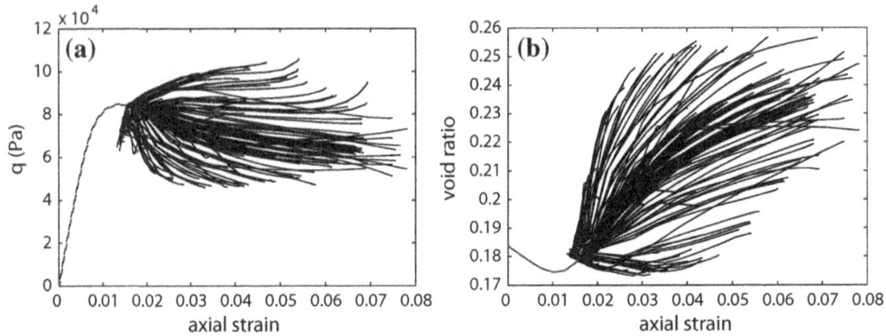

Fig. 1 FEM simulation of a biaxial test. Bifurcation of **a** deviatoric stress, and **b** void ratio are shown for different Gauss points

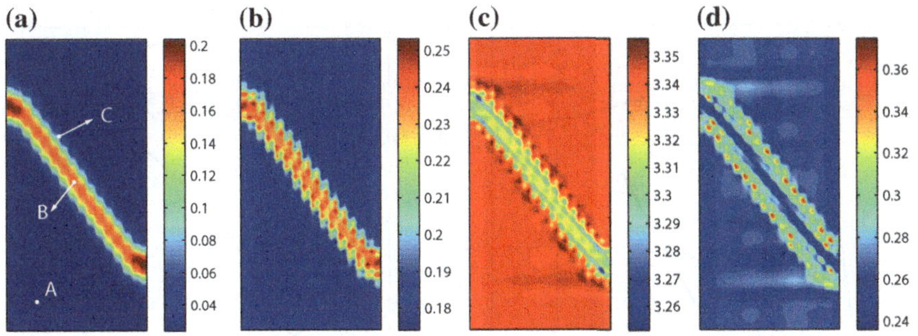

Fig. 2 Localized fields of: **a** deviatoric strain, **b** void ratio, **c** coordination number, and **d** fabric anisotropy, at global axial strain of 2%. The representative points are identified on (**a**)

reached; a state which has been previously seen to be connected to the isostatic limit of the granular assembly [4].

A typical localized mode of the failure is confirmed in Fig. 2a and b with both the deviatoric and volumetric strains localized within a shear band. Figure 2c and d illustrate the remarkable advantage of such a purely micromechanical constitutive model displaying the evolution of microvariables and the important role it can play in instability analysis. Interestingly, the fields of coordination number and fabric anisotropy also follow the localization pattern of strain in general, while the fabric anisotropy decreases to its critical value only within the shear band which is in accordance with previous DEM studies [6]. Compared to the strain localization pattern in Fig. 2a and b, the fields of coordination number and fabric anisotropy show a wider transition zone around the shear band that hints towards a different loading path for these point. This is further investigated by probing the response of three representative points inside (A), outside (B), and on the boundary (C) of the shear band.

Figure 3 shows that, for a material point inside the shear band (point B), the stress-strain response as well as the microsructure steadily tend toward the critical state, while the material outside the shear band (point A) undergoes unloading. At the boundaries of the shear band (point C), however, the material experiences a transition in softening rate, as more clearly seen in Fig. 3a and b. Despite the decrease in deviatoric stress at these boundaries, the local microvariables do not tend towards the critical state, but instead move along the SES surface [4] with slight change in fabric anisotropy. As such, the drop in deviatoric stress at the boundaries of shear band is mainly due to the drop in mean stress, p, and hence decrease of average interparticle force.

It should be noticed that similar value of fabric anisotropy inside and outside the shear band in Fig. 2d is merely a coincidence. The reason is that the loading process has stopped at global axial strain of 2% resulting in partial unloading at point A, as evidenced in Fig. 3a. Indeed, the trends in Fig. 3d indicate that the decrease in fabric anisotropy at points A and B are due to different mechanisms and had the

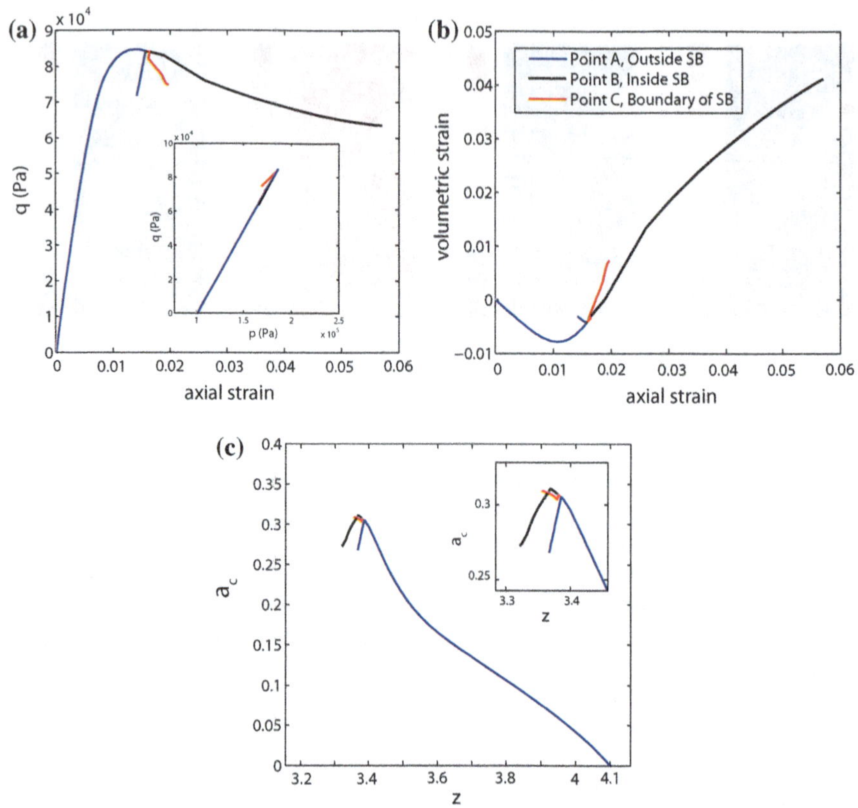

Fig. 3 Stress-strain and microstructural response of the material points inside (*A*), outside (*B*), and on the boundary (*C*) of the shear band (see Fig. 2). Graphs show: **a** stress-strain and stress path, **b** volumetric response, and **c** the evolution of fabric anisotropy versus coordination number

loading been continued, the fabric anisotropy outside the shear band would have kept decreasing, as opposed to within the shear band where the fabric anisotropy would asymptotically tend to the critical state value.

4 Conclusions

The study presents a novel multiscale framework for formulating a constitutive model for granular materials based on purely micromechanical considerations. The expressions for global stress and strain in terms of microvariables such as coordination number and fabric anisotropy, together with consistency relations describing the interrelated evolutions of microvariables have been used to arrive at a rate-independent incrementally linear constitutive model. A simple version of such models has been developed for the case of 2D granular media under proportional loading conditions, and has been subsequently implemented into a finite elements code to simulate a biaxial test. Being predicated on microstructural descriptions, the model provide the opportunity to investigate instability characteristics of granular media from a micromechanical point of view. A post-peak bifurcation is shown to take place only when the state of material reaches the "stable evolution state" surface [4]. The distribution of coordination number and fabric anisotropy within the sample follow the strain localization pattern with the microstructure approaching the critical state only inside the shear band. A transition zone was identified at boundaries of the localized zone where rate of softening changes under combined stress continuity and strain discontinuity condition. Interestingly, the microstructural characteristics are found not to approach the critical state within this transition zone, but instead move along the SES surface as the mean stress decreases locally.

Acknowledgements This work was funded by the Natural Sciences and Engineering Research Council of Canada and Foundation Computer Modelling Group.

References

1. Rothenburg, L., Bathurst, R. J.: Analytical study of induced anisotropy in idealized granular materials. Geotechnique **39**(4), 601–614 (1989)
2. Rothenburg, L., Bathurst, R. J., Dusseault, M. B.: Micromechanical ideas in constitutive modelling of granular materials. Powders and Grains, Biarez, Gourves (eds.) (1989)
3. Pouragha, M., Wan, R.: Onset of structural evolution in granular materials as a redundancy problem. Granular Matter, **18**(3), 1–13 (2016)
4. Pouragha, M., Wan, R.: On elastic deformations and decomposition of strain in granular media (in preparation)
5. Pouragha, M., Wan, R.: Strain in granular media: probabilistic approach to Dirichlet tessellation. Journal of Engineering Mechanics, **143**(1), (2017) C4016002
6. Wan, R., Pouragha, M.: Fabric and connectivity as field descriptors for deformations in granular media. Continuum Mechanics and Thermodynamics, **27**(1–2), 243 (2015)
7. Pouragha, M., Wan, R.: Non-dissipative structural evolutions in granular materials within the small strain range. International Journal of Solids and Structures, **110**, 94–105 (2017)

Part VII
Hans Muhlhaus Tribute

Interplay Between Symmetry and Anti-symmetry in the Evolution of Localized Buckle Patterns

M.K. Paesold, T.J. Dodwell and G.W. Hunt

Abstract A model of geological folding comprising a thin elastic beam supported by a nonlinear viscoelastic (Kelvin-Voigt) material is subjected to a slow rate of applied end-shortening. The description reduces to the nonlinear Swift-Hohenberg partial differential equation (PDE), supplemented by a constraint condition. A modified one degree-of-freedom Galerkin description is introduced, built by adopting the evolving modeshapes of the corresponding statical equilibria at the same state of compression. An *evolutionary energy landscape* is described, formed by plotting total potential energy against the single degree of freedom and the end-shortening. Comparisons of the reduced system with numerical solutions of the full PDE are found to be in good qualitative agreement for slow rates of applied end-shortening.

1 Introduction

Following the pioneering work of Biot [1], folded buckle patterns in structural geology have traditionally been assumed to be periodic. A beam in viscous medium, for example, would be expected to display the *dominant wavelength*, suggested by linear analysis as that growing most rapidly over time. In contrast, some elastic buckling problems display spatially localised solutions, driven by material and geometric nonlinearities [2–4]. In this, much progress has developed around stability of a compressed beam supported by a nonlinear elastic foundation, with carefully selected

M.K. Paesold
Qintiq, 417 St Kilda Road, Melbourne, VIC 3004, Australia
e-mail: martin.paesold@gmail.com

T.J. Dodwell
College of Engineering, Mathematics & Physical Sciences, University of Exeter, Exeter, Devon EX4 4PY, UK
e-mail: T.Dodwell@exeter.ac.uk

G.W. Hunt (✉)
Department of Mechanical Engineering, University of Bath, Bath BA2 7AY, UK
e-mail: g.w.hunt@bath.ac.uk

E. Papamichos et al. (eds.), *Bifurcation and Degradation of Geomaterials with Engineering Applications*, Springer Series in Geomechanics and Geoengineering, DOI 10.1007/978-3-319-56397-8_65

properties to mimic either softening and/or stiffening characteristics in the supporting medium. Synergy between these two separate developments then naturally led to study of viscoelastic systems with the same nonlinearities [5].

Here, a beam supported by a nonlinear viscoelastic (Kelvin-Voigt) material under slow end-compression is described by the nonlinear Swift-Hohenberg PDE, together with a constraint condition. The foundation has an elastic part that first softens and then re-stiffens. This reflects many purely elastic formulations [6–9], where the equilibrium states are known to comprise two alternative forms of localised *snaking* path, one spatially symmetric and the other anti-symmetric, that emerge from the flat state at a critical bifurcation point. These paths are generally disconnected, but are linked by *ladders* of non-symmetric equilibria, bringing with them regions of bi-stability. For the dynamical PDE, interest then naturally falls on the transitions between these two attractors as the system evolves.

To explore the qualitative aspects further, a modified one degree-of-freedom Galerkin description is introduced, built on the evolving modeshapes of the statical equilibrium states. Total potential energy is then plotted against the single degree of freedom and the end-shortening to provide an *evolutionary energy landscape*. Movement on the surface, governed by a viscous gradient-flow process in one dimension and a constant flow rate in the other, describes the dynamics. Changing patterns of minima and maxima in the (constrained) energy then bring about possibilities for dynamical bifurcation and related reversals in the direction of flow.

2 The Model

We consider the model for dynamic localized folding seen in Fig. 1, comprising an infinitely long inextensible elastic beam, supported by a nonlinear viscoelastic foundation and subjected to end-shortening at a constant rate R. Deformation is characterised by vertical displacement of the centreline $u(x, t)$, where x is arc-length measured along the beam, and $t > 0$ is time. The shortening at time t is given by

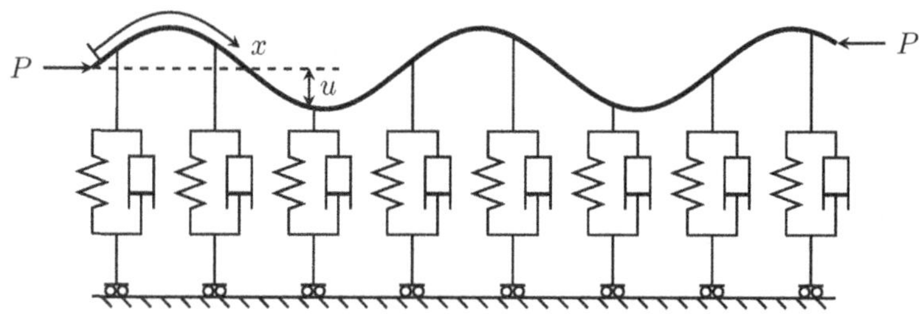

Fig. 1 Strut supported by nonlinear springs and linear dashpots in parallel. Axial load P would normally be accompanied by bending moments and shear forces at points of application (not shown)

$$\Delta = \frac{1}{2}\int_{-\infty}^{\infty} u'^{\,2}\,\mathrm{d}x = Rt, \tag{1}$$

to first order, where primes denote differentiation with respect to x. For moderately-large deflections the bending energy is $\frac{1}{2}B\int u''^2\,\mathrm{d}x$, where B is the beam bending stiffness. The Winkler foundation, comprising a nonlinear spring and linear dashpot in parallel, provides a strictly local and vertical total resistive force,

$$f(u) = f_e(u) + f_v(u) = k_1 u - k_2 u^3 + k_3 u^5 + \eta \dot{u}, \tag{2}$$

where $(\dot{\ })$ denotes differentiation with respect to time t. The negative cubic and positive quintic coefficients mean that, as displacements into the foundation grow, the resisting stiffness initially drops and then increases [7].

We introduce the Lagrangian

$$\mathscr{L}(x,t,u) = \frac{1}{2}Bu''^2 + \frac{1}{2}k_1u^2 - \frac{1}{4}k_2u^4 + \frac{1}{6}k_3u^6 + \frac{1}{2}\eta\dot{u}u - \frac{1}{2}Pu'^2. \tag{3}$$

Elastic strain energy at time t is given by the functional

$$E(t,u) = \int_{-\infty}^{\infty} \frac{1}{2}Bu''^2 + \frac{1}{2}k_1u^2 - \frac{1}{4}k_2u^4 + \frac{1}{6}k_3u^6\,\mathrm{d}x, \tag{4}$$

whereas

$$D(t,u) = \frac{1}{2}\eta\int_0^t\int_{-\infty}^{\infty} \dot{u}u\,\mathrm{d}x\,\mathrm{d}s, \quad \text{and} \quad W(t,u) = \frac{1}{2}\int_0^t\int_{-\infty}^{\infty} P(s)u'^2\,\mathrm{d}x\,\mathrm{d}s, \tag{5}$$

are the energy dissipated in the dashpot and the work done by the load up to time t, respectively. Evolution of the system is described by the Euler-Lagrange equation,

$$\frac{\partial\mathscr{L}}{\partial u} - \frac{\partial}{\partial t}\left(\frac{\partial\mathscr{L}}{\partial\dot{u}}\right) + \frac{\partial^2}{\partial x^2}\left(\frac{\partial\mathscr{L}}{\partial u''}\right) - \frac{\partial}{\partial x}\left(\frac{\partial\mathscr{L}}{\partial u'}\right) = 0, \tag{6}$$

subject to appropriate initial, essential and natural boundary conditions. This can be physically interpreted as a vertical force balance on $\mathrm{d}x$, at time t. Appropriate differentiation of (3) yields the constrained nonlinear fourth-order PDE,

$$Bu'''' + Pu'' + k_1u - k_2u^3 + k_3u^5 + \eta\dot{u} = 0, \quad \text{subject to} \quad \frac{1}{2}\int_{-\infty}^{\infty} u'^2\,\mathrm{d}x = Rt. \tag{7}$$

We can reduce this to the non-dimensional form,

$$\dot{u} = -\left(u'''' + pu'' + u - u^3 + \alpha u^5\right) \quad \text{and} \quad \frac{1}{2}\int_{-\infty}^{\infty} u'^{\,2}\,\mathrm{d}x = \rho t, \tag{8}$$

by the following transformation: $x \mapsto (B/k_1)^{1/4}x$; $t \mapsto (\eta/k_1)t$; $u \mapsto \sqrt{k_1/k_2}u$: where $p(t) = P(t)/\sqrt{Bk_1}$; $\alpha = k_1k_3/k_2^2$; and $\rho = R\eta k_2/k_1^2\sqrt{B/k_1}$. We note that the final system is in fact a two parametric group in (α, ρ), as the load $p(t)$ is a free variable directly imposed by the shortening constraint.

3 Static Equilibrium—Snakes and Ladders

Before seeking solutions to PDE (8), it is useful to review the associated stationary states, using the fourth-order reversible ordinary differential equation (ODE) in x:

$$u'''' + pu'' + u - u^3 + \alpha u^5 = 0. \tag{9}$$

with the same 'cubic-quintic' foundation force characteristic: $f_e(u) = u - u^3 + \alpha u^5$. This is known to exhibit a Hamiltonian-Hopf bifurcation from the unbuckled state into a periodic buckling mode at $p = 2$ [3, 4]. We compute such solutions in AUTO [10] over $x \in [-L, L]$, starting arc-length continuation close to the bifurcation point and seeding it with the eigenmode in two configurations, one symmetric about $x = 0$ (cosine) and the other anti-symmetric (sine). For convenience, pinned ($u = u'' = 0$) rather than homoclinic [11] boundary conditions are chosen, the difference being numerically inconsequential for a long beam [12]. End-shortening Δ is chosen as the continuation parameter, with the load p in Eq. (9) being regarded as free.

Figure 2 shows a typical load–end-shortening bifurcation diagram and corresponding solution shapes. This shows the classic *snakes-and-ladders* scenario of a pair of *snaking* equilibrium solutions, one symmetric in x about its mid-point and the other anti-symmetric, each fluctuating in load as Δ increases. The equilibrium shapes themselves are all homoclinic, with a central region that grows with Δ; these are the initial stages of a heteroclinic connection to a periodic state at the Maxwell load [7]. Along the snakes the folded profiles are each a function of both x and Δ, so they change shape but also grow in amplitude as Δ increases. The snaking paths are connected at bifurcation points by *ladders* [8, 9], comprising states of transition

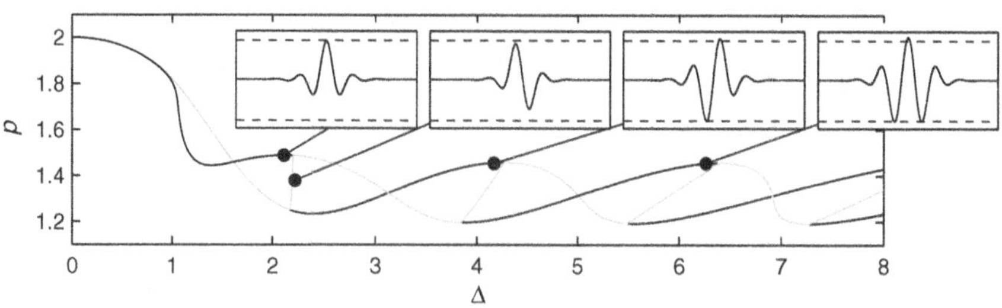

Fig. 2 Snakes and ladders equilibrium paths for $\alpha = 0.3$. Stable and unstable states are shown as *black* and *grey* respectively

between symmetry and anti-symmetry; for a recent account of such behaviour see [13]. Unlike the response under controlled load where limit points have a part to play, stability here is only lost or gained at the bifurcation points.

4 Evolution of Transient Folding Patterns

4.1 Finite Element (FE) Procedure for Constrained Gradient Flow

Equation (8) can be solved as a constrained gradient flow problem over a large-but-finite domain $X := [-L, L]$, discretized into N nodes $x_i = ih - L$ where $h = 2L/N$ and $i = 0, 1, \ldots, N$. The weak form is found by multiplying by a suitable test function v, integrated over the domain X and then by parts, to give

$$\int_X \dot{u}v\,\mathrm{d}x = -\int_X u''v''\,\mathrm{d}x + p\int_X u'v'\,\mathrm{d}x - \int_X f_e(u)v\,\mathrm{d}x \quad \text{such that} \quad \frac{1}{2}\int_X u'^2\,\mathrm{d}x = \rho t. \quad (10)$$

By approximating u and v as piecewise cubic functions spanned by the cubic FE shape functions $\phi_i(x)$, this constrained time-dependent variational equation converts to a system of differential algebraic equations (DAEs) of index-1 given by

$$\begin{bmatrix} \underline{U}^T C & 0 \\ A & 0 \end{bmatrix} \begin{bmatrix} \underline{\dot{U}} \\ \dot{p} \end{bmatrix} = \begin{bmatrix} R \\ -(B - pC)\underline{U} - \underline{D} \end{bmatrix}. \quad (11)$$

Here the matrices are defined as follows

$$A_{ij} = \int_X \phi_i\phi_j\,\mathrm{d}x, \quad B_{ij} = \int_X \phi_i''\phi_j''\,\mathrm{d}x, \quad C_{ij} = \int_X \phi_i'\phi_j'\,\mathrm{d}x, \text{ and } D_i = \int_X f_e(\underline{U})\phi_i\,\mathrm{d}x. \quad (12)$$

where the nonlinear functional defined by D_i is computed using a high order Gaussian quadrature rule. To produce an index 1 DAE, we differentiate the constraint equation in time and impose the constraint as $\int_X u'u'_T\,\mathrm{d}x = \rho$. Full details of the FE procedure can be found for a similar problem in [14, Sect. 6.1].

Up to now the formulation is general, without consideration of boundary conditions. We impose homoclinic boundary conditions at each end of the finite domain by setting end nodal values (u_1, u_1', u_N, u_N'), such that the linearisation of (6) i.e.

$$u(x) = A_1 e^{\xi_1 x} + A_2 e^{-\xi_1 x} + A_3 e^{\xi_2 x} + A_4 e^{-\xi_2 x} \quad (13)$$

is satisfied in the homoclinic tails [11]. Here ξ_1 and ξ_2 are the characteristics of the linear equation $\xi^4 + p\xi^2 + 1 = 0$. As $x \to -\infty$, solutions must remain bounded and therefore $A_2 = A_4 = 0$. This leads to the *outset* condition,

$$u_1 = A_1 e^{-\xi_1 L} + A_3 e^{-\xi_2 L} \quad \text{and} \quad u_1' = A_1 \xi_1 e^{-\xi_1 L} + A_3 \xi_2 e^{-\xi_2 L}. \tag{14}$$

Solving for A_1 and A_3 enables solutions for u_1 and u_1' to be computed. The process is repeated at $x = x_N = L$, where the *inset* condition ($A_1 = A_3 = 0$) is imposed.

4.2 Numerical Experiments

The dynamic behaviour of (11) under different rates of applied end-shortening ρ is next investigated. An error convergence study indicates that a uniform FE mesh with $N = 70$ over length $L = 30$ is suitably accurate and efficient. The quintic coefficient of (8) was taken as $\alpha = 0.3$, allowing comparisons with the stationary solutions of Fig. 2 as computed in Sect. 3. Figure 3 shows a series of numerical solutions for increasing rates of loading from top to bottom: $\rho = 10^{-4}$, 10^{-3} and 10^{-2} respectively. To avoid the system becoming locked in the trivial equilibrium state, all runs are seeded by an incremental displacement into a symmetric localised shape.

First we make some general observations:

- For rates of end-shortening $\rho < 10^{-4}$ say, the system is *quasi-static*. Jumps between near-equilibrium states occur immediately or soon after stability is lost.
- Increasing ρ mean that solutions tend to drift from the static state, and can also delay the unstable jumps so they occur with increasing as well as decreasing load.
- Behaviour at high rates is dominated by the dynamics. Significantly, a high rate can lead to a jump being by-passed, so the system can remain in a symmetric or anti-symmetric state even when its stationary counterpart has become unstable.

Figure 3 suggests a change in rate causes a dynamical jump to drift, but it is also possible for it to change suddenly at a dynamical bifurcation. Figure 4 shows solution paths for two rates, 10^{-7} apart, found via a root searching algorithm, bounding the critical rate $\rho^* \simeq 1.6135 \times 10^{-4}$. This is explored phenomenologically later in Fig. 6.

5 Evolutionary Galerkin Procedure and Energy Landscape

To help interpret the dynamics it is useful to model the system with just a single degree of freedom. We thus decompose the description into linear combinations of the symmetric and anti-symmetric stationary modes at the given $\Delta = \rho t$, as follows,

$$\tilde{u}(x,t) = q_s(t)\psi_s(x,t) + q_a(t)\psi_a(x,t). \tag{15}$$

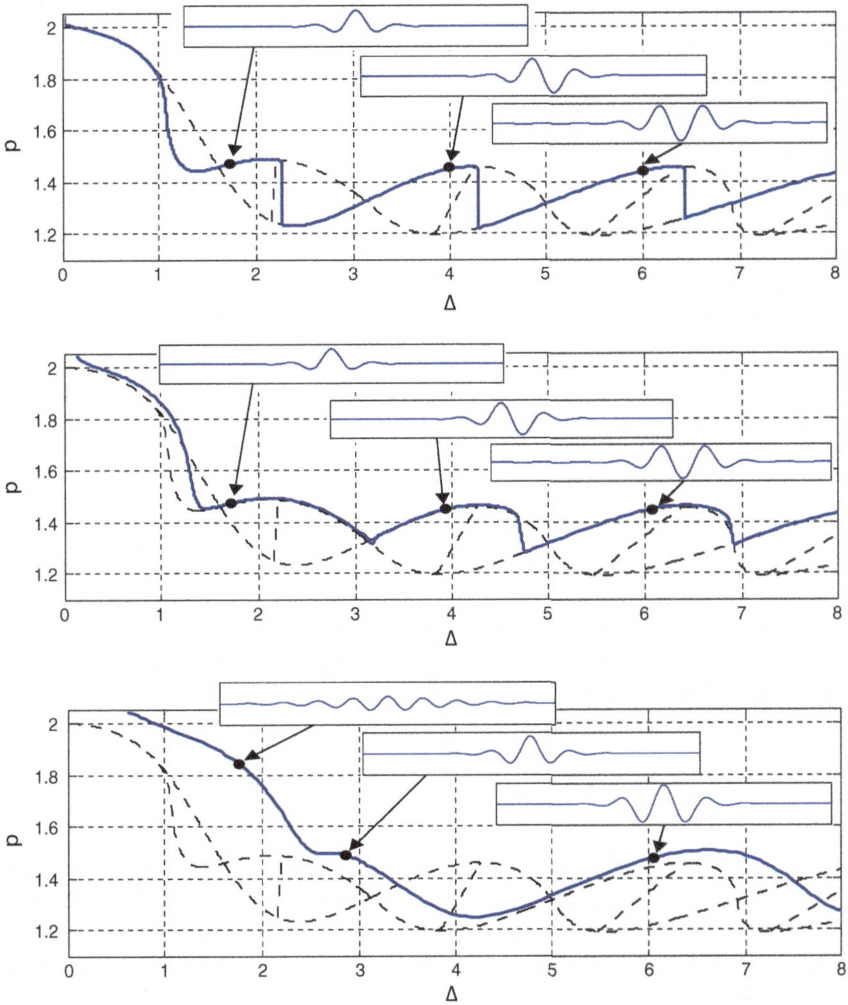

Fig. 3 Load p against end-shortening Δ at rates $\rho = 10^{-4}$ (*top*), 10^{-3} (*middle*), 10^{-2} (*bottom*), compared against stationary solutions of Fig. 2. *Inset plots* show solutions profiles at positions indicated

where ψ_i are the mode shapes of Fig. 2, with amplitudes q_i. End-shortening is now,

$$\Delta = \rho t = \frac{1}{2}\int_X \tilde{u}'^2\,\mathrm{d}x = \frac{1}{2}q_s^2\int_X \psi_s'^2\,\mathrm{d}x + \frac{1}{2}q_a^2\int_X \psi_a'^2\,\mathrm{d}x \tag{16}$$

at time t, where X is the (long) domain over which the modeshapes are computed. Since $\frac{1}{2}\int \psi_s'^2\mathrm{d}x$ and $\frac{1}{2}\int \psi_a'^2\mathrm{d}x$ compute to the same value independently, q_s and

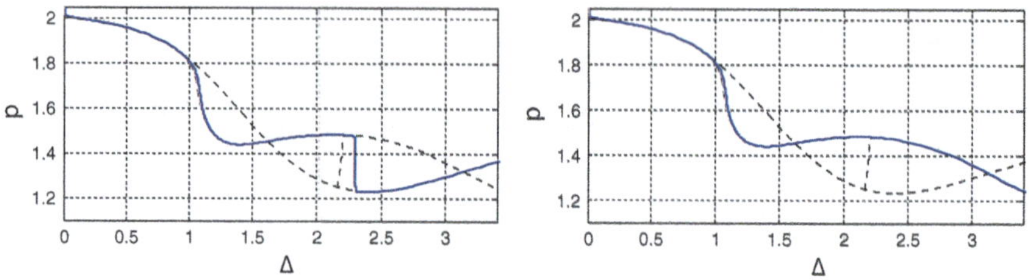

Fig. 4 Numerical evidence of a dynamical bifurcation at a critical rate $\rho^* \simeq 1.6135 \times 10^{-4}$, at which solution branching occurs at $\Delta^* \simeq 2.274$. *Plots* shows Δ against p of the dynamic paths *blue* (*solid*) and the primary stationary states (*thin-dashed*) for $\rho = 1.613 \times 10^{-4}$ (*left*); $\rho = 1.614 \times 10^{-4}$ (*right*)

q_a lie on the unit circle. Thus $q_a = \sqrt{1 - q_s^2}$ and the system has a single degree of freedom. Energy levels can then be computed directly from (4).

Substituting (15) into (8) gives

$$\dot{q}_s\psi_s + q_s\dot{\psi}_s + \dot{q}_a\psi_a + q_a\dot{\psi}_a = -\big(q_s\psi_s'''' + q_a\psi_a'''' + p(q_s\psi_s'' + q_a\psi_a'') + f_e(q_a\psi_a + q_s\psi_s)\big). \tag{17}$$

Evolution equations for the q_i are found by multiplying by the corresponding ψ_i and integrating over the domain X. These, along with the constraint equation, lead to the following governing equations, written in matrix form,

$$\begin{bmatrix} A & 0 \\ 2t\underline{Q}^T & 0 \end{bmatrix} \begin{bmatrix} \underline{\dot{Q}} \\ \dot{p} \end{bmatrix} = -\begin{bmatrix} (B - pC)\underline{Q} + \underline{D} + \frac{1}{2}\dot{A}\underline{Q} \\ \underline{Q}^T\underline{Q} - 1 \end{bmatrix} \tag{18}$$

where $\underline{Q} = [q_s;\ q_a]$. A, B, C and D are defined by (12) but with the ϕ_i replaced by ψ_i. Load p is fixed by the constraint condition (16) which, after differentiating with respect to time yields the bottom row in (18). We note, in contrast to the FE formulation, the appearance of momentum-like terms due to variations of the shape functions with time.

The constrained gradient flow equation (8) is thus reduced to three DAEs of index-1, which again can be solved using MATLAB's inbuilt function `ode23s`. To avoid the system sitting unrealistically on an energy maximum, zero values of q_s or q_a are replaced by the small non-zero quantity $+10^{-12}$, the positive sign ensuring that solutions are all found in the first quadrant of the unit circle. Typical dynamical outcomes are seen in Fig. 6 later.

Figure 5 shows contours of the change in energy from the symmetric state, as computed from (4), at different but constant values of Δ. This is plotted against the polar angle θ on the unit circle ($q_s = \cos\theta$, $q_a = \sin\theta$), with the darker regions representing higher energies. Sections at constant Δ are shown on the right. For small Δ, as seen in the bottom and middle slices, the symmetric form is preferred while the anti-symmetric form changes from a global maximum to a local minimum. As Δ is increased, the symmetric form first relinquishes its status as the global minimum

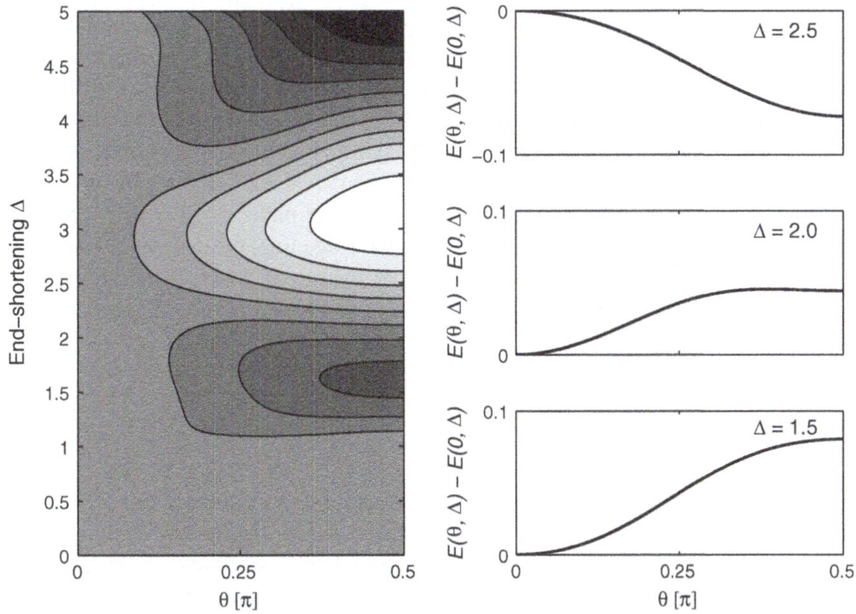

Fig. 5 *Left* energy landscape. *Right* three sections at different Δ values, that embrace a region of bistability and illustrate transfer of stability from symmetry to anti-symmetry

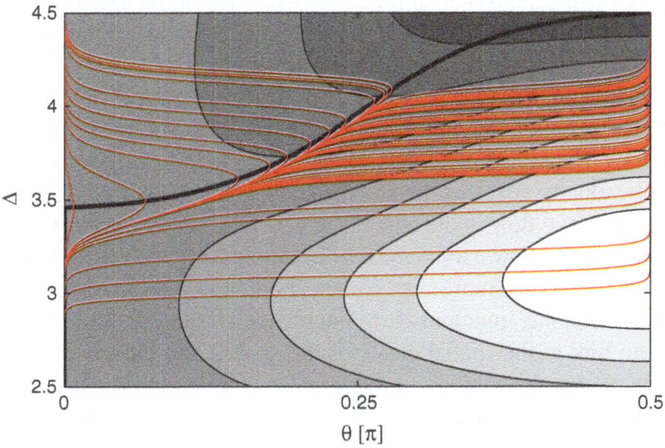

Fig. 6 Dynamical paths from a start close to the symmetric shape at different rates. The ridge line of equilibria associated with the second ladder of Fig. 2 is shown in *black*

(somewhere around $\Delta = 2.18$), and then evolves into a local maximum as seen at the top. When symmetric and anti-symmetric minima coexist the system is bistable, and we note the appearance then of a third equilibrium state in the form of a local maximum; this reflects the first ladder of Fig. 2.

As seen in Figs. 3 and 4, typical dynamical responses of the PDE can be subject to *dynamical bifurcations*. The reduced view of Fig. 5 is useful in describing how this comes about. Figure 6 shows a set of runs at differing rates on the landscape of Fig. 5, all starting from the same position close to the symmetric state. After venturing into the mixed region the runs either return to the symmetric state, or veer off to the anti-symmetric one. The critical rate $\rho^* \simeq 8.2077455491 \times 10^{-4}$ defines the transition, and the difference between the two closest diverging paths is $\Delta\rho = 10^{-15}$. Separation occurs close to, but not coincident with, the ridge line, the difference being due to the (small) velocity component provided by the rate ρ. Starts from closer to the symmetric state bifurcate at a lower critical rate and closer to the ridge line.

6 Concluding Remarks

We have presented two models for the nonlinear visco-elastic system of Fig. 1, a multi degree-of-freedom FE formulation, and a single degree-of-freedom evolutionary Galerkin procedure based on static modal shapes obtained from the path-following routine AUTO [10]. According to the associated dynamical responses, the two models show good qualitative agreement at low rates of end-shortening. The pure symmetric and anti-symmetric snaking equilibrium paths and associated modeshapes are the same, but the ladder equilibrium states of Fig. 2, being based on these pure forms, fail to match precisely. The first ladder extends over the range $2.157 < \Delta < 2.198$ for the FE model, for example, whereas for the Galerkin it covers $1.90 < \Delta < 2.35$. Better accuracy could be obtained with the addition of further shapes derived from the ladders, but this would be at the expense of simplicity.

The single degree-of-freedom Galerkin model is shown to be a useful phenomenological tool. The reduction allows the dynamics to be seen as movement on a 2D surface, with velocity governed by a constant rate of end-shortening on one dimension and visco-elastic flow in the other. The benefit is neatly demonstrated by Fig. 6, which shows visually how a dynamical bifurcation associated with the energy maxima of the ladder equilibrium states comes about, and thus provides the key ingredient for many chaotic situations. It is interesting to see that in this instance there is an intimate link to the spatial chaos of the snakes-and-ladders scenario. We are of course aware that the solution of a nonlinear PDE is not exactly expressible as a linear combination of two stationary states, but the numerical evidence suggests that the stationary states are such strong attractors (in terms of the energy surface) that the process does offer a fair and highly illustrative new interpretation.

References

1. Biot, M.A.: Mechanics of Incremental Deformations. Wiley, New York (1965)
2. Potier-Ferry, M.: Amplitude modulation, phase modulation and localization of buckling patterns. In: Thompson, J.M.T., Hunt, G.W. (eds.) Collapse: The Buckling of Structures in Theory and Practice, pp. 149–159. Cambridge University Press, Cambridge (1983)
3. Hunt, G.W., Bolt, H.M., Thompson, J.M.T.: Structural localization phenomena and the dynamical phase-space analogy. Proc. R. Soc. Lond. A **425**, 245–267 (1989)
4. Champneys, A.R., Toland, J.F.: Bifurcation of a plethora of multi-modal homoclinic orbits fo autonomous Hamiltonian systems. Nonlinearity **6**, 665–772 (1993)
5. Hunt, G.W., Mühlhaus, H.-B., Whiting, A.I.M.: Evolution of localized folding for a thin elastic layer in a softening visco-elastic medium. Pure Appl. Geophys. **146**(2), 229–252 (1996)
6. Woods, P.D., Champneys, A.R.: Heteroclinic tangles in the unfolding of a degenerate Hamiltonian Hopf bifurcation. Phys. D **129**(3–4), 147–170 (1999)
7. Hunt, G.W., Peletier, M.A., Champneys, A.R., Woods, P.D., Ahmer Wadee, M., Budd, C.J., Lord, G.J.: Cellular buckling in long structures. Nonlinear Dyn. **21**(1), 3–29 (2000)
8. Burke, J., Knobloch, E.: On snakes and ladders: localized states in the Swift-Hohenberg equation. Phys. Lett. A **360**(6), 681–688 (2007)
9. Dawes, J.H.P.: The emergence of a coherent structure for coherent structures: localized states in nonlinear systems. Philos. Trans. R. Soc. Lond. A **368**, 3551–3565 (2010)
10. Doedel, E.J., Champney, A.R., Fairgrieve, T.E., Kuznetsov, Yu.K., Sanstede, B., Wang, X.: Auto07: continuation and bifurcation software for ordinary differential equations (with homcont). Tecnical Report, Concordia University (2007)
11. Hunt, G.W., Dodwell, T.J., Hammond, J.: On the nucleation and growth of kink and shear bands. Philos. Trans. R. Soc. Lond. A **371**, 20120431 (2013)
12. Hunt, G.W., Wadee, M.K.: Comparative Lagrangian formulations for localized buckling. Proc. R. Soc. Lond. A **434**, 485–502 (1991)
13. Kao, Hsien-Ching, Beaume, Cédric, Knobloch, Edgar: Spatial localization in heterogeneous systems. Phys. Rev. E **89**, 012903 (2014). Jan
14. Budd, C.J., Chakhchoukh, A.N., Dodwell, T.J., Kuske, R.: Chevron folding patterns and heteroclinic orbits. Phys. D **330**, 32–46 (2016)

Coupled Phenomena and Instability in Fault Zones

Jean Sulem and Ioannis Stefanou

Abstract In the last 20 years, it has been recognized that multiphysics couplings play a major role in earthquake nucleation and seismic slip. On the basis of field observations, laboratory experiments and advanced modeling, the detailed study of coupled processes has brought some new insights in the understanding of earthquake mechanics that are reviewed in this paper.

Keywords Faults · Earthquakes · Multiphysics couplings · Strain localization

1 Introduction

The study of earthquake mechanisms has shown that earthquakes appear to be the result of a frictional instability and occur by sudden slippage along a pre-existing fault or at plate interface. The thickness of the slip zones of major recent earthquakes have been identified in several drilling programs in active faults. All the field observations confirm that slip during individual earthquake is localized along extremely thin slip zones; for example: 1 mm for the principal slip zone (PSZ) in Nojima fault (Kobe earthquake, Japan, 1995, [1]) and few mm for the PSZ in Chelungpu fault (Chi Chi earthquake, Taiwan, 1999, [2]).

The seismic cycle is commonly modelled as a stick-slip mechanism on the basis of a simple conceptual model of a 'spring-slider system' [3] where the spring represents the elasticity of a crustal block under tectonic remote loading in which a narrow fault zone is embedded and sheared. Stick-slip instability is possible as soon as the softening rate of the frictional resistance at the interface between the block and the fault is larger than the stiffness of the spring. Thus a key factor for the

J. Sulem (✉) · I. Stefanou
Laboratoire Navier/CERMES, Ecole des Ponts ParisTech, IFSTTAR, CNRS, Université Paris-Est, 77455 Marne la Vallée, France
e-mail: jean.sulem@enpc.fr

E. Papamichos et al. (eds.), *Bifurcation and Degradation of Geomaterials with Engineering Applications*, Springer Series in Geomechanics and Geoengineering, DOI 10.1007/978-3-319-56397-8_66

nucleation of seismic slip is how the shear resistance of the fault weakens with slip. In the following, we will examine various weakening mechanisms of fault resistance during seismic slip. Considering the extreme thinness of the localized slip zone, it is expected that thermal weakening processes might be of major importance [4, 5].

2 Weakening Mechanisms During Seismic Slip

The fault zone mineralogy plays a major role in the acceleration of the slip process. We generally distinguish between fault gouge for which the friction resistance increases with slip rate as observed in the presence of abundant clay minerals and fault gouge which are velocity weakening as commonly observed in limestone units. However, even if clay rich zones may prevent from sudden seismic rupture because they exhibit velocity strengthening at low slip velocity, they can dramatically weaken at high slip velocity, so that a rupture propagating from depth might not be arrested in clay-rich sediments. The Tohoku earthquake in 2011 in Japan with a magnitude of 9.1 is an example of a large rupture that propagated up to the surface through a fault zone composed primarily of clays and that was believed to be seismically stable.

It has been proposed that slip weakening of seismic faults is caused by thermally-activated processes triggered by localized frictional heating and high temperatures attained in the slip zone [6]. Among the weakening mechanisms that affect the fault resistance, the drop of the friction coefficient with slip and/or slip rate, or with temperature is related to different physical processes. Among them, flash heating at micro-asperity contacts [6, 7] appears to be a general weakening mechanism for most rocks at the early stages of slip, once a critical slip rate in the range of 0.01–0.1 m/s is overcome. Lubrication of the fault plane, in relation with the formation of a macroscopic melt layer or a 'gel-like' layer in wet silica rich fault zones, and the production of nano-particles during slip [8] is also responsible for the dynamic drop in friction. Considering the ubiquitous presence of fluids in fault zones, thermal pressurization of the pore fluid within the fault core by frictional heating reduces the effective normal stress acting on the fault and hence the shear strength [6, 9–11]. Thermo-poro-mechanical couplings due to shear heating can be also associated to chemical effects such as dehydration of minerals or decomposition of carbonates [12–14]. Such reactions induce two competing effects: a direct increase in pore pressure because they these reactions release fluid in the system and a limit in temperature increase because are endothermic so that part of the frictional heat is actually absorbed by the chemical reactions. On the other hand they can form a mineral assemblage stronger (reaction hardening) or weaker (reaction weakening) than the original material. Although quite scarce, the presence of pseudotachylites which are melted rock has been reported in fault zones [1, 15] with intrusion patterns indicating that the melts behaved like a viscous fluid under high pressure that could fracture the rock.

Evidence of the occurrence of thermal decomposition of hydrous minerals during seismic slip with a release of water, and of carbonates with a release of CO_2 has been reported in many field observations of active crustal faults. Observations of thermal decomposition of minerals have also been reported in exhumed faults. For example Collettini et al. [16] have documented an outcrop exhumed from 2 to 3 km depth in the Spoleto thrust fault in Central Italy. A principal slip zone with a thickness of 0.3–1 mm has been identified. The accumulated displacement was estimated between 5 and 10 km. This slip zone shows evidence of calcite decarbonation and illite-smectite dehydration.

3 Modelling Weakening Mechanisms During Seismic Slip

In order to highlight the dominant phenomena that occur during seismic slip, we consider a rapidly deforming and infinitely long layer of thickness h consisting of fluid-saturated material.

The 1D model is justified by the fact that the length scales in the direction parallel to the fault, over which the thermo-poro-mechanical fields vary, are much larger than in the direction normal to it. In this 1D-model the velocity components, $u_x(z, t)$, and $u_z(z, t)$ in the direction parallel and normal to the fault respectively depend only on the time since the onset of slip and of the position z in the direction normal to the band. An overall slip-rate V is imposed in the x-direction, as shown in Fig. 1. Inside such a shear-band the pore pressure p and the temperature T are assumed to be functions only of time t and of the position z in the direction normal to the band.

The governing equations of the system are the balance laws of linear momentum, mass and energy. The balance of linear momentum leads to constant shear and total normal stresses inside the layer. The fluid mass balance equation is the sum of several terms: the diffusion term, the thermal pressurization term, the chemical effect of mineral decomposition corresponding to the mass of the fluid released in the system by the reaction and also the term corresponding to the effect on pore

Fig. 1 Model of a of fault zone as an infinite layer under uniform shear strain rate

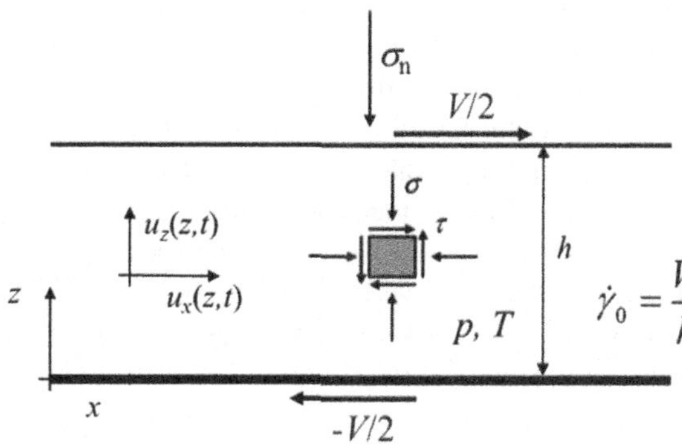

pressure of inelastic porosity change (compaction or dilatancy and also porosity change due to the mineral decomposition):

$$\frac{\partial p}{\partial t} = c_{hy} \frac{\partial^2 p}{\partial x_2^2} + \Lambda \frac{\partial T}{\partial t} + \frac{1}{\rho_f \beta^*} \frac{\partial m_d}{\partial t} - \frac{1}{\beta^*} \frac{\partial n^p}{\partial t} - \frac{1}{\beta^*} \frac{\partial n_d}{\partial t} \tag{1}$$

In Eq. (1), c_{hy} is the hydraulic diffusivity, $\Lambda = n(\lambda_f - \lambda_n)/\beta^*$ is the undrained thermal pressurization coefficient, where λ_f is the pore fluid thermal expansion coefficient and λ_n is the thermal expansion coefficient of the pore volume, and n is the porosity of the rock. β^* is the storage capacity, m_d is the mass per unit volume of the fluid released in the system by mineral decomposition, $\partial n^p/\partial t$ is the rate of inelastic porosity change and $\partial n_d/\partial t$ is the rate of porosity change due to mineral decomposition. Typical values for Λ range from 0.1 to 1 MPa/°C [17].

The energy balance is the sum of the diffusion term, the source term corresponding to the frictional heat and also the term corresponding to the energy consumed in the reaction (with a sign minus here for an endothermic reaction):

$$\frac{\partial T}{\partial t} = c_{th} \frac{\partial^2 T}{\partial x_2^2} + \frac{1}{\rho C} \tau \dot{\gamma}_0^p - \frac{1}{\rho C} \Delta H_T \frac{\partial m_d}{\partial t} \tag{2}$$

In Eq. (2), c_{th} is the thermal conductivity, ρC is the specific heat, $\tau \dot{\gamma}_0^p$ is the plastic work, which is assumed to be entirely converted into heat and ΔH_T is the enthalpy change per unit mass of the dehydration mineral. For endothermic reactions, $\Delta H_T > 0$. The shear stress τ is proportional to the effective Terzaghi stress: $\tau = f(\sigma_n - p_0)$ where f is the friction coefficient. Part of the frictional energy is absorbed in the chemical reaction, which limits the temperature increase due to shear heating. Sulem and Famin [13] have proposed the first theoretical study on the mechanical effect of calcite thermal decomposition ($CaCO_3$ (calcite) → CaO (*lime*) + CO_2 (carbon dioxyde)) on a fault layer of given thickness during seismic slip. The effect of porosity change due to the solid decomposition on the permeability has been accounted for assuming a Carman Kozeny porosity-permeability relationship. Typical results are shown in Fig. 2: (i) the endothermic chemical reaction limits the co-seismic temperature rise to a value which is controlled by the reaction kinetics; (ii) pore pressure increase is first due to thermal pressurization and then accelerates at the onset of the reaction up to a maximum and then decreases due to the reduction of solid volume and of the fluid diffusion process (pore pressure pulse). This results in a dynamic initial weakening of the shear strength followed by a re-strengthening.

Several investigations have shown that the ultracataclastic gouge zones forming the fault core have a much lower permeability (<10^{-19} m^2) than that in the surrounding damage zone [18, 19]. In the above computations, the initial permeability of the slip zone was taken equal to 10^{-19} m^2. The friction coefficient was taken equal to 0.1 in accordance with high velocity friction experiments that show that during coseismic slip the friction coefficient is low [8, 20].

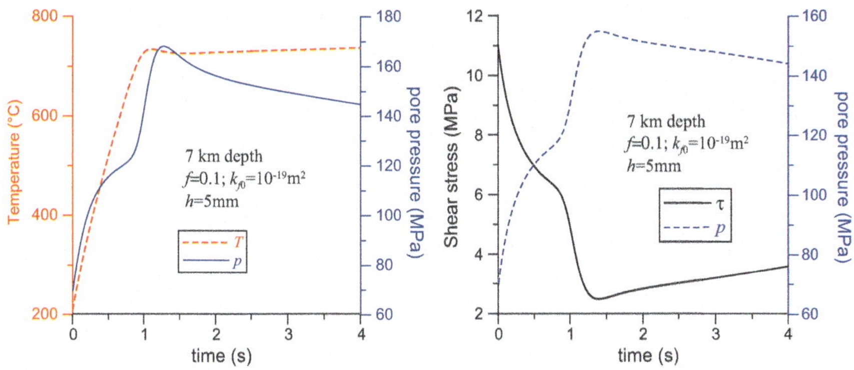

Fig. 2 Effect of thermal decomposition of calcite on a homogeneous layer at 7 km depth ($T_0 = 210\,°C$, $p_0 = 70\,MPa$, $\sigma_n = 180\,MPa$), uniformly sheared at a constant strain rate (imposed slip velocity $V = 1$ m/s)

The observation that endothermic chemical reactions might buffer the temperature of fault zones during seismic slip leads to the conclusion that the activation of other weakening mechanisms such as melt lubrication might be inhibited. High temperature rise limitation by thermal decomposition would provide another explanation to the notorious absence of positive heat flow anomaly on active crustal faults such as San Andreas [9]: a large part of the heat produced by friction would be consumed by endothermic reactions.

Shear heating, thermal pressurization and thermal decomposition of carbonates are weakening mechanisms that can also explain large landslides. The Heart Mountain landslide of northwest Wyoming is the largest known sub-aerial landslide on Earth. This Eocene age landslide covers more than 3000 km^2 with a very long runabout of more than 45 km [21]. An intriguing question is how such a massive volume of rock could have moved across a basal surface with an average regional dip of only 2°. Field observations and experiments on rocks taken from the landslide have shown that since the shear zone of the Heart Mountain slide is located within a dolomite layer, thermal decomposition and release of CO_2 induced by flash heating occurs, allowing a huge upper plate rock to slide over a 'cushion' of pressurized material [22, 23].

4 Chemical Weakening and Earthquake Nucleation

An interesting situation corresponds to shear localization in a chemically weakening material. This case is relevant for understanding the nucleation of intermediate and deep earthquakes within subduction zones. As discussed by Green [24], deep earthquakes have been a paradox since their discovery in the 1920s. The combined increase of pressure and temperature with depth precludes brittle failure or frictional sliding beyond a few tens of kilometers. Nevertheless, earthquakes still occur in subduction zones to 700 km. Growing evidence suggests that the great majority of subduction zone earthquakes shallower than 400 km are initiated by breakdown of hydrous phases and that deeper ones probably initiate as a shearing instability associated with breakdown of metastable olivine to its higher-pressure polymorphs. In the case of serpentinite dehydration, the reaction products may be weaker. This is the case for the dehydration of lizardite for which the produced ultra-fine grained olivine is weaker than the serpentinite aggregates [25]:

$$5\,\underset{\text{lizardite}}{Mg_3Si_2O_5(OH)_4} \rightarrow \underset{\text{talc}}{Mg_3Mg_4Si_4O_{10}(OH)_{10}} + 6\,\underset{\text{olivine}}{Mg_2SiO_4} + 9\,H_2O \tag{3}$$

This leads to a self-lubrication of the fault by the newly formed material. The intrinsic reaction-weakening process assumed here is thus an interesting possibility for dehydration-induced earthquakes at intermediate depths. Brantut and Sulem [26] have assumed a simple evolution law for the friction coefficient f in order to describe chemical weakening during dehydration:

$$f = f_0 + a\ln(\dot{\gamma}/\dot{\gamma}_0) - b\mu \tag{4}$$

where f_0 is a reference friction coefficient for a reference strain rate $\dot{\gamma}_0$ and a and b are positive constitutive parameters. This friction law is similar to the commonly used Rate and Stae Friction (RSF) laws where the 'state' is identified as the reaction extent μ. The reaction rate is assumed to be of first order following an Arrhenius law. It can be linearized close to the critical temperature T_c at which the reaction is triggered:

$$\frac{\partial\mu}{\partial t} \approx \begin{cases} 0 & \text{if } T < T_c \\ c_T(T - T_c) - c_\mu \mu & \text{if } T \geq T_c \end{cases} \tag{5}$$

The linear stability analysis of undrained adiabatic shearing of a layer considering the above dehydration reaction is presented in detail in [26, 27]. It gives the following expression for the critical wavelength (smallest wave length for which the growth coefficient of the instability is positive):

$$\lambda_{cr}^{ch} = 2\pi\sqrt{\frac{ac_{th}}{\dot{\gamma}_0}\frac{\rho C}{b\tau_0}\frac{c_\mu}{c_T}} \tag{6}$$

In order to illustrate this analysis, we use the parameter values given in Table 1. They are taken from [26] for lizardite dehydration of a layer at 30 km depth, which is sheared at a very low strain rate of $10^{-6}\ s^{-1}$. In Fig. 3, the growth coefficient s of the instability mode is plotted in terms of the perturbation wave length. The smallest wave length for which s becomes positive is given by Eq. (6).

Table 1 Parameter values for lizardite dehydration at a depth of around 30 km (from [26])

Quantity	Value
Friction coefficient, f_0	0.6
Rate strengthening parameter, a	0.002
Reaction weakening parameter, b	0.5
Specific heat capacity, ρC	2.7 MPa °C^{-1}
Thermal dependency of the chemical kinetics, c_T	2.58×10^{-7} °C^{-1} s^{-1}
Depletion dependency of the chemical kinetics, c_μ	2.12×10^{-6} s^{-1}
Initial shear stress, τ_0	240 MPa
Nominal strain rate, $\dot{\gamma}_0$	10^{-6} s^{-1}
Thermal pressurization coefficient, Λ	0.5 MPa °C^{-1}
Thermal diffusivity, c_{th}	10^{-6} m^2 s^{-1}
Hydraulic diffusivity, c_{hy}	10^{-6} m^2 s^{-1}

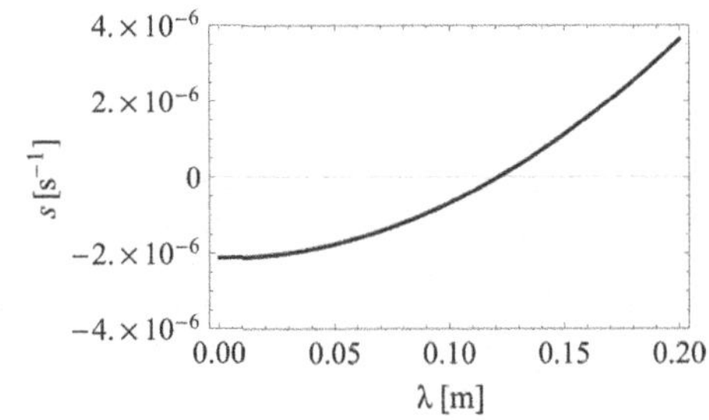

Fig. 3 Growth coefficient (Lyapunov exponent) in terms of perturbation wavelength

It is worth investigating the evolution of the localization zone in a simple case of a gouge layer of 5 m thickness. An initial small perturbation of the shear strain field is imposed with a wavelength equal to the gouge thickness. According to Fig. 3, this wave length is unstable. In particular, the critical wave length given by (6) is 0.12 m.

Figure 4a shows the evolution of the shear stress and of the shear strain rate in the middle of the localization zone. The numerical problem is discretized in space with finite differences and integrated in time using an adaptive scheme with Mathematica© ODE solver. Figure 4b presents the profile of the shear strain at various time instants showing the localization and the de-localization process when the reaction is depleted. Note that in this example, a very low strain rate corresponding to a subducting slab was assumed. For this strain rate and due to partial dehydration (the material in the middle is depleted but not in the adjacent zones) chemical traveling waves are triggered and travel towards the boundaries of the gouge layer. Figure 4c depicts the temperature increase due to shearing in the middle of the shear band. It is worth mentioning that after a point the (endothermic) chemical reaction buffers the temperature rise. Figure 4d shows the evolution of the chemical reaction and the pore pressure increase in the middle of the shear band.

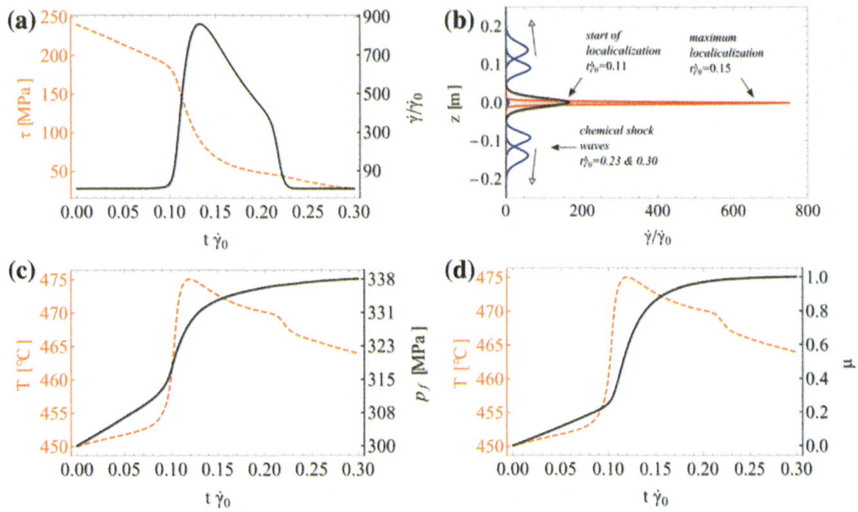

Fig. 4 Strain localization due to chemical softening in a 5 m thick dehydrating gouge layer: **a** Shear stress drop in the middle of the shear band and shear strain evolution. The system localizes to a narrow band because of dehydration, but then it delocalizes due to depletion; **b** Evolution of shear band localization—profile of shear strain rate; **c** Temperature and pore pressure increase in the middle of the shear band. The chemical reaction buffers temperature rise; **d** Evolution of the chemical reaction in the middle of the shear band. At $t\dot{\gamma}_0 \simeq 0.1$, the reaction effects become important and the material is rapidly depleted in the middle of the gouge ($\mu \rightarrow 1$) until $t\dot{\gamma}_0 \simeq 0.3$. Due to partial depletion, chemical shock waves are triggered and travel towards the boundaries of the gouge

5 Conclusion

We have presented the framework of localized failure in earthquake rupture as the result of instability of the deformation process. Frictional heating, thermal weakening, rock comminution and fault lubrication by the production of nano-particles, pore fluid pressurization and thermal decomposition of minerals are weakening mechanisms can trigger strain localization and acceleration of the seismic slip. These endothermic chemical reactions also influence the energy budget of the system as they limit the coseismic temperature rise and inhibit other thermal weakening mechanisms such as melting. The actual shear band thickness plays a major role in the energy budget of the system as it controls the feedback of the dissipative terms in the energy balance equation. It is known that it is necessary to introduce a localization limiter in order to obtain a finite thickness for the localized shear zone. Rate independent constitutive models for a classical (Cauchy) continuum lead to an infinitesimally thin zone. On the contrary, a rate dependent friction law or a Cosserat framework lead to a finite value for the critical wave length of the perturbation, above which, homogeneous deformation is unstable [27].

A challenging issue is the modelling of coupled thermo-chemo-hydro-mechanical phenomena with evolution of the microstructure of the material through various mechanical and chemical processes.

References

1. Otsuki, K.: Fluidization and melting of fault gouge during seismic slip: Identification in the Nojima fault zone and implications for focal earthquake mechanisms. J. Geophys. Res. **108** (B4), 2192 (2003)
2. Kuo, L.W., Hsiao, H.C., Song, S.R., Sheu, H.S., Suppe, J.: Coseismic thickness of principal slip zone from the Taiwan Chelungpu fault Drilling Project-A (TCDP-A) and correlated fracture energy. Tectonophysics **619–620**, 29–35 (2014)
3. Scholz, C.H.: The Mechanics of Earthquakes and Faulting, 2nd edn. Cambridge University Press, Cambridge (2002)
4. Rice, J.R., Rudnicki, J.W., Platt, J.D.: Stability and localization of rapid shear in fluid-saturated fault gouge: 1. Linearized stability analysis. J. Geophys. Res. Solid Earth (2014)
5. Platt, J.D., Rudnicki, J.W., Rice, J.R.: Stability and localization of rapid shear in fluid-saturated fault gouge, 2. Localized zone width and strength evolution. J. Geophys. Res. Solid Earth n/a–n/a (2014)
6. Rice, J.R.: Heating and weakening of faults during earthquake slip. J. Geophys. Res. Solid Earth **111**(5) (2006)
7. Beeler, N.M., Tullis, T.E., Goldsby, D.L.: Constitutive relationships and physical basis of fault strength due to flash heating. J. Geophys. Res. **113**(B1), B01401 (2008)
8. Di Toro, G., Han, R., Hirose, T., De Paola, N., Nielsen, S., Mizoguchi, K., Ferri, F., Cocco, M., Shimamoto, T.: Fault lubrication during earthquakes. Nature **471**(7339), 494–498 (2011)
9. Lachenbruch, A.H., Sass, J.H.: Heat flow and energetics of the San Andreas Fault Zone. J. Geophys. Res. **85**(B11), 6185 (1980)
10. Mase, C.W., Smith, L.: Effects of frictional heating on the thermal, hydrologic, and mechanical response of a fault. J. Geophys. Res. **92**(1), 6249 (1987)
11. Sulem, J., Lazar, P., Vardoulakis, I.: Thermo-poro-mechanical properties of clayey gouge and application to rapid fault shearing. Int. J. Numer. Anal. Methods Geomech. **31**(3), 523–540 (2007)
12. Brantut, N., Sulem, J., Schubnel, A.: Effect of dehydration reactions on earthquake nucleation: Stable sliding, slow transients, and unstable slip. J. Geophys. Res. **116**(B5), 1–16 (2011)
13. Sulem, J., Famin, V.: Thermal decomposition of carbonates in fault zones: slip-weakening and temperature-limiting effects. J. Geophys. Res. **114**(B3), 1–14 (2009)
14. Veveakis, M., Poulet, T., Alevizos, S.: Thermo-poro-mechanics of chemically active creeping faults: 2. Transient considerations. J. Geophys. Res. Solid Earth **119**(6), 4583–4605 (2014)
15. Di Toro, G., Nielsen, S., Pennacchioni, G.: Earthquake rupture dynamics frozen in exhumed ancient faults. Nature **436**(August), 1009–1012 (2005)
16. Collettini, C., Viti, C., Tesei, T., Mollo, S., Collettini, C., Viti, C., Tesei, T., Mollo, S., Nazionale, I., Murata, V.: Thermal decomposition along natural carbonate faults during earthquakes. Geology (2013)
17. Ghabezloo, S., Sulem, J.: Stress dependent thermal pressurization of a fluid-saturated rock. Rock Mech. Rock Eng. **42**(1), 1–24 (2008)
18. Wibberley, C.A.J., Shimamoto, T.: Internal structure and permeability of major strike-slip fault zones : the Median Tectonic Line in Mie Prefecture, Southwest Japan. J. Struct. Geol. **25** (2003)

19. Sulem, J., Vardoulakis, I., Ouffroukh, H., Boulon, M., Hans, J.: Experimental characterization of the thermo-poro-mechanical properties of the Aegion Fault gouge. Comptes Rendus Geosci. **336**(4–5), 455–466 (2004)
20. Di Toro, G., Goldsby, D.L., Tullis, T.E.: Friction falls towards zero in quartz rock as slip velocity approaches seismic rates. Nature **427**(January), 774–777 (2004)
21. Hauge, T.A.: The Heart Mountain detachment, northwestern Wyoming; 100 years of controversy. In: Snoke, A.W., Steidtmann, J.R., Roberts, S.M. (eds.) Geology of Wyoming: Memoir. Geological Survey of Wyoming, pp. 530–571. Laramie, WY (1993)
22. Goren, L., Aharonov, E., Sparks, D., Toussaint, R.: Pore pressure evolution in deforming granular material: a general formulation and the infinitely stiff approximation. J. Geophys. Res. **115**(B9), B09216 (2010)
23. Mitchell, T.M., Smith, S.A.F., Anders, M.H., Di Toro, G., Nielsen, S., Cavallo, A., Beard, A. D.: Catastrophic emplacement of giant landslides aided by thermal decomposition: Heart Mountain, Wyoming. Earth Planet. Sci. Lett. **411**, 199–207 (2015)
24. Green, H.W.: Shearing instabilities accompanying high-pressure phase transformations and the mechanics of deep earthquakes. Proc. Natl. Acad. Sci. U.S.A. **104**(22), 9133–9138 (2007)
25. Rutter, E.H., Brodie, K.H.: Experimental 'sytectonic' dehydration of serpentinite under conditions of controlled pore water pressure. J. Geophys. Res. **93**(B5), 4907 (1988)
26. Brantut, N., Sulem, J.: Strain localization and slip instability in a strain-rate hardening, chemically weakening material. J. Appl. Mech. **79**(3), 31004 (2012)
27. Sulem, J., Stefanou, I.: Thermal and chemical effects in shear and compaction bands. Geomech. Energy Environ. **6**, 4–21 (2016)

Modelling Grain Damage Under Plane Strain Compression Using a Micro-polar Continuum

Erich Bauer, Linke Li and Mohammadkeya Khosravi

Abstract A novel concept for modelling the influence of grain damage on the reduction of the mean grain diameter and consequently on the compaction behaviour and incremental stiffness is proposed. Herein the constant solid hardness in previous models is replaced by a pressure dependent stiffness factor. The change of grading of the grain sizes caused by grain damage is reflected in a simplified manner by a reduction of the mean grain diameter. In order to distinguish the effect of the loading path on the evolution of grain damage the evolution equations for the solid hardness and the mean grain diameter are decomposed into parts reflecting the influence on isotropic and deviatoric loading. The mean grain diameter is embedded as the internal length in the micro-polar hypoplastic model. Numerical simulations of the influence of grain damage on the compaction behaviour are carried out for biaxial compression tests. The results show that for the case of shear strain localization the reduction of the stiffness factor and the mean grain diameter is more pronounced in the shear band than outside of this zone.

1 Introduction

The mechanical behaviour of granular materials is mainly dictated by the grain size distribution, the grain shape, the surface roughness, the initial density and the loading path. Under higher stresses the mechanical behaviour is also influenced by grain damage in the form of grain abrasion and grain fragmentation. The amount of grain damage does not only depend on the pressure level but it is also influenced by the loading path. Grain damage leads to a change of the grading of the grain sizes and as a consequence to a reorientation of grains within the grain skeleton towards a denser state. Only little experimental data is available in the literature, so that numerical simulations are a possibility to gain deeper insight into the mechanisms of grain damage.

E. Bauer (✉) · L. Li · M. Khosravi
Institute of Applied Mechanics, Graz University of Technology, 8020 Graz, Austria
e-mail: erich.bauer@tugraz.at

E. Papamichos et al. (eds.), *Bifurcation and Degradation of Geomaterials with Engineering Applications*, Springer Series in Geomechanics and Geoengineering, DOI 10.1007/978-3-319-56397-8_67

For the constitutive modelling of the influence of grain damage under large monotonic shearing a concept based on a micro-polar continuum was recently proposed by Bauer et al. [1] and Bauer [2]. The numerical investigations show, that for shearing under nearly constant mean pressure, micro-polar effects like abrasion and grain crushing caused by particle rotations and curvatures are the dominant causes of grain damage in zones of pronounced shear strain localization. In the present paper an extended concept with the focus on the influence of the amount of stress on grain damage is considered. The additional compaction behaviour caused by grain damage is reflected using a pressure dependent stiffness parameter as a key parameter for modelling the reduction of the incremental stiffness. In order to distinguish the effect of the loading path on grain damage the evolution equations for the reduction of the stiffness parameter and the mean grain diameter are decomposed into terms reflecting the influence of the isotropic and deviatoric part of the stress tensor. The concept proposed for the stress dependent reduction of the stiffness parameter and the mean grain diameter is implemented in the micro-polar hypoplastic model by Huang et al. [3]. In order to demonstrate the influence of particle damage numerical simulations of biaxial compression tests are performed.

2 Compaction Behaviour and Reduction of the Mean Grain Diameter Under Increasing Mean Pressure

To model the change of the current void ratio e of a cohesionless granular material for a monotonic increase of the mean pressure, $p = -\sigma_{ii}/3$, the following isotropic compression law was proposed by Bauer [4]:

$$e = e_0 \exp[-(3p/h_s)^n]. \tag{1}$$

Herein e_0 denotes the initial void ratio for $p = 0$. h_s is the so-called solid hardness, which is defined for the grain assembly, where the compression curve in a semi-logarithmic representation shows the point of inflection. Parameter n is related to the inclination of the compression curve (Fig. 1a). It is obvious that h_s is a key parameter for describing the compaction behaviour of the granular material under isotropic compression. Equation (1) can approximate the behaviour for an unlimited pressure range and thus it also reflects the history and the effect of grain damage on the compaction behaviour. Depending on the strength of the grains the value for the solid hardness ranges between 10 MPa and 300 MPa. Isotropic compression tests with such high stresses are not usually standard in soil mechanics laboratories, so that h_s is commonly correlated to the compression curve obtained from a high pressure oedometer test. Numerical simulations indicate that for high stresses the difference between isotropic and oedometric compression is usually small and can be neglected for practical applications.

If the maximum stress in the compression experiment is lower than h_s, the adaptation of Eq. (1) to the experimental data usually results in a much higher value

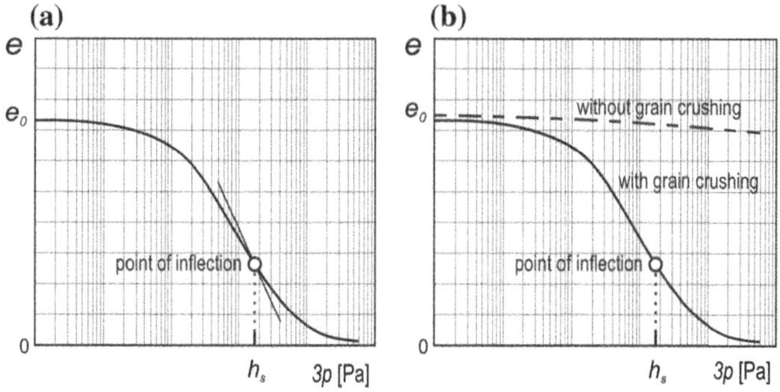

Fig. 1 Illustration of monotonic compression: **a** compression law of Eq. (1), **b** behaviour with and without grain crushing

of h_s. For Karlsruhe sand, for instance, a solid hardness of $h_s = 5800$ MPa was reported to reflect the experimental data for a low pressure range [5], while a value of $h_s = 190$ MPa was obtained for a high pressure test [4]. The higher h_s is relevant for the pressure range where grain crushing is negligibly small, while the lower h_s is related to a pressure range where grain crushing becomes dominant. In this context it is of interest to discuss the results of DEM simulations by Laufer [6], which are illustrated in a simplified manner in Fig. 1b. For the assumption of unbreakable particles, the compaction behaviour is small and determined by the properties of the contact points and the reorientation of the particles into a denser state. However, when particle crushing is taken into account, the DEM simulation of monotonic compression up to a large pressure range shows a behaviour similar to the one observed in real experiments. These results suggest that it makes sense to replace the constant solid hardness h_s in Eq. (1) with the pressure dependent stiffness parameter, h_{sp}. To this end the following evolution equation for a decrease of the stiffness parameter with an increase of the mean pressure is proposed:

$$\dot{h}_{sp} = -(\dot{p}/p)(3p/b_{op})^{n_{op}} h_{sp}. \tag{2}$$

Herein, $\dot{h}_{sp}$ and $\dot{p}$ denote the rates of h_{sp} and p, respectively, and b_{op} and n_{op} are constitutive constants. Integration of Eq. (2) leads to the pressure dependent stiffness parameter $h_{sp} = h_{sp}(p)$. Replacing the constant solid hardness in Eq. (1) with the stress dependent stiffness parameter h_{sp} one obtains:

$$e = e_0 \exp\left[-(3p/h_{sp})^n\right]. \tag{3}$$

For $b_{op} = 457$, $n_{op} = -0.31$, an initial void ratio of $e_o = 1.01$ for $p = 100$ Pa and an initial stiffness parameter of $h_{so} = 5800$ MPa, the reduction of the void ratio with an increase of the mean pressure is shown in Fig. 2a together with experimental results obtained by Bauer [4].

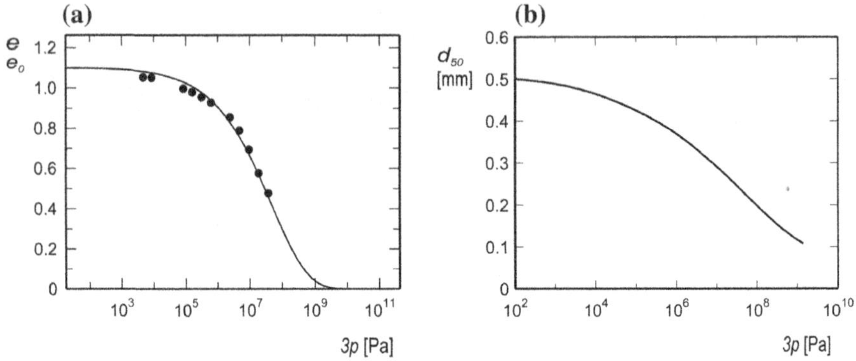

Fig. 2 Influence of the pressure level p on the reduction of **a** the void ratio e (*dots* are experimental data by Bauer [4]); **b** the mean grain diameter d_{50}

In the following the influence of grain damage on the change of the grain size distribution is taken into account in a simplified manner by reducing the mean grain diameter d_{50}. To this end the following evolution equation is proposed:

$$\dot{d}_{50p} = -b_{50p} n_{50p} (\dot{p}/p)(3p/h_{sp})^{n_{50p}} d_{50p}. \tag{4}$$

Herein, the change of the mean grain diameter is also related to the current value of h_{sp}. b_{50p} and n_{50p} are constitutive constants which can be calibrated based on compression tests. As h_{sp} is higher for lower pressures, the change of the mean grain diameter is also smaller within the lower pressure range.

3 Reduction of h_s and d_{50} Under Deviatoric Loading

To model the influence of deviatoric loading on grain damage the following evolution equation is proposed for the rate of the stiffness parameter:

$$\dot{h}_{sd} = -(b_{sd}/h_{sd})\left|\dot{\sigma}^{*symm}/b_{od}\right|\left|\sigma^{*symm}/b_{od}\right| h_{sd}, \tag{5}$$

and for the rate of the mean grain diameter:

$$\dot{d}_{50d} = -b_{d50d}\left|\dot{\sigma}^{*symm}/b_{50od}\right|\left|\sigma^{*symm}/b_{50od}\right| d_{50d}. \tag{6}$$

Herein the norm of the stress deviator, σ^{*symm}, is computed from the symmetric part of the stress tensor. b_{sd}, b_{d50d} and b_{50od} are constitutive constants.

4 Micro-polar Description of Grain Damage

In order to take into account the effect of abrasion and damage caused by the rotation resistance of grains on the reduction of h_s and d_{50} hypothetical evolution equations were recently proposed within the framework of a micro-polar continuum [1, 2]. With respect to Eqs. (2–6) these evolution equations for h_s^* and d_{50} can be extended as follows:

$$\dot{h}_s^* = -\left[b_\kappa\|\dot{\boldsymbol{\kappa}}\| + b_\omega\left\|\dot{\omega}_3^c - \dot{\omega}_3\right\| + \frac{\dot{p}}{p}\left(\frac{3p}{b_{op}}\right)^{nop} + \frac{b_{sd}}{h_s^*}\frac{|\dot{\sigma}^{*symm}||\sigma^{*symm}|}{b_{od}^2}\right]h_s^*, \tag{7}$$

and

$$\dot{d}_{50} = -\left[b_\kappa\|\dot{\boldsymbol{\kappa}}\| + b_\omega\left\|\dot{\omega}_3^c - \dot{\omega}_3\right\| + b_{50p}n_{50p}\frac{\dot{p}}{p}\left(\frac{3p}{h_s^*}\right)^{n_{50p}} + b_{d50d}\frac{|\dot{\sigma}^{*symm}||\sigma^{*symm}|}{b_{50od}^2}\right]d_{50}. \tag{8}$$

Herein d_{50} and h_s^* are the current values of these state quantities. $\|\dot{\boldsymbol{\kappa}}\|$ denotes the norm of the rate of the curvature, and $\left\|\dot{\omega}_3^c - \dot{\omega}_3\right\|$ denotes the norm of the difference of the rate of the micro- and macro-rotations. Factors b_κ and b_ω are constitutive constants. Equations (7) and (8) are embedded into the micro-polar hypoplastic model by Huang et al. [3]. For the extended model the evolution equations for the objective stress rate, the objective couple stress and of the rate of the void ratio, $\dot{e}$, read:

$$\overset{\circ}{\boldsymbol{\sigma}} = f_s\left[\hat{a}^2\dot{\boldsymbol{\varepsilon}}^c + (\hat{\boldsymbol{\sigma}}:\dot{\boldsymbol{\varepsilon}}^c + \hat{\boldsymbol{\mu}}:\dot{\bar{\boldsymbol{\kappa}}})\hat{\boldsymbol{\sigma}} + f_d(\hat{\boldsymbol{\sigma}} + \hat{\boldsymbol{\sigma}}^d)\sqrt{\hat{a}^2\|\dot{\boldsymbol{\varepsilon}}^c\|^2 + a_m^2\|\dot{\bar{\boldsymbol{\kappa}}}\|^2}\right], \tag{9}$$

$$\overset{\circ}{\boldsymbol{\mu}} = d_{50}f_s\left[a_m^2\dot{\bar{\boldsymbol{\kappa}}} + \left(\hat{\boldsymbol{\sigma}}:\dot{\boldsymbol{\varepsilon}}^c + \hat{\boldsymbol{\mu}}:\dot{\bar{\boldsymbol{\kappa}}} + 2f_d\sqrt{\hat{a}^2\|\dot{\boldsymbol{\varepsilon}}^c\|^2 + a_m^2\|\dot{\bar{\boldsymbol{\kappa}}}\|^2}\right)\hat{\boldsymbol{\mu}}\right], \tag{10}$$

$$\dot{e} = (1+e)\dot{\boldsymbol{\varepsilon}}^c : \dot{\boldsymbol{\varepsilon}}^c. \tag{11}$$

Herein, $\hat{\boldsymbol{\sigma}} = \boldsymbol{\sigma}/tr\boldsymbol{\sigma}$ denotes the normalized non-symmetric Cauchy stress tensor, $\hat{\boldsymbol{\sigma}}^d = \hat{\boldsymbol{\sigma}} - \mathbf{I}/3$ its deviator, $\hat{\boldsymbol{\mu}} = \boldsymbol{\mu}/(d_{50}tr\boldsymbol{\sigma})$ the normalized couple stress tensor, $\dot{\varepsilon}^c$ the rate of deformation tensor, and $\dot{\bar{\boldsymbol{\kappa}}} = d_{50}\dot{\boldsymbol{\kappa}}$ denotes the normalized rate of the curvature tensor. Factors f_s and f_d are functions of the current void ratio e, the pressure dependent limit void ratios, e_i and e_d, and the critical void ratio e_c [4]. The scalar $\hat{a}$ is related to critical stress states, i.e. to the critical friction angle φ_c, and a_m is related to the rotation resistance of particles as outlined in more detail in Huang et al. [7].

Table 1 Constitutive parameters used for numerical simulations

h_{so} (Mpa)	n	d_{50} (mm)	e_{io}	e_{do}	e_{co}	α	β	φ_C (°)
5800	0.4	0.5	1.2	0.51	0.82	0.15	1.05	30
a_m	n_{op}	b_{50p}	n_{50p}	b_{sd} (Pa)	b_{od} (Pa)	b_{50d}	b_{50od}	b_{op} (Pa)
0.8	−0.31	1.05	0.2	10^{-5}	1.0	10^{-10}	1.0	457

Numerical investigations show that the first two terms in Eqs. (7) and (8) are only significant when pronounced shear strain localization takes place [1]. As the focus of the present paper is on investigating the influence of the mean pressure and the deviatoric stress deviator, the first two terms in Eqs. (7) and (8) are not considered in the numerical simulations in Sect. 5, i.e. parameters b_κ and b_ω are set to zero. The values of the constitutive parameters used for numerical simulations are summarized in Table 1.

5 Numerical Simulation of Biaxial Compression

For numerical simulations with the finite element program ABAQUS the present micro-polar hypoplastic model is implemented into the version originally developed by Huang [3]. The specimen with a height of 12 cm and a width of 4 cm is discretized by 3702 four-node plane strain Cosserat elements with linear shape functions for displacements and micro-polar rotations. For the initial state a random distribution of the void ratio and an initial stress of −100 Pa is assumed (Fig. 3a). In the second loading step the vertical and lateral stress is increased up to -10^5 Pa. Then the lateral stress is kept constant and the specimen is vertically compressed.

The results shown in Fig. 3b–d are obtained for a vertical displacement of the top surface of 1.4 cm. Shear strain localization is clearly visible in Fig. 3b where the darker area indicates a lower value of the void ratio, i.e. a denser state of the material. It is also obvious that the initial random distribution of the void ratio (Fig. 3a) is almost swept out within the localized zone (Fig. 3b). The active shear band thickness is approximately 12 times of the mean grain diameter. This value is in a good accordance with the observed thickness during biaxial compression experiments with dense Karlsruhe sand [8] and the analysis by Mühlhaus and Vardoulakis [9]. In this context it can be mentioned that the numerical simulations by Huang and Bauer [7] indicate that the active thickness of the shear band is not a material constant and may also be influenced on the loading history. As a result of an increase of the isotropic and deviatoric stress the value of h_s decreases form 5800 MPa to an average value of 240 MPa (Fig. 3c). In the zone of intense shear strain localization, the minimum value is 202 MPa and represented in the contour plot by the darker area. For d_{50} a reduction from 0.5 mm to an average value of 0.4 mm is predicted for the entire specimen (Fig. 3d). It is clearly visible that the reduction of h_s and d_{50} is more pronounced within the localized zone. At the

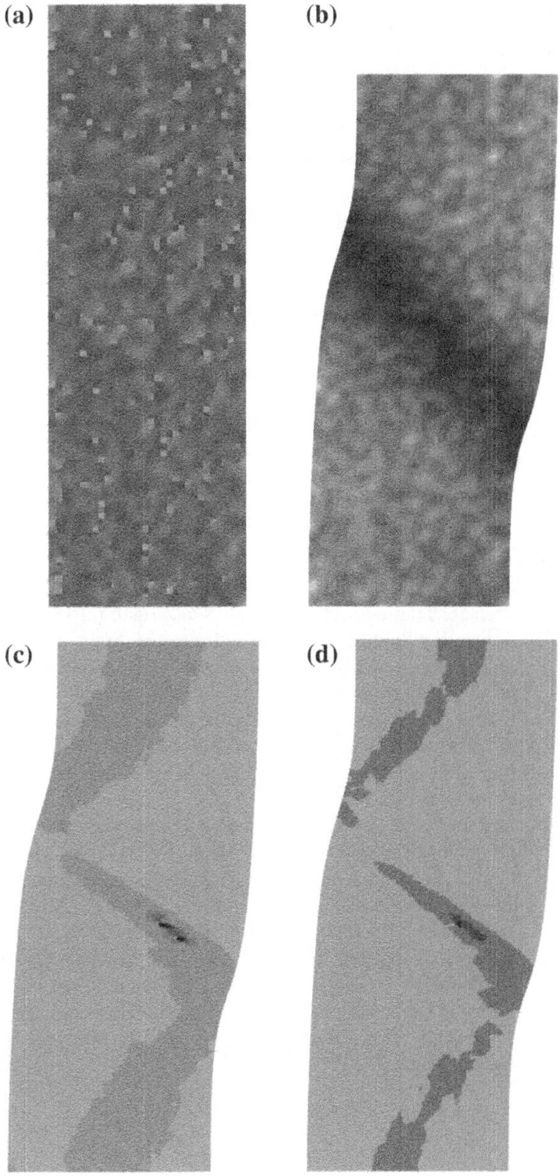

Fig. 3 Contour plot of the **a** initial random distribution of the void ratio e [1.15, 0.93]; **b–d** the void ratio e [1.01, 0.73], stiffness parameter h_s^* [243, 209 MPa], mean grain diameter d_{50} [0.41, 0.39 mm] for a lateral stress of − 0.1 MPa and a vertical compression of 1.4 cm

beginning of shear localization the reduction of this quantities starts in the middle of the shear band and is therefore not homogeneous along the shear band. This can be explained by a higher mean pressure in the middle of the deformed specimen. With a decrease of d_{50} also the active shear band thickness decreases.

6 Conclusions

In this paper a new concept is proposed to model the influence of the loading path on grain damage and compaction behaviour using a micro-polar continuum description. To this end corresponding evolution equations for modelling the reduction of the stiffness parameter and the mean grain on the loading history are introduced and implemented into a micro-polar hypoplastic model. Numerical simulations of biaxial compression tests show that the effect of grain damage leads to a significant reduction of the resistance to compaction. With a decrease of the mean grain diameter the thickness of the zone of shear strain localization is also reduced.

References

1. Bauer, E., Li, L., Huang, W.: Hypoplastic constitutive modelling of grain damage under plane shearing. In: Chau, K.T., Zhao, J. (eds.) Bifurcation and Degradation of Geomaterials in the New Millennium. Springer Series in Geomechanics and Geoengineering, pp. 181–187. Springer International Publishing Switzerland (2015)
2. Bauer, E.: Simulation of the influence of grain damage on the evolution of shear strain localization. In: Albers B., Kuczma M. (eds.) Continuous Media with Microstructure, vol. 2, pp. 231–244. Springer International Publishing Switzerland (2016)
3. Huang, W., Nübel, K., Bauer, E.: Polar extension of a hypoplastic model for granular materials with shear localization. Mech. Mater. **34**, 563–576 (2002)
4. Bauer, E.: Calibration of a comprehensive hypoplastic model for granular materials. Soil Found. **36**(1), 13–26 (1996)
5. Herle, I., Gudehus, G.: Determination of parameters of a hypoplastic constitutive model from properties of grain assemblies. Mech. Cohes.-Frict. Mater. **4**, 461–486 (1999)
6. Laufer, I.: Grain crushing and high-pressure oedometer tests simulated with the discrete element method. Granul. Matter **17**(3), 389–412 (2015)
7. Huang, W., Bauer, E.: Numerical investigations of shear localization in a micro-polar hypoplastic material. Int. J. Numer. Anal. Methods Geomech. **27**, 325–352 (2003)
8. Vardoulakis, I.: Scherfugenbildung in Sandkörpern als Verzweigungsproblem. Publication Series of the Institute of Soil Mechanics, University of Karlsruhe, p. 70 (1977)
9. Muehlhaus, H.B., Vardoulakis, I.: The thickness of shear bands in granular materials. Geotechnique **37**, 271–283 (1987)

Opening of Mode I Fractures in Geomaterials with Rotating Particles: Small-Scale Cosserat Continuum Approach and Its Verification

E. Pasternak, A.V. Dyskin, M. Esin and Y. Xu

Abstract Geomaterials often contain constituents (e.g., grains, blocks) capable of rotating. Modelling of such geomaterials requires the use of Cosserat continua. A Cosserat continuum possesses characteristic lengths, which distinguish it from the scale-independent classical continuum. For geomaterials with grains these characteristic lengths are of the order of grain size, which restricts the use of continuum modelling to scales (e.g., crack lengths) macroscopic with respect to the grain size and to the Cosserat characteristic lengths. This automatically brings us to the pseudo-Cosserat continuum with constrained microrotations leading to what we call a small-scale Cosserat continuum. We consider a 2D Mode I crack in such a continuum and determine its characteristics, in particular the opening. We then verify this approach using physical modelling and a direct numerical simulation based on discrete element modelling.

1 Introduction

Geomaterials often consist of cemented grains or slightly cemented and unbounded constituents (e.g., clay, sand) thus possessing highly nonhomogeneous microstructure. Furthermore the constituents (particles) can rotate under loading and thus involve additional, rotational degrees of freedom. This complicates the modelling of mechanical behaviour of geomaterials especially fracture propagation therein. In this situation continuum modelling is attractive since it permits skipping the mechanical and geometric details of the constituents and the binder concentrating instead on average behaviour. Such a simplification is especially important

A.V. Dyskin (✉) · Y. Xu
School of Civil, Environmental and Mining Engineering,
University of Western Australia, Crawley, Australia
e-mail: arcady.dyskin@uwa.edu.au

E. Pasternak · M. Esin
School of Mechanical and Chemical Engineering, University of Western Australia,
Crawley, Australia

E. Papamichos et al. (eds.), *Bifurcation and Degradation of Geomaterials with Engineering Applications*, Springer Series in Geomechanics and Geoengineering, DOI 10.1007/978-3-319-56397-8_68

for fracture modelling as it allows one to neglect the complex interaction between the stress concentration and the particle movement at the crack tip.

The use of continua for modelling inherently heterogeneous and even discontinuous materials such as geomaterials is based on the *hypothesis of separation of scales* (e.g., [1]): if l_{micro} is the characteristic size of the material microstructure (e.g., the largest particle dimension) and L is the characteristic length of the variations of the external fields (e.g., the smallest size of the geometry of the region modelled or the fracture length or the wave length in dynamic problems), then there exists an intermediate size H such that $l_{micro} \ll H \ll L$. Using this size the continuum can be formed which refers to the fields (e.g., stress, moment stress, rotation, strain) obtained by averaging the corresponding quantities in the geomaterial over volume elements of size H. This continuum will be refereed to as an *equivalent continuum* and the volume element of size H is called the *representative volume element* (*RVE*).

While the construction of an equivalent continuum explicitly involves the size of the representative volume element, the continuum description does not contain H. This can only occur when the above double inequality is taken to the limit, which corresponds to retaining only the main term of the double asymptotics

$$l_{micro}/H \to 0, \quad H/L \to 0 \tag{1}$$

This is a mathematical representation of the hypothesis of separation of scales.

Geomaterials with constituents (e.g., grains) capable of rotating possess three rotational degrees of freedom on top of three translational degrees of freedom characterising conventional materials. Modelling of geomaterials with rotational degrees of freedom requires the introduction of high order continua, first of all the micropolar or Cosserat continuum [2–4]. Opposite to the scale independent classical continua, the Cosserat continuum introduces internal length scales. They emerge from the difference in units of the elastic moduli relating the stress and strain and the moduli relating the moment stress and the rotation gradients (curvature-twist). The units differ by a factor of length squared, which gives rise to a set of characteristic lengths that can be constructed from the possible pairs of moduli.

Theoretically, the Cosserat characteristic lengths can assume any values. However, as shown in [5–9] for Cosserat continua constructed to model particulate materials consisting of spherical particles connected to each other by 3 translational and 3 rotational springs that the Cosserat lengths are of the order of the particle size. Therefore, the Cosserat lengths obtained are of the order of l_{micro} and hence cannot enter the continuum description, which is the main term of asymptotics (1). Subsequently, due to the separation of scales condition, the model of particulate material considered in [5–9] corresponds to double asymptotics

$$l_k/H \to 0, \quad H/L \to 0 \tag{2}$$

where l_k are the Cosserat characteristic lengths. This asymptotics leads to the Cosserat continuum with constrained rotations (coupled stress continuum), which now becomes a legitimate model of a particulate material rather than a mere simplification of a general Cosserat continuum. In order to emphasise that it is not an approximation, a term *small-scale Cosserat continuum* was introduced. Based on this a model of a crack in a particulate material was developed and a criterion of its propagation based on the concentration of moments stress was proposed.

In this paper we first prove the asymptotics (2) from general dimensional analysis and then consider experimental and numerical verification of the model [5–9].

2 Separation of Scales in Cosserat Continuum Modelling of Geomaterials

Consider a geomaterial with rotating constituents. Let its microstructure be characterised by microscopic lengths λ_1, λ_2, …; l_{micro} being their maximum, moduli E_1, E_2, … and Poisson's ratios ν_1, ν_2, …. Then the dimensional analysis gives that the Cosserat characteristic lengths l_k are

$$l_k = l_{micro} f_k\left(\frac{\lambda_1}{l_{micro}}, \frac{\lambda_2}{l_{micro}}, \ldots, \frac{E_1}{E_{\max}}, \frac{E_2}{E_{\max}}, \ldots, \frac{\nu_1}{\nu_{\max}}, \frac{\nu_2}{\nu_{\max}}, \ldots\right) \tag{3}$$

where $E_{\max} = \max(E_1, E_2, \ldots)$; $\nu_{\max} = \max(\nu_1, \nu_2, \ldots)$. It is presumed that the Poisson's ratios are not negative; otherwise we choose the Poisson's ratio of maximum absolute value as $\nu_{\max}$. Figure 1 gives an example of two grains bonded by a cement layer. Here $\lambda_1 = l_{micro} = a$, $\lambda_2 = l_c$, $\lambda_3 = h$.

Now, since f_k are constants of the material, the asymptotics (1) leads to asymptotics (1). Therefore the obtained small-scale Cosserat continuum is a natural model of a material with microstructure with rotational degrees of freedom as long as only the main term of asymptotics (1) is retained.

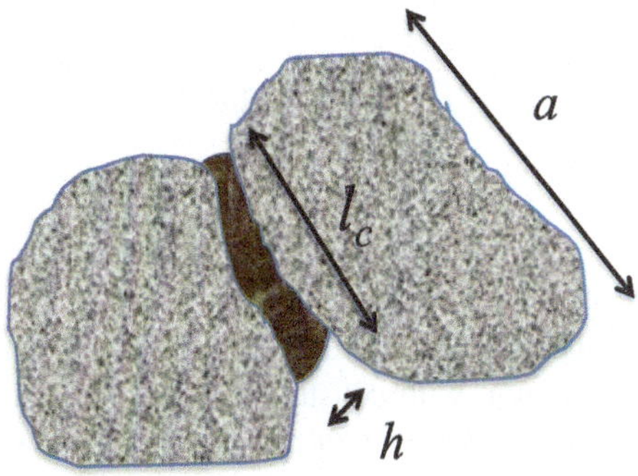

Fig. 1 A fragment of a particulate material with two grains bonded by a cement layer

3 Wedge Crack in Small-Scale Cosserat Continuum

Consider a crack in two-dimensional isotropic Cosserat continuum in plane (x_1, x_2). It has only three degrees of freedom represented by two dimensional displacement (u_1, u_2) and rotation φ_3. Subsequently the following deformation measures are introduced: strain and rotation gradients (e.g., [10])

$$\gamma_{ii} = \frac{\partial u_i}{\partial x_i}, \quad \gamma_{12} = \frac{\partial u_1}{\partial x_2} + \varphi_3, \quad \gamma_{21} = \frac{\partial u_2}{\partial x_1} - \varphi_3, \quad \kappa_{i3} = \frac{\partial \varphi_3}{\partial x_i}, \quad i = 1, 2 \tag{4}$$

The equations of equilibrium and the constitutive law read

$$\frac{\partial \sigma_{ji,j}}{\partial x_j} = 0, \quad i, j = 1, 2, \quad \frac{\partial \mu_{13}}{\partial x_1} + \frac{\partial \mu_{23}}{\partial x_2} + \sigma_{12} - \sigma_{21} = 0 \tag{5}$$

$$\begin{aligned} &\sigma_{11} = (2\mu + \lambda)\gamma_{11} + \lambda\gamma_{22}, \quad \sigma_{12} = (\mu + \alpha)\gamma_{12} + (\mu - \alpha)\gamma_{21} \\ &\sigma_{21} = (\mu + \alpha)\gamma_{21} + (\mu - \alpha)\gamma_{12}, \quad \sigma_{22} = \lambda\gamma_{11} + (2\mu + \lambda)\gamma_{22} \\ &\mu_{13} = B\kappa_{13}, \quad \mu_{23} = B\kappa_{23} \end{aligned} \tag{6}$$

Here λ and μ are the Lamé constants, α is the Cosserat shear modulus, B is the modulus relating moment stress and rotation gradient; it has units of Pa/m^2.

By substituting (4), (5) into (6) the Lamé equations are obtained

$$\begin{cases} (\lambda + 2\mu)\left(\frac{\partial^2 u_1}{\partial x_1^2} + \frac{\partial^2 u_2}{\partial x_1 \partial x_2}\right) + (\mu + \alpha)\left(\frac{\partial^2 u_1}{\partial x_2^2} - \frac{\partial^2 u_2}{\partial x_1 \partial x_2}\right) + 2\alpha \frac{\partial \varphi_3}{\partial x_2} = 0 \\ (\lambda + 2\mu)\left(\frac{\partial^2 u_1}{\partial x_1 \partial x_2} + \frac{\partial^2 u_2}{\partial x_2^2}\right) + (\mu + \alpha)\left(\frac{\partial^2 u_2}{\partial x_1^2} - \frac{\partial^2 u_1}{\partial x_1 \partial x_2}\right) - 2\alpha \frac{\partial \varphi_3}{\partial x_1} = 0 \\ l_2^2\left(\frac{\partial^2 \varphi_3}{\partial x_1^2} + \frac{\partial^2 \varphi_3}{\partial x_2^2}\right) - \varphi_3 + \frac{1}{2}\left(\frac{\partial u_2}{\partial x_1} - \frac{\partial u_1}{\partial x_2}\right) = 0 \end{cases} \tag{7}$$

which contain only one length parameter $l_2^2 = B/(4\alpha)$.

Asymptotics (2) leads to $l_2 \to 0$ and eventually to the following set of equations constituting the isotropic small-scale Cosserat continuum:

$$\begin{cases} (\lambda + 2\mu)\left(\frac{\partial^2 u_1}{\partial x_1^2} + \frac{\partial^2 u_2}{\partial x_1 \partial x_2}\right) + \mu\left(\frac{\partial^2 u_1}{\partial x_2^2} - \frac{\partial^2 u_2}{\partial x_1 \partial x_2}\right) = 0 \\ (\lambda + 2\mu)\left(\frac{\partial^2 u_1}{\partial x_1 \partial x_2} + \frac{\partial^2 u_2}{\partial x_2^2}\right) + \mu\left(\frac{\partial^2 u_2}{\partial x_1^2} - \frac{\partial^2 u_1}{\partial x_1 \partial x_2}\right) = 0 \\ \varphi_3^0 = \frac{1}{2}\left(\frac{\partial u_2}{\partial x_1} - \frac{\partial u_1}{\partial x_2}\right) \end{cases} \tag{8}$$

These equations formally coincide with equations for a 2D isotropic Cosserat continuum with constrained rotations.

The importance of this feature of the small-scale Cosserat continuum is that the first two equations are simply the conventional Lamé equations for a classical 2D elastic plane. Subsequently, if the boundary conditions are classical, that is specified in terms of displacements or tractions then the displacement field can be determined

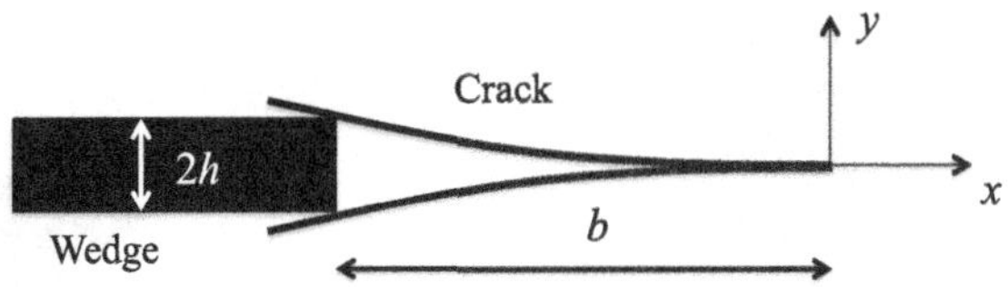

Fig. 2 Wedge crack

in a usual way without involving the rotational degree of freedom. After that, the field of rotations can be found using the third equation of (8) and from it the moment stress field can be calculated, using the constitutive Eq. (6).

A comment should be made here. According to (2) the asymptotics considered corresponds to $l_2/H \to 0$, where H is the characteristic length of redistribution of stress/strain fields. In other words, H is the scale of the modelling. Subsequently, the asymptotics of small-scale Cosserat continuum can be implemented by $H \to \infty$, rather than $l_2 \to 0$. Therefore, modulus $B \sim l_2^2$ is not small and hence the moment stresses associated with (constrained) rotations can be large and capable of inducing local failure.

The above consideration explains the following algorithm for solving the problem of a wedge crack in a small-scale Cosserat continuum (see [8] for detail): (i) solve the problem for a conventional wedge crack in isotropic elastic plane; (ii) determine the rotation and moment stress fields using (6), (8).

Implementing this algorithm we start with a semi-infinite crack in an infinite plane opened by a thin rigid rectangular wedge with thickness of $2h$ at a distance b from the crack tip, Fig. 2. Due to the symmetry, only the upper half plane $y > 0$ is considered (here we use x and y instead of x_1 and x_2).

For the plane stress problem (for the plane strain problem E should be replaced with $E/(1-\nu^2)$ and ν with $\nu/(1-\nu)$) the displacement field (u, v) is [11]:

$$\begin{cases} 2Gu = (1-2\nu)\mathrm{Re}\overline{Z} - y\mathrm{Im}Z \\ 2Gv = 2(1-\nu)\mathrm{Im}\overline{Z} - y\mathrm{Re}Z \end{cases}, \quad \varphi = \frac{1}{2}\left(\frac{\partial v}{\partial x} - \frac{\partial u}{\partial y}\right) \tag{9}$$

where Z is the Westergaard complex stress function, which for this crack is:

$$Z(z) = \frac{Eh}{\pi}\frac{1}{\sqrt{z(z+b)}}, \quad z = x + iy, \; Z = \frac{d}{dz}\overline{Z}, \quad \overline{Z}(z) = \frac{Eh}{\pi}\sinh^{-1}\sqrt{\frac{z}{b}} \tag{10}$$

Figure 3 shows the displacement and rotation fields for the upper half plane, $y > 0$ for $b = 62$ mm, $2h = 10.8$ mm, $\nu = 0.3$. As can be seen there is singularity in the rotation field at the crack tip ($x = 0$, $y = 0$) and at the point of the wedge ($x = -b$). For the purpose of model verification the paper focuses only on the rotations.

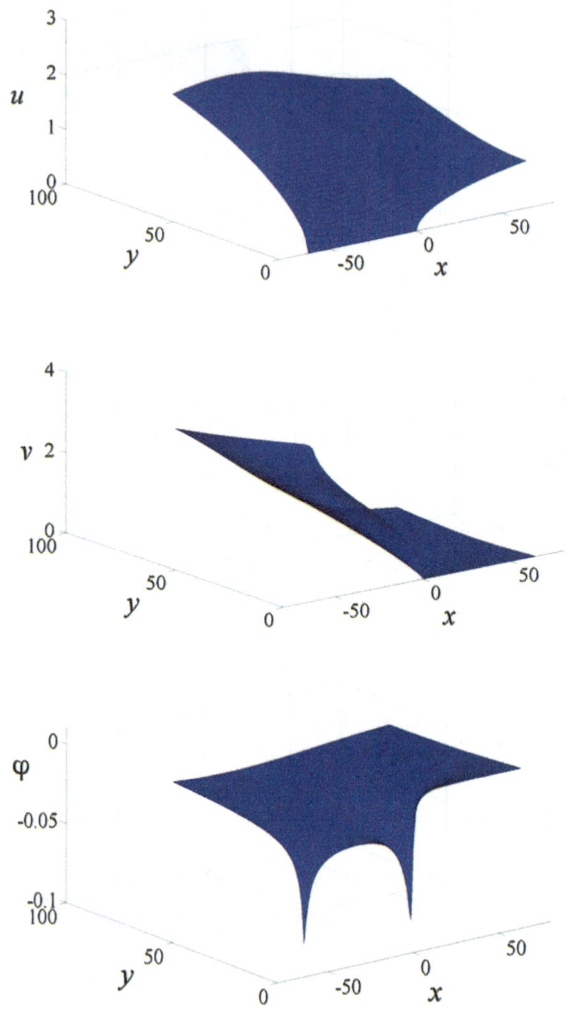

Fig. 3 Displacement (u, v) and rotation φ fields created by the crack

4 Wedge Crack: Physical and Discrete Element Models

Experimental verification of the proposed small-scale Cosserat continuum approach will be based on a 2D physical model of particulate material [12]. The model consists of an assembly of steel discs placed between two plexiglass walls. The assembly was restricted from the left and right sides by aluminium plates positioned between the walls. The disc rotations were measured using digital image correlation. To this end the discs were covered with speckles. The images taken by Canon 5D Mark II with Sigma 70 mm macro lens were processed in VIC-2D software to

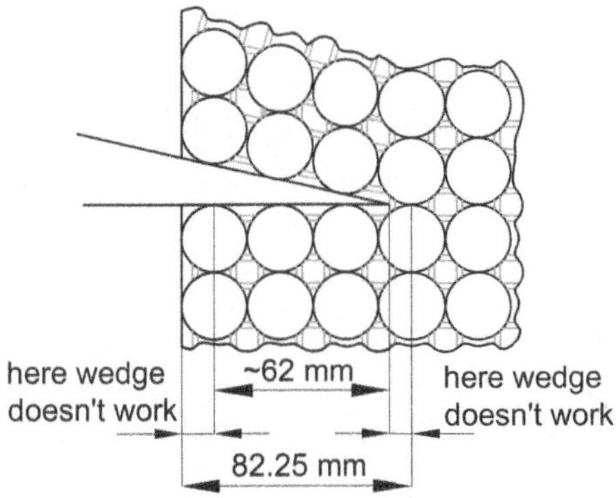

Fig. 4 The fracture (cut bonds between three pairs of the discs) opened by a wedge covering three discs

determine the disc rotations. A proprietary algorithm was developed which averages the obtained rotations exported from VIC-2D within each area of interest.

In the experiment reported a square packing of disks was used, Fig. 4. The reason for this choice of assembly geometry was dictated by the presence of straight boundary between the discs such that the crack was easy to model simply by removing the bonds between the particles. The technically simplest loading was performed through a lateral wedge of the angle 5°, Fig. 4.

5 Discrete Element Model

The Itasca Particle Flow Code^{2D} (PFC2D) realising the discrete element method (DEM) was used alongside the physical model. The linear parallel bond (PB) represented the silicone sealant between the discs, while the linear contact model simulated the behaviour of the un-glued pairs of discs. Some parameters of the model were taken from a previous study, namely the normal and shear stiffnesses of wall-to-disc and disc-to-disc contact, critical shear and normal damping ratio, friction coefficient of wall-to-disc and disc-to-disc contact [12]. The newly included micro-properties—normal and shear stiffnesses of PB model were estimated using the sealant's mechanical properties and the geometry of gluing area as $\bar{k}_n = E/L = 0.08\,\text{MPa/mm}$, $\bar{k}_s = E/(2(1+\nu)L) = 0.03\,\text{MPa/mm}$, where E is the Young's modulus of the silicone sealant, L is a length of the bond, $\nu = 0.3$ is the Poisson's ratio. Since fracture propagation is not considered, the parallel bond tensile and shear strength are set to large values of 10 GPa. Since in the experiment powdered graphite was used as a lubricant, the friction coefficient between the wedge and the discs was set to a small value of 0.01.

The PFC solution scheme is based on the explicit finite-difference method, which requires specifying the time step. We used the time step of 0.0005 s; reducing the time step did not change the results.

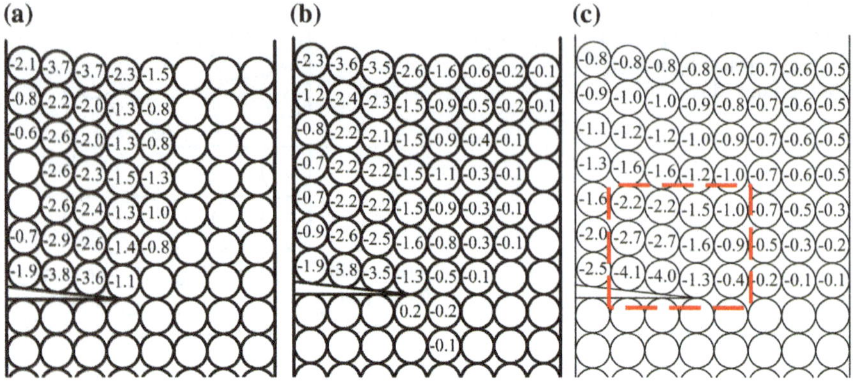

Fig. 5 Comparison between rotations calculated/measured in different models: **a** physical model; **b** DEM model and **c** analytically calculated. The number inside each disc indicates the angle of rotation in degrees (negative angles mean clockwise rotation). The *dash-lined rectangle* indicates a zone where analytical solution and experimental and numerical results are in reasonable agreement

Figure 5 shows the comparison of disc rotations in the assembly during the physical experiment (Fig. 5a), DEM model (Fig. 5b) and the analytical model (Eqs. (9), (10)) at the stage when the wedge reaches the third pair of the discs (Fig. 5c). The numbers in each disc show the rotation (in degrees). Negative values denote the clockwise direction. Rotations in both directions smaller than the measurement error of 0.3° are not shown. As the wedge thickness is relatively small, the discs above the wedge rotate slightly (maximum magnitude is 3.8° in Fig. 5a, b and 4.1 in Fig. 5c), the discs near the right wall (slider) practically do not rotate.

During movement the wedge hardly contacts the discs along the upper fracture surface (except of the disc at the left boundary of the assembly). This shows that the opening of the Mode I fracture is determined by the wedge thickness at the left edge of the physical model, which justifies the use of straight wedge in analytical calculations. Furthermore, despite very approximate nature of the analytical model it gives acceptable results near the crack surfaces, which is the area where the knowledge of rotations is most important as the rotation-induced moment stresses cause failure and the resulting fracture pattern (see [9, 13] for details).

6 Conclusions

2D physical experiment and discrete element modelling (DEM) of mechanical behaviour of an idealised slightly cemented granular material with a pre-existing fracture features particle rotations associated with the fracture opening. Subsequently, the homogenisation of the particulate material should involve the Cosserat theory to account for the rotational degree of freedom. The obtained Cosserat continuum has the Cosserat lengths commensurate with the particle size and hence become infinitesimal in the continuum. This corresponds to the asymptotics of small Cosserat lengths; the main asymptotic term (*Small-scale Cosserat continuum*) is given by the standard equations of elasticity with the rotational field obtained from the displacement field and the moment stresses determined from the rotation gradients using the Cosserat constitutive equations. In particular, the rotations caused by opening of a fracture obtained using this approach show adequate correspondence to both experimentally measured and DEM calculated rotations.

Acknowledgements The authors acknowledge the financial support from the UWA Near Miss Scheme 2015.

References

1. Hunter, S.C.: Mechanics of Continuous Media. Willey, New York, London, Sydney, Toronto (1976)
2. Cosserat, E., Cosserat, F.: Théorie des corps déformables. A. Herrmann et Fils, Paris (1909)
3. Toupin, R.A.: Perfectly elastic materials with couple stresses. Arch. Ration. Mech. Anal. **11**, 385–414 (1962)
4. Mindlin, R.D., Tiersten, H.F.: Effects of couple-stresses in linear elasticity. Arch. Ration. Mech. Anal. **11**, 415–448 (1962)
5. Dyskin, A.V., Pasternak, E.: Rotational mechanism of in-plane shear crack growth in rocks under compression. In: Potvin, Y., Carter, J., Dyskin, A.V., Jeffrey, R. (eds.) 1st Southern Hemisphere International Rock Mechanics Symposium SHIRMS, pp. 111–120. Australian Centre for Geomechanics (2008)
6. Dyskin, A.V., Pasternak, E.: Cracks in Cosserat continuum—macroscopic modelling. In: Maugin, G.A., Metrikine, A.V. (eds.) Mechanics of Generalized Continua: One Hundred Years After the Cosserats. Springer Series Advances in Mechanics and Mathematics, pp. 35–42. Springer, New York (2010)
7. Pasternak, E., Dyskin, A.V.: Intermediate asymptotics for scaling of stresses at the tip of crack in Cosserat continuum. In: 12th International Conference Fracture ICF12, Paper T40.014, Ottawa (2009)
8. Dyskin, A.V., Pasternak, E.: Asymptotic analysis of fracture propagation in materials with rotating particles. Eng. Fract. Mech. **150**, 1–18 (2015)
9. Dyskin, A.V., Pasternak, E., Esin, M.: Multiscale rotational mechanism of fracture propagation in geomaterials. Philos. Mag. **95**, 3167–3191 (2015)
10. Nowacki, W.: Theory of Micropolar Elasticity. Springer, Wien (1970)
11. Tada, H., Paris, P.C., Irwin, G.R.: The Stress Analysis of Cracks Handbook, 3rd edn. ASME Press, New York (2000)
12. Pasternak, E., Dyskin, A.V., Esin, M., Hassan, G.M., MacNish, C.: Rotations and pattern formation in granular materials under loading. Philos. Mag. **95**, 3122–3145 (2015)
13. Esin, M., Dyskin, A.V., Pasternak, E., Xu, Y.: Mode I crack in particulate materials with rotational degrees of freedom. Eng. Fract. Mech. (2017) (in print). doi:10.1016/j.engfracmech.2016.12.0240013-7944

Strain Localization with Rate Dependent Models Versus Cosserat Continuum: An Illustrative Example

Ioannis Stefanou and Jean Sulem

Abstract A simple example of adiabatic shearing of a rock layer under constant shear stress is considered in order to investigate and juxtapose two different modeling frameworks concerning strain localization and shear band thickness. The first framework is the Cauchy continuum with a rate dependent constitutive law (viscoplasticity). The second modeling framework is Cosserat elastoplasticity. Cosserat continuum is a special case of higher order continua. It is shown that the conditions for shear band triggering have a similar mathematical form, even though the starting point is different from a physical point of view.

1 Introduction

The simple example of adiabatic shearing of a rock layer under constant shear stress and in inelasticity is considered in this chapter. The purpose of this example is to investigate and juxtapose two different modeling frameworks of at least, at first approximation, different physics, and compare the conditions for which shear band localization takes place in each case.

The first framework is the Cauchy continuum with rate-dependent constitutive law (viscoplasticity). Rate dependent constitutive and in general visco-plastic constitutive laws in the frame of Cauchy continuum are frequently used in the literature as they lead to finite thickness shear band formation. The relation between viscosity and shear band thickness (and consequently material length scale) has been discussed in several publications (e.g. [1]).

The second modeling framework is Cosserat elastoplasticity. Cosserat continuum is a special case of higher order continua, which are also called generalized continua (for a classification of the most common higher order continua we refer to [2]). Cauchy continuum is a special case of the Cosserat continuum, if one neglects

I. Stefanou (✉) · J. Sulem
Laboratoire Navier, CNRS (UMR 8205), ENPC, IFSTTAR, Université Paris-Est, Champs-sur-Marne, France
e-mail: ioannis.stefanou@enpc.fr

E. Papamichos et al. (eds.), *Bifurcation and Degradation of Geomaterials with Engineering Applications*, Springer Series in Geomechanics and Geoengineering, DOI 10.1007/978-3-319-56397-8_69

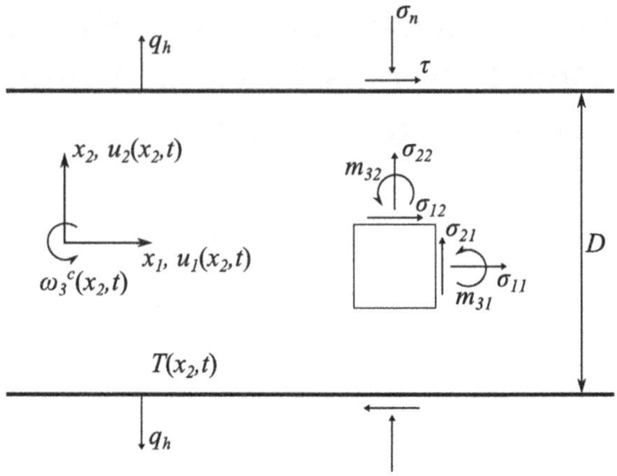

Fig. 1 Shearing of a rock layer: Cosserat rotational degree of freedom ω and couple stresses

the additional rotational degrees of freedom that the latter has and their conjugate in energy generalized stresses.

Thermal softening is taken into account as a destabilizing mechanism that may lead to shear band localization. The complexity of the chosen constitutive laws and of the multiphysical couplings considered is kept to a minimum degree in order to reveal the salient futures of each framework and highlight their similarities and their differences as far as strain localization is concerned. For a more detailed modeling in the frame of Cosserat continuum involving thermo-poro-chemo-mechanical couplings and more elaborate constitutive laws for the rock material, the reader is referred to [3–7]. The reader is referred to [8] for the Cauchy rate dependent framework under thermo-poro-chemo-mechanical couplings.

The thickness of the rock layer considered is D and constant normal and shear stresses are applied at its boundaries as depicted in Fig. 1. Initially, the layer is considered to be in a state of homogeneous shear deformation.

In both models it is assumed that all the plastic work is converted to heat and that Fourier's law is applicable. Under these assumptions, the heat equation is written in indicial notation as follows:

$$\frac{\partial T}{\partial t} = c_{th}\frac{\partial^2 T}{\partial x_2^2} + \frac{1}{\rho C}\sigma_{ij}\dot{\gamma}^p_{ij} + \frac{1}{\rho C}m_{ij}\dot{\kappa}^p_{ij} \tag{1}$$

where, T is the temperature inside the layer (the flux is denoted as q_h), c_{th} is the thermal diffusivity, ρC the specific heat, σ_{ij}, $\dot{\gamma}^p_{ij}$ are respectively the non-symmetric stress and plastic deformation rate tensors and m_{ij}, κ_{ij} the couple stress and plastic curvature rate tensors respectively. A common assumption is to consider that the layer is invariant in the x_1 and x_3 directions (infinite layer hypothesis). Consequently the derivatives in these directions vanish. Repeated indices indicate summation and $i,j = 1,2,3$. $(.)_{,i}$ denotes derivation in the ith direction, i.e. $(.)_{,i} = \frac{\partial}{\partial x_i}(.)$, and $(\dot{.})$ is the time derivative. Small deformations are considered and the slip event is sufficiently rapid in order to justify adiabatic conditions at the boundaries of the layer.

2 Cauchy Continuum with Rate Dependent Constitutive Law

Let's assume a simple rate dependent constitutive law for the shear stress at a point inside the shear layer:

$$\sigma_{12} = \tau_0 + H\dot{\gamma}_{12} + \xi(T - T_s) \tag{2}$$

where H is a mechanical hardening parameter (positive), ξ a thermal softening parameter (negative), T_s a reference temperature and τ_0 the shear stress at steady state and reference temperature.

For a Cauchy continuum the linear momentum balance is:

$$\sigma_{ij,j} = 0 \tag{3}$$

Inertia terms and body forces are neglected in this example. The angular momentum balance imposes the symmetry of the stress tensor, $\sigma_{ij} = \sigma_{ji}$. For a Cauchy continuum, $\gamma_{ij} = u_{i,j}$, where u_i is the displacement in the ith direction.

At steady state $T = T^* = T_s$, $\sigma_{12} = \overset{*}{\sigma}_{12} = \tau_0$, $\sigma_{22} = \overset{*}{\sigma}_{22} = \sigma_0$, $\dot{\gamma}_{12} = \dot{\overset{*}{\gamma}}_{12} = 0$ and $\dot{T}^* = 0$. This state will be stable as long as any perturbation does not grow in time. By perturbing the temperature and displacements fields at steady state, $T = T^* + \tilde{T}$, $u_i = u_i^* + \tilde{u}_i$) and by neglecting higher order terms, Eqs. (1), (2) and (3) become:

$$\begin{aligned} &\tilde{\sigma}_{12} = H\dot{\tilde{\gamma}}^p_{12} + \xi\tilde{T}; \quad \frac{\partial \tilde{\sigma}_{12}}{\partial x_2} = 0; \quad \frac{\partial \tilde{\sigma}_{22}}{\partial x_2} = 0 \\ &\frac{\partial \tilde{T}}{\partial t} = c_{th}\frac{\partial^2 \tilde{T}}{\partial x_2^2} + \frac{1}{\rho C}\overset{*}{\sigma}_{12}\dot{\tilde{\gamma}}^p_{12} \end{aligned} \tag{4}$$

The perturbations $\tilde{T}$, $\tilde{u}_i$ should fulfill the boundary conditions of the rock layer. Equation (4) together with the boundary conditions $\left.\frac{\partial \tilde{T}}{\partial z}\right|_{z=\pm\frac{D}{2}} = 0$, $\tilde{\sigma}_{12}\left(z = \pm\frac{D}{2}\right) = 0$ and $\tilde{\sigma}_{22}\left(z = \pm\frac{D}{2}\right) = 0$ form a linear system of partial differential equations which admits solutions of the type $\tilde{u}_i = U_i e^{st} \sin\frac{2\pi}{\lambda}z$, $\tilde{T} = Te^{st}\cos\frac{2\pi}{\lambda}z$, where s is the so-called growth coefficient and $\lambda = \frac{D}{N}$, $N = 1, 2, 3, \ldots$ Replacing the perturbations into (4) we obtain: $s = -\frac{\xi\tau_0}{H\rho C} - \frac{4\pi^2 c_{th}}{\lambda^2}$. The system is unstable when $s > 0$ or, equivalently, when the wavelength of the perturbation is bigger than a critical wavelength λ_{cr}^{rd}: $\lambda > \lambda_{cr}^{rd} = 2\pi\sqrt{\frac{Hc_{th}\rho C}{-\xi\tau_0}}$.

3 Cosserat Elastoplasticity

Compared to the classical Cauchy continuum, Cosserat continuum is equipped with additional degrees of freedom, i.e. the Cosserat rotations. The rotational degrees of freedom are conjugate in energy with moments, the so-called couple stresses. For a Cosserat continuum, the angular momentum balance leads to the additional equation for the couple stresses:

$$m_{ij,j} + \varepsilon_{ijk}\sigma_{kj} = 0 \quad (5)$$

Body forces and moments as well as inertia and micro-inertia terms are also neglected in this paragraph. It is worth mentioning that due to the presence of the couple stresses the stress tensor is not symmetric, i.e. $\sigma_{ij} \neq \sigma_{ji}$. ε_{ijk} is the Levi-Civita symbol. The generalized strains of the Cosserat continuum are $\gamma_{ij} = u_{i,j} + \varepsilon_{ijk}\omega_k^c$ and the curvatures are $\kappa_{ij} = \omega_{i,j}$. An elastic perfectly plastic constitutive behavior with thermal softening is assumed in this example. More advanced Cosserat constitutive models such as the Mühlhaus-Vardoulakis Cosserat plasticity model [9] might be used, but the advantage of this simple model is that analytical derivations can be performed, which permits a convenient comparison with the above rate dependent model. By analogy with the Cauchy rate dependent model presented in the previous paragraph, the yield surface is defined as:

$$F = \sigma_{(12)} - \tau_0 - \xi(T - T_s) \leq 0 \quad (6)$$

where $\sigma_{(ij)}$ denotes the symmetric part of the stress tensor. In this way the same shear stress limit and thermal softening with the Cauchy model is retrieved if one neglects the rate dependent term in Eq. (2). The strains and curvatures of the Cosserat medium are split in elastic and plastic parts $\gamma_{ij} = \gamma_{ij}^{el} + \gamma_{ij}^{pl}$, $\kappa_{ij} = \kappa_{ij}^{el} + \kappa_{ij}^{pl}$ as usual. Nevertheless, because of the chosen yield surface (Eq. (6)) the plastic curvatures are zero and therefore they do not contribute to the heat equation (Eq. (1)). In a centrosymmetric, linear elastic isotropic Cosserat medium, the stresses are related to the generalized elastic deformation measures according to the following constitutive relations [10]:

$$\begin{aligned} \sigma_{ij} &= K\gamma_{kk}^{el}\delta_{ij} + 2G\left(\gamma_{(ij)}^{el} - \frac{1}{3}\gamma_{kk}^{el}\delta_{ij}\right) + 2\eta_1 G\gamma_{[ij]}^{el} \\ m_{ij} &= 4GR^2\left(\kappa_{(ij)}^{el} + \eta_2\delta_{ij}\kappa_{kk}^{el}\right) + 4GR^2\eta_3\kappa_{[ij]}^{el} \end{aligned} \quad (7)$$

where K is the bulk modulus, G is the shear modulus, η_1, η_2, η_3 are positive material constants and R is an internal length parameter, which here is identified to the mean radius of the grains of the Representative Volume Element (RVE). $\gamma_{(ij)}$ and $\gamma_{[ij]}$ denote respectively the symmetric and anti-symmetric part of γ_{ij}. The Cosserat shear modulus, which expresses the stiffness related to the relative rotation of the particle (e.g. of a grain) with respect to the macro-rotation of the continuum (e.g. assemblage of grains) is defined as $G_c = \eta_1 G$.

At steady state we have a Cauchy continuum under homogeneous shear. In particular, $\dot{T}^* = 0$, $T = T^* = T_s$, $\sigma_{(12)} = \sigma^*_{(12)} = \tau_0$, $\sigma_{[12]} = \sigma^*_{[12]} = 0$, $m_{32} = m^*_{32} = 0$ and $\sigma_{22} = \sigma^*_{22} = \sigma_0$. This state will be stable as long as any perturbation does not grow in time. The temperature, the displacement and the rotation fields at steady state are perturbed ($T = T^* + \tilde{T}$, $u_i = u^*_i + \tilde{u}_i$, $\omega^c_3 = \omega^{c*}_3 + \tilde{\omega}^c_3$) as in the Cauchy case. The perturbations $\tilde{T}$, $\tilde{u}_i$ and $\tilde{\omega}^c_3$ have to fulfill the boundary conditions of the rock layer as in the Cauchy continuum case and additionally $\tilde{m}_{32}\left(z = \pm\frac{D}{2}\right) = 0$. A linear system is then formed which admits solutions of the form:$\tilde{u}_i = U_i e^{st} \sin\frac{2\pi}{\lambda} z$, $\tilde{\omega}^c_3 = \Omega_3 e^{st} \cos\frac{2\pi}{\lambda} z$, $\tilde{T} = T e^{st} \cos\frac{2\pi}{\lambda} z$. The critical growth coefficient is then: $s = -\frac{16G\pi^4 R^2 \rho C c_{th}}{\lambda^2\left(4G\pi^2 R^2 \rho C + \left(8\pi^2 R^2 + \lambda^2\right)\xi\tau_0\right)}$ where we set $G_c = G$ for simplicity. The system is unstable when $s > 0$ or, equivalently when the wavelength of the perturbation is bigger than a critical wavelength λ^{Cos}_{cr}: $\lambda > \lambda^{Cos}_{cr} = 2\pi\sqrt{\frac{R^2(G\rho C + 2\xi\tau_0)}{-\xi\tau_0}} \approx 2\pi\sqrt{\frac{R^2 G\rho C}{-\xi\tau_0}}$. The Lyapunov exponent is singular for λ^{Cos}_{cr}. This is a mathematical artifact due to the absence of inertia and micro-inertia (see [3] for more details). However, here we focus on the onset of the instability. For typical values of the shear modulus, the applied shear stress at the boundary, the thermal softening parameter and specific heat, it holds $G\rho C \gg \xi\tau_0$.

4 Conclusions

Even though both frameworks are based on different constitutive assumptions and micro-mechanisms, the resemblance of the expressions for the critical wavelength impels an analogy between the hardening parameter of the viscoplastic model H and the Cosserat internal length, which here is chosen equal to the mean grain radius: $Hc_{th} \sim R^2 G$. The hardening parameter H can be measured experimentally for a given rock and it generally decreases during shearing together with the size of the grains and the shear modulus, which also decrease due to important shearing and comminution. It is worth mentioning that the term $R^2 G$ represents the rolling stiffness of the grains, which, in comparison with the classical Cauchy continuum, rigidifies the system in the same way that the viscous term in the rate dependent friction law does. If we take the example of a highly granulated fault gouge with a grain size of 10 μm and assuming a shear modulus $G = 300$ MPa, then for

$c_{th} = 1\,\mathrm{mm}^2/\mathrm{s}$, the hardening parameter H is equal to $H = 0.03\,\mathrm{MPa\,s}$, which is in agreement with experimental measurements [11, 12]. Consistently, we observe the similar role of the diffusion length and of the Cosserat internal length in the control of the thickness of the localization zone.

References

1. Wang, W.M., Sluys, L.J., De Borst, R.: Interaction between material length scale and imperfection size for localisation phenomena in viscoplastic media. Eur. J. Mech.—A/Solids **15**(3), 447–464 (1996)
2. Godio, M., Stefanou, I., Sab, K., Sulem, J.: Dynamic finite element formulation for Cosserat elastic plates. Int. J. Numer. Methods Eng. **101**(13), 992–1018 (2015)
3. Sulem, J., Stefanou, I., Veveakis, E.: Stability analysis of undrained adiabatic shearing of a rock layer with Cosserat microstructure. Granul. Matter **13**(3), 261–268 (2011)
4. Veveakis, E., Stefanou, I., Sulem, J.: Failure in shear bands for granular materials: thermo-hydro-chemo-mechanical effects. Géotech. Lett. **3**, 31–36 (2013)
5. Veveakis, E., Sulem, J., Stefanou, I.: Modeling of fault gouges with Cosserat continuum mechanics: influence of thermal pressurization and chemical decomposition as coseismic weakening mechanisms. J. Struct. Geol. **38**, 254–264 (2012)
6. Godio, M., Stefanou, I., Sab, K., Sulem, J.: Multisurface plasticity for Cosserat materials: plate element implementation and validation. Int. J. Numer. Methods Eng. **108**(5), 456–484 (2016)
7. Mühlhaus, H.B., Vardoulakis, I.: The thickness of shear bands in granular materials. Géotechnique **37**(3), 271–283 (1987)
8. Rice, J.R., Rudnicki, J.W., Platt, J.D.: Stability and localization of rapid shear in fluid-saturated fault gouge: 1. Linearized stability analysis. J. Geophys. Res. Solid Earth **119**(5), 4311–4333 (2014)
9. Vardoulakis, I., Sulem, J.: Bifurcation Analysis in Geomechanics. Blackie, Glascow (1995)
10. Vardoulakis, I.: Lecture Notes on Cosserat Continuum Mechanics with Application to the Mechanics of Granular Media, Jan 2009
11. Blanpied, M.L., Lockner, D.A., Byerlee, J.D.: Frictional slip of granite at hydrothermal conditions. J. Geophys. Res. **100**(B7), 13045 (1995)
12. Chester, F.M., Higgs, N.G.: Multimechanism friction constitutive model for ultrafine quartz gouge at hypocentral conditions. J. Geophys. Res. **97**(B2), 1859 (1992)

A Breakage Diffusion Model for Strength Softening Rock

Hans Muhlhaus and Lutz Gross

Abstract A generalised version of the gradient plasticity model by Muhlhaus and Aifantis [4] is proposed. As an example, simple shear with a weak zone of reduced cohesive strength around the center of the shear layer is considered. Details of the formulation, implementation and numerical and analytic results are presented. The results include unexpected elements, which are interpreted in terms of a partial (analytic) solution of the governing equations.

1 Introduction

The need for strain gradient theories of plasticity arises mainly in the context of strain softening materials [1] where the gradient terms provide an internal length scale avoiding mesh dependency of the numerical solution beyond the usual discretisation error. In geotechnical engineering, specifically in connection with mining applications, the ability of computational models to realistically capture strain/strength softening in numerical simulations is of crucial importance for usefulness of numerical simulations.

2 A Strength Softening Model

Our main aim is the development of a set of equations that describe inelastic behavior of brittle rock including the decrease of cohesive strength after a critical stress level is reached as defined by a yield criterion. For our formulation we choose the simplest model applicable, a Drucker–Prager type yield criterion in combination

H. Muhlhaus · L. Gross (✉)
School of Earth Sciences, The University of Queensland, St Lucia, QLD 4072, Australia
e-mail: l.gross@uq.edu.au

H. Muhlhaus
e-mail: h.muhlhaus@uq.edu.au

E. Papamichos et al. (eds.), *Bifurcation and Degradation of Geomaterials with Engineering Applications*, Springer Series in Geomechanics and Geoengineering, DOI 10.1007/978-3-319-56397-8_70

with a non-associated Prandtl–Reuss flow rule. The following notations apply: The components of the stress tensor are designated as σ_{ij}, the deviatoric stress reads $\sigma'_{ij} = \sigma_{ij} + p\delta_{ij}$, where $p = -\frac{1}{3}\sigma_{kk} = -\frac{1}{3}(\sigma_{11} + \sigma_{22} + \sigma_{33})$ is the pressure. With $\tau = \sqrt{\frac{1}{2}\sigma'_{ij}\sigma'_{ij}}$ the Drucker-Prager yield criterion is written as

$$F = \tau - \alpha p - \tau_Y \leq 0 \tag{1}$$

with friction parameter α and cohesive strength τ_Y. With the above assumptions, the constitutive relationships for the stretching D_{ij} are obtained as

$$\dot{\sigma}_{ij} = 2G\, D_{ij} + (K - \frac{2}{3})D_{kk}\delta_{ij} - \dot{\gamma}\,(G\frac{\sigma'_{ij}}{\tau} + \beta\, K\delta_{ij}) \tag{2}$$

where $\dot{\gamma}$ is the equivalent plastic strain rate, β is the dilatancy function, and G and K are the shear and bulk moduli, respectively. We also have to obey the secondary conditions $\dot{\gamma} \geq 0$ and $F \cdot \dot{\gamma} = 0$. The cohesive strength τ_Y is a function of the plastic strain γ in the form

$$\tau_Y = g(\gamma) \tag{3}$$

where g is the strain hardening/softening function. For the simple bi-linear model this takes the form

$$\tau_Y = g(\gamma) = max(g_0 - h \cdot \gamma, g_{res}) \tag{4}$$

where g_0 is the initial strength, g_{res} is the final strength and h is the hardening/softening modulus. It is well known that softening (i.e. $h > 0$) leads to localization and because of the absence of an internal length scale strong mesh sensitivity of the solution of the discretized equations follows. To regularize the situation we modify relationship (4) and thus the yield criterion (1) by adding a non-local, flux term [4]

$$\tau_Y = g(\gamma) + q_{j,j} \quad \text{with flux} \quad q_j = -\kappa\gamma_{,j} \tag{5}$$

where κ is a positive constant. In (3) and (5) the strain considers the contribution of the inelastic deformation mechanisms within the representative volume element (RVE) while the non-local term, i.e. flux q_j, considers the interactions of the mechanisms between neighboring RVEs. One can write $\kappa = G \cdot l^2$ where parameter l is interpreted as the characteristic length scale of the inelastic mechanisms, e.g. when crack spacing is treated as a material parameter. The negative sign of the plastic strain gradient term in (5) guarantees that the non-local term always stabilizes the deformation, independent of the sign of the derivative.

3 Numerical Scheme

We use a simple, tangential integration time scheme with time step size $dt > 0$. The current stress is set to $\sigma_{ij} = \sigma_{ij}^- + \Delta\sigma_{ij}$ where the stress increment $\Delta\sigma_{ij} = dt \cdot \dot{\sigma}_{ij}$ is obtained from Eq. (2) as

$$\Delta\sigma_{ij} = 2G\,\Delta\varepsilon_{ij} + (K - \frac{2}{3}G)\Delta\varepsilon_{kk}\delta_{ij} - \Delta\gamma\,(G\frac{\sigma_{ij}^{'-}}{\tau^-} + \beta\,K\delta_{ij}) \tag{6}$$

where $\Delta\gamma = \Delta t \cdot \dot{\gamma}$ and $\Delta\varepsilon_{ij} = \Delta t \cdot D_{ij}$ are the plastic strain and the strain tensor increments. The stress increment needs to fullfill conservation of momentum

$$-(\sigma_{ij}^- + \Delta\sigma_{ij})_j = 0 \tag{7}$$

which defines a system of partial differential equation for the displacement increment Δu_i and the plastic strain $\Delta\gamma$. To complete the system we need an additional equation which is derived from the yield condition (1). First we observe that

$$p = p^- - KD_{kk} + \beta K\Delta\gamma \quad \text{and} \quad \tau = \tau^- + \frac{G}{\tau^-}\sigma_{ij}^{'-}\Delta\varepsilon_{ij} - G\,\Delta\gamma\,. \tag{8}$$

Under the assumption that the yield condition is met in the current time step we obtain from identities (8):

$$\Delta\tau_Y = \tau_Y - \tau_Y^- = F^- + \frac{G}{\tau^-}\sigma_{ij}^{'-}\Delta\varepsilon_{ij} + \alpha K\Delta\varepsilon_{kk} - (G + \alpha\beta K)\Delta\gamma\,. \tag{9}$$

On the other hand the incremental form of the modified model for the cohesive strength (5) is given in the form

$$\Delta\tau_Y = h_T^-\,\Delta\gamma - (\kappa\,\Delta\gamma_{,j})_{,j} \tag{10}$$

with tangential softening modulus $h_T^- = g'(\gamma^-)$. Combining Eqs. (9) and (10) leads to

$$(h_T^- + G + \alpha\beta K)\,\Delta\gamma - (\chi\kappa\,\Delta\gamma_{,j})_{,j} - \chi(\frac{G}{\tau^-}\sigma_{ij}^{'-}\Delta\varepsilon_{ij} + \alpha K\Delta\varepsilon_{kk}) = \chi F^- \tag{11}$$

where the function χ marks regions in the domain in which the yield condition $F = 0$ is reached. We use

$$\chi = \begin{cases} 0 & \text{if } \hat{F} < 0 \\ 1 & \text{otherwise} \end{cases} \tag{12}$$

with $\hat{F}$ providing a predictor for the yield condition:

$$\hat{F} = \tau - \alpha p - \tau_Y^- = F^- + \frac{G}{\tau^-}\sigma_{ij}'^- D_{ij} + \alpha K D_{kk} \; . \tag{13}$$

Notice that Eq. (11) implies $\Delta\gamma = 0$ in the elastic region where $F < 0$ which in particular enforces $\Delta\gamma$ to be zero on the boundary of the plastic region $F = 0$ if this boundary is not an external boundary. Starting from $\chi = 0$ the coupled system of PDEs (7) and (11) is solved iteratively in each time step where χ is updated according to (12) and (13). Convergence is detected when changes in the displacement and plastic strain increments are small which is typically reached after three steps. Then the cohesive strength is updated using identity (9) which in comparison to Eq. (10) provides the advantage that no second order derivatives need to be evaluated. It is pointed out that in the absence of the diffusion term in Eq. (11) (i.e. $\kappa = 0$) the plastic strain increment $\Delta\gamma$ can be eliminated from the momentum Eq. (7) [3]. For this case no boundary conditions for regions in the plastic state are enforced on the $\Delta\gamma$ and as a consequence – in contrast to the case $\kappa > 0$ – $\Delta\gamma$ can become discontinuous at the boundaries between elastic and plastic regions.

The solution scheme has been implemented in python using the finite element solver esys-escript, see [2, 5].

4 Simple Shear

We consider a simple shear parallel to the horizontal $x = x_1$ axis. All fields depend on the vertical $z = x_2$ coordinate only. We define H as thickness of the layer ($-\frac{H}{2} \leq z \leq \frac{H}{2}$). We assume that the horizontal displacement is kept at zero at the bottom of the domain $z = -\frac{H}{2}$ and is slowly increased at a constant velocity v_1 at the top of the domain $z = \frac{H}{2}$ over time. No external pressure is applied and it is set $\alpha = \beta = 0$. We also assume incompressibility which makes the shear components ε_{12} and σ_{12} the only non-zero entries in the strain and stress tensor, respectively. To meet the momentum equation $\sigma_{12} = \tau$ needs to be constant. A weak zone is introduced by reducing the initial cohesive strength g_0 over a zone of thickness P at the center of the domain ($-\frac{P}{2} \leq z \leq \frac{P}{2}$). In this zone it is $\dot{\tau} = \dot{\tau}_Y = \dot{\sigma}_{12}$ and hence we can derive from Eq. (5)

$$\dot{\sigma}_{12} = -h\,\dot{\gamma} - \kappa\dot{\gamma}_{,zz} \tag{14}$$

In addition the boundary conditions $\dot{\gamma}(\frac{P}{2}) = \dot{\gamma}(-\frac{P}{2}) = 0$ need to be fulfilled if $P < H$. The solution is given as

$$\dot{\gamma} = \frac{\dot{\sigma}_{12}}{h}\left(\frac{\cos(\lambda z)}{\cos(\lambda\frac{P}{2})} - 1\right) \text{ with } \lambda = \sqrt{\frac{h}{\kappa}} \tag{15}$$

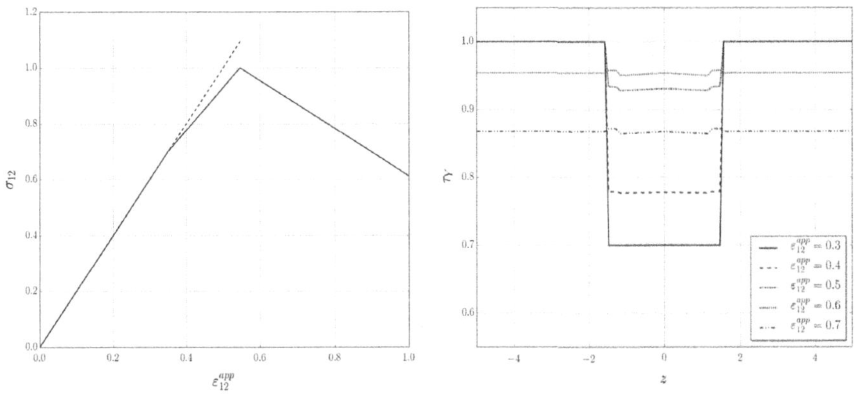

(a) Stress-Strain (b) Cohesive strength

Fig. 1 External strain ε_{12}^{app} versus stress σ_{12}, which is constant across the domain. The *straight line* shows the elastic stress-strain relation. Part **b** shows the cohesive strength τ_y as function of depth z for selected values of ε_{12}^{app}

for $-\frac{P}{2} \leq z \leq \frac{P}{2}$. The term within the brackets remains non-negative if

$$\lambda \frac{P}{2} < \frac{\pi}{2} \quad \text{or} \quad \frac{P^2\, h}{\pi^2} < \kappa\,. \tag{16}$$

As $\dot{\gamma} \geq 0$ $\dot{\sigma}_{12}$ must be positive. It can be shown that under condition (16) such a $\dot{\sigma}_{12}$ does in fact exist but can otherwise not be guaranteed (details are left out here). Note that here the internal length through the diffusion term controls the shape of the plastic strain within the weak zone but does control its width P.

Figure 1a shows the accumulated shear stress σ_{12} versus the applied total shear strain $\varepsilon_{12}^{app} = \frac{v_1 \cdot t}{H}$. We use $H = 10$, $G = 1$, $h = 0.3$, $\kappa = 1$, $P = 3$ and a displacement increment of 0.02 at the top boundary $z = \frac{H}{2}$. The initial strength g_0 is set to $0.5G$ within the weak zone and to G elsewhere. The history of the cohesive strength τ_Y is presented in Fig. 1b showing an initially increasing τ_Y in the weak zone until a constant value across the domain is reached at $\varepsilon_{12}^{app} \approx 0.52$. At this point the entire domain enters the plastic state, see Fig. 1a. Initially the plastic strain γ is accumulating in the weak zone only but is increasing through out the domain once global softening has kicked, see Fig. 2a, b. The corresponding behavior is observed for the shear strain ε_{12} as shown in Fig. 3a, b. After global softening has started ε_{12} is increasing at a constant rate across the domain where the rate equals the externally applied shear strain rate. Initially ε_{12} is increasing faster within the weak zone resulting in a reduction of the shear strain rate in the elastic zone. At the same time the shear stress σ_{12} is increasing in accordance to (15) but diverts from the initial elastic change, see Fig. 1a. At the point global plasticity is kicking in shear stress starts to decrease. This is consistent with Eq. (14) which in this situation simplifies to $\dot{\sigma}_{12} = -h\dot{\gamma} \leq 0$.

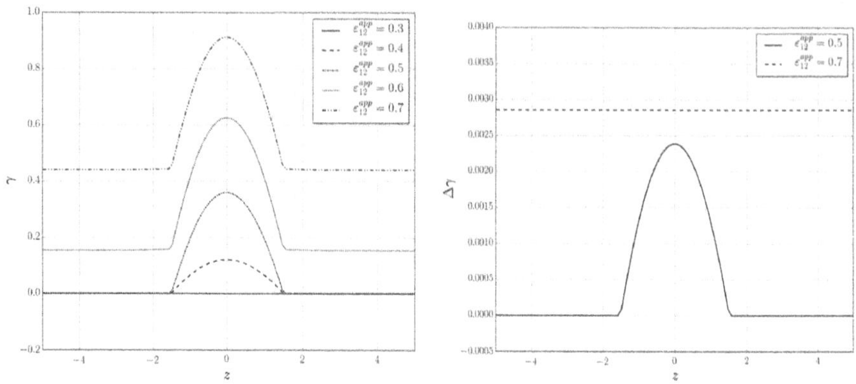

(a) Accumulated plastic strain (b) Incremental plastic strain

Fig. 2 Accumulated and incremental plastic strain as function of z for selected values of ε_{12}^{app}

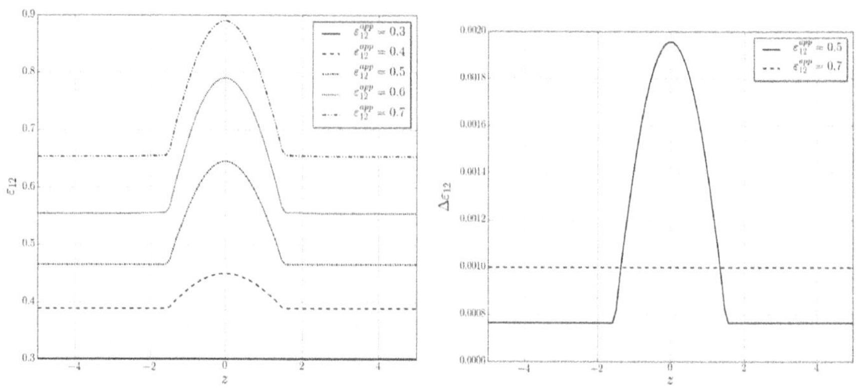

(a) Accumulated shear strain (b) Incremental shear strain

Fig. 3 Accumulated and incremental shear strain as function of z for selected values of ε_{12}^{app}

5 Conclusions

The solution of problems involving strain softening are non-unique in general. We have presented a solution of a simple shear problem which differs in some aspects from what has been presented in the literature. The differences are owed mainly to a specific choice of the strength parameters and the magnitude of the coefficient of the gradient hardening term. The solution is of interest, we believe, in its own right but also as a benchmark problem for finite element implementations.

Acknowledgements The support by Dr. R. Puscasu and Prof. G. Chitombo of the Bryan Mining Centre of the University of Queensland is gratefully acknowledged. The second author likes to acknowledge the support by AuScope NCRIS.

References

1. Bažant, Z.P., Belytschko, T.B., Chang, T.: Continuum theory for strain-softening. J. Eng. Mech. **110**(12), 1666–1692 (1984)
2. Gross, L., Bourgouin, L., Hale, A.J., Muhlhaus, H.B.: Interface modeling in incompressible media using level sets in escript. PEPI **163**, 23–34 (2007)
3. Gross, L., Mühlhaus, H., Thorne, E., Steube, K.: A new design of scientific software using Python and XML. Pure Appl. Geophys. **165**(3), 653–670 (2008)
4. Mühlhaus, H.B., Alfantis, E.: A variational principle for gradient plasticity. Int. J. Solids Struct. **28**(7), 845–857 (1991)
5. Schaa, R., Gross, L., Du Plessis, J.: PDE-based geophysical modelling using finite elements: examples from 3D resistivity and 2D magnetotellurics. J. Geophys. Eng. **13**, S59–S73 (2016)

Numerical Simulations of Debris Flows and Its Application in Hazard Assessments

K.T. Chau

Abstract As over seventy percent of the land of Hong Kong is mountainous, rainfall-induced debris flows are not uncommon in Hong Kong. The objective of this study is to incorporate numerical simulations of debris flows with GIS to identify potential debris flow hazard areas. To illustrate this approach, the proposed methodology is applied to Tsing Shan and Leung King Estate in Tuen Mun, Hong Kong. A three-dimensional Digital Elevation Model (DEM) of the terrain and the potential debris-flow sources were generated by using GIS to provide the required terrain and flow source data for the numerical simulations. A theoretical model by Takahashi et al. [9] improved by incorporating a new erosion initiation criterion was used for simulating the runout distances of debris flows. The well-documented 1990 Tsing Shan debris flow, was used to calibrate most of the flow parameters needed for computer simulations. Based on the simulation results, a potential hazard zone was identified and presented by using GIS. Our proposed hazard map was thus determined by flow dynamics and a deposition mechanism through computer simulations without using any so-called expert opinions.

1 Introduction

Over seventy percent of the 1100 km^2 of land in Hong Kong is hilly, and usable flat land is very scarce. Reclamation within the Victoria Harbor has been banned in recent years. The population of Hong Kong increased from 2.2 million in 1953 to 7.3 million in 2016. Natural hillsides have been transformed into residential and commercial. The risk of debris flow, rockfall and landslides in Hong Kong has been increasing tremendously because the rapid development of rural areas next to steep terrain [1, 4–7].

K.T. Chau (✉)
Department of Civil and Environmental Engineering, The Hong Kong Polytechnic University, King's Park, Hong Kong
e-mail: cektchau@polyu.edu.hk

E. Papamichos et al. (eds.), *Bifurcation and Degradation of Geomaterials with Engineering Applications*, Springer Series in Geomechanics and Geoengineering, DOI 10.1007/978-3-319-56397-8_71

The main objective of this paper is to summarize the numerical model proposed by Chau and Lo [2] and to provide some new results. Although there has been some criticism on such approach, we believe that theory based numerical simulation is superior to empirical approach in hazard estimation (e.g. [3]). Ideally, a reliable debris flow hazard map should carry "appropriate" weights from historical landslide events, from geomorphological analysis, and from mechanical or dynamical analysis of slides, falls, and flows of the earth mass. Since all three aspects of hazard analysis involve a large amount of factual, geological and simulated data, the use of computer or information technology is crucial to the success of such analysis. The present study uses the geographical information systems (or GIS) to incorporate the dynamics of debris flow.

2 Debris Flow Model

In the literature, there is no theoretical flow model that can simulate debris flow over a three-dimensional terrain, that at the same time takes into consideration of bed erosion and deposition. The only available models that can simulate three-dimensional debris flow approximately are the two-dimensional depth average models (e.g. [9]). Most of the existing models did not incorporate erosion and deposition mechanism; only the models by Takahashi et al. [9] and Ghilardi et al. [8] incorporated the possibility of erosion and deposition. In this study, we have adopted the model proposed by Takahashi et al. because it has been available for us to use (private communication, Takahashi and Nakagawa). Without going into the details, we should emphasize a major limitation of the model by Takahashi et al. [9]. That is, the critical slope gradient for the onset of erosion is assumed constant in the model. This is only an approximation since the critical gradient would naturally depend on the streampower of the flow.

In the Takahashi et al. [9] model, the mixture is assumed to consist of three components, water, fine solid particles, and coarse solid particles. The content of solids is represented by volumetric concentrations of solid particles in the mixture (c_{sp}), of coarse solid particles in the mixture (C_{cp}), of fine solid particles in mixture (c_{fp}), and of fine solid particles in interstitial fluid (c_{if}). They are defined as:

$$c_{sp} = \frac{V_s}{V_T}, \quad c_{cp} = \frac{V_c}{V_T}, \quad c_{fp} = \frac{V_f}{V_T}, \quad c_{if} = \frac{V_f}{V_i} \tag{1}$$

The flows of the fine and coarse particles through a small control volume satisfy the following continuity equations [9],

$$\frac{\partial c_{cp}}{\partial t} + \frac{\partial c_{cp}M}{\partial x} + \frac{\partial c_{cp}N}{\partial y} = ic^*_{cp}, \quad \frac{\partial c_{fp}}{\partial t} + \frac{\partial c_{fp}M}{\partial x} + \frac{\partial c_{fp}N}{\partial y} = ic^*_{fp} \tag{2}$$

where M and N are the average fluxes over flow depth h along the x- and y-directions respectively. The velocity of gain or loss of the solid particles is denoted by i. The "*" denotes the eroded or deposited debris mixture at the bed, what all the mixture concentrations on the left hand side must be given. The momentum equations of the debris mixtures flowing along the x- and y-directions are respectively [9]:

$$\frac{\partial M}{\partial t}+\beta\frac{\partial u_0 M}{\partial x}+\beta\frac{\partial v_0 M}{\partial y}=gh\sin\theta_{bx}-gh\cos\theta_{bx}\frac{\partial(z_b+h)}{\partial x}-\frac{\tau_{bx}}{\rho_T} \quad (3)$$

$$\frac{\partial N}{\partial t}+\beta\frac{\partial u_0 N}{\partial x}+\beta\frac{\partial v_0 N}{\partial y}=gh\sin\theta_{by}-gh\cos\theta_{by}\frac{\partial(z_b+h)}{\partial y}-\frac{\tau_{by}}{\rho_T} \quad (4)$$

where g is the gravitational constant (9.81 m/s^2); z_b is the deposit thickness; u_o and v_o are the velocities along x- and y-directions respectively; β is a momentum correction factor; ρ^T is the equivalent density of the debris mixture; θ_{bx} and θ_{by} are the tangents at the bed along the x- and y-directions respectively; and τ_{bx} and τ_{by} are the base shear resistances along the x- and y-directions respectively.

A new erosion initiation criterion is expressed in terms of the minimum energy gradient, than $\theta_{\min}$ (i.e. a natural slope with energy gradient θ less than $\theta_{\min}$ will suffer no erosion) [2]:

$$\theta_{\min}=\sin^{-1}\left\{\frac{c_{sp}G_s\rho_w\left(\int_{P_f}^{1}\omega_r dP\right)/(1-P_f)}{F\rho_m\sqrt{u_0^2+v_0^2}}\left(\frac{\sigma-2\rho_m}{\sigma-\rho_m}\right)\right\} \quad (5)$$

where $c_{sp}, F, G_s, \sigma, \rho_w, P_f, P, u_0, v_0, \rho_m$ and ω_r are solid concentration in the flow, excess fraction of the streampower available for erosion (typically 0.1–0.2), specific gravity of the solid particles, density of the solid particles, density of water, percentage of fine solid (i.e. particle size less than 63 μm), percentage of solid mass in the fluid with a certain particle size d, average flow velocity along the x-direction, average flow velocity along y-direction, density of the muddy water (i.e. the interstitial fluid), and relative settling velocity of particle of size d respectively. The deposition velocity can be calculated from [2]:

$$i=K\sqrt{gh}\sin^{3/2}\theta\left[1-\frac{\sigma-\rho_i}{\rho_i}c_{cp}\left(\frac{\tan\alpha}{\tan\theta}-1\right)\right]^{1/2}\left(\frac{\tan\alpha}{\tan\theta}-1\right)(c_{ca}-c_{cp})\frac{h}{d} \quad (6)$$

where $K, g, \sigma, \rho_i, h, d, \alpha$ and c_{ca} are a numerical constant (typically 0.06), the gravitational constant, the density of the interstitial fluid, the density of the debris material, the flow depth, the mean diameter of solid particles, the dynamic frictional angle, and the equilibrium solid concentration. Equation (6) is obtained by assuming that erosion is caused by the dynamic action of the shear stress on the bed by the interstitial fluid of the overlying sediment-laden flow. This erosion process is assumed to continue as long as the entrained-solids is less than the equilibrium value c_{ca}. As shown in Eq. (6) that when c_{cp} increases to c_{ca}, the erosion velocity i diminishes to zero.

3 Numerical Simulation for Debris Flow in Hong Kong

To calibrate the parameters of the present numerical model, back analysis will be conducted for the 1990 Tsing Shan debris flow event. The 1990 Tsing Shan debris flow occurred on the eastern flank of Tsing Shan on 11 September 1990 and is the largest debris flow in the recorded history of Hong Kong (see Fig. 1). It was estimated that a total of 19 000 m^3 of debris deposited down slope. The total travel distance is about 1 km. Although this event is relatively well-documented, similar to most other debris flows reported elsewhere no discharge histrograph was measured. We adopt the discharge histrograph recorded at Mount. St. Helens and scale it down match to the total volume of debris reported for the 1990 Tsing Shan debris flow. Regarding the solid (both fine and coarse) concentrations in the discharge, we use the solid concentrations adopted for the Horadani debris flow simulation (private communication, Takahashi and Nakagawa). The results of the simulation are shown in Figs. 1 and 2. Figure 1 shows the 3-D simulation result together with an aerial photo of the same slope at Tsing Shan after the debris flow. It is clear that the results are similar. A plan view of the observed and predicted deposition is shown in Fig. 2. Considering the roughness of the contour-based DEM, the comparison is considered reasonable. The parameters, especially those related to the criteria for erosion and deposition, interpreted from this back analysis will be used for numerical analysis of debris flow for other parts of Hong Kong.

Next, we conduct a debris flow hazard analysis for Leung King Estate. Note that Leung King Estate appeared in the background in Fig. 1. Field trip to this area, in conjunction with information extracted from aerial photographs, revealed landslides

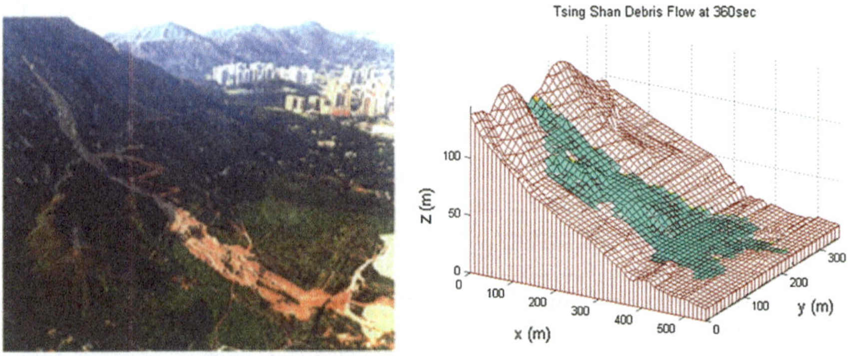

Fig. 1 Numerical simulation of the 1990 Tsing Shan debris flow

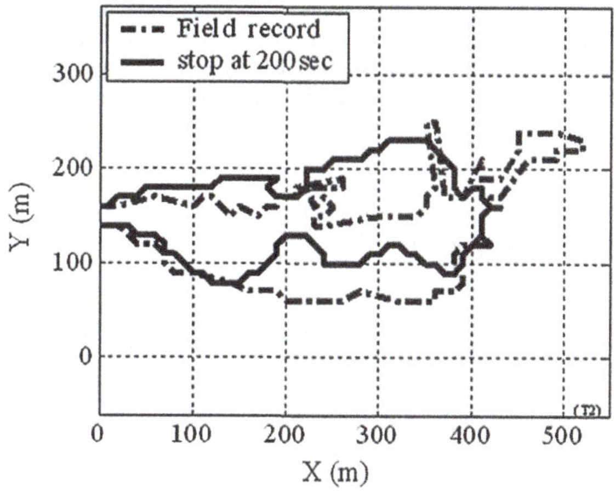

Fig. 2 Numerical simulation of the 1990 Tsing Shan debris flow

and debris flows have occurred in the past in the mountain range next to Leung King Estate.

Without going into the details, we will show in Fig. 3 that a hazard map was generated by superimposing the results of a number of numerical simulations. There are three areas on the map that the deposition may be up to 10 m (back color zones). The southern "black zone" does not arrive the buildings (mainly appears as outlines of Y shape and cross shape), the middle "black zone" stopped at the gabion wall built after the 2000 event, and the northern "black zone" touched the buildings and is a zone that remedial measures needed to be done.

In this paper, we illustrate that fluid-dynamics-based numerical simulation incorporated with GIS provides a powerful tool in debris flow hazard mapping.

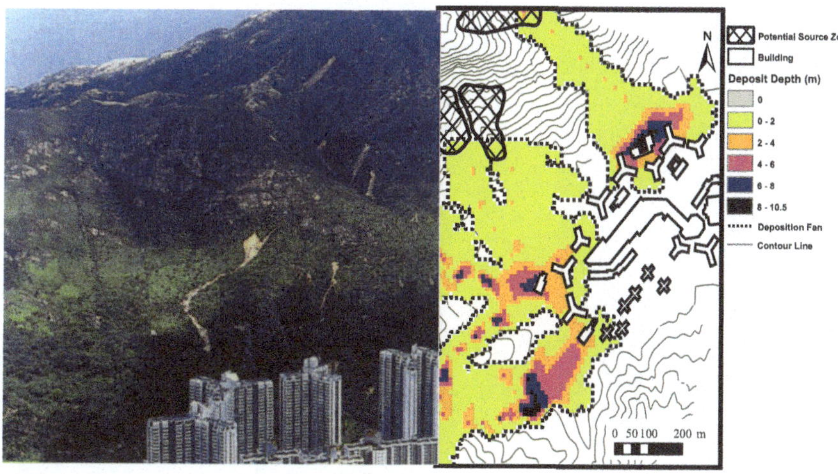

Fig. 3 Debris flow hazard map for Leung King Estate

Acknowledgements This paper was written under the support of PolyU research project "Theory, Analysis and Monitoring of Landslides and Debris Flows". The author is grateful to his former student K.H. Lo in carrying out the numerical simulations. This paper is dedicated to the 70th birthday of Professor Hans Muhlhaus, who is my mentor and collaborator at CSIRO when I visited in 1994.

References

1. Chau, K.T., Chan, J.E.: Regional bias of landslide data in generating susceptibility maps using logistic regression: case of Hong Kong Island. Landslides **2**(4), 280–290 (2005)
2. Chau, K.T., Lo, K.H.: Hazard assessment of debris flow for Leung King Estate of Hong Kong by incorporating GIS with numerical simulations. Nat. Hazards Earth Syst. Sci. **4**(1), 103–116 (2004)
3. Chau, K.T., Lo, K.H.: Author's reply to discussion on "Hazard assessment of debris flows for Leung King Estate of Hong Kong by incorporating GIS with numerical simulations" by S. Parry, M. E. Ruse and S. J. Williamson. Nat. Hazards Earth Syst. Sci. **5**, 25–27 (2005)
4. Chau, K.T., Wong, R.H.C., Liu, J., Lee, C.F.: Rockfall hazard analysis for Hong Kong based on rockfall inventory. Rock Mech. Rock Eng. **36**(5), 383–408 (2003)
5. Chau, K.T., Tang, Y.F., Wong, R.H.C.: GIS based rockfall hazard map for Hong Kong. Int. J. Rock Mech. Mining Sci. **41**(3), 530 (2004)
6. Chau, K.T., Sze, Y.L., Fung, M.K., Wong, W.Y., Fong, E.L., Chan, L.C.P.: Landslide hazard analysis for Hong Kong based on landslide inventory and GIS. Comput. Geosci. **30**(4), 429–443 (2004)
7. Chau, K.T., Wong, R.H.C., Ho, K.W., Mak, L.M.: Rockfall field test at Tung Shan Terrace, Hong Kong. In: Thirty Years of Slope Safety Practice in Hong Kong, GEO, CEDD, HKSAR Government, pp. 226–231 (2007)
8. Ghilardi, P., Natale, L., Savi, F.: Modeling debris flow propagation and deposition. Phys. Chem. Earth **26**(9), 651–656 (2001)
9. Takahashi, T., Nakagawa, H., Harada, T., Yamashiki, Y.: Routing debris flows with particle segregation. J. Hydraul. Res. **118**(11), 1490–1507 (1992)

Customized Coarse Models for Highly Heterogeneous Materials

T.J. Dodwell, A. Sandhu and R. Scheichl

Abstract Using a toy model for subsurface flow in highly heterogeneous materials, we demonstrate a methodology for building customized coarse models from generalized eigenvalue problems in overlapping subdomains. We show that this methodology allows to efficiently build accurate multiscale models with very few macroscale degrees of freedom for cases where classical computational homogenization methods break down. Such methods show great potential for modelling highly heterogeneous materials with properties that vary on small length scales. Such problems are typical in many engineering applications, for example subsurface flow or composite materials.

Keywords Generalized multiscale finite elements · Upscaling · Porous media · Eigenvalue problems

1 Introduction

We consider mathematical models represented by elliptic partial differential equations (PDEs) with high contrast coefficients that vary over small length scales relative to the macroscale dimension. Such models naturally arise in many engineering applications, for example composite materials or flow within porous media. In this paper, we consider a toy model for porous media given by the scalar elliptic PDE

$$-\nabla \cdot (k(\vec{x})\nabla u(\vec{x})) = 1 \quad \forall \vec{x} \in \Omega := [0,1]^2 \tag{1}$$

T.J. Dodwell (✉) · A. Sandhu
College of Engineering, Mathematics and Physical Sciences, University of Exeter, Exeter, UK
e-mail: t.dodwell@exeter.ac.uk

A. Sandhu
e-mail: as909@exeter.ac.uk

R. Scheichl
Department of Mathematical Sciences, University of Bath, Bath, UK
e-mail: r.scheichl@bath.ac.uk

E. Papamichos et al. (eds.), *Bifurcation and Degradation of Geomaterials with Engineering Applications*, Springer Series in Geomechanics and Geoengineering, DOI 10.1007/978-3-319-56397-8_72

subject to homogeneous Dirichlet boundary conditions $u(\vec{x}) = 0$ for all $\vec{x} \in \partial\Omega$. In porous media flow, $u(\vec{x})$ denotes fluid pressure and $k(\vec{x})$ is a spatially varying permeability field. For the numerical examples within this paper we take $k(\vec{x})$ to be the random field generated from a single sample of a log-normal random field with exponential covariance function, shown in Fig. 1 (left). Such fields are not differentiable, but they are Hölder continuous with exponent t, for any $t < 1/2$.

Equation 1 can be solved using finite elements (FE), for which we seek the approximate solution $u_h \in V_h$, the space of piece linear functions on a grid $\mathcal{T}_h$, which satisfies the variational equation

$$\int_\Omega k(\vec{x})\nabla u_h \cdot \nabla v_h \, d\vec{x} + \int_\Omega v_h \, d\vec{x} = 0 \quad \forall v_h \in V_h. \tag{2}$$

Capturing the fine scale details, arising from the variations in $k(\vec{x})$ at the small length scales, is computationally expensive since we require a sufficiently small mesh size (h). It is therefore, natural to try and build efficient multiscale methods which upscale the microscale information to a coarse/lower dimensional FE space V_H, which still captures the local microscale information of $k(\vec{x})$.

A natural question arises. What is a good choice for the coarse space V_H? It is easy to construct examples where the coarse model will provide a poor solution if V_H has insufficient degrees of freedom. For example, let V_H be the span of piecewise linear finite elements on a coarse grid characterized by a mesh size $H \gg h$. Consider a block of rock (Fig. 1 (right)) spanned by a single coarse linear finite element, which

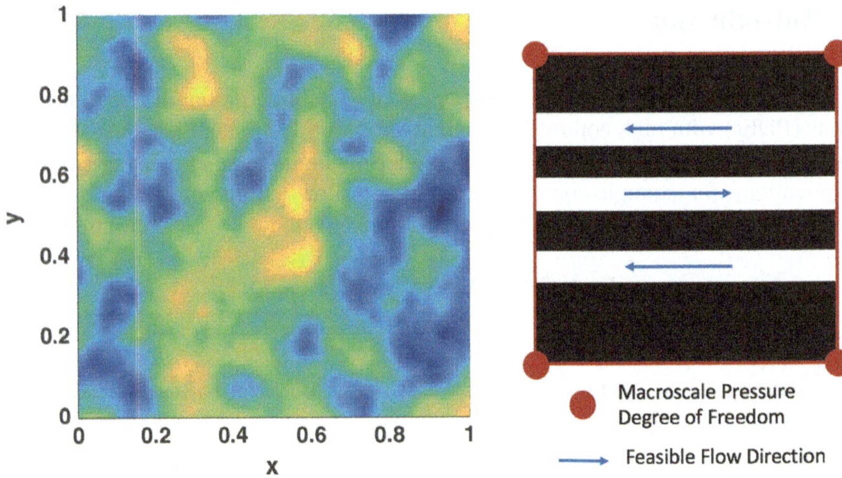

Fig. 1 *Left* Permeability field generated from a single realization of a log-normal random field, characterized by a small correlation length and high contrast (colors plotted on a log scale). *Right* A block of rock with three high permeability channels, spanned by a single macroscale bilinear finite element with 4 degrees of freedom

contains three channels of high permeability rock (white) surrounded by low permeability regions (black). One possible flow configuration is the fluid flows in one direction in two channels and the opposite direction in the third (as indicated by the arrows in Fig. 1 (right). For a single quadrilateral bilinear finite element the horizontal flow can only vary linearly in the vertical direction. Therefore this local switching of the flow is not a *mode* which is captured on the macroscale. For particular boundary conditions, such a coarse space would give non-representative results of the true microscale solution.

In such cases, additional degrees of freedom need to be added to V_H. Particularly in continuum mechanics this has led to the development of higher order continuum models, for example the Cosserat or strain-gradient continuum [1, 2]. Such models often provide accurate coarse scale descriptions of the microscale although their development often requires significant physical intuition and must be built on a case-by-case basis. In this paper, we describe a type of generalized multiscale FE method [3, 4] whereby the macroscale model is custom built from a few local eigenmodes computed on overlapping subdomains. These coarse spaces have also been shown to provide excellent preconditioners for two-level Schwarz method [5]. In this contribution we focus on their use as a multiscale method and provide step-by-step details an toy example of this exciting new multiscale method.

2 Customized Coarse Models for Highly Heterogeneous Models

Let V_h be the set of functions spanned by the basis of piece-wise linear finite element functions $\{\phi_j(\vec{x})\}_{j=1}^{n:=\dim(V_h)}$ on a grid $\mathcal{T}_h$, which is characterized by the small grid size h. The indices j will denote the jth degree of freedom, and so the FE solution of (2) in V_h can be represented as $u_h = \sum_{j=1}^{n} u_j\phi_i(\vec{x})$. This converts the variational problem (2) to the system of linear equations

$$\mathbf{K}\vec{u} = \vec{f} \tag{3}$$

where $\mathbf{K}_{ij} = \int_\Omega k(\vec{x})\nabla\phi_i \cdot \nabla\phi_j \, d\vec{x}$ is the stiffness matrix, $f_i = \int_\Omega \phi_i \, d\vec{x}$ the load vector and $\vec{u} = [u_1, u_2, \ldots, u_n]^T$ the solution vector.

Starting with a non-overlapping partition of Ω into N subdomains made up of disjoint sets of elements $\Omega = \cup_{j=1}^{N}\Omega_j'$, we extend each of the subdomains Ω_j' by O layers to achieve an overlapping partition $\Omega = \cup_{j=1}^{N}\Omega_j$. The part of Ω_j that is overlapped by neighbouring domains is denoted by the set

$$\Omega_j^\circ := \{\vec{x} \in \Omega_j \, : \, \exists\, i \neq j \text{ such that } \vec{x} \in \Omega_i\}.$$

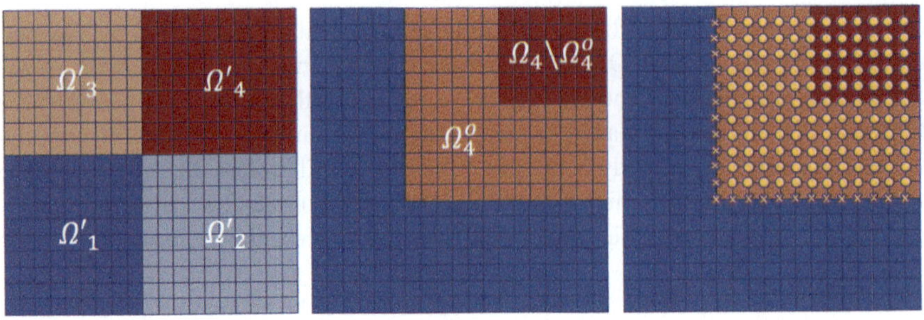

Fig. 2 *Left* Domain Ω split into $N = 4$ non-overlapping subdomains Ω'_j. *Middle* Overlapping subdomain Ω_4 (*red* and *orange*) with the subregion Ω_4° (*orange*), the overlapping region generated once each subdomain is extended by three layers. *Right Circles* mark $k \in \text{dof}(\Omega_4)$ and *crosses* mark $k \in \overline{\text{dof}}(\Omega_4)\backslash\text{dof}(\Omega_4)$

Furthermore the subset of indices k for which ϕ_k is supported in Ω_j is defined by

$$\text{dof}(\Omega_j) := \{k : 0 \le k \le n \text{ such that } \text{supp}(\phi_k) \subset \Omega_j\}, \tag{4}$$

whilst those degrees of freedom *active* in Ω_j are denoted by the set

$$\overline{\text{dof}}(\Omega_j) := \{k : 0 \le k \le n \text{ such that } \text{supp}(\phi_k) \cap \Omega_j \neq \emptyset\}. \tag{5}$$

The definition of each of these sets can perhaps be better understood with an example, as shown in Fig. 2.

The aim is to construct global multiscale basis functions from local eigenfunctions. These local eigenfunctions are FE functions restricted to a subdomain Ω_j. It is therefore natural to define the function space

$$V_h(\Omega_j) := \{v|_{\Omega_j} : v \in V_h\}$$

and the further restriction to those FE functions with support on Ω_j i.e.

$$V_{h,0}(\Omega_j) := \{v \in V_h(\Omega_j) : \text{supp}(v) \subset \Omega_j\}.$$

A function $v \in V_{h,0}(\Omega_j)$ can be extended to V_h by padding it with zeros. This extension operator is defined by $R_j^T : V_{h,0}(\Omega_j) \to V_h$. Its transpose $R_j : V_h \to V_{h,0}(\Omega_j)$ defines a restriction operator of any function $v \in V_h$ to $V_{h,0}(\Omega_j)$. In practice, the extension operator can be encoded in a sparse matrix $\mathbf{R}_j^T$ of 0's and 1's mapping local degrees of freedom (nodal values) to the their global counterparts.

However, the local eigenfunctions are in $V_h(\Omega_j)$ and so they are in general nonzero on the boundary of Ω_j. It is important to take care when extending these local eigenfunctions to global multiscale basis functions of V_h. To patch such (overlapping) functions together, we scale the values at shared nodes (that lie in several subdomains) using a *partition of unity* operator. In that way they can be more gradually brought to zero outside the subdomain Ω_j.

First, we start by defining, for each global degree of freedom k, the number $\xi_k := |\{j : k \in \mathrm{dof}(\Omega_j)\}|$ of subdomains in which ϕ_k is supported. Then, for each $j = 1, \ldots, N$, the local partition of unity operator applied to a function $v = \sum_{k\in\overline{\mathrm{dof}(\Omega_j)}} v_k\phi_k \in V_h(\Omega_j)$ is defined as

$$\Xi_j(v) := \sum_{k\in\mathrm{dof}(\Omega_j)} \frac{1}{\xi_k} v_k \phi_k|_{\Omega_j}. \tag{6}$$

This operator maps $V_h(\Omega_j) \to V_{h,0}(\Omega_j)$. Given this set of local partition of unity operators, we can reconstruct any global FE function $\Phi \in V_h$ from the local parts $\Phi|_{\Omega_j}$ as follows

$$\Phi = \sum_{j=1}^{N} R_j^T \Xi_j(\Phi|_{\Omega_j}). \tag{7}$$

In practice, the partition of unity operator Ξ_j is implemented as a matrix multiplication with a diagonal matrix $\mathbf{X}^{(j)}$ acting on the vector of nodal values $\vec{w^{(j)}}$ i.e. $\mathbf{X}^{(j)}\vec{w^{(j)}}$. The diagonal entries of $\mathbf{X}^{(j)}$ are $\mathbf{X}^{(j)}_{kk} = \xi_k^{-1}$, for $k \in \mathrm{dof}(\Omega_j)$.

A robust coarse space V_H that captures enough of the fine scale variation can now be constructed by calculating, for each $j = 1, \ldots, N$, the first m eigenfunctions $w^{(j,i)} \in V_h(\Omega_j)$, $i = 1, \ldots, m$, of a suitable generalised eigenproblem. In matrix form, the corresponding eigenproblem for the coefficient vector $\vec{w}^{(j,i)}$ of $w^{(j,i)}$ is

$$\mathbf{K}_{\Omega_j}\vec{w}^{(j,i)} = \lambda^{(j,i)}\left(\mathbf{X}^{(j)}\mathbf{K}_{\Omega_j^\circ}\mathbf{X}^{(j)}\right)\vec{w}^{(j,i)} \tag{8}$$

where $\mathbf{K}_{\Omega_j}$ and $\mathbf{K}_{\Omega_j^\circ}$ denote the stiffness matrices (with Neumann boundary conditions) on $V_h(\Omega_j)$ and on $V_h(\Omega_j)$, respectively. Once these eigenvectors are computed on each subdomain, the coarse space can be constructed as

$$V_H := \mathrm{span}\{R_j^T \Xi_j(w^{(j,\ell)}) : \ell = 1, \ldots, m;\ j = 1, \ldots, N\}.$$

The mapping of coarse mode coefficients in V_H to the original fine scale space V_h can be encoded by the mapping $\mathbf{R}_H^T$, a matrix of size $\dim(V_h) \times Nm$, with columns $\mathbf{R}_j^T\mathbf{X}_j\mathbf{w}_j^{(i)}$ for $j = 1, \ldots, N$ and $i = 1, \ldots, m$. Finally we reformulate the original variational problem (2) in the customized coarse space so that

$$\mathbf{K}_H\vec{U} = \vec{F} \tag{9}$$

where $\mathbf{K}_H = \mathbf{R}_H\mathbf{K}\mathbf{R}_H^T$, $\vec{U}$ is the solution vector which contains the coefficients of the coarse modes and $\vec{F}$ the load vector with entries $\vec{F}_{m(j-1)+i} = \vec{f}^T\mathbf{R}_j^T\mathbf{X}_j\vec{w}_j^{(i)}$.

Importantly the computation of the eigenvectors on each subdomain are independent, and the assembly of $\mathbf{K}_H$ and $\vec{F}$ require only nearest-neighbour (subdomain) communication. Therefore this coarse model can be assembled efficiently in parallel. In effect, we reduce the problem from one large solve to N independent (parallel) small eigenvalue problems and one inexpensive coarse solve.

3 Numerical Results and Concluding Remarks

We now test this method on the toy problem (2) with permeability field $k(\vec{x})$ taken to be a single realization of a random field define by the log-normal distribution shown in Fig. 1 (left). The fine mesh $\mathcal{T}_h$ is defined on a uniform square grid with $h = 1/200$ (i.e. 40,000 elements and 36,601 degrees of freedom), the domain is split into $N = 16$ square non-overlapping domains Ω_j', each with 2500 elements, and then overlapped by O layers to form an overlapping partition $\{\Omega_j\}$.

Figure 3 shows the first 5 eigenvectors $\vec{w}^{(6,i)}$, $i = 1, \ldots, 5$, of subdomain Ω_6 (with $O = 5$). We note that the smallest eigenvalue is zero and the corresponding eigenvector is constant on Ω_6. The same holds for any subdomain Ω_j that contains no degrees of freedom on the exterior boundary of Ω. In Fig. 4, we first compare the fine scale solution $\vec{u}_h$ computed on the fine grid $\mathcal{T}_h$ (left) visually with the coarse approximation $\vec{U}_H$ with an overlap of $O = 5$ and $m = 20$ (middle). Defining the error as $\varepsilon = \|\vec{u}_h - \vec{R}_H^T\vec{U}_H\|_2/\|\vec{u}_h\|_2$, this coarse model with over 100 times less degrees of freedom computes the solution with $\varepsilon = 0.034$. Finally in Fig. 4 (right), we study how the error ε in the coarse model relative to the fine scale model varies with the number m of eigenmodes per subdomain, for various sizes of overlap O.

Remark: We note that here we only applied the extension operator $\vec{R}_H^T$ to demonstrate the good agreement of $\vec{R}_H^T\vec{U}_H$ with the actual fine scale solution u_h. This reconstruction on the fine grid would not be generally done, since its cost is of the same order as an efficient iterative solver for finding u_h itself. In practice, it often suffices to project the pressure field onto a coarse FE space or to recover the fine scale solution in localized regions of interest.

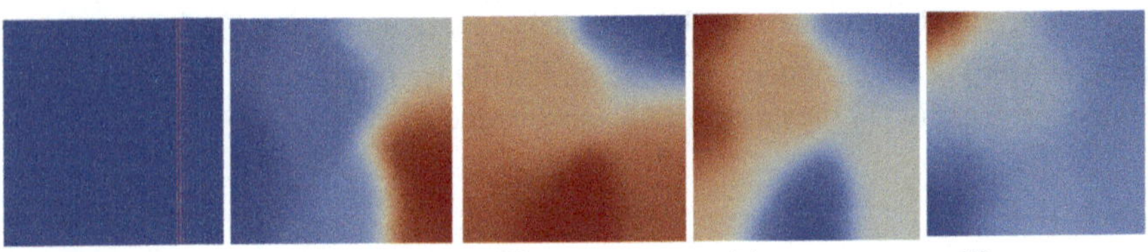

$\lambda_6^{(1)} = 0.0$ $\lambda_6^{(2)} = 0.0193$ $\lambda_6^{(3)} = 0.0511$ $\lambda_6^{(4)} = 0.0937$ $\lambda_6^{(5)} = 0.2056$

Fig. 3 Eigenvectors associated with the lowest five eigenvalues for Ω_6 with $O = 5$

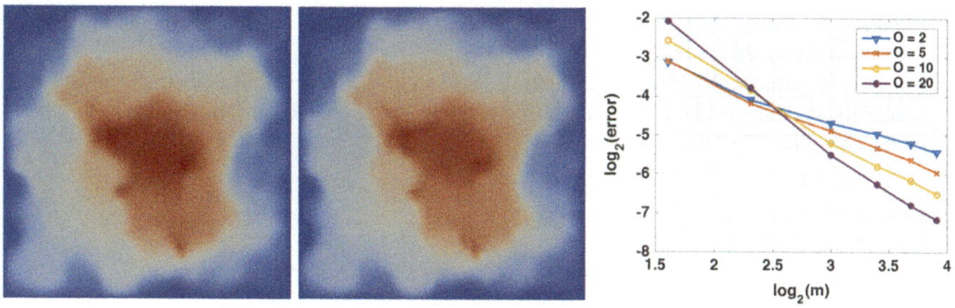

Fig. 4 (*Left*) Fine scale solution $\vec{u}_h$ (*Middle*) Bespoke Coarse space solution $\vec{R}_H^T\vec{U}_H$ with $m = 20$ and $O = 5$ (*Right*) Plot of $\log(m)$ against $\log_2(\epsilon)$ for various values of O

In this short paper, we have demonstrated the power of using local spectral information to build bespoke generalized multiscale FE spaces [4, 5]. This method allows the construction of coarse models with a fraction of the degrees of freedom of their fine scale counterparts, which still achieve accurate solutions with microscale information. For the particular toy model presented, the degrees of freedom are reduced by a factor of over 100, to achieve a solution within 3% of the fine scale solution. The methods show particular promise for applications with material parameters that vary over a range of length scales, have a complex distribution and/or demonstrate high contrast. For such cases, the relevant macroscale modes can be unclear, and the use of local eigenfunctions on subdomains in the coarse space construction appears natural.

We have not demonstrated the computational savings of this method, since we only tested it on small toy problems with $\sim 3.6 \times 10^4$ degrees of freedom. For such cases, even the fine scale problem can be solved efficiently on a single processor with a good direct solver. We observed a factor 2–3 speed up for an error of 3%. However, for 3D problems with large numbers of degrees of freedom ($\dim(V_h) > 10^7$) we expect to see significant gains. In effect, we substitute one large solve for N independent (parallel) small eigenvalue problems and one inexpensive coarse solve. Further gains can be expected for more structured high contrast materials (e.g. composites materials), since in such cases only representative local eigenproblems need to be solved. In our current work, we are exploring the application of these methods to 3D linear elasticity (a vector-valued elliptic PDE).

References

1. Muhlhaus, H.B.: Continuum models for layered and blocky rock. In: Comprehensive Rock Engineering, Vol. 2: Analysis and Design Methods. Pergamon Press, Oxford (1991)
2. Dodwell, T.J.: Internal wrinkling instabilities in layered media. Philos. Mag. **95**, 3225–3243 (2015)
3. Hou, T.Y., Wu, X.H.: A multiscale finite element method for elliptic problems in composite materials and porous media. J. Comput. Phys. **134**, 168–189 (1997)

4. Efendiev, Y., Galvis, J., Hou, T.Y.: Generalized multiscale finite element methods (GMsFEM). J. Comput. Phys. **251**, 116135 (2013)
5. Spillane, N., Dolean, V., Hauret, P., Nataf, F., Pechstein, C., Scheichl, R.: Abstract robust coarse spaces for systems of PDEs via generalized eigenproblems in the overlaps. Numerische Mathematik **126**, 741–770 (2014)

Physical and Mathematical Modeling of Erosion Processes Across the Scales

A. Scheuermann, H.-B. Muehlhaus, S. Galindo-Torres, H.M.D. Harshani, M. Aminpour, T. Bittner, P. To, M. Gholami-Korzani, D. Pedroso, L. Li and L. Gross

Abstract Internal erosion is the interaction of micro-scale hydraulic and mechanic processes taking place within the soil structure leading to macro-scale damages causing failures of whole constructions. While the geometrical and hydraulic conditions on the macro-scale supporting erosion are well investigated, the processes involved in the onset and continuation of erosion, the conditions influencing its temporal evolution are not well understood. The investigation of the processes involved in internal erosion requires a bijective or one-to-one approach with simultaneously implemented physical experiments and computational modelling on both relevant scales, micro- or pore-scale and macro- or continuum-scale. The aim of this approach is to accomplish an improved understanding of the underlying physics to be able to transfer this new knowledge into computational models enabling the solving of problems on the technical scale.

1 Introduction

Erosion is the process of particle transport within the soil structure induced by hydro mechanical forces leading either to the mixing of different fractions of soils or the washout of particles out of the structure. Erosion takes place in the hidden within the soil structure which is why it is referred to internal erosion [1]. To study erosion, numerical models have been developed on both, micro- and macro-scale. The rigorous development of these models requires validation based on experiments with well-defined hydraulic and mechanic boundary conditions and with

A. Scheuermann (✉) · S. Galindo-Torres · H.M.D. Harshani · M. Aminpour · T. Bittner · M. Gholami-Korzani · D. Pedroso · L. Li
The School of Civil Engineering, The University of Queensland, St Lucia, Australia
e-mail: a.scheuermann@uq.edu.au

H.-B. Muehlhaus · L. Gross
School of Earth Science, The University of Queensland, St Lucia, Australia

P. To
James Cook University, Townsville City, Australia

E. Papamichos et al. (eds.), *Bifurcation and Degradation of Geomaterials with Engineering Applications*, Springer Series in Geomechanics and Geoengineering, DOI 10.1007/978-3-319-56397-8_73

observations of the overall reactions in terms of discharge and deformation. Special attention is given to the observation of the state variables governing the underlying processes, such as the porosity.

In the following, numerical and experimental investigations and their results from a common project of the authors are presented and discussed against the background of improving the understanding of the temporal evolution of erosion.

2 The Continuum Scale

The most important parameter governing the process of erosion and which is changed by erosion is the porosity. An increase in porosity increases the hydraulic conductivity, which enhances the dislodgement of particles and consequently a further increase in porosity. Changes in hydraulic conductivity based on porosity can be described using the well-known Kozeny-Carman Equation [2]. Furthermore, porosity influences the resistance of the particles against dislodgement. It is always the interplay between hydraulic conditions and changes in structural and mechanical conditions which needs to be described in computational models and observed in experiments to provide the full picture of the overall phenomenon.

For being able to numerically simulate erosion, the dislodgement of particles needs to be introduced in the mathematical model. But, dislodged particles do not simply disappear. They rather transfer from a solid state in rest to a fluidized state. There are different approaches suggested in literature to describe this transition process which have in common to consider the fluidized solid phase as an additional liquid phase besides of water, and to use the flow velocity as the parameter driving the transformation process [3]. As a result, the mass balance equations for particles in rest and fluidized particles include a mass transformation rate being positive when erosion takes place and negative for the deposition of particles.

Based on this consideration of a three-phase continuum model consisting of water, solid phase in rest and solid phase in fluidization, Vardoulakis has developed a numerical model for describing the pre- and post-failure processes leading to the production of sand during suffusion [4]. The same approach of a three-phase continuum model was chosen by Scheuermann et al. [5] to simulate contact erosion due to horizontal water flow parallel to the interface of fine particles to coarse particles. Alternatively, one can also assume that dislodged particles are simply removed from the domain without simulating the flow of fluidized particles. Mercier et al. [6] have simulated the Hole Erosion Test based on this assumption. Changes in the shape of the hole were then simulated using the Level-Set method.

The Level-Set method was also used by Muehlhaus et al. [7] for simulating changes in porosity in a fluidized bed leading to the preferential growth of the porosity front in form of finger-like structures. Similar observations have been made during experiments and numerical simulations using Discrete Element Modelling (DEM) simulations coupled with the Lattice Boltzmann Method (LBM) for simulating water flow (see [8] and Fig. 1).

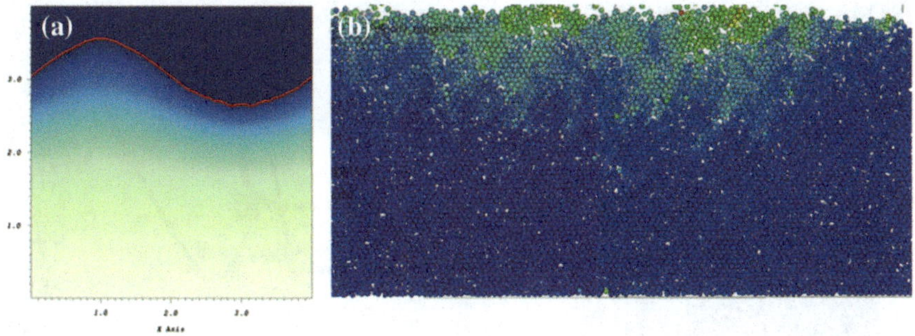

Fig. 1 Comparison between numerical results using **a** a continuum model to simulate the growth of a porosity front in a fluidized bed [7] and **b** results from DEM/LBM simulations [8]

Electromagnetic measurement methods, such as Time Domain Reflectometry (TDR) or frequency-based methods are most suitable for determining changes in porosity during experiments also as a profile [9]. As the dielectric permittivity of water (80) dominates the permittivity of the soil/water mixture, changes in permittivity can directly correlated with changes in porosity [10]. This approach is currently used to observe porosity distributions during contact erosion tests [11]. Figure 2a shows the column used for erosion experiments. The column is designed to form a coaxial line cell consisting of an inner conductor and an outer one. The sample forms the dielectric between both conductors. Furthermore, a calibration set-up (Fig. 2b) was developed to provide calibration for dielectric measurements and further constitutive relationships such as for the shear strength of a granular medium depending on the hydraulic gradient (Fig. 3).

3 The Micro-scale

Micro-scale simulations are frequently considered as a substitute for experimental investigations. For instance, simulations using Discrete Element Models (DEM) are used to derive general constitutive relationships [12] or to investigate the structural composition of granular media [13]. And flow simulations using the Lattice Boltzmann Method (LBM) are implemented to scrutinize the validity of flow laws for porous media [14]. Nevertheless, especially in engineering sciences some proof of validation of micro-scale models is expected to make sure that these models are capable of reproducing experimental investigations before using them for parametric studies and extrapolation.

The application of DEM and LBM allows the user to observe every desirable detail during the numerical simulation. Naturally, there are limitations if aspects on even smaller scales are considered, such as the effect of attracting and repelling surface forces on particles. But nevertheless, DEM allows to observe every motion of particles and the force created between them. And with LBM pressure, velocity

Fig. 2 **a** Column erosion tests with simultaneous observation of porosity. **b** Calibration set-up for investigating dielectric, mechanic and hydraulic parameters of fluidized granular media [11]

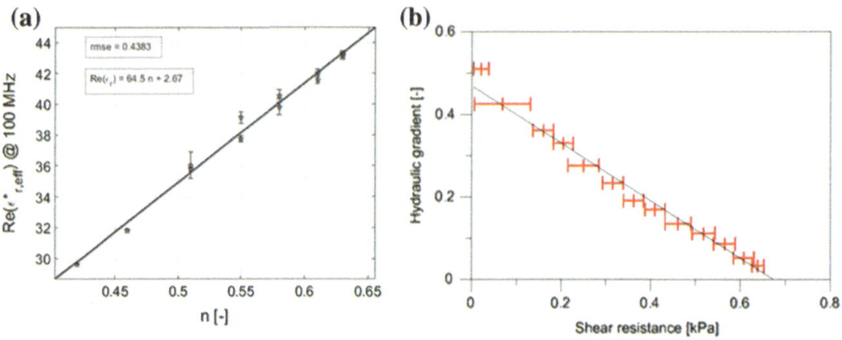

Fig. 3 **a** Relationship between porosity and dielectric permittivity at 100 MHz. **b** Relationship between shear strength and applied hydraulic gradient of a granular medium before fluidization

and moving direction of any liquid in every LBM cell can be captured. These possibilities frequently tempt users to try to look into parameters, which cannot be observed in physical experiments.

In the case of erosion, one problem considered with coupled DEM/LBM simulations concerned the onset of erosion. Focus was put in the question which parameter first shows the onset of erosion, and what first visible manifestation of erosion occurs in terms of movement of particles. With this question in mind, simulations on contact erosion have been conducted [15] and compared to experimental investigations implemented with glass beads [10]. The parameter used to quantitatively compare the results from simulation and experiment was the porosity. This comparison between experiments and simulation showed a satisfactory agreement. However, with respect to the parameter indicating the onset of erosion, several parameters including rotation of particles indicated the beginning of erosion. In contrast to the common understanding, it was not the movement of base particles in flow direction showing the first visual manifestation of particle movement. It was the penetration of coarse filter particles into the bed of fines indicating the onset of erosion which has been also observed in experiments.

Particle Imaging Velocimetry (PIV) is a well-known and suitable experimental method for investigating the pore-scale flow. For this method it is necessary to use particles and a liquid which are refractive index matched to create an overall transparent medium. This can be achieved by using for example glass beads in combination with oil or glycerin [16]. Seeding particles are then used as tracer to mark the liquid, which can be illuminated in a light sheet usually created by a laser system. A series of pairs of pictures taken at a known time interval using a high-speed camera are then used to quantify the flow field in the considered pore space. In order to simplify experimental procedures, a new set-up was developed [17]. This set-up involved a light emitting diode (LED) based light source for creating the light sheet. Furthermore, water was used as liquid in combination with hydro-gel beads to create a more adaptable set-up. The combination of LBM simulations and experiments using PIV allowed the direct comparison of flow patterns (Fig. 4) and the statistical analysis of flow velocities in porous media (Fig. 5).

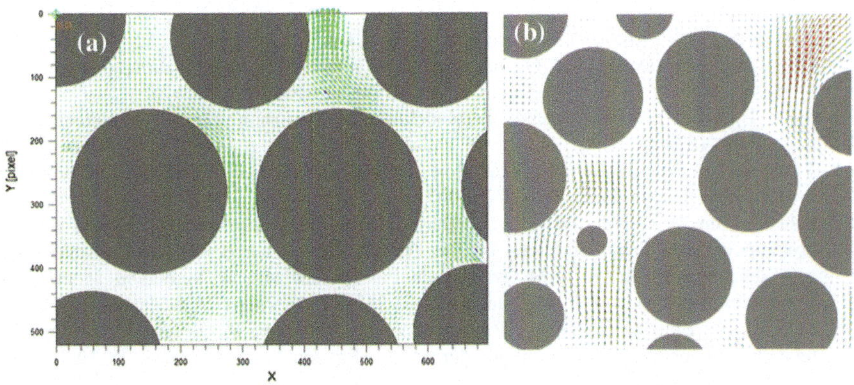

Fig. 4 Flow patterns measured with PIV (*left*) and simulated using LBM (*right*)

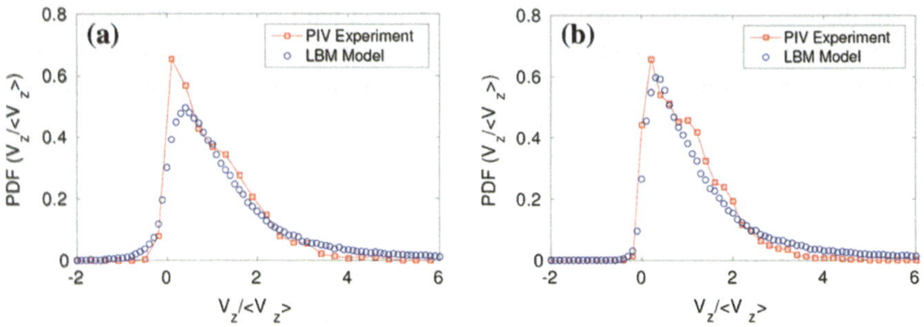

Fig. 5 Flow velocity in vertical direction for Re = 0.8 (*left*) and Re = 2.2 (*right*)

4 Discussion and Conclusions

The presented bijective approach of integrating the development of mathematical models and experimental set-ups optimizes the outcome of numerical tools to be developed. Frequently, parameters are involved in mathematical models, which cannot be observed using conventional experimental set-ups. At the same time, experiments also frequently produce results, which cannot be explained based on macroscopic observations only. In both cases, unconventional observation methods are required, which provide additional information of internal, mostly structural changes of the sample during testing. These observation tools, in form of tomographic methods, visual methods or geophysical methods, are available and need to be integrated in experimental set-ups. This contribution shows, as we hope, the great advantage of combining the developments of mathematical and experimental tools for developing numerical solutions for engineering problems.

References

1. Bonelli, S. (ed.): Erosion of Geomaterials. Wiley (2012)
2. Chapuis, R.P., Aubertin, M.: On the use of the Kozeny Carman equation to predict the hydraulic conductivity of soils. Can. Geotech. J. **40**(3), 616–628 (2003)
3. Steeb, H., Diebels, S., Vardoulakis, I.: Modeling internal erosion in porous media. Geotech. Spec. Publ. **157** (2007)
4. Vardoulakis, I.: Fluidisation in artesian flow conditions: hydromechanically unstable granular media. Géotechnique **54**(3), 165–177 (2004)
5. Scheuermann, A., Vardoulakis, I., Papanastasiou, P., Stavropoulou, M.: A sand erosion problem in axial flow conditions on the example of contact erosion due to horizontal groundwater flow. In: IUTAM Symposium on Theoretical and Numerical Methods in Continuum Mechanics of Porous Materials, pp. 169–175. Springer Netherlands (2001)
6. Mercier, F., Bonelli, S., Golay, F., Anselmet, F., Philippe, P., Borghi, R.: Numerical modelling of concentrated leak erosion during Hole Erosion Tests. Acta Geotech. **10**(3), 319–332 (2015)

7. Muhlhaus, H., Gross, L., Scheuermann, A.: Sand erosion as an internal boundary value problem. Acta Geotech. **10**(3), 333–342 (2015)
8. Harshani, H.M.D., Galindo-Torres, S.A., Scheuermann, A., Muhlhaus, H.B.: Micro-mechanical analysis on the onset of erosion in granular materials. Philos. Mag. **95** (28–30), 3146–3166 (2015)
9. Scheuermann, A.: Determination of porosity distributions of water saturated granular media using spatial time domain reflectometry (spatial TDR). Geotech. Test. J. **35**(3), 441–450 (2012)
10. Scheuermann, A., Muehlhaus, H.B., Bittner, T., Bieberstein, A.: Measurement of porosity distributions during erosion experiments using spatial time domain reflectometry (spatial TDR). In: 6th International Conference on Scour and Erosion (ICSE-6), pp. 94–101 (2012)
11. Bittner, T., Bore, T., Wagner, N., Karlovsek, J., Scheuermann, A.: Experimental investigation of the dielectric properties of soil under hydraulic loading. Meas. Sci. Technol. (2016)
12. Andrade, J.E., Avila, C.F., Hall, S.A., Lenoir, N., Viggiani, G.: Multiscale modeling and characterization of granular matter: from grain kinematics to continuum mechanics. J. Mech. Phys. Solids **59**(2), 237–250 (2011)
13. To, H.D., Torres, S.A.G., Scheuermann, A.: Sequential sphere packing by trilateration equations. Granul. Matter (2016)
14. Galindo-Torres, S.A., Scheuermann, A., Li, L.: Numerical study on the permeability in a tensorial form for laminar flow in anisotropic porous media. Phys. Rev. E **86**(4), 046306 (2012)
15. Galindo-Torres, S.A., Scheuermann, A., Mühlhaus, H.B., Williams, D.J.: A micro-mechanical approach for the study of contact erosion. Acta Geotech. **10**(3), 357–368 (2015)
16. Budwig, R.: Refractive index matching methods for liquid flow investigations. Exp. Fluids **17** (5), 350–355 (1994)
17. Harshani, H.M.D., Galindo-Torres, S.A., Scheuermann, A., Muehlhaus, H.: Experimental study of porous media using hydro-gel beads and LED based PIV. Meas. Sci. Technol. (accepted 11 Nov 2016) (2016)

Modelling of Shear Localization During Granular Flow Within Non-local Hypoplasticity Using Material Point Method

Pawel Hajko and Jacek Tejchman

Abstract The paper presents some 2D simulation results of shear localization during advanced granular flow. The sand behaviour was simulated using the finite element method based on hypoplasticity enhanced by a characteristic length of micro-structure by means of a non-local theory. In order to avoid the mesh distortion, the material point method was used. A satisfactory agreement between numerical and experimental results was achieved.

1 Introduction

Shear localization is a fundamental phenomenon in granular materials. Thus, it is of primary importance to take it into account while modelling the granular behaviour [1]. Localization under shear occurs either in the interior domain in the form of a spontaneous shear zone as a single zone or as a pattern of zones. It may be also created at interfaces in the form of an induced single zone where structural members interact and stresses are transferred from the surrounding granular body to the structure. When modelling shear zones within continuum mechanics, three basic conditions have to be fulfilled [1, 2]: (a) the material behaviour has to be described with a realistic constitutive model, (b) the constitutive model has to be enhanced by a characteristic length of micro-structure in order to obtain mesh-independent results and (c) the excessive mesh distortion has to be avoided by means of a suitable description of the material motion.

The paper deals with the description of 2D quasi-static shear localization in cohesionless sand during plane strain compression and confined silo flow between two parallel walls. In order to realistically capture the sand behaviour, a hypoplastic

P. Hajko · J. Tejchman (✉)
Faculty for Civil and Environmental Engineering, Gdańsk University of Technology, Gdańsk-Wrzeszcz, Narutowicza 11/12, 80-952 Gdańsk, Poland
e-mail: tejchmk@pg.gda.pl

P. Hajko
e-mail: pawhajko@pg.gda.pl

E. Papamichos et al. (eds.), *Bifurcation and Degradation of Geomaterials with Engineering Applications*, Springer Series in Geomechanics and Geoengineering, DOI 10.1007/978-3-319-56397-8_74

constitutive model was used. To properly capture the width of shear zones, the model was enhanced by a characteristic length of micro-structure by means of a non-local theory. The Material Point Method was used which was suitable for problems with an excessive mesh distortion and entanglement during advanced granular flow [3]. Numerical results were compared to corresponding experiments.

2 Material Point Method

The Material Point Method was used to simulate the motion of granular bodies. We took advantage of the explicit MPM implemented in the open program UINTAH 1.6.0 (developed at Utah University). In this approach, the continuum is discretized by material points whose motion is traced by a background arbitrary mesh. In this method the material is not connected with the finite element mesh. This approach deals instead of finite elements with material points which carry the information needed for the analysis (mass, momentum, stresses, constitutive parameters). The material points pass the required information to the mesh in each time step. The time step data are resolved on the mesh level and given back to the material points. The material point positions, stresses and constitutive parameters are next updated. The equation of the virtual work in MPM is the following [3]:

$$\int_{\Omega} \varrho\left(a_i w_i + \frac{1}{\varrho}\sigma_{ij} w_{ij}\right) dx = \int_{\Omega} \varrho b_i w_i dx + \int_{\Gamma_\sigma} t_i w_i ds + \int_{\Gamma_c} \sigma_{ij} n_j w_i ds, \quad (1)$$

where Ω describes the region of the analyzed continuum at the time $t \in [0, T]$, $\partial\Omega$ is the boundary of the region Ω where the displacements Γ_u, stresses Γ_σ and contacts Γ_c are given, σ_{ij} is the Cauchy stress tensor, t_i denotes the surface traction, b_i is the vector of body forces, ϱ represents the mass density, a_i is the vector of accelerations, w_i is the weighting function and n_j denotes the unit vector outwardly normal to the boundary region $\partial\Omega$. The entire mass of the continuum sub-region was concentrated in representative material points. The density field around each material point was expressed as:

$$\varrho(x) = \sum_{z=1}^{N} M_z \delta(x - X_z), \quad (2)$$

where M_z is the mass and X_z is the position of z-th material point, $\delta(x)$ denotes the Dirac delta function with dimension of the inverse of volume and N is the number of material points. Each time increment included two steps (Lagrangian step and convective step). In the Lagrangian step, the calculations were carried out in a similar way as in the usual FEM. In the convective step the material points velocities were mapped to the computational mesh.

3 Constitutive Model for Granular Materials

Hypoplastic constitutive models, which have been developed at Karlsruhe University, are of the rate type [1]. They are capable of describing a number of significant properties of granular materials: non-linear stress-strain relationship, dilatant and contractant behaviour, pressure dependence (barotropy), density dependence (pycnotropy), dependence on the deformation rate direction and material softening. A further feature of hypoplastic models is the inclusion of critical states, i.e. states in which a grain aggregate can continuously be deformed at constant stress and constant volume. In contrast to elasto-plastic models, a decomposition of deformation components into elastic and plastic parts, the formulation of a yield surface, plastic potential, flow rule and hardening rule are not needed. Moreover, both the coaxiality (understood as a coincidence of the directions of the principal stresses and principal plastic strain increments) and stress-dilatancy rule are not assumed in advance. The hallmark of these models is their simple formulation and procedure for determining material parameters with standard laboratory experiments. The material parameters are related to granulometric properties, viz. size distribution, shape, angularity and hardness of grains. The following representation of the general constitutive equation was used [4, 5]

$$\overset{\circ}{\sigma}_{ij} = f_s\left[L_{ij}(\hat{\sigma}_{kl}, d_{kl})\right] + f_d N_{ij}(\hat{\sigma}_{ij})\sqrt{d_{kl}d_{kl}}, \quad (3)$$

where $\overset{o}{\sigma}_{ij}$ is the Jaumann stress rate tensor, $\hat{\sigma}_{ij}$ denotes the normalized stress, L_{ij} and N_{ij} are the isotropic tensorial functions, d_{kl} is the rate of deformation, f_s , f_d represent the stiffness and density factor. The hypoplastic FE-calculations were carried out with the non-local modulus of the deformation rate d^*. The modulus of the deformation rate in Eq. 3, $d = \sqrt{d_{kl}d_{kl}}$, was treated non-locally [1, 2, 6]:

$$d^*(\mathbf{x}) = \frac{\int_V \omega(||\mathbf{x}-\boldsymbol{\xi}||)\, d(\boldsymbol{\xi})\mathrm{d}\boldsymbol{\xi}}{\int_V \omega(||\mathbf{x}-\boldsymbol{\xi}||)\mathrm{d}\boldsymbol{\xi}} \quad \text{with} \quad \omega(x) = \frac{1}{l_c\sqrt{\pi}} e^{-(r/l_c)^2}, \quad (4)$$

where: d^*—the non-local modulus of deformation, V—the volume of the body, $\mathbf{x}$—the coordinates of the considered (actual) point, $\boldsymbol{\xi}$—the coordinates of the surrounding points and ω—the weighting function, l_c—the characteristic length of micro-structure and r—the distance between two material points. A non-local approach may be used independently of the motion formulation. The characteristic length of micro-structure l_c was taken as 1.5 mm based on comparative calculations of the thickness of a shear zone in the interior of the granular specimen of Karlsruhe sand during a plane strain compression test [6]. The specimen material constants were assumed as for Karlsruhe sand [1]: $e_{i0} = 1.3$, $e_{d0} = 0.51$, $e_{c0} = 0.82$, $\phi_c = 30°$, $h_s = 190$ MPa, $\beta = 1$, $n = 0.5$ and $\alpha = 0.3$.

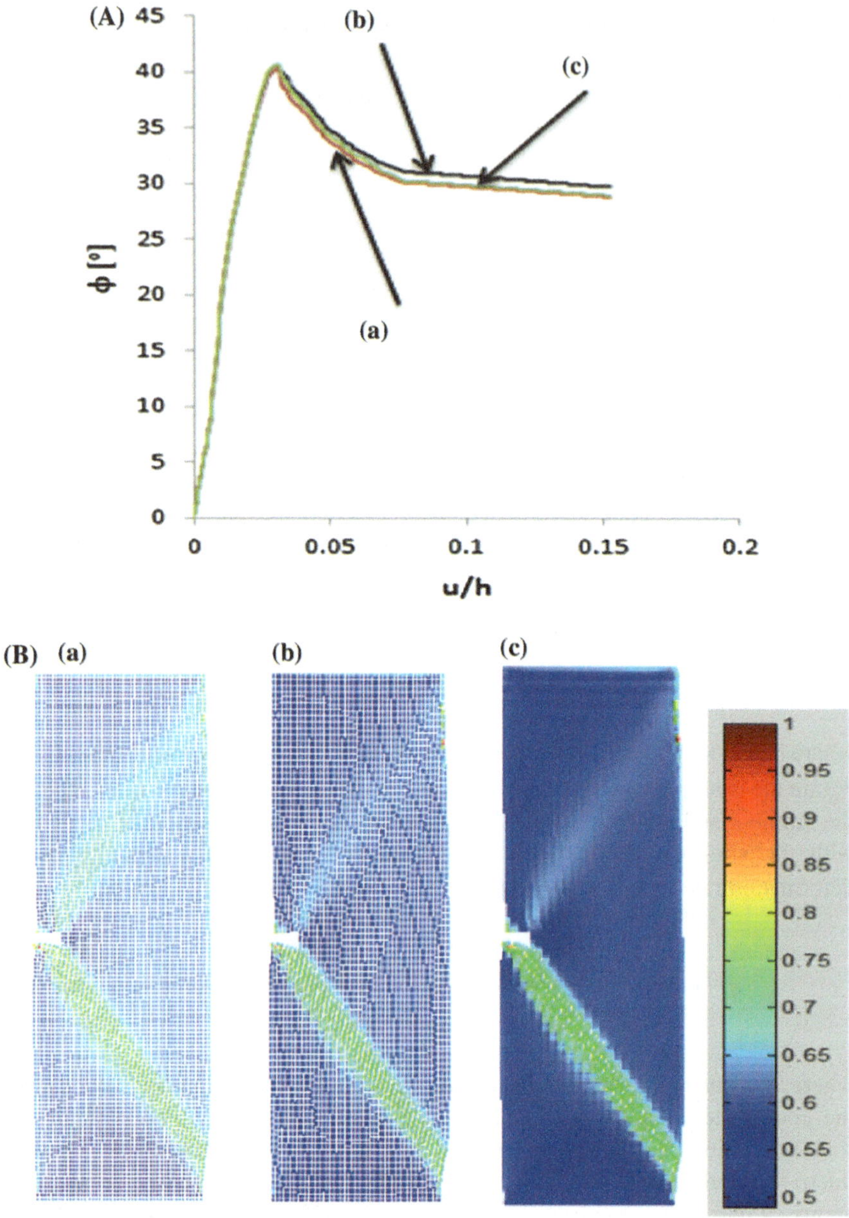

Fig. 1 FE results for plane strain compression: **A** evolution of mobilized internal friction angle ϕ versus normalized vertical displacement of *top* boundary *u/h* and **B** distribution of void ratio *e* at residual state (with scale attached) for different mesh of material points: **a** 22 × 112, **b** 44 × 112 and **c** 44 × 170 (*h*—specimen height)

4 Numerical Results with MPM

In the first step, two different BVPs with shear localization were simulated with a moderate mesh distortion to check the capability of MPM. Initially plane strain compression of the initially dense sand specimen was numerically simulated with the size of 140 × 40 mm^2. The deformation was induced in the specimen by prescribing a constant displacement along the top. Gravity was omitted. The lateral pressure was 200 kPa and initial void ratio was $e_o = 0.6$. In order to induce a shear zone, the imperfection of the size 5 × 5 mm^2 in the form of a higher void ratio ($e_o = 0.9$) was inserted.

Figure 1 shows the evolution of mobilized internal friction angle versus the normalized vertical top displacement and the distribution of void ratio in the specimen for the different number of material points. During deformation initially 2 shear zones occurred, starting from the initial imperfection. Later a single shear zone dominated. The thickness of the shear zone was 5 × l_c and its inclination against the bottom was about 54°. The results were in agreement with experiments [1] and did not depend on the mesh of material points (Fig. 1).

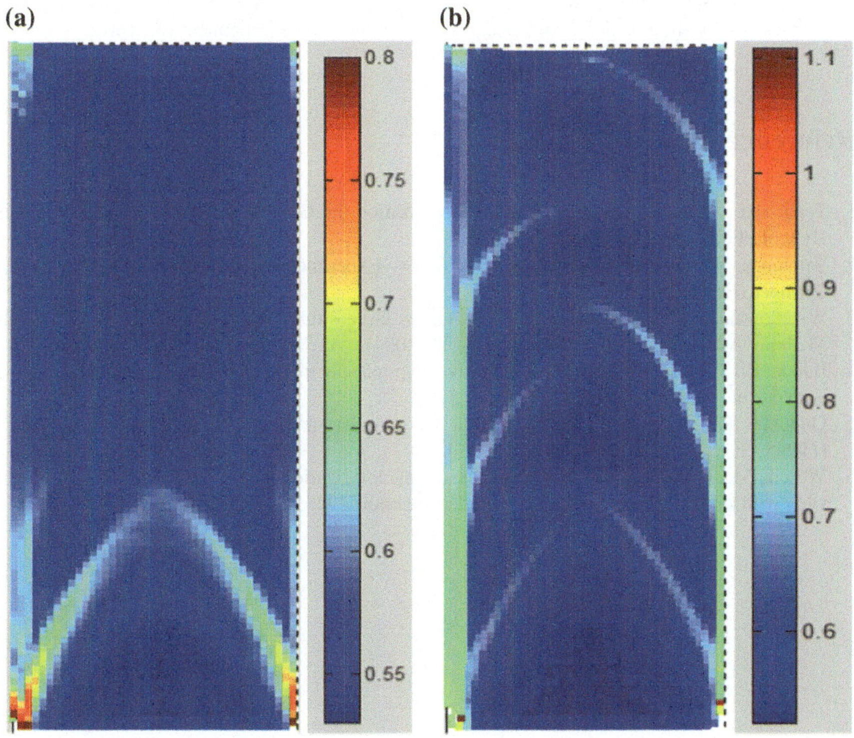

Fig. 2 Distribution of void ratio e (with scale attached) in sand after silo bottom displacement u: **a** u = 2 mm and **b** u = 4 mm

Later confined granular plane strain flow was simulated in a silo with a slowly movable bottom (height of 0.5 m and width 0.2 m). The silo walls were rigid and very rough (the wall friction angle was 45° [1]). The initial void ratio was e_o = 0.55. The results of the shear zone formation in the silo based on the distribution of void ratio are shown in Fig. 2.

The results indicate that dilatant shear zones occurred along the very rough walls and inside the flowing sand (zone pattern of a parabolic shape). The outcomes are in agreement with the silo experiments [6].

5 Conclusions

The FE results show that MPM based on a hypoplastic constitutive model with non-local softening properly captures shear localization in granular bodies during advanced deformation (with a moderate mesh distortion). The numerical results were in agreement with the experimental outcomes. They were insensitive of the mesh of material points.

Acknowledgments We acknowledge the support from the Grant WND-POIG.01.03.01-00-099/12-01 financed by the Polish National Centre for Science and Development (NCBiR).

References

1. Tejchman, J.: FE Modeling of Shear Localization in Granular Bodies with Micro-Polar Hypoplasticity. Springer, Heidelberg (2008)
2. Tejchman, J.: Confined Granular Flow in Silos—Experiments and Numerical Investigations. Springer, Heidelberg (2013)
3. Więckowski, Z.: The material point method in large strain engineering problems. Comput. Methods Appl. Mech. Eng. **193**, 4417–4438 (2004)
4. Bauer, E.: Calibration of a comprehensive hypoplastic model for granular materials. Soils Found. **36**(1), 13–26 (1996)
5. Gudehus, G.: A comprehensive equation for granular materials. Soils Found. **36**(1), 1–12 (1996)
6. Wójcik, M., Tejchman, J.: Modeling of shear zone localization during confined granular flow in silos with non-local hypoplasticity. Powder Technol. **192**, 298–310 (2009)

Instability of Geomaterials Caused by Transitional Negative Stiffness

Arcady Dyskin and Elena Pasternak

Abstract A mechanism of instability of geomaterials in compression is proposed based on the phenomenon of transitional negative stiffness associated with the process of crack/fracture formation. This phenomenon acts only for a very short time as the fracture is being formed. The negative stiffness cracks increase the Poisson's ratio and thus are capable of making nearly incompressible material elastically unstable. A geomaterial can reach the limiting value of the incremental Poisson's ratio due to dilatancy observed in compression. While dilatancy is a non-linear phenomenon, incrementally, at the time of fracture formation it is elastic and hence can be made unstable by the transitional negative stiffness.

1 Introduction

Compressive loading of geomaterials leads to their instability and failure when the peak stress is reached. An important characteristic feature of instability in compression is the dependence of its manifestation upon the stiffness of the loading frame: if the frame is soft, the complete (and often violent) failure is observed when the peak is reached, while if the frame is sufficiently stiff the post-peak softening phase is exhibited whereby the descending branch of the stress-strain curve can be registered. In other words, sufficiently stiff frame, that is the frame that, in the process of loading, releases insufficient amount of elastic energy to cause instability of the

A. Dyskin (✉)
School of Civil, Environmental and Mining Engineering, University of Western Australia, Crawley, Australia
e-mail: arcady.dyskin@uwa.edu.au

E. Pasternak
School of Mechanical and Chemical Engineering, University of Western Australia, Crawley, Australia

E. Papamichos et al. (eds.), *Bifurcation and Degradation of Geomaterials with Engineering Applications*, Springer Series in Geomechanics and Geoengineering, DOI 10.1007/978-3-319-56397-8_75

sample permitting post-peak regime, while a soft frame is incapable of stabilisation (e.g., [1]). Such dependence of the loading frame stiffness is similar to the behavior of an element with negative stiffness, which is unstable on its own (since the elastic energy is not positive definite), but can be stabilised if the encompassing system (loading frame in this case) can store enough energy to make the total energy of the system with the negative stiffness element positive definite [2–6].

This similarity is supported by the negative values of the deformation modulus in the (stable) post-peak regime, when the displacement/strain increases. Subsequently the mechanics of immediate failure (instability) or post-peak softening (stability) can be based on the development of negative stiffness elements in geomaterial in the process of its compressive loading.

A mechanism of negative stiffness recently put forward is the rotation of non-spherical constituents (e.g., grains, blocks) in the presence of compressive load [2–6]. The effect comes from the moment equilibrium of a non-spherical constituent about a contact point whereby the resistance to rotation associated with the compressive force decreases as the rotation progresses. While in the case of granular matter or blocky rock mass this mechanism and the associated instabilities are apparent [7–11], in rock the rotation of grains requires development of considerable damage capable of reducing the matrix resistance to grain rotation. However, the needed degree of damage does not always develop.

In this paper we consider another mechanism—a mechanism that creates negative stiffness only momentarily. It is the *transitional negative stiffness* characteristic of the cracks being formed. While the negative stiffness only lasts for a very short time, we assume that this time interval is sufficient to cause instability owing to the fact that the velocity of crack growth is considerably smaller than the velocities of waves whose propagation eventually brings the sample to equilibrium after the crack formation. (Indeed, the maximum crack growth speed is half of the Rayleigh wave speed and this needs some time to be reached).

2 Transitional Negative Stiffness

Consider an unbounded elastic material under uniform load σ_{ij}^0 and suppose that a crack is formed. Assume, for the sake of simplicity, that the crack is planar and disc-like of radius a. Introduce a local coordinate set $x_1' \, x_2' \, x_3'$ with axis x_3' directed perpendicular to the crack plane. Let the applied stress in the local coordinate frame have components σ_{ij}'. We also assume for now that the normal stress component is not negative such that the crack does not get closed. (In real cracks there is always some initial aperture small enough to be neglected when the displacement discontinuity is determined, but sufficient to prevent the crack closure, at least under load of low magnitude). Under this stress the crack develops displacement discontinuity, $\Delta u_i'$, distributed over the crack plane A. Hereafter we are only interested in the integral of displacement discontinuity over the crack plane, which reads (e.g. [12]):

$$V_i' = \iint_A \Delta u_i' dS = \frac{8\pi}{3}\frac{1-\nu_0^2}{E_0}a^3 c_{(i)}\sigma_{i3}', \tag{1}$$

where E_0 and ν_0 are the Young's modulus and Poisson's ratio of the material the crack is in, a is the crack radius. For a conventional disc-like crack

$$c_{(1)} = c_{(2)} = \frac{4}{\pi}\frac{1}{2-\nu_0}, \quad c_{(3)} = \frac{2}{\pi}. \tag{2}$$

For the normal component of the displacement discontinuity, V_i' it is just the volume of crack opening; two other components can be called "shear volumes".

It is easy to show that the energy change, associated with the crack opening is

$$E_{opening} = \frac{1}{2}V_i'\sigma_{i3}' = \frac{4\pi}{3}\frac{1-\nu_0^2}{E_0}a^3 c_{(i)}\sigma_{i3}'\sigma_{i3}', \tag{3}$$

where the summation over index i is presumed.

Now consider the energy change, associated with the crack formation. At the moment of formation of the crack, the corresponding stress components acting on the plane of future crack, σ_{i3}', is equal to the material strength. After the crack is formed, these stress components vanish. In other words, σ_{i3}' can be thought of as stresses holding the crack faces together thus effectively removing its presence. In the process of crack formation, when the opening commences, σ_{i3}' is applied to the crack faces in the direction opposite to their displacement thus producing negative work. Assuming that stress components σ_{i3}' reduce to zero linearly with the crack displacements we arrive to the negative energy associated with crack formation

$$E_{formation} = -\frac{1}{2}V_i'\sigma_{i3}' = -\frac{4\pi}{3}\frac{1-\nu_0^2}{E_0}a^3 c_{(i)}\sigma_{i3}'\sigma_{i3}'. \tag{4}$$

Therefore we can introduce negative stiffness associated with crack formation:

$$k_i = -\frac{8\pi}{3}\frac{1-\nu_0^2}{E_0}a^3 c_i. \tag{5}$$

This negative stiffness acts only during the short time of crack formation and for that reason will be called the *transitional negative stiffness*.

3 Momentary Change in Effective Elastic Moduli Associated with Transitional Negative Stiffness

Consider now macroscopic (effective) elastic moduli the (geo) material temporarily acquires during the duration of crack formation. In order to accomplish this we use the theory of effective characteristics (e.g. [12]) and assume that the crack concentration is small enough such that the crack interaction can be neglected (indeed, the probability that during the duration of the crack formation other cracks can be formed not far from each other is negligible).

Central for calculating the effective characteristics is expression (1), which reflects the traditional crack opening. The energy associated with the opening is given by (3). The energy associated with the crack formation (4) is the same but with reverse sign. Therefore in order to find the effective characteristics caused by transitional negative stiffness one can use the conventional equations for calculating the effective characteristics of material with non-interacting cracks and revert the sign of the terms responsible for the effect of cracks.

For the case of disc-like cracks the conventional theory of effective characteristics gives (e.g. [12]):

$$E = E_0\left[1 - \frac{16}{45}(10 - 3\nu_0)\frac{1-\nu_0^2}{2-\nu_0}\mathrm{v}\right], \quad \nu = \nu_0\left[1 - \frac{16}{15}(3 - \nu_0)\frac{1-\nu_0^2}{2-\nu_0}\mathrm{v}\right] \tag{6}$$

Here E_0 and ν_0 are the Young's modulus and Poisson's ratio of the material (between the cracks) and

$$\mathrm{v} = N\langle a^3\rangle \ll 1, \tag{7}$$

N is the number of cracks per unit volume and <.> denotes averaging over all cracks.

Reverting signs in (6) we arrive at the expression for *transitional effective moduli*, that is the moduli the material temporarily show during the crack formation:

$$E = E_0\left[1 + \frac{16}{45}(10 - 3\nu_0)\frac{1-\nu_0^2}{2-\nu_0}\mathrm{v}\right], \quad \nu = \nu_0\left[1 + \frac{16}{15}(3 - \nu_0)\frac{1-\nu_0^2}{2-\nu_0}\mathrm{v}\right] \tag{8}$$

These equations suggest that both the effective Young's modulus and Poisson's ratio increase. Yet this increase is short leaved, after the cracks are formed the material assumes the effective moduli, which are given by Eq. (6), that is, as expected, smaller than the moduli of the material.

4 Elastic Instability Associated with Transitional Negative Stiffness

Examining Eq. (8) one can see that the transitional negative stiffness increases both the Young's modulus and Poisson's ratio (because of the 'plus' in (8)). This increase is however marginal (since the increase is proportional to v <<1) and temporary. While the increase in the Young's modulus does not affect the material stability, the increase in the Poisson's ratio can throw it beyond the upper boundary $\nu = 0.5$, when ν_0 is close enough to 0.5, which will render the material unstable.

Suppose $\nu_0 = 0.5 - \varepsilon$, where $\varepsilon \ll 1$ is a small number. Then the concentration of forming cracks needed to bring the effective Poisson's ratio to the upper boundary $\nu = 0.5$ and thus make the geomaterial unstable is

$$\mathrm{v} > (3/16)\varepsilon + O(\varepsilon^2). \tag{9}$$

Thus the critical value of the concentration of forming cracks marking the geomaterial unstable is proportional to ε and hence can be really small as long as the Poisson's ratio of the geomaterial reached without these cracks is close to the limit of incompressibility 0.5. Usually the geomaterials have Poisson's ration far from 0.5, however under compressive load they show dilatancy, that is the increase in the relative volume of the sample thus reaching the incompressibility, whereby the incremental Poisson's ratio can reach 0.5.

A comment is due here. The phenomenon of dilatancy is non-elastic as it involves crack growth in compression (e.g., [13]) and hence energy dissipation. However the process of crack growth in compression is stable and slow compared to the process of crack formation. Therefore during the action of the effect of transitional negative stiffness the geomaterial can be considered elastic. Hence, as soon as the transitional effective Poisson's ratio reaches 0.5, the geomaterial instantaneously loses stability, presumably in the form of strain localisation.

5 Conclusions

We consider a mechanism of instability of geomaterials constituted by what we call a *transitional negative stiffness*. This is a momentary effect produced by developing fractures (or broken bonds), which for a very short time of energy release act as *negative stiffness elements*. After the fracture is formed it immediately becomes a conventional fracture with usual positive stiffness/compliance characteristics. We demonstrate that the forming fractures even in low concentrations are capable of momentarily bringing the Poisson's ratio to 0.5 if the geomaterial itself had Poisson's ratio (or incremental Poisson's ratio) close to 0.5; the latter is characteristic of compressive loading of geomaterials due to the phenomenon of dilatancy.

Acknowledgments The authors acknowledge financial support from the UWA Near Miss Scheme 2015.

References

1. Cook, N.G.W.: The failure of rock. Int. J. Rock Mech. Min. Sci. **2**, 389–403 (1965)
2. Dyskin, A.V., Pasternak, E.: Effective anti-plane shear modulus of a material with negative stiffness inclusions. In: Papanastasiou, P., Papamichos, E., Zervos, A., Stavropoulou, M. (eds.) 9th HSTAM10, Limassol, Cyprus 12–14 July, 2010, Vardoulakis mini-symposia—Wave Propagation, paper, vol. 116, pp. 129–136 (2010)
3. Dyskin, A.V., Pasternak, E.: Negative stiffness: is thermodynamic defeated? In: Das, R., John, S. (eds.) Proceeding of the 8th Australasian Congress of Applied Mechanics ACAM 8, Melbourne, pp. 1–7 (2014)
4. Dyskin, A.V., Pasternak, E.: Elastic composite with negative stiffness inclusions in antiplane strain. Int. J. Eng. Sci. **58**, 45–56 (2012)
5. Pasternak, E., Dyskin, A.V.: Dynamic instability in geomaterials associated with the presence of negative stiffness elements. In: Chau, K.-T., Zhao, J. (eds.) Bifurcation and Degradation of Geomaterials in the New Millennium. Springer Series in Geomechanics and Geoengineering. Springer, Heidelberg, pp. 155–159 (2015)
6. Pasternak, E., Dyskin, A.V., Esin, M.: Wave propagation in materials with negative Cosserat shear modulus. Int. J. Eng. Sci. **100**, 152–161 (2016)
7. Dyskin, A.V., Pasternak, E.: Friction and localisation associated with non-spherical particles, In: Bonelli, S., Dascalu, C., Nicot, F. (eds.) Advances in Bifurcation and Degradation in Geomaterials. Proceedings of 9th International Workshop on Bifurcation and Degradation in Geomaterials. Springer, pp. 53–58 (2011)
8. Dyskin, A.V., Pasternak, E.: Rock mass instability caused by incipient block rotation, In: Qian, Q., Zhou, Y. (eds.) Harmonising Rock Engineering and the Environment. Proceedings of 12th International Congress on Rock Mechanics. CRC Press, Balkema, pp. 201–204 (2012)
9. Dyskin, A.V., Pasternak, E.: Rock and rock mass instability caused by rotation of non-spherical grains or blocks. In: Rock Engineering and Technology for Sustainable Underground Construction. Proceedings of Eurock 2012, paper 102P (2012)
10. Dyskin, A.V., Pasternak, E.: Mechanical effect of rotating non-spherical particles on failure in compression. Phil. Mag. **92**(28–30), 3451–3473 (2012)
11. Dyskin, A.V., Pasternak, E.: Bifurcation in rolling of non-spherical grains and fluctuations in macroscopic friction. In: Holzapfel, G.A., Ogden, R.W. (eds.) ESMC 2012, 8th European Solid Mechanics Conference, Graz, Austria, p. 2 (2012)
12. Salganik, R.L.: Mechanics of bodies with many cracks. Mech. Solids **8**(4), 135–143 (1973)
13. Germanovich, L.N., Salganik, R.L., Dyskin, A.V., Lee, K.K.: Mechanisms of brittle fracture of rock with multiple pre-existing cracks in compression. Pure Appl. Geophys. (PAGEOPH) 143(1/2/3), 117–149 (1994)

DEM Modelling of Instability of Particulate Materials. Transitional Negative Stiffness

Yuan Xu, Elena Pasternak and Arcady Dyskin

Abstract We show that broken bonds can act as negative stiffness elements, but during a very short time. If at the time of bond breakage the geomaterial was in an almost incompressible state due to for instance dilatancy in compression, the breakage leads to material instability that starts at the peak stress.

1 Introduction

Quasi-static loading of geomaterials in compression leads to either instantaneous failure when the loading frame is not stiff enough or, in displacement-controlled loading, to strain localisation resulting in formation of shear-type fracture. The strain localisation stage is characterised by post-peak softening that is the descending branch of the stress-strain curve. The negative slope of the stress-strain curve in post-peak softening and the fact that the mere presence of this stage requires high stiffness of the loading frame suggest the involvement of negative stiffness elements of a sort. It was discussed in [1–3] that while the negative stiffness element is unstable due to the fact that its potential energy is not positive-definite, the material stability can still be ensured if the loading is provided by a loading frame whose stiffness is sufficient to make the potential energy positive definite.

A number of mechanisms engineered to exhibit, under displacement-controlled loading, negative stiffness were proposed (e.g. [4, 5]). The only natural mechanism of negative (shear) stiffness and instability of particulate and granular materials suggested in [1–3, 6–12] is based on the rotation of non-spherical (non-circular) particles in the presence of compression whereby a descending branch of the shear

Y. Xu · A. Dyskin (✉)
School of Civil, Environmental and Mining Engineering,
University of Western Australia, Perth, Australia
e-mail: arcady.dyskin@uwa.edu.au

E. Pasternak
School of Mechanical and Chemical Engineering,
University of Western Australia, Perth, Australia

E. Papamichos et al. (eds.), *Bifurcation and Degradation of Geomaterials with Engineering Applications*, Springer Series in Geomechanics and Geoengineering, DOI 10.1007/978-3-319-56397-8_76

force-displacement curve is exhibited with the slope proportional to the magnitude of compressive force.

It is the essence of this mechanism that the particles are not spherical; spherical (circular in 2D) particle does not produce the effect of negative stiffness. Yet direct discrete element modelling involving only spherical or circular particles [13, 14] demonstrates the overall post-peak softening, which indicates the presence of negative stiffness elements associated neither with the particle shape nor rotation. It was suggested in [15] that the effect of negative stiffness can be momentarily (for a very short time) attributed to fracture production (so-called *transitional negative stiffness* [15]), the reason being that the cohesive forces resisting the crack formation act in the directions opposite to the crack face displacement thus producing negative work. Here we demonstrate that the bond breakage also creates the effect of transitional negative stiffness. Following [16] we use 2D numerical simulations using the Particle Flow Code ^{2D} (PFC ^{2D}).

2 Transitional Negative Stiffness of a Separate Link and Instability

In order to investigate the mechanics of transitional negative stiffness we made a simple DEM model, Fig. 1. We use a PFC 2D built-in contact model Linear Parallel Bond model (PB). In order to keep our model simple we assume the radius of the each disc of 1 m. The only contact normal stiffness activated is set to 100 N/m. The top and bottom walls move towards each other to compress the particles. The left

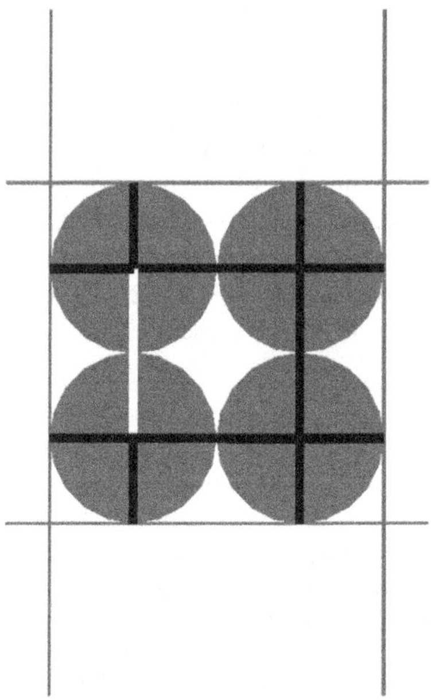

Fig. 1 The model

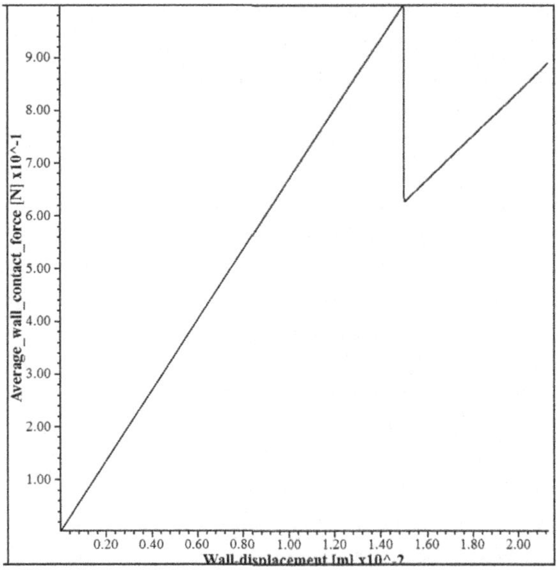

Fig. 2 Force-displacement curve

and right walls are set to keep all particles in contact. During loading the average force exerted on the top and bottom walls is monitored. Once the force reaches a set value of 1 N, the normal stiffness of the white contact (the link between the two discs on the left) is reduced to 10 N/m to simulate bond breakage. The force-displacement curve, Fig. 2, shows the corresponding force drop. A zoom in the region of the force drop, Fig. 3 shows that the curve exhibits negative slope. We

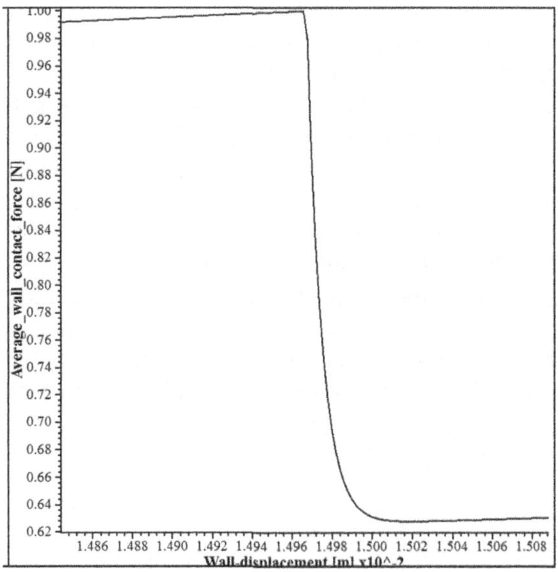

Fig. 3 Zoom at the force drop

shall note here that micro parameters in PFC do influence the macro properties however there is no definitive relationships between them. Hence the value we use for bond stiffness is only for the purpose of keeping the model simple.

This negative stiffness effect is momentary. It is present only during the *transitional time* when the sample reaches equilibrium after bond breakage. We extracted the data from PFC and confirmed that the drop in the force-displacement curve covers 270 computational steps. In order to gauge the extent of these 270 steps we found the number of computational steps needed for numerical stabilisation, i.e. the absence of further change of the forces in the walls. By applying a step loading to the model (with the original stiffness of the bonds) it was found that it took 580 computational steps to reach the numerical convergence. Hence the force drop requires less computational steps than the number of steps for numerical stabilisation. Hence we observed the *transitional negative stiffness*.

The transitional negative stiffness acts during a very short time such that the concentration of negative stiffness elements at any time is very low. Subsequently, the loss of stability associated with reaching critical concentration of negative stiffness elements [3, 11] cannot be expected. It was hypothesised in [15] that the fracture or strain localisation are initiated by elastic instability caused by transitional negative stiffness. The mechanics of the instability is based on the increase in Poisson's ratio caused by negative stiffness fractures [15] (Fig. 5 shows that Poisson's ratio increases as the axial stress peaks) such that if the geomaterial is already in a nearly incompressible state the transitional negative stiffness brings the Poisson's ratio to its limit. This limit (0.5 in 3D isotropic case) makes the elastic energy loose its positive definiteness leading to the instability.

The near incompressible state in geomaterials can be reached due to dilatancy produced in compression. Of course, the phenomenon of dilatancy is non-elastic as it involves crack growth in compression [17] and the associated energy dissipation. Assuming that the process of crack growth in compression is stable and slow compared to the process of crack formation one can conclude that during the time of the transitional negative stiffness the geomaterial can be considered elastic. As soon as the transitional effective Poisson's ratio reaches its limiting value (0.5 for the 3D isotropic material) the geomaterial instantaneously looses stability.

This hypothesis is verified in the next section using 2D discrete element modelling. The assembly of discs is isotropic. It should however be noted that while the assembly is isotropic in the plane, it is characterised by infinite stiffness in the third direction. This, in effect, corresponds to the transversal-isotropic material with plane of isotropy coinciding with the computational plane. Therefore the volumetric strain is the sum of only two strain components, $\varepsilon_x + \varepsilon_y$ in coordinate set (x, y) drawn in the plane of modelling. It is easy to show that the instability of such material is reached when the (incremental) Poisson's ratio equals 1.

3 Simulation of Instability in Compression

The simulation is conducted according to the following procedure: generating initial compact sample, installing bond, loading sample to isotropic compressive stress state of 10 MPa and conducting confined compressive test until failure of the sample. The initial sample contained 4131 particles. Particles of radius ranging from 0.3 to 0.42 mm were randomly generated in a box of dimension of 31.7 mm by 63.4 mm surrounded by four walls. The density of the particles was 2630 kg/m^3 and the porosity of the sample was 16%.

For the bond we used the PFC 2D Linear Parallel Bond model (PB). It consists of two sets of springs: the linear springs and parallel springs both characterised by normal and shear stiffnesses. The bonds breaks when either tensile strength σ_t or shear strength τ_t of the bond is exceeded.

In many numerical studies conducted using PFC, the tensile strength and shear strength of PB were set to be close (e.g. [16]). However as demonstrated in [18] local tensile failure caused by moment stress can act as a mechanism of global failure of particulate materials. To avoid this we set the shear bond strength 7 times higher than the tensile bond strength to eliminate shear failure. The Young's modulus for both linear and parallel bonds was E = 62GPa, the stiffness ratio $k_n/k_s = 2.5$. The tensile and shear strengths were $\sigma_t = 157$ MPa and $\tau_t = 1099$ MPa.

In PFC we cannot obtain macro-properties of the virtual sample directly from properties of particle bonds. Gradual modification of the bond properties has to be made in order to let the sample reach realistic macro-properties however this can be time consuming. Hence in our study we took the bond parameters as well as particle density, radius and porosity from the existing literature [16].

In simulations, we firstly loaded the samples to isotropic compressive stress of 10 MPa. Then the top and bottom walls move towards each other at a rate of 0.05 m/s. While this loading rate looks very high, PFC simulations are based on the dynamic mode governed by Newton's second law, the time step in each calculation cycle is chosen to be infinitely small (e.g., we used 10^{-9} s). Hence for a static analysis this rate can be considered as slow enough. During the entire simulation lateral stress was kept at 10 MPa by a simulated servo-control mechanism.

The bonds fail when the sum of tensile stresses acting on the bond cross-section caused by translational movements and relative rotations of bonded particles (moment stress) reaches the tensile strength. The virtual sample at failure and the stress-strain curve are shown in Fig. 4. Post peak softening is obtained and shear failure mode is observed.

The number of computational steps required to reach numerical equilibrium is found to be 100,000. This was used as window to compute incremental Poisson's ratio. Figure 5 shows incremental Poisson's ratio and axial stress respectively. It can be seen that the peak axial stress corresponds to the incremental Poisson's ratio reaching 1. Therefore the transitional negative stiffness induced by bond breakage can be responsible for the global instability of the particulate material.

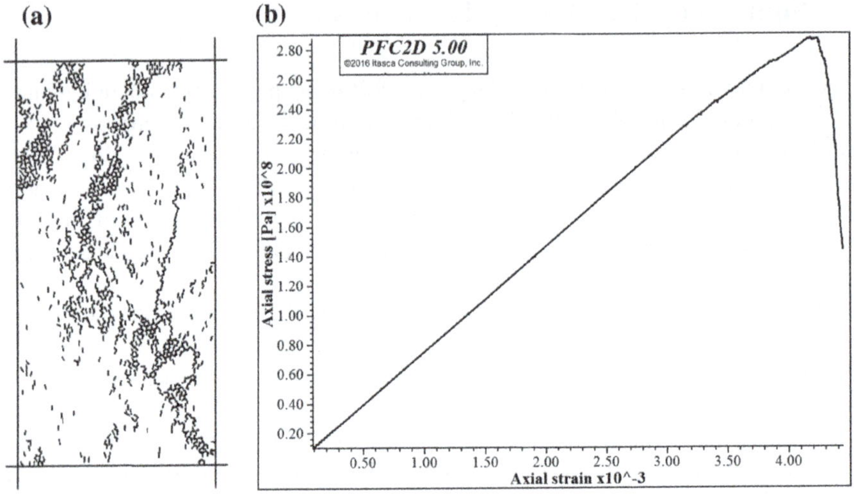

Fig. 4 PFC modelling results. **a** Broken bonds at failure. **b** Stress-strain curve

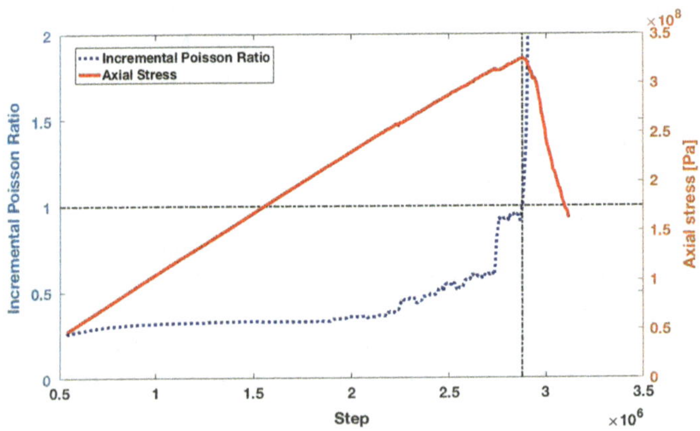

Fig. 5 Incremental poisson's ratio and axial stress versus time step

4 Conclusions

The discrete element simulation confirms that the bond breakage can induce the effect of negative stiffness albeit for a short time. The mechanics of instability associated with the *transitional negative stiffness* is in increasing Poisson's ratio of a nearly incompressible material to its limit corresponding to full incompressibility. In 2D discrete element simulation the limit of Poisson's ratio is 1. The discrete element simulation of sample failure in compression shows that the peak stress and the onset of instability just correspond to incremental Poisson's ratio equal 1.

References

1. Dyskin, A.V., Pasternak, E.: Effective anti-plane shear modulus of a material with negative stiffness inclusions. In: Papanastasiou, P., Papamichos, E., Zervos, A., Stavropoulou, M. (eds.) 9th HSTAM10, Limassol, Cyprus, Vardoulakis mini-symposia—Wave Propagation, pp 129–136 (2010)
2. Dyskin, A.V., Pasternak, E.: Negative stiffness: Is thermodynamic defeated? In: Das, R., John, S. (eds.) 8th Australasian Congress of Applied Mechanics ACAM 8, Melbourne, vol. 1, p. 7 (2014)
3. Dyskin, A.V., Pasternak, E.: Elastic composite with negative stiffness inclusions in antiplane strain. Int. J. Eng. Sci. **58**, 45–56 (2012)
4. Bažant, Z.P., Cedolin, L.: Stability of structures: elastic, inelastic, fracture, and damage theories. Oxford University Press, Oxford (1991)
5. Thompson, J.M.T., Hunt, G.W.: A general theory of elastic stability. Wiley, London (1973)
6. Pasternak, E., Dyskin, A.V.: Dynamic instability in geomaterials associated with the presence of negative stiffness elements. In: Chau, K.-T., Zhao, J. (eds.) Bifurcation and Degradation of Geomaterials in the New Millennium, pp. 155–159. Springer Series in Geomechanics and Geoengineering, Springer, London (2015)
7. Pasternak, E., Dyskin, A.V., Esin, M.: Wave propagation in materials with negative Cosserat shear modulus. Int. J. Eng. Sci. **100**, 152–161 (2016)
8. Dyskin, A.V., Pasternak, E.: Friction and localisation associated with non-spherical particles. In: Bonelli, S., Dascalu, C., Nicot, F. (eds.) Advances in bifurcation and degradation in geomaterials. In: Proceedings of 9th International Workshop on Bifurcation and Degradation in Geomaterials, pp. 53–58. Springer (2011)
9. Dyskin, A.V., Pasternak, E.: Rock mass instability caused by incipient block rotation. In: Qian, Q., Zhou, Y. (eds.) Harmonising rock engineering and the environment. In: Proceedings of 12th International Congress on Rock Mechanics, pp. 201–204. CRC Press, Balkema (2012)
10. Dyskin, A.V., Pasternak, E.: Rock and rock mass instability caused by rotation of non-spherical grains or blocks. In: Proceedings of Eurock 2012 Rock Engineering & Technology for Sustainable Underground Construction. Paper 102P (2012)
11. Dyskin, A.V., Pasternak, E.: Mechanical effect of rotating non-spherical particles on failure in compression. Phil. Mag. **92**(28–30), 3451–3473 (2012)
12. Dyskin, A.V., Pasternak, E.: Bifurcation in rolling of non-spherical grains and fluctuations in macroscopic friction. In: Holzapfel, G.A., Ogden, R.W. (eds.) ESMC 2012, 8th European Solid Mechanics Conference, Graz, Austria, 2012, p. 2 (2012)
13. Hazzard, J.F., Young, R.P., Maxwell, S.C.: Micromechanical modeling of cracking and failure in brittle rocks. J. Geophys. Res. Solid Earth **105**, 16683–16697 (2000)
14. Potyondy, D.O., Cundall, P.A.: A bonded-particle model for rock. Int. J. Rock Mech. Min. Sci. **41**, 1329–1364 (2004)
15. Dyskin, A.V., Pasternak, E.: Instability of geomaterials caused by transitional negative stiffness. This volume (2017)
16. Potyondy, D.O., Cundall, P.A., Lee, C.A.: Modelling rock using bonded assemblies of circular particles. In: 2nd North American Rock Mechanics Symposium, American Rock Mechanics Association, Montreal, Quebec, Canada, 1996, pp. 1937–1944 (1996)
17. Germanovich, L.N., Salganik, R.L., Dyskin, A.V., Lee, K.K.: Mechanisms of brittle fracture of rock with multiple pre-existing cracks in compression. PAGEOPH **143**(1/2/3), 117–149 (1994)
18. Dyskin, A.V., Pasternak, E.: Asymptotic analysis of fracture propagation in materials with rotating particles. Eng. Fract. Mech. **150**, 1–18 (2015)

The Effect of Constriction in Hydraulic Fracturing

Junxian He, Elena Pasternak, Arcady Dyskin, Maxim Lebedev and Boris Gurevich

Abstract Fractures in geomaterials e.g., hydraulic fractures often contain bridges—parts of unbroken material connecting the opposite faces of the fracture distributed over the fracture and constricting its opening. Our laboratory experiments demonstrate that the bridges can even hold the sample cut through by a fracture together, in one piece. We model such a fracture as a crack with Winkler layer whose stiffness is controlled by the bridge geometry and distribution. The model shows that short constricted fractures are insignificantly different from the conventional cracks; only large fractures, i.e. the fractures whose size is of the order of the characteristic scale of the bridge constriction are affected. The constricted fractures have the opening and the Mode I stress intensity factor bound as the fracture dimensions proportionally increase, which distinguish them from the conventional cracks where both the opening and the stress intensity factors tend to infinity as the crack size increases.

1 Introduction

Fractures and cracks in geomaterials often contain bridges that are parts of unbroken material connecting the opposite faces of the fracture. In fractures opened and driven by putting fluid under pressure, such as in hydraulic fracturing, the bridges are perceived as obstacles for fluid flow, similarly to the effect of the roughness of crack surfaces [1]. Bridging is often observed (in different materials)

J. He · A. Dyskin (✉)
School of Civil, Environmental and Mining Engineering,
University of Western Australia, Perth, Australia
e-mail: arcady.dyskin@uwa.edu.au

E. Pasternak
School of Mechanical and Chemical Engineering,
University of Western Australia, Perth, Australia

M. Lebedev · B. Gurevich
Department of Exploration Geophysics, Curtin University, Perth, Australia

E. Papamichos et al. (eds.), *Bifurcation and Degradation of Geomaterials with Engineering Applications*, Springer Series in Geomechanics and Geoengineering, DOI 10.1007/978-3-319-56397-8_77

near the crack front (crack tips in 2D) and considered as part of the fracture process zone underpinning its special mechanical behaviour (e.g. [2–5]). However, if the bridges are distributed over the whole fracture they could constrict its opening (we presume that the fracture is of Mode I). Hereafter such a fracture will be called *fracture with constricted opening (HFCO).*

2 Evidence of Distributed Bridges in Physical Model of Hydraulic Fracture

In order to demonstrate the existence of distributed bridging we initiated hydraulic fractures in laboratory using a modelling material: mortar with water to cement mass ratio of 0.4 and sand (particle size below 0.15 mm) to cement ratio of 1. The fractures were produced in cylindrical samples of 38 mm diameter and 76 mm length with a 4 mm diameter model borehole. The fracturing fluid (glycerol of viscosity 1410 cp) was injected from the top surface through a stainless steel tube of 10 mm in length and 3 mm outer diameter. The steel tube also played a role of casing leaving a 40 mm open section of the model borehole uncased.

The fluid pressure was increased incrementally from initial 400 psi with a step of 100 psi till 1000 psi was reached; after that the pressure increment was reduced to 50 psi. The flow rate in each step was 0.06 ml/min. The pressure was being increased until the fracture was created. The formation of macroscopic fracture was indicated by a dramatic increase of flow rate and sudden pressure drop.

Before and after hydraulic fracturing test, the X-ray micro-CT scanning was performed on each sample using the Xradia Versa XRM-500. The resolution of the scanned image was 33.72 micron per pixel.

The fracturing pressure for sample 1 and sample 2 was 1840 psi and 1450 psi respectively. In sample 1, two major vertical fractures were observed, Fig. 1 (see [6] for more detail). Investigation of the x-ray images showed that the length of the rock bridges was about 2 mm. The bridges of similar kind could be observed at different locations of the sample 1. In sample 2, two vertical and one horizontal fractures were created, Fig. 1. A fragment of the x-ray micro-CT scan of sample 2 at the (lower) resolution of 33.72 micron per pixel is presented in Fig. 2. It shows the bridges that connect the opposite faces of the fracture. Similar bridges can also be observed in other scan slices at different locations of the hydraulic fracture.

Despite the fractures traversing the whole samples, Fig. 1, the samples were still in one piece even after cutting out the top parts with the casing glued in. This indicates that the bridges are capable of holding the opposite surfaces of the fracture thus effecting fracture constriction.

Another hydraulic fracture test was conducted using a real rock (Metasediment). The fracture again traversed the sample but it was still in one piece. Also a force-displacement curve, Fig. 3 obtained from the direct tensile test on rock sample in the direction normal to fracture plane permitted the estimation of the total bridge stiffness: $k = 110$ N/mm.

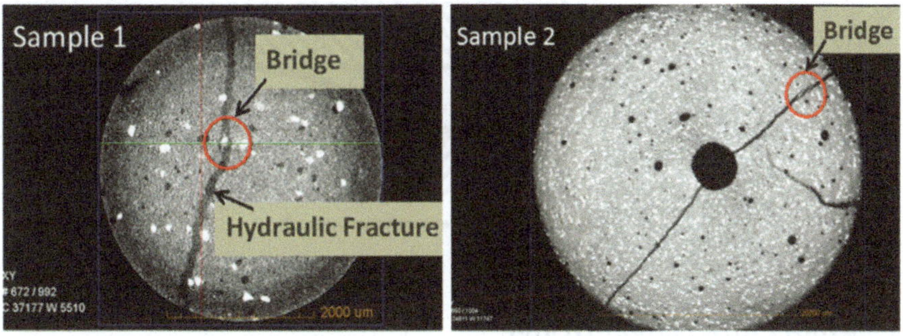

Fig. 1 X-ray CT images of samples 1 and 2 with hydraulic fractures

Fig. 2 Micro-CT image of a bridge

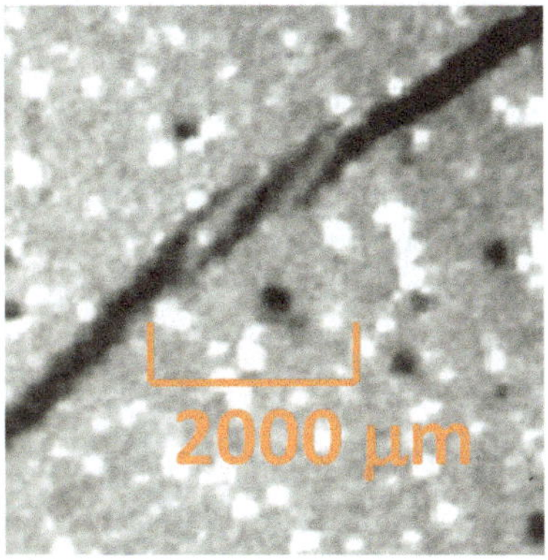

Fig. 3 Tensile force-displacement curve of rock

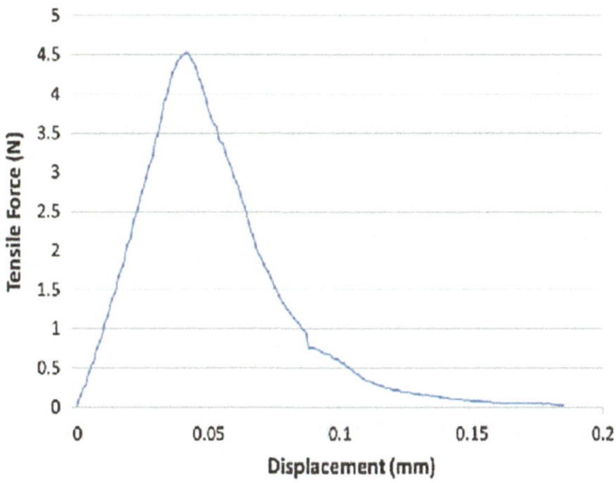

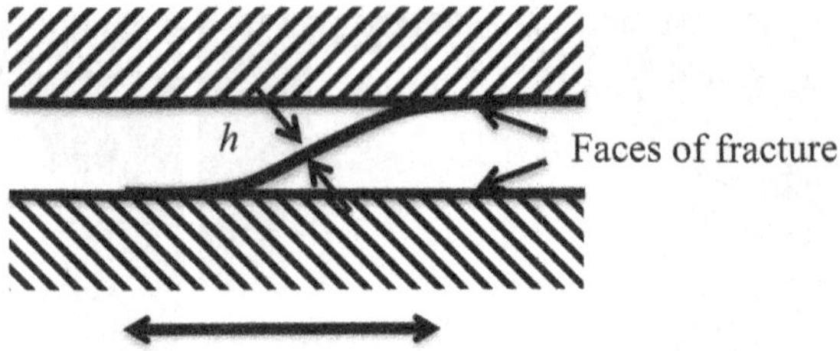

Fig. 4 Model of a bridge as a beam of size *b* and thickness *h*

3 Model of a Single Bridge

The configuration of the bridges suggests the model of the constriction effect by representing the bridge as a beam of thickness *h* deformed in bending, Fig. 4. Assume the beam having cross-sectional area *h* by *b*, where *h* is the beam thickness. Examination of Fig. 2 reveals that the bridge cross-section is not uniform with one end being thinner than the other. We will approximately model this feature by assuming that one end of the beam is clamped, while the other end is pinned. For such a beam the *equivalent stiffness* (the coefficient relating the displacement and the applied force) is (e.g. [7])

$$k_{eq} = Eh^3/\left(12b^2\right), \tag{1}$$

where *E* is the Young's modulus of the bridge material. This will be approximately taken as stiffness of a single bridge.

Let *N* be the total number of bridges distributed over area *A* of the fracture. Then the equivalent stiffness of all the bridges in the fracture is:

$$k = Nk_{eq} = ENh^3/\left(12b^2\right). \tag{2}$$

From here, knowing the measured stiffness, $k = 110$ N/mm, average bridge thickness, $h = 0.05$ mm, length $b = 0.5$ mm and that the total number of bridges $N = 77$ were observed in the fracture one can back calculate the rock Young's modulus, $E = 34$ GPa. This value is within the order of magnitude accuracy compared to the value of static modulus of approximately 74 GPa estimated from the measured dynamic modulus of 88 GPa for this rock [8].

4 Crack with Constricted Opening

Taking into account the number of bridges distributed over the whole fracture rather than only the fracture process zone, its opening can be modelled at the scale macroscopic with respect to the bridge scale (maximum of bridge length and the distance between the bridges). At this scale the joint action of bridges can be

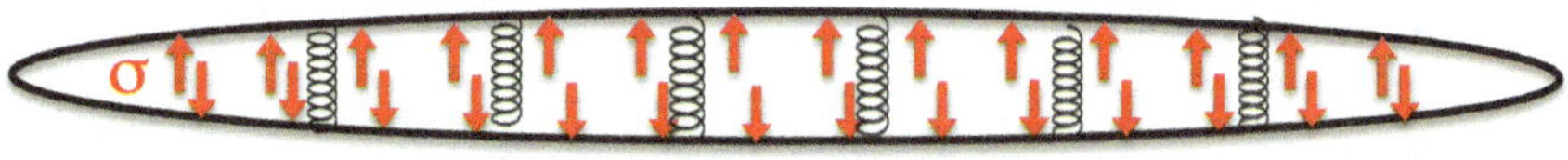

Fig. 5 Crack with Winkler layer (shown by *springs*) opened by internal pressure σ (*arrows*)

homogenised, which leads to modelling the crack with constricted opening as a crack with Winkler layer under load (Fig. 5). The Winkler layer is characterised by effective stiffness k_W relating displacement discontinuity Δu with normal stress, $\sigma = k_W \Delta u$. Here $k_W = k/A$, where A is the area over which the bridges are distributed. Using (2) one obtains:

$$k_w = EN_A h^3/(12b^2) \cong Eh^3/(12L^2 b^2), \tag{3}$$

where $N_A = N/A$, the number of bridges per unit area, L is the average distance between the centres of the bridges.

Consider, as an example, a disc-like crack of radius R with Winkler layer under uniform pressure p [9]. In the limit $R \to \infty$ the opening at the crack centre (displacement discontinuity) and the Mode I stress intensity factor read:

$$\Delta u = u^+ - u^- \xrightarrow[R \to \infty]{} 4(1-\nu^2)p/k_w, \quad K_I \xrightarrow[R \to \infty]{} p\sqrt{2E}/\left[\pi\sqrt{k_w(1-\nu^2)}\right] \tag{4}$$

where E and ν are the Young's modulus and Poisson's ratio of the material around the crack (e.g., rock), u^+ and u^- are displacements of the opposite crack faces at the centre. It is seen that both the crack opening (displacement discontinuity) and the Mode I stress intensity factor are bounded even for infinite crack radius. This is the effect of bridge constriction. Note, for conventional cracks both the opening and the stress intensity factor tend to infinity with the radius increase (e.g. [10]).

The model of the crack with constricted opening is characterised by 2 dimensional parameters. It is the Young's modulus (more precisely the ratio $E/(1-\nu^2)$) representing resistance of the elastic material and the stiffness of Winkler layer, k_W. As these parameters have different units, Pa and Pa/m it is possible to construct a parameter of a unit of length, λ, which plays a role of the characteristic length of the crack with constricted opening

$$\lambda = E/k_w = b^2 L^2/(2h^3) \tag{5}$$

Therefore the constricting effect of bridges is important when $R > \lambda$, i.e. only for large fractures. On the other hand when the fracture is small, $R << \lambda$ the bridge effect can be neglected (the fracture contains too few bridges).

5 Conclusions

The well-known phenomenon of bridging in fractures has significance far beyond the obstruction of fluid flow into the hydraulic fractures or forming the fracture process zone. The bridges are capable of constricting the fracture opening and subsequently bounding the value of Mode I stress intensity factor. This is only important for large fractures, i.e. fractures whose size is of the order of the characteristic scale of the bridge constriction λ or larger. This characteristic size is determined by the average size, b, thickness, h, of bridges and the average distance, L, between the centres of the bridges. The fractures that are much smaller than the characteristic scale are not affected by the constriction; for them the traditional crack models can be applied. The constricted opening and the stress intensity factor can be estimated using the model of crack with Winkler layer. The constricted fractures have the opening and the stress intensity factors bound as the fracture dimensions proportionally increase, which distinguish them from the conventional cracks where both the opening and the stress intensity factors tend to infinity as the crack size increases. The proposed model is more accurate than the traditional ones and can provide more adequate analysis of extended fractures in geomaterials and, in particular, the design and monitoring of extensive hydraulic fractures.

Acknowledgements We acknowledge the financial support from the Australian Research Council linkage project LP 120200797, Australian Worldwide Exploration (AWE) limited and Norwest Energy NL Companies. Assistance of Tim Sercombe (UWA) with setting up the tensile tests is greatly appreciated. Discussion with Andy Bunger from University of Pittsburgh is acknowledged. AD and EP acknowledge support from the UWA Near Miss Scheme.

References

1. van Dam, D.B., de Pater, C.J.: Roughness of hydraulic fractures: The importance of in-situ stress and tip processes. SPE Annual Technical Conference and Exhibition. Society of Petroleum Engineers (1999). doi:10.2118/56596-MS
2. Swanson, P.L., Fairbanks, C.J., Lawn, B.R., Mai, Y.-W., Hockey, B.J.: Crack-interface grain bridging as a fracture resistance I, mechanism in ceramics: I. Experimental study on alumina. J. Am. Ceramic Soc. **70**(4), 279–289 (1987)
3. Hu, X., Wittmann, F.H.: An analytical method to determine the bridging stress transferred within the fracture process zone: I. General theory. Cement Concr. Res. **21**(6), 1118–1128 (1991)
4. Hallam, S.D., Last, N.C.: Geometry of hydraulic fractures from modestly deviated wellbores. J. Petrol. Technol. **43**, 742–748 (1991)
5. Li, Q., Xing, H., Liu, J., Liu, X.: A review on hydraulic fracturing of unconventional reservoir. Petroleum **1**, 8–15 (2015)
6. Dyskin, A.V., Pasternak, E., He, J., Lebedev, M., Gurevich, B.: The role of bridge cracks in hydraulic fracturing. In: Kotousov, A., Ma, J. (eds.) Proceedings of the 10th International Conference on Structural Integrity and Failure (SIF2016), Adelaide, Australia, Paper #6 (2016)

7. Tedesco, J.W., McDougal, W.G., Ross, C.A.: Structural Dynamics: Theory and Applications. Pearson (1988)
8. Mockovčiaková, A., Pandula, B.: Study of the relation between the static and dynamic moduli of rocks. Metalurgija **42**(1), 37–39 (2003)
9. Shifrin, E.I.: The I-Mode crack whose sides interact according to a linear law. Mech. Solids **23**(5), 91–97 (1988)
10. Green, A.E., Sneddon, I.N.: The distribution of stress in the neighbourhood of a flat elliptical crack in an elastic solid. In: Mathematical Proceedings of the Cambridge Philosophical Society, vol. 46, pp. 159–163 (1950)

Erratum to: Propagation and Evolution of Strain Localization in Clay

Dunja Perić and Xingdong Wu

Erratum to: Chapter 'Propagation and Evolution of Strain Localization in Clay' in: E. Papamichos et al. (eds.), *Bifurcation and Degradation of Geomaterials with Engineering Applications*, Springer Series in Geomechanics and Geoengineering, DOI 10.1007/978-3-319-56397-8_3

In the original version of the book, incorrect author name "Xindong Wu" has to be corrected to read as "Xingdong Wu" in Chapter 3. The erratum chapter and the book have been updated with the change.

The updated online version of this chapter can be found at
https://doi.org/10.1007/978-3-319-56397-8_3

E. Papamichos et al. (eds.), *Bifurcation and Degradation of Geomaterials with Engineering Applications*, Springer Series in Geomechanics and Geoengineering, DOI 10.1007/978-3-319-56397-8_78

Author Index

E. Papamichos et al. (eds.), *Bifurcation and Degradation of Geomaterials with Engineering Applications*, Springer Series in Geomechanics and Geoengineering, DOI 10.1007/978-3-319-56397-8

The manufacturer's authorised representative in the EU is Springer Nature Customer Service Centre GmbH, Europaplatz 3, 69115 Heidelberg, Germany. If you have any concerns regarding our products, please contact ProductSafety@springernature.com

Printed and bound by CPI Group (UK) Ltd, Croydon, CR0 4YY
22/09/2026
02221052-0005